REVIEWS IN MINERALOGY AND GEOCHEMISTRY

Volume 63   2006

NEUTRON SCATTERING IN EARTH SCIENCES

EDITOR

Hans-Rudolf Wenk
University of California
Berkeley, California

COVER FIGURE CAPTIONS: *Top left:* View of the Spallation Neutron Source at Oak Ridge in Tennessee. *Top right:* Magnetic structures of hematite (see Fig. 7, Chapter 6). *Bottom left:* Design of the new multi-detector diffractometer PEARL at ISIS, used mainly for high *P-T* applications. *Bottom right:* View of the Los Alamos Neutron Science Center in New Mexico.

Series Editor: **Jodi J. Rosso**

GEOCHEMICAL SOCIETY
MINERALOGICAL SOCIETY OF AMERICA

SHORT COURSE SERIES DEDICATION

Dr. William C. Luth has had a long and distinguished career in research, education and in the government. He was a leader in experimental petrology and in training graduate students at Stanford University. His efforts at Sandia National Laboratory and at the Department of Energy's headquarters resulted in the initiation and long-term support of many of the cutting edge research projects whose results form the foundations of these short courses. Bill's broad interest in understanding fundamental geochemical processes and their applications to national problems is a continuous thread through both his university and government career. He retired in 1996, but his efforts to foster excellent basic research, and to promote the development of advanced analytical capabilities gave a unique focus to the basic research portfolio in Geosciences at the Department of Energy. He has been, and continues to be, a friend and mentor to many of us. It is appropriate to celebrate his career in education and government service with this series of courses.

Reviews in Mineralogy and Geochemistry, Volume 63

Neutron Scattering in Earth Sciences

ISSN 1529-6466
ISBN 978-0-939950-75-1

COPYRIGHT 2006

THE MINERALOGICAL SOCIETY OF AMERICA
3635 CONCORDE PARKWAY, SUITE 500
CHANTILLY, VIRGINIA, 20151-1125, U.S.A.
WWW.MINSOCAM.ORG

VOLUME DEDICATION

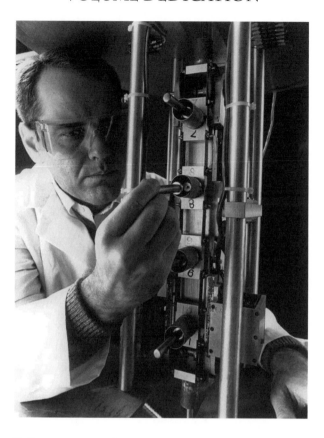

Dr. James D. Jorgensen passed away September 7, 2006. Jim was instrumental in the development of modern neutron powder diffraction, especially at spallation neutron sources. Building on earlier work by Brugger and Worlton at the Idaho Falls reactor, Jim championed high-pressure neutron powder diffraction at the Argonne National Laboratory, first at reactor sources and then at the ZING-P and the Intense Pulsed Neutron Source. His leadership and scientific contributions, while a member of the Materials Science Division and as leader of the Neutron and X-ray Scattering Group at Argonne, were recognized by a number of awards including the 1991 Warren Diffraction Physics Award of the American Crystallographic Association. His early high-pressure investigations of ice and silica had a profound effect on our views of the compression mechanism in minerals, and his papers on superconductivity became citation classics. We humbly dedicate this volume to him, his service to the community of neutron scientists, and his contributions across a broad range of basic scientific research. His mentorship, friendship and support will be greatly missed.

NEUTRON SCATTERING *in* EARTH SCIENCES

63   *Reviews in Mineralogy and Geochemistry*   63

FROM THE SERIES EDITOR

The review chapters in this volume were the basis for a two day short course on *Neutron Scattering in Earth Sciences* held in Emeryville, California, U.S.A. (December 7-8, 2006) prior to the American Geophysical Union Meeting in San Francisco, California.

Rudy Wenk, the editor of this volume, did an excellent job organizing this volume. He met deadlines (well ahead of schedule). Even when he was "out of the office," he was always available for answering questions, etc. Thank you Rudy for all your hard work! I also want to thank all the authors who willingly perservered through copyright permissions, figure revisions, equation modifications, and many many many e-mails from me.

Any supplemental material and errata (if any) can be found at the MSA website *www.minsocam.org*.

Jodi J. Rosso, Series Editor
West Richland, Washington
September 2006

PREFACE

For over half a century neutron scattering has added valuable information about the structure of materials and many experiments have been performed. Contrary to X-rays that have quickly become a standard laboratory technique and are available to all modern researchers in physics, chemistry, materials and earth sciences, neutrons have been elusive and reserved for specialists. A primary reason is that neutron beams, at least so far, are only produced at large dedicated facilities with nuclear reactors and accelerators and access to those has been limited. Yet there are a substantial number of experiments that use neutron scattering as the following summary from some large facilities documents:

Approved User Proposals

Facility	1995	2005
ILL	697	696
ISIS	633	581
IPNS	92	138
Lujan	136	306

These numbers are lower than for synchrotron facilities such as ESRF (781 for 2005) and APS (1083 for 2005) but not by a large amount. This is rather surprising since neutron scattering is so much weaker than X-ray scattering and experiments take much longer. Clearly, Europe has been dominating in neutron research with three times as many experiments than in the U.S., at only the two largest facilities. But usage has stabilized, whereas in the U.S. there has been a steady increase over the last decade and new users come increasingly from fields such

1529-6466/06/0063-0000$05.00   DOI: 10.2138/rmg.2006.63.0

as materials and earth sciences. With the addition of the Spallation Neutron Source (SNS) in Oak Ridge, the second target station at ISIS and J-PARC in Japan, opportunities for neutron experiments are greatly expanded and it is necessary to attract and prepare users from applied sciences to take advantage of neutron scattering.

While earth science users are still a small minority, neutron scattering has nevertheless contributed valuable information on geological materials for well over half a century. Important applications have been in crystallography (e.g. atomic positions of hydrogen and Al-Si ordering in feldspars and zeolites, Mn-Fe-Ti distribution in oxides), magnetic structures, mineral physics at non-ambient conditions and investigations of anisotropy and residual strain in structural geology and rock mechanics. Applications range from structure determinations of large single crystals, to powder refinements and short-range order determination in amorphous materials. Zeolites, feldspars, magnetite, carbonates, ice, clathrates are just some of the minerals where knowledge has greatly been augmented by neutron scattering experiments. Yet relatively few researchers in earth sciences are taking advantage of the unique opportunities provided by modern neutron facilities. The goal of this volume, and the associated short course by the Mineralogical Society of America held December 7-9 in Emeryville/Berkeley CA, is to attract new users to this field and introduce them to the wide range of applications. As the following chapters will illustrate neutron scattering offers unique opportunities to quantify properties of earth materials and processes.

Focus of this volume is on scientific applications but issues of instrumental availabilities and methods of data processing are also covered to help scientists from such diverse fields as crystallography, mineral physics, geochemistry, rock mechanics, materials science, biomineralogy become familiar with neutron scattering. A few years ago European mineralogists spearheaded a similar initiative that resulted in a special issue of the European Journal of Mineralogy (Volume 14, 2002). Since then the field has much advanced and a review volume that is widely available is highly desirable. At present there is really no easy access for earth scientists to this field and a more focused treatise can complement Bacon's (1955) book, now in its third edition, which is still a classic.

The purpose is to provide an introduction for those not yet familiar with neutrons by describing basic features of neutrons and their interaction with matter as well illustrating important applications. The volume is divided into 17 Chapters. The first two chapters introduce properties of neutrons and neutron facilities, setting the stage for applications. Some applications rely on single crystals (Chapter 3) but mostly powders (Chapters 4-5) and bulk polycrystals (Chapters 15-16) are analyzed, at ambient conditions as well as low and high temperature and high pressure (Chapters 7-9). Characterization of magnetic structures remains a core application of neutron scattering (Chapter 6). The analysis of neutron data is not trivial and crystallographic methods have been modified to take account of the complexities, such as the Rietveld technique (Chapter 4) and the pair distribution function (Chapter 11). Information is not only obtained about solids but about liquids, melts and aqueous solutions as well (Chapters 11-13). In fact this field, approached with inelastic scattering (Chapter 10) and small angle scattering (Chapter 13) is opening unprecedented opportunities for earth sciences. Small angle scattering also contributes information about microstructures (Chapter 14). Neutron diffraction has become a favorite method to quantify residual stresses in deformed materials (Chapter 16) as well as preferred orientation patterns (Chapter 15). The volume concludes with a short introduction into neutron tomography and radiography that may well emerge as a principal application of neutron scattering in the future (Chapter 17).

As was mentioned, we hope to attract many new users to this field and this will stimulate new research. Thus if this volume sparks new applications and becomes soon obsolete the goal has been achieved.

It is my privilege to acknowledge the contributions by authors and their enthusiasm and perseverance, as well as patience with editorial requests. Jodi Rosso has been an outstanding copy editor and without her such a coherent presentation could never have been accomplished. Also important were reviews, mostly by contributors to the volume, but also other experts, including Bjoern Clausen, Stephen Covey-Crump, Eberhard Lehmann, and Hongwu Xu.

The workshop, organized by the Mineralogical Society of America, was subsidized by DOE, Office of Basic Research, by NSF through the COMPRES Program, the Office of the President of the University of California, the Lujan Center of LANSCE, Los Alamos, and the Spallation Neutron Source at Oak Ridge and this generous support is highly appreciated.

Rudy Wenk
Dept. Earth and Planetary Science
University of California at Berkeley

September 2006

FREQUENTLY USED SYMBOLS

b	scattering length
c	velocity of light
e	elementary charge
F_{hkl}	nuclear unit-cell structure factor
$g(\mathbf{r})$	atomic pair distribution function
$G(\mathbf{r}, t)$	time-dependent pair-correlation function
$G_s(\mathbf{r}, t)$	self time-dependent pair-correlation function
$g(r)$	static pair-distribution function
$\mathbf{G}$	vector in reciprocal space
h	Planck's constant $= 6.626 \times 10^{-34}$ J s
$\hbar$	$= h/2\pi$
k_B	Boltzmann constant
$\mathbf{k}, \mathbf{k}'$	initial and final wave vectors of neutron
$I(\mathbf{Q}, t)$	intermediate function
$I_s(\mathbf{Q}, t)$	self intermediate function
$\mathbf{I}$	spin angular momentum of nucleus
M	mass of atom
m_n	mass of neutron
N	number of nuclei in scattering system
$\mathbf{p}$	$\mathbf{p} = \hbar\mathbf{k}$; momentum of the neutron
$\mathbf{Q}$	$\mathbf{Q} = \mathbf{k} - \mathbf{k}'$; scattering vector
$\mathbf{q}$	wave vector of normal mode (variable in reciprocal space)
$\mathbf{r}_j$	position of jth nucleus
$\mathbf{r}$	position of neutron; generalized position of nucleus
$S(\mathbf{Q}, \omega)$	scattering function
$S_i(\mathbf{Q}, \omega)$	incoherent scattering function
$S(\mathbf{Q})$	structure factor
T	absolute temperature
t	time variable
V	volume of crystal
v	velocity of neutron
$<v(0){\cdot}v(t)>$	velocity autocorrelation function
$\delta(x)$	Dirac delta function
θ	scattering angle
λ	wavelength of neutron
μ_n	magnetic dipole moment of neutron
μ_N	nuclear magneton
μ_B	Bohr magneton
$\rho(\mathbf{r}, t)$	particle density operator
ρ	mean number density

Continued on following page

FREQUENTLY USED SYMBOLS (CONT.)

Φ	neutron flux, incident neutrons cm^{-2} s^{-1}
$\Phi(v)$	velocity distribution of incident neutron flux
ρ_0	number density of atoms (nuclei) in the system of N atoms
σ	total number of neutrons scattered per second/Φ
σ_{tot}	total scattering cross-section
σ_{coh}	$= 4\pi(\bar{b})^2$
σ_{tot}	$= 4\pi\left\{\overline{b^2} - (\bar{b})^2\right\}$ total scattering
σ_a	absorption cross-section
Ψ_k, $\Psi_{k'}$	initial and final wave functions of neutron
ω	defined by $\hbar\omega = E - E'$

NEUTRON SCATTERING FACILITIES AND ABBREVIATIONS USED IN TEXT

Forschungsneutronenquelle Heinz Maier-Leibnitz (**FRM II**), Garching/Munich, Germany
http://wwwnew.frm2.tum.de/en.html
Institut Laue Langevin (**ILL**), Grenoble, France
http://www.ill.fr/
Intense Pulse Neutron Source (**IPNS**), Argonne IL, United States
http://www.pns.anl.gov/
ISIS at the Rutherford Appleton Laboratory, Chilton, United Kingdom
http://www.isis.rl.ac.uk/
Joint Institute for Nuclear Research (**JINR**), Dubna, Russia
http://www.jinr.dubna.su/
Japan Proton Accelerator Research Complex (**J-PARC**), Tokai, Japan
http://j-parc.jp/index-e.html
Laboratoire Leon Brillouin (**LLB**), Saclay, France
http://www-llb.cea.fr/index_e.html
Los Alamos Neutron Science Center (**LANSCE**), Los Alamos NM, United States
http://lansce.lanl.gov/
Spallation Neutron Source (**SNS**), Oak Ridge TN, United States
http://www.sns.gov/

NEUTRON SCATTERING *in* EARTH SCIENCES

63   *Reviews in Mineralogy and Geochemistry*   **63**

TABLE OF CONTENTS

1   Introduction to Neutron Properties and Applications

John B. Parise

2   Neutron Production, Neutron Facilities and Neutron Instrumentation

Sven C. Vogel, Hans-Georg Priesmeyer

3   Single-crystal Neutron Diffraction: Present and Future Applications

Nancy L. Ross, Christina Hoffmann

4   Neutron Rietveld Refinement

Robert B. Von Dreele

5   Application of Neutron Powder-Diffraction to Mineral Structures

Karsten Knorr, Wulf Depmeier

6   Neutron Diffraction of Magnetic Materials

Richard J. Harrison

7   Neutron Powder Diffraction Studies of Order-Disorder Phase Transitions and Kinetics

Simon A. T. Redfern

8   Time-resolved Neutron Diffraction Studies with Emphasis on Water Ices and Gas Hydrates

Werner F. Kuhs, Thomas C. Hansen

9   High Pressure Studies

John B. Parise

10   Inelastic Scattering and Applications

Chun-Keung Loong

11   Analysis of Disordered Materials Using Total Scattering and the Atomic Pair Distribution Function

Thomas Proffen

12   Structure of Glasses and Melts

Martin C. Wilding, Chris J. Benmore

13 Neutron Scattering and Diffraction Studies of Fluids and Fluid-Solid Interactions

David R. Cole, Kenneth W. Herwig
Eugene Mamontov, John Z. Larese

14

Small-Angle Neutron Scattering and the Microstructure of Rocks

Andrzej P. Radlinski

15   Neutron Diffraction Texture Analysis

Hans-Rudolf Wenk

16   Internal Stresses in Deformed Crystalline Aggregates

Mark R. Daymond

17	Applications of Neutron Radiography and Neutron Tomography

Bjoern Winkler

Reviews in Mineralogy & Geochemistry
Vol. 63, pp. 1-25, 2006
Copyright © Mineralogical Society of America

Introduction to Neutron Properties and Applications

John B. Parise

*Department of Geosciences, Chemistry Department
and Center for Environmental Molecular Sciences
Stony Brook University
Stony Brook, New York, 11794-2100, U.S.A.*
e-mail: John.Parise@stonybrook.edu

INTRODUCTION

Neutrons are powerful probes of nuclear and magnetic structure of condensed matter, and of lattice dynamics. Many of the unique properties of the neutron (shown in Tables 1 and 2) have no equivalent in X-ray studies (Bacon 1962). Neutrons are important in locating light atoms such as hydrogen and lithium and to study the diffusion of hydrogenous molecules. With some exceptions, they easily distinguish neighboring elements in the periodic table, such as Mg/Al and Si/Al for example, which are indistinguishable with X-rays. Neutrons are particularly well suited to investigations taking advantage of isotope substitution, since different isotopes (^{1}H/^{2}H) have very different scattering power. Indeed the inherent differences in the X-ray and neutron scattering processes mean that the neutron scattering length, the equivalent of the X-ray scattering factor, can be negative for certain isotopes, or near zero. This leads to possibilities unique to neutron experiments such as the use of null scattering materials (Table 2) like alloys of Ti and Zr, and vanadium metal. TiZr and V are ubiquitous at neutron sources, and are used to make neutron transparent sample holders and for environmental cells. This property also allows contrast matching studies of nano-materials in mixtures of H_2O-D_2O, for example. The overall scattering length for H_2O is negative and that of D_2O is positive; by mixing the two forms of water, average scattering lengths in between those of H_2O and D_2O are created. These mixtures can then be used to match the average scattering length, contrast match, particles immersed in this fluid. In another example of the power of neutron scattering, the determination of magnetic structure and site magnetic moments is routine with neutron scattering techniques, because absolute measurements of intensity are straightforward. No magnetic structures have been solved with powder X-ray scattering and no magnetic moments have been measured.

With the advent of very bright synchrotron X-ray sources, and the higher energy resolution now available at these facilities, some of the "traditional" neutron experiments, such as inelastic and magnetic scattering *can* be carried out with X-ray scattering techniques. Other examples include resonant X-ray scattering experiments to discriminate between sites occupied by adjacent elements or oxidation states, enabling distinction between say Fe^{2+} and Fe^{3+} (Materlik et al. 1994). When dealing with elements lighter than say manganese, the energy of the X-ray absorption edges is so low that micro-absorption and a limited data range conspire to limit the advantages of X-ray scattering. Both radiations have their unique properties—there are no neutron experiments equivalent to the ubiquitous X-ray Absorption Spectroscopy (XAS) but when working with the most common element in the universe, hydrogen, neutrons provide the most information; clearly neutron and X-ray techniques are complementary rather than competitive. An increasing number of facilities and users see advantages in combining the results obtained from both sources, and this justifies situating synchrotron X-ray storage rings and neutron sources in close geographic proximity. There are now several sites where

1529-6466/06/0063-0001$05.00

DOI: 10.2138/rmg.2006.63.1

Table 1. Neutrons have both particle-like and wave-like properties.

Mass: $m_n = 1.675 \times 10^{-27}$ kg
Charge = 0
Spin = ½
Magnetic dipole moment: $\mu_n = 1.913\ \mu_N$
Nuclear magneton: $\mu_N = 5.051 \times 10^{-27}$ J T^{-1}

Relationship between velocity (v), kinetic energy (E), wave vector (**k**), wavelength (λ), temperature (T) of the neutron.

$$E = \tfrac{1}{2}m_n v^2 = \mathbf{k}_B T = (h\mathbf{k}/2\pi)^2/(2m_n) = (\hbar^2 k^2)/(2m_n)$$
$$\mathbf{k} = 2\pi/\lambda = m_n v/(h/2\pi) = m_n v/\hbar$$
Units: E (meV); **k** (Å^{-1}); λ (Å); v (m/sec); T (K);

Approximate values for ranges of E and corresponding T and λ

	Energy (E/meV)	Temp (T/K)	Wavelength (λ/Å)
Cold	0.1 - 10	1 - 120	30 - 3
Thermal	5 - 100	60 - 1000	4 - 1
Hot	100 - 500	1000 - 6000	1 - 0.4

* Adapted from Squires (1996) and the "Neutron Data Booklet" published by ILL, Grenoble (Dianoux and Lander 2002)

the X/N synergy is available on the same campus; ESRF/ILL at Grenoble, France, APS/IPNS at Argonne, U.S.A. and Diamond/ISIS at Chilton, United Kingdom.

This chapter provides some of the important properties of the neutron and an introduction the mathematical basis for neutron scattering. More detailed explanations of the applications are provided in subsequent chapters. The applications and instruments to carry them out depend on how neutrons are produced, at either reactor or spallation sources and these aspects are covered explicitly in Vogel and Priesmeyer (2006, this volume). For other details see the neutron portal *http://neutron.neutron-eu.net/n_links* for links to new and existing facilities for neutron research.

PROPERTIES OF THE NEUTRON

It is important to understand those properties that make the neutron such a unique probe for condensed matter research. A neutron is an elementary particle found in the nucleus of atoms. As all particles of this length scale (protons, electrons, photons etc.), it has both particle (e.g., mass) and wave properties (e.g., wavelength). Some of the more important of the properties of the neutron are summarized in Table 1. Four properties are of particular note.

Charge

The neutron has zero net charge and so interacts very weakly with matter, penetrating deeply into a sample. This allows the use of large, geologically relevant, rock samples for absorption studies (Winkler 2006, this volume). The neutron also easily penetrates sample enclosures such as low temperatures cryostats and pressure cells. Further, there is no Coulomb interaction with the electronic charge cloud and neutrons interact directly with the nuclei of atoms. Although for most elements the absorption cross section is small, there are important exceptions and these are discussed below along with the nuclear scattering process.

Magnetic moment

Although the neutron carries no net charge, and so does not interact with the charge of the electron, it does have a magnetic moment, a spin, that interacts with the *unpaired* electron spins.

Table 2. Bound* scattering lengths, b (fm) and cross-sections for selected isotopes and for selected naturally occurring isotopic mixtures of the elements; σ (barns; 1 barn =100 fm^2). Z: atomic number; A: mass number; I: spin of nuclear ground state; C: natural abundance; b_{coh}, b_{inc}: coherent and incoherent scattering lengths; σ_{coh}, σ_{inc}: coherent and incoherent cross-sections; σ_a: absorption cross section for 2.2 km s^{-1} neutrons.

Element	Z	M	I	C (%)	b_{coh}	b_{inc}	σ_{coh}	σ_{inc}	σ_a
H	1				−3.74		1.76	79.9	0.33
		1	½	99.98	−3.74	25.22	1.76	79.9	0.33
		2	1	0.02	6.67	4.03	5.59	2.04	0.00
Li	3				−1.9		0.45	0.83	70.5
		6	1	7.5	2−0.3i	−1.9+0.3i	0.51	0.46	940
		7	³/₂	92.5	−2.22	−2.32	0.62	0.68	0.05
C	6				6.64		5.55	0.00	0.00
O	8				5.803		4.23	0.0008	0.0002
Mg	12				5.375		3.63	1.62	0.06
Al	13				3.449	0.256	1.49	0.008	0.23
Si	14				4.149		2.16	0.004	0.17
Ca	20				4.70		2.78	0.05	0.43
Ti	22				−3.44	1.49	2.76		6.1
Mn	25	55	⁵/₂	100	−3.73	1.79	1.75	0.4	13.3
Fe	26				9.54		11.44	0.39	2.56
Ni	28				10.3		13.3	5.2	4.49
		58	0	68.27	14.4	0	26.1	0	4.6
		62	0	3.59	-8.7	0	9.5	0	14.5
Zr	40				7.16		6.44	0.16	0.19
Cd	48				5.1−0.7i		3.3	2.4	2520
Au	79	197	³/₂	100	7.63	−1.84	7.32	0.43	98.65
Pb	82				9.40		11.11	0.00	0.17
U	92				8.42		8.90	0.00	7.57

*In the *free state* nuclei recoil upon neutron impact; cross-sections related to atomic mass number (M) by $\sigma_{free} = (A/A + 1)^2 \sigma_{bound}$. For H, $\sigma_{free} = \frac{1}{4} \sigma_{bound}$, with the difference between the two cross-sections decreasing rapidly for heavier nuclei.

In magnetic atoms the strength of the interaction is comparable to that of the nuclear interaction and the scattered intensity associated with magnetic effects is comparable to the scattering from the nuclei. Neutron beams can be prepared which contain a single angular momentum state, either spin up (+½) or spin down (−½). These spin-polarized neutrons have unique applications in determining complex magnetic structures from single crystals, separating magnetic from nuclear scattering, and isolating incoherent scattering from the total scattering. Neutron scattering has contributed enormously to our understanding of magnetism (Von Dreele 2006, this volume).

Wave-like nature and Bragg (elastic) scattering

The neutron was discovered in 1932, and soon after, in 1936 it was shown that neutrons could be diffracted (see Bacon 1975) in an experiment that used a low flux Ra–Be neutron source. The geometry of scattering is as it is for X-ray diffraction, with the conditions required summarized in Bragg's law:

$$\lambda = 2 \, d_{hkl} \sin\theta$$

Here λ is the wavelength of the neutron or X-ray, d_{hkl} is spacing between planes in the crystalline solid with Miller index *hkl,* and θ is the scattering angle, half the angle between the incident and diffracted beams, 2θ. We will return to a more detailed discussion of geometry for scattering experiments at reactor and spallation sources when we summarize some of the important application later in this chapter. The equivalent energy for a particular wavelength can be deduced from the de Broglie equation, $\lambda = h/mv$, where h is Plank's constant and m and v are the mass and velocity of the particle, respectively.

The neutron's mass and inelastic scattering

The mass of the neutron is relatively large (Table 1) and this has several important consequences. One consequence has important implication for the production of neutrons with a particular energy distribution. Atoms of course have a finite mass and they can gain or lose energy when a neutron, or any kind of radiation, collide with them. Free neutrons have a life time of ~15 minutes and can be generated by fission, at reactor sources or by spallation, where energetic protons impinge on a heavy metal target (mercury for example) and spall off neutrons. The energetic neutrons produced by these processes can be slowed down, or moderated, by collisions with atoms of similar mass, such as hydrogen or deuterium, atoms with about the same mass as the neutron, in substances such as water, methane and polyethylene. The moderator produces a Maxwellian distribution of neutron energies (Fig. 1) in an appropriate range (1–100 meV) for studying a wide variety of dynamical phenomena in solids and liquids. By varying the type and

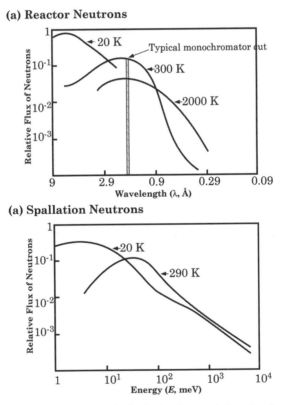

Figure 1. (a) The intensity versus wavelength distribution for a typical neutron beam emerging from a reactor and brought into equilibrium with moderators at the temperatures indicated. The wavelength band selected by the monochromator is indicated. (b) Energy distribution, on the same scale as (a), for neutrons moderated at a spallation neutron source. Adapted from *http://www.mrl.ucsb.edu/~pynn/Lecture_2_Facilities.pdf*

temperature of the moderator, and allowing the neutrons to come to thermal equilibrium, the shape and Maxwellian is varied (Fig. 1). At reactor sources a narrow range of wavelengths (energies) is selected with a crystalline monochromator while at a spallation neutron source the distribution of energies (Fig. 1b) impinges on the sample. It is conventional to state the relationship between a neutron with energy E with a temperature T as $E = \frac{1}{2}mv^2 = k_B T$. Other relationships are summarized in Table 1 along with typical ranges of energy, temperature and wavelength. The wavelengths corresponding to the root mean square velocities of neutrons in equilibrium at temperatures of 100, 300 and 500 K are approximately 2.51, 1.45 and 1.13 Å, respectively and are quite suitable for scattering experiments. Most experiments of interest to geologist are carried out with "thermal" neutrons; those neutrons coming from a moderator near room temperature with kinetic energies and wavelengths (Table 1, Fig. 1) similar to typical excitation processes, such as lattice vibrations, and interatomic distances, respectively. However other important applications require "cold" (longer λ, lower E) neutrons or "hot" neutrons (Table 1, Fig. 1) that are produced by passing neutron beams into say solid methane and hot graphite, respectively.

Inelastic scattering. Another important consequence of the neutron's mass is its ability to interact strongly with lattice vibrations. Neutron scattering can be either elastic, involving no energy transfer to the atoms, or it can be inelastic, involving transfer of energy to or from the scattering system (see below). Atomic motions in a solid can be analyzed into a spectrum of thermal vibrations, which travel as waves through the solid. The energy associated with these waves is quantized and each quantum of energy is regarded as a "phonon," i.e., particle-like entities representing the quantized elastic waves. This is by direct analogy to the theory of radiation where a quantum of energy is regarded as a "photon." When neutrons, or X-rays for that matter, are inelastically scattered the phonons are generated or annihilated as the neutrons lose or gain energy, respectively. Studies of the phonons in solids are of particular value, and are particularly easy, with neutron beams because the phonon energies are about the same as the kinetic energy of thermal neutrons (10–100 meV). Loss of the energy of the phonon will produce a big change in the velocity or energy of the neutron; typically a modest and easily achievable energy resolution of $\Delta E/E \sim 10\%$ is sufficient to obtain useful results. This is very different from the case of inelastic scattering by X-rays where, because of the large probe energy—about 10^4 eV for X-rays—comparable observations may require $\Delta E/E \sim 10^{-7}$; this resolution is achievable at modern synchrotron X-ray facilities (Hill et al. 1996). The aim of the measurements of inelastic scattering is to determine the phonon spectrum in the solid and to determine the dispersion law, which describes the dependence of the frequency ω on the wave-vector k, numerically equivalent to $2\pi/\lambda$ (Table 1), for any direction in the crystal.

The interactions of the neutron beam with matter are complex, and even the theory of the interaction between a neutron and the nucleus of an atom is still incomplete. However the unique manner in which neutrons scatter from matter gives rise to important applications. We will now look in some detail at that process before summarizing applications and highlighting two of the more important, Bragg scattering and inelastic scattering.

THE SCATTERING OF NEUTRONS BY ATOMS

In neutron scattering experiments, one exposes a sample to a beam of neutrons (Fig. 2) and measures the intensity of the scattered beam as a function of scattering angle and the wavelength (λ) of the probe. The intensity data are related to the structure and/or dynamics of the sample; the positions of the atoms constituting the sample in space ($\mathbf{r}$) and time (t). The analysis of the scattered intensity in terms of a structural model is central to modern condensed matter science. These statements are true for any scattering experiment—neutrons, electrons or X-rays—and the approach taken in describing the interactions below are fundamentally common to any scattering experiment.

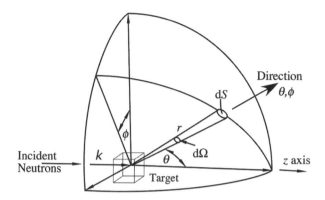

Figure 2. Geometry of the scattering experiment; adapted from Squires (1996)

Scattering cross-sections

The quantity actually measured in any scattering experiment is the cross-section. The effective area presented by a nucleus to an incident neutron is termed the *cross-section* (σ) in barns: 1 barn = 10^{-24} cm^2. Following the formalism of Squires (1996) consider a beam of energy E and a flux Φ (neutrons cm^{-2}sec^{-1}) incident on a target sample (Fig. 2). The neutrons interact with the collection of atoms in the target, the scattering system. If we set up a counter in a particular direction and can discriminate the energy of the neutrons after they encounter the scattering system, E', the result is expressed in terms of the cross-section. We specify the geometry in polar coordinates with the direction of the scattered neutrons given by θ and ϕ. If σ is the total number of neutrons scattered per second normalized to incident flux (Φ),the partial differential cross-section is defined by

$$\frac{d^2\sigma}{d\Omega\, dE'} = \frac{\text{neutrons scattered sec}^{-1} \text{ into } d\Omega \text{ in direction } \theta, \phi \text{ with final energy between } E' \text{ and } E' + \text{dE}}{\Phi\, d\Omega\, dE'}$$

Suppose we do not analyze the energy of the scattered neutrons but count all neutrons scattered into the solid angle $d\Omega$ in the direction θ, ϕ. We would measure the *differential cross-section* per unit solid angle

$$\frac{d\sigma}{d\Omega} = \frac{\text{neutrons scattered sec}^{-1} \text{ into } d\Omega \text{ in the direction } \theta, \phi}{\Phi\, d\Omega} = \int \frac{d^2\sigma}{d\Omega\, dE'} dE'$$

The *total cross-section* σ_{tot} is given by

$$\sigma_{tot} = \int \frac{d\sigma}{d\Omega} d\Omega$$

The information on structure and dynamics is contained in the scattering cross-sections, which relates a sample dependent property (the probability of a particular scattering event occurring) to a real experimental value (the number of detected counts in a detector) and this is discussed further below. The cross-sections can be calculated, if we know all the properties of the scattering system. The experimentally measured intensity, I, is related to the cross-section, modified by geometric factors such as polarization and absorption and appropriately scaled.

$$I = A\frac{d\sigma}{d\Omega}$$

Differences between X-ray and neutron cross-sections. The scattering of neutrons by atomic nuclei is described by the atomic scattering length b, related to the total cross-section by the expression $\sigma_{tot} = 4\pi b^2$, derived below. Unlike the total scattering of X-rays, which can be treated theoretically (Waller and Hartree 1929), there is no suitable theory of nuclear forces we can use to predict b from other properties of the nucleus; the neutron scattering lengths are parameters that are determined experimentally. Other important differences between X-ray and neutron scattering include the dependence of scattering on the atomic number, isotope and scattering vector. X-rays interact with electrons in an electron cloud of comparable size to the wavelength of the X-rays; the X-ray scattering process can be thought of as involving absorption of the X-ray photon, accompanied by the excitation of the electronic system, immediately followed by de-excitation and re-radiation of an X-ray photon. The X-ray scattering power scales with the number of electrons (the atomic number, Z) and falls off at higher scattering angle. Since the size of the electron cloud is comparable to the X-ray wavelength, intra-atomic interference effects are important, and the X-ray atomic form factor is thus strongly dependent on angle. Neutrons, in contrast, interact with atomic nuclei via the nuclear force. The scattering process can be regarded as the momentary capture of the incoming neutron by a nucleus, and then re-emission of the neutron. The nuclear force ranges over ~10^{-13} m while typical thermal neutron wavelengths (Table 1) are of order 10^{-10} m, so scattering is "point-like" involving no dependence on scattering angle. Further, the strength of the interaction is different for different isotopes of the same atom and does not vary regularly as it does for X-ray scattering where it increases regularly with Z. The irregular variation of the neutron-nuclear interaction can be used to advantage when studying elements with adjacent Z-numbers Further, the phase change in the scattered neutron wave can be of the opposite sign, as it is in the case of ^{1}H and ^{2}H (D, deuterium), for example. The magnetic form factor, a measure of the strength of the interaction between the magnetic moment of the neutron and the unpaired electrons in the electron shell giving rise to magnetic scattering, *does* depend on scattering angle and is of comparable magnitude to the nuclear interaction. We will return to a discussion of scattering lengths, and their implications for scattering experiments, once we introduced some concepts of neutron scattering.

Sample scattering amplitude

Consider a scattering system consisting of a single nucleus at the origin (Fig. 3), *fixed* in position so no kinetic energy is transferred to the neutron (the magnitude of wave vector **k**, momentum, is unchanged by the scattering). The incident neutrons are then represented by the wave function

$$\Psi_{incident} = \exp(ikz) \tag{1}$$

The elastic scattering is spherically symmetrical and the wave function of the scattered neutron at point r can we written as

$$\Psi_{scatter} = -\frac{b}{r}\exp(ikr) \tag{2}$$

where b is the constant scattering length, independent of the angles θ, ϕ (Fig. 2). The minus sign in Equation (2) is inserted to ensure that most nuclei have a positive value of b, implying a phase change of 180° between incident and scattered waves, just as there is for X-ray scattering.

Resonant neutron scattering. We can distinguish two types of isotope. In the first type the scattering is complex, having both real and imaginary components, and varying rapidly with the energy of the neutron, especially in the vicinity of a resonance, in a manner analogous to anomalous X-ray scattering (Materlik et al. 1994). The resonance phenomenon in neutron anomalous scattering is associated with the formation of a compound nucleus consisting of the original nucleus plus the incident neutron, which has energy close to an excited state. The

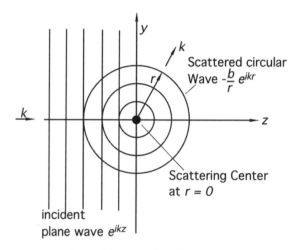

Figure 3. Scattering by a fixed nucleus.

neutron scattering length of these nuclei can be written as $b_0 + b' + ib'' = b + ib''$ (Table 2). The correction terms b' and b'' are strongly wavelength dependent and in favorable case b'/b_0 and b''/b_0 can be of order 10 whereas they are small fractions in X-ray anomalous scattering. This raises the possibility of the crystallographic use of the phenomenon for substances containing such light elements as ^6Li and ^{10}B that produce no sizeable resonant effect with X-rays. Examples of other nuclei displaying this behavior are ^{103}Rh, ^{113}Cd, ^{157}Gd and ^{176}Lu. In some cases the phenomenon is used for structure solution (Singh and Ramasesh 1968a,b; Flook et al. 1977). Since the imaginary part of the scattering length corresponds to absorption, these nuclei strongly absorb neutrons. As a consequence, lightweight machinable boron nitride collimation, cadmium and gadolinium oxide "paint" are ubiquitous shielding materials at neutron sources. Lead, as used for shielding X-Rays and Gamma Radiation, is almost transparent for neutrons (Table 2).

One consequence of neutron capture close to the energy of an isolated resonance is the possibility of nuclear resonance spectroscopy, proposed recently as a technique for the noninvasive determination of sample temperature in experiments at high pressures (Stone et al. 2005). The technique relies on the measurement of the thermally induced Doppler broadening of neutron absorption resonances.

For most nuclei the compound nucleus is not formed near an excited state. The scattering length is independent of energy and is a real quantity. The value of b will depend on the particular isotope and the spin-state of the compound nucleus. The neutron has spin ½ (Table 1). For a nucleus of spin I the spin of the compound nucleus is $|I \pm ½|$. Each spin has its own value of b. If the spin of the nucleus is zero then, of course, the compound nucleus can only have spin ½. Otherwise each spin states has its own value of b. The values of b vary erratically with both atomic number (Z) and isotope (N) (Bacon 1975; Squires 1996; Dianoux and Lander 2002) and will be discussed in some additional detail below.

As described by Squires (1996) the differential cross-section $d\sigma/d\Omega$ for scattering can be readily calculated using the expressions for $\Psi_{incident}$ and $\Psi_{scatter}$ in Equations (1) and (2). If v is the velocity of the neutron, which is the same before and after scattering, the number of neutrons passing through the area dS per second (Fig. 2) is:

$$vdS\left|\Psi_{scatter}\right|^2 = vdS\frac{b^2}{r^2} = vb^2 d\Omega$$

The number of incident neutrons passing through unit area is

$$\Phi = v\left|\Psi_{incident}\right|^2 = v$$

Now, recalling the definition of the total cross-section

$$\frac{d\sigma}{d\Omega} = \frac{vb^2 d\Omega}{\Phi d\Omega} = b^2 = \text{constant}$$

so

$$\sigma_{tot} = 4\pi b^2$$

Scattering by more than one nucleus with k = k′

In the special case where the incident and scattered radiation have the same energy, that is the scattering is elastic, derivation of the scattering cross section follows straightforwardly. Consider the situation of scattering from two sites containing the same nuclei depicted in Figure 4; we calculate the phase difference between the two scattering centers. The incident and diffracted beams are written as before in the form $\exp(i\mathbf{k}\cdot r)$ and $\exp(i\mathbf{k}'\cdot r)$, respectively. Here $\mathbf{k}$ and $\mathbf{k}'$ are the wave-vectors, related to the wavelength of the wave, $|\mathbf{k}| = 2\pi/\lambda$. The scattering vector $\mathbf{Q}$ is the vector difference

$$\mathbf{Q} = \mathbf{k}' - \mathbf{k}$$

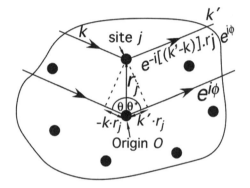

Figure 4. Scattering of a beam of radiation from two particles separated by vector $\mathbf{r}_j$. The wave-vectors of he incoming and scattered beams are $\mathbf{k}$ and $\mathbf{k}'$.

In this case the scattering is elastic and no energy, and hence no momentum $\hbar\mathbf{k}$ is exchanged between wave and scatterer,

$$|\mathbf{k}'| = |\mathbf{k}| = 2\pi/\lambda$$

The total phase difference, the sum of the path length differences between the site at the origin and that at distance $\mathbf{r}_j$ from it (Fig. 4) is $2\pi r_j(\cos\theta^* + \cos\theta)/\lambda$, which, from the definitions above, is represented by the dot product, $(\mathbf{k}' - \mathbf{k})\cdot\mathbf{r}_j$ or $\mathbf{Q}\cdot\mathbf{r}_j$. If we assume for the moment that the amplitude of scattering is unity for all nuclei, the wave scattered by the site at the origin is of the form

$$\Psi_o = \exp(i\phi)$$

where ϕ is a phase term. The wave scattered from the site $\mathbf{r}_j$ from the origin will be out of phase by the amount calculated above ($\mathbf{Q}\cdot\mathbf{r}_j$) and will be of the form

$$\Psi_r = [\exp(-i\mathbf{Q}\cdot\mathbf{r}_j)\exp(i\phi)]$$

The total scattering from these two sites would be the sum of the two waves

$$\Psi_{scatter} = \Psi_o + \Psi_r = \exp(i\phi)[1 + \exp(-i\mathbf{Q}\cdot\mathbf{r}_j)] \tag{3}$$

and the amplitude of the scattered beam is modified by the phase factor $[1 + \exp(-i\mathbf{Q}\cdot\mathbf{r}_j)]$.

In deriving this result, only the possibility of single scattering was considered (kinematic approximation), which is appropriate when the scattering is weak, as it is for X-rays and neutrons in interacting with powders. In the case of highly perfect large single crystals multiple scattering can be important and the scattering process is best described in terms of the dynamical theory of diffraction (Cowley 1975).

The treatment above can be generalized to a collection of N sites with unitary scattering lengths (Fig. 4) by adding phase shifts for each site r_j ($j = 1,...,N$) with respect to the origin

$$\Psi(\mathbf{Q}) = \sum_j \exp(-i\mathbf{Q} \cdot \mathbf{r}_j) \tag{4}$$

This *scattering amplitude* is sometimes referred to as F(**Q**). The *intensity* of the scattered beam will be proportional to the square of this value. Also note that when we need to take account of the situation when scattering occurs from nuclei with different scattering lengths and that the positions r_j are not static

$$\Psi(\mathbf{Q},t) = \frac{1}{<b>} \sum_j b_j \exp\left(-i\mathbf{Q} \cdot \mathbf{r}_j(t)\right)$$

where $r_j(t)$ defines the position of the jth nucleus as a function of time (t), b_j is the scattering amplitude of the jth nucleus, and $<b>$ represents a compositionally average scattering length. In the following discussion we suppress the variable t, for the time being.

The expression for the scattered wave Equations (3) and (4) look like a Fourier transform: the Fourier transform of a function $f(x)$ is given by

$$F(\mathbf{Q}) = \sum_{j=-\infty}^{\infty} f(x_j) \exp(-i\mathbf{Q} \cdot x_j)$$

The total scattering cross-section is then related to the Fourier transform of the atomic position vector. The structural information is contained in the phase of the exponential factor for $\Psi(\mathbf{Q})$. Thus, if we know this scattering amplitude, including its phase, we can determine the atomic structure exactly, by taking the inverse Fourier transformation. However, we do not directly measure the scattered amplitude but only the intensity of the diffracted beam, which is directly related to the square of the magnitude of $\Psi(\mathbf{Q})$.

$$\frac{d\sigma_{coh}}{d\Omega} = \frac{\langle b \rangle^2}{N} |\Psi(\mathbf{Q})|^2 = \frac{1}{N} \sum_{j,j'} b_{j'} b_j \exp[-i\mathbf{Q} \cdot (\mathbf{r}_{j'} - \mathbf{r}_j)] \tag{5}$$

$$I(Q) = \frac{d\sigma_{coh}}{d\Omega} + \frac{d\sigma_{inc}}{d\Omega}$$

$$S(Q) = \frac{I(Q)}{\langle b \rangle^2}$$

where the subscript *coh* and *inc* refer to coherent and incoherent scattering described in detail below. In measuring I, all phase information in $\Psi(\mathbf{Q})$ is lost[1].

---

[1] There are several designations for $S(\mathbf{Q})$ in the literature: for example, the *total scattering structure function*, or often just *structure function* (though this terminology is at times used for the time-dependent function ($S(\mathbf{Q},\omega)$ described below); in the inelastic scattering community $S(\mathbf{Q},\omega)$ is the *dynamic structure factor or structure function*; in the liquid and glass community $S(\mathbf{Q})$ is called the structure factor, while in the crystallographic community the term structure factor scales the amplitude, Ψ_s, and often is represented as F_{hkl}; there will be some confusion added by the desire of certain communities to normalize $S(\mathbf{Q})$ with respect to $<b^2>$ or $<b>^2$]

The result above can be related to a general Fourier analysis[2]. In the static approximation the particle density $\rho(\mathbf{r})$ of the scattering system is given as the sum of delta functions, each of which represents a single point particle at position $\mathbf{r}_j$:

$$\rho(\mathbf{r}) = \rho_0 g(\mathbf{r}) \propto \sum_j b_j \delta(\mathbf{r} - \mathbf{r}_j)$$

where ρ_0 is the number density of nuclei and $g(\mathbf{r})$ is the static *atomic pair distribution function*. The Dirac δ-function has the usual properties, viz. its value is zero unless the position vector $\mathbf{r}$ coincides with $\mathbf{r}_j$ and the integral of this function is unity. The Fourier transform of the delta function above is obtained by performing the integral over ranges $\mathbf{r} \neq \mathbf{r}_j$ and $\mathbf{r} = \mathbf{r}_j$.

$$\int_{-\infty}^{\infty} \delta(\mathbf{r} - \mathbf{r}_j) \exp(-i\mathbf{Q} \cdot \mathbf{r}) d\mathbf{r} = \int_{\mathbf{r} \neq \mathbf{r}_j} \delta(\mathbf{r} - \mathbf{r}_j) \exp(-i\mathbf{Q} \cdot \mathbf{r}) d\mathbf{r} + \int_{\mathbf{r} = \mathbf{r}_j} \delta(\mathbf{r} - \mathbf{r}_j) \exp(-i\mathbf{Q} \cdot \mathbf{r}) d\mathbf{r}$$

$$= \exp(-i\mathbf{Q} \cdot \mathbf{r}_j)$$

Thus the Fourier transform of the particle density is given as

$$\rho(\mathbf{Q}) = \int \rho(\mathbf{r}) \exp(-i\mathbf{Q} \cdot \mathbf{r}) \, d\mathbf{r} = \sum_j b_j \int \delta(\mathbf{r} - \mathbf{r}_j) \exp(-i\mathbf{Q} \cdot \mathbf{r}) \, d\mathbf{r} = \sum_j b_j \exp(-i\mathbf{Q} \cdot \mathbf{r}_j)$$

This is of the same form as the scattering amplitude $\Psi(\mathbf{Q})$ given above in Equation (4). The nuclear density is said to be the "Fourier transform" of the scattering amplitude or structure factor and the scattering amplitude is the Fourier transform of the nuclear density. Thus the scattered intensity is given by

$$I(\mathbf{Q}) = \langle \rho(\mathbf{Q})\rho(-\mathbf{Q}) \rangle = \sum_{j,j'} b_{j'} b_j \exp[-i\mathbf{Q} \cdot (\mathbf{r}_{j'} - \mathbf{r}_j)]$$

General law for neutron scattering

The static approximation is valid provided energy transfers are small and the quantum state of the scattering system does not change in the scattering process (Van Hove 1954). In this case we describe the differential cross section in terms of simple density distribution functions for the particles. For most condensed matter studied with X-rays the static approximation is valid. The time for oscillation is of the order of $t_0 = 10^{-12}$ s. Interatomic spacings are of order 1 Å $= 10^{-10}$ m. So for X-rays traveling at the speed of light, c, the time taken to traverse the distance between interatomic sites is $t_1 \sim a/c \sim 10^{-18}$ s. Thus $t_1 \ll t_0$ and the static approximation is valid. For thermal neutrons $v \sim 10^3$ ms^{-1}, and $t_1 \sim 10^{-13}$ s. Thus thermal neutrons potentially gives information on the correlation between sites in *space and time*, over a wide range of time scales, making it a useful tool for the study of condensed matter. It also means that what we detect in the scattering experiment (Fig. 2) depends on whether we integrate over energy or not.

---

[2] A brief description of the Fourier Transform is given in Appendix B of Squires (1996) and in standard college and crystallographic texts. Briefly: if $f(x)$ is a function defined as

$$f(x) = \int_{-\infty}^{\infty} g(\mathbf{k}) \exp(ikx) d\mathbf{k}$$

then $g(k)$ is said the be the Fourier transform of $f(x)$ with $g(k)$ given explicitly by

$$g(\mathbf{k}) = \frac{1}{2\pi} \int_{-\infty}^{\infty} f(x) \exp(-ikx) dx$$

Van Hove (1954) provided a suitably generalized pair distribution function $G(\mathbf{r},t)$ depending on a space vector $\mathbf{r}$ and a time interval t, and a detailed study of this function in for a number of real systems. It turns out the generalized pair-distribution function $G(\mathbf{r},t)$, to which neutron scattering gives direct experimental access, is a very natural extension of the static pair distribution function $g(\mathbf{r})$. A general expression for the partial differential cross-section for any assembly of nuclei is given in Chapter 2 of Squires (1996) and summarized elsewhere (Pynn 1990). The analysis treats the scattering process in terms of perturbation theory, utilizing Fermi's pseudo potential $V(\mathbf{r})$ for the neutron-nucleus interaction to show the partial differential cross-section is given by:

$$\frac{d^2\sigma}{d\Omega dE'} = \frac{\mathbf{k}'}{\mathbf{k}} \frac{1}{2\pi\hbar} \sum_{j,j'} b_j b_{j'} \int_{-\infty}^{\infty} \left\langle \exp\{-i\mathbf{Q}\cdot\mathbf{r}_{j'}(0)\}\exp\{i\mathbf{Q}\cdot\mathbf{r}_j(t)\}\right\rangle \times \exp(-i\omega t)dt \qquad (6)$$

where $\omega = \omega_{inc} - \omega_{sc}$ is the change in frequency ($= 0$ for elastic scattering) between the incoming and scattered neutron beams; recall that $E = \hbar\omega$. The integral over t essentially gives a thermal average with the angle brackets denoting an average over all possible thermodynamic states of the sample. The result is presented here without derivation. This formalism is complicated and its evaluation is non-trivial (Squires 1996; Van Hove 1954). By approximating the quantum mechanical operators to classical (vector) approximations, it forms the basis for important neutron scattering techniques. Also note Squires (1996) gives the quantum mechanical relationships that are not obeyed by classical mechanics, those that are non-commuting and the prescriptions needed to calculate various functions exactly from the atomic properties of the scattering system. Treating the system as one that obeys classical mechanics the sum over atomic sites can be rewritten as

$$\sum_{j,j'} b_j b_{j'} \left\langle \exp\{-i\mathbf{Q}\cdot[\mathbf{r}_{j'}(0)-\mathbf{r}_j(t)]\}\right\rangle = \sum_{j,j'} b_j b_{j'} \int_{-\infty}^{\infty} \delta\left\langle\left(\mathbf{r}-[\mathbf{r}_{j'}(0)-\mathbf{r}_j(t)]\right)\right\rangle \exp(-i\mathbf{Q}\cdot\mathbf{r})\, d^3\mathbf{r}$$

If we make the simplifying assumption that all scattering lengths are the same ($b = b_j = b_{j'}$) the right hand side becomes

$$Nb^2 \int_{-\infty}^{\infty} G(\mathbf{r},t)\exp(-i\mathbf{Q}\cdot\mathbf{r})\, d^3\mathbf{r}$$

$$\text{where } G(\mathbf{r},t) = \frac{1}{N}\sum_{j,j'} \delta\left\langle\left(\mathbf{r}-[\mathbf{r}_{j'}(0)-\mathbf{r}_j(t)]\right)\right\rangle$$

The delta function now is zero except when nucleus j' at time zero and j at time t are separated by vector $\mathbf{r}$. The delta-function is summed over all possible pairs of atoms and so $G(\mathbf{r},t)$, the *time dependent pair correlation function*, is the probability of one nucleus being at the origin *and* a nucleus being at position $\mathbf{r}$ at time t, and describes how the correlation between two nuclei evolves with time.

Coherent and incoherent scattering

For neutrons the scattering lengths in Equation (6), $b_j b_{j'}$, depend on the nuclear isotope and on the nuclear spin relative to the neutron. The cross-section we measure is very close to the cross-section averaged over all systems with the same number of nuclear positions and motions but with different distributions of b's. Equation (6) then becomes

$$\frac{d^2\sigma}{d\Omega dE'} = \frac{\mathbf{k}'}{\mathbf{k}} \frac{1}{2\pi\hbar} \sum_{j,j'} \overline{b_j b_{j'}} \int_{-\infty}^{\infty} \left\langle \exp\{-i\mathbf{Q}\cdot\mathbf{r}_{j'}(0)\}\exp\{i\mathbf{Q}\cdot\mathbf{r}_j(t)\}\right\rangle \exp(-i\omega t)dt \qquad (7)$$

where the over-bar signifies an average scattering length for nucleus j. On the assumption of no correlation between b values of different nuclei

$$\overline{b_j b_{j'}} = (\overline{b})^2, j \neq j' \qquad \text{and} \qquad \overline{b_j b_{j'}} = \overline{b^2}, j = j'$$

and this allows separation of Equation (7) into the following equations (Squires 1996)

$$\left(\frac{d^2\sigma}{d\Omega dE'}\right)_{coh} = \frac{\sigma_{coh}}{4\pi}\frac{k'}{k}\frac{1}{2\pi\hbar}\sum_{j,j'}\int_{-\infty}^{\infty}\left\langle\exp\{-i\mathbf{Q}\cdot\mathbf{r}_{j'}(0)\}\exp\{i\mathbf{Q}\cdot\mathbf{r}_j(t)\}\right\rangle\exp(-i\omega t)\,dt \qquad (8)$$

$$\left(\frac{d^2\sigma}{d\Omega dE'}\right)_{inc} = \frac{\sigma_{inc}}{4\pi}\frac{k'}{k}\frac{1}{2\pi\hbar}\sum_{j}\int_{-\infty}^{\infty}\left\langle\exp\{-i\mathbf{Q}\cdot\mathbf{r}_{j}(0)\}\exp\{i\mathbf{Q}\cdot\mathbf{r}_j(t)\}\right\rangle\exp(-i\omega t)\,dt \qquad (9)$$

where $\sigma_{coh} = 4\pi(\overline{b})^2$, $\sigma_{inc} = 4\pi\{\overline{b^2}-(\overline{b})^2\}$.

This separation has important and useful consequences, although the forms of these equations are still not very useful. The coherent scattering (Eqn. 8) depends on the correlation between positions of the same (j, j) and different (j, j') nuclei at different times (t). This gives rise to interference effects, Bragg scattering and phonon scattering, for example, and is the scattering the system would give if all the scattering lengths were equal to b. The incoherent scattering depends only on the correlation between positions of the same nucleus at different times and arises from the random distribution of the deviations of the scattering lengths in the scattering system from their mean values. The incoherent scattering does not give rise to interference effects but we anticipate now that it might be useful in studies of diffusion, since it arises from correlations of the same nucleus at different times.

The values of the coherent and incoherent scattering lengths for different elements and isotopes (Fig. 5 and Table 2) do not vary in any obviously systematic way (Bacon 1975). Most of the values of b are positive (Sears 1995) and there are important and useful exceptions (Table 2). The elements Mn and Fe differ by one in atomic number and so, far from absorption edges, are difficult to distinguish with X-rays (Fig. 5). The difference in their coherent neutron scattering lengths make studies of site occupancy distributions straightforward using neutron diffraction. Because several isotope pairs (^{1}H/^{2}H; ^{6}Li/^{7}Li; ^{58}Ni/^{62}Ni) have positive and negative values of b_{coh} they can be combined in appropriate mixtures to give $<b_{coh}> = 0$. Further, machinable alloys, such as mixtures of Zr and Ti (Table 2, Fig. 4) are used to make vessels with $<b_{coh}> \sim 0$ that are suitable as sample containers that contribute little or no coherent parasitic scattering to scattering experiments. There is no equivalent to these types of studies possible with X-rays.

An example of the power of isotope substitution, in combination with NMR and X-ray synchrotron techniques, is the study of short and long range order in $LiNi_{0.5}Mn_{0.5}O_2$, a potential Li-ion battery positive electrode material, with a structure related ot that of α-$NaFeO_2$ (Breger et al. 2005). Three samples were studied: ^{6}Li(NiMn)$_{0.5}O_2$, ^{7}Li(NiMn)$_{0.5}O_2$ and ^{7}Li(NiMn)$_{0.5}O_2$ enriched with ^{62}Ni so that the resulting $<b>_{coh}$ of Ni atoms is null. Calculations of the pair distributions function, g($\mathbf{r}$) (see Proffen 2006, this volume, and the description below) revealed considerable local distortions in the layers that were not captured in the refinements of local structure models. Isolation of the specific site-site interactions by contrast matching out various components of the structure allowed the testing of models of cation ordering (Breger et al. 2005). Low energy X-rays of about 8 keV will allow us to follow changes in the oxidation sate of Ni in these materials, but only the unique contrast matching possible with neutrons allows us to isolate contributions from the metal components. This simplifies considerably the pair distribution function and discriminates between possible models to fit this function.

Coherent scattering: definition of $I(Q,t)$, $G(r,t)$ and $S(Q,\omega)$

The various properties of interest for the scattering system, atomic positions for example, are related to the scattering cross-sections. This relationship is expressed in terms of *correlation*

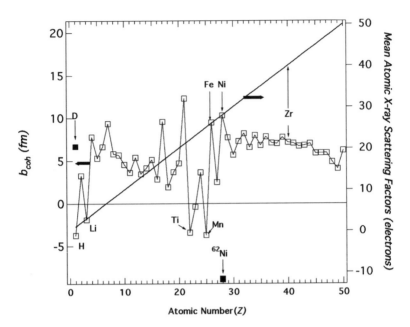

Figure 5. Irregular variation of coherent neutron scattering length (*b*, open squares, left axis) and regular monotonic increase in mean atomic X-ray scattering factor at $\theta = 0$ (line, right axis) with atomic number. Variation of *b* among isotopes of the same element also occurs and is illustrated in selected case of ^2H (deuterium) and ^{62}Ni by the filled squares. Neutron data taken from *http://www.ncnr.nist.gov/resources/n-lengths/list.html*.

functions, a tool that describes on average how random variables at two different points in space or time vary with respect to one-another. An excellent introduction to correlation functions and their use in scattering is provided in several introductory texts (Giacovazzo 1992). The correlation functions appropriate to neutron scattering provide insight into the physical significance of the terms that occur in the scattering cross-sections. For reference, we can now define the following functions (Squires 1996):

repeating expression (8) for the coherent scattering cross section

$$\left(\frac{d^2\sigma}{d\Omega dE'}\right)_{coh} = \frac{\sigma_{coh}}{4\pi}\frac{k'}{B}\frac{1}{2\pi\hbar}\sum_{jj'}\int_{-\infty}^{\infty}\left\langle\exp\{-i\mathbf{Q}\cdot\mathbf{r}_{j'}(0)\}\exp\{-i\mathbf{Q}\cdot\mathbf{r}_j(t)\}\right\rangle\times\exp(-i\omega t)\,dt$$

we define the intermediate function $I(\mathbf{Q},t)$

$$I(\mathbf{Q},t) = \frac{1}{N}\sum_{j,j'}\left\langle\exp\{-i\mathbf{Q}\cdot\mathbf{r}_{j'}(0)\}\exp\{i\mathbf{Q}\cdot\mathbf{r}_j(t)\}\right\rangle \qquad (10)$$

where *N* is the number of nuclei in the scattering system. The term *intermediate scattering function* is sometimes used and this expression include the scattering lengths, $b_{j'}b_j$

The functions $G(\mathbf{r},t)$ and $S(\mathbf{Q},t)$ are defined as

$$G(\mathbf{r},t) = \frac{1}{(2\pi)^3}\int I(\mathbf{Q},t)\exp\{-i(\mathbf{Q}\cdot\mathbf{r})\}\,d\mathbf{Q} \qquad (11)$$

and

$$S(\mathbf{Q},\omega) = \frac{1}{2\pi\hbar} \int I(\mathbf{Q},t) \exp(-i\omega t) \, dt \qquad (12)$$

From the inverse relations for Fourier transforms

$$I(\mathbf{Q},t) = \int G(\mathbf{r},t) \exp(i\mathbf{Q}\cdot\mathbf{r}) \, d\mathbf{r}$$

$$I(\mathbf{Q},t) = \hbar \int S(\mathbf{Q},\omega) \exp(i\omega t) \, d\omega$$

thus

$$G(\mathbf{r},t) = \frac{\hbar}{(2\pi)^3} \int S(\mathbf{Q},\omega) \exp\{-i(\mathbf{Q}\cdot\mathbf{r}) - \omega t\} \, dQ \, d\omega$$

$$S(\mathbf{Q},\omega) = \frac{1}{2\pi\hbar} \int G(\mathbf{r},t) \exp\{i(\mathbf{Q}\cdot\mathbf{r}) - \omega t\} \, dr \, dt$$

$G(\mathbf{r},t)$ is the *time-dependent pair-correlation function* of the scattering system. $S(\mathbf{Q},\omega)$ is the *structure* function, *dynamical structure function, coherent scattering function*, also referred to as the *scattering function* (Squires 1996) or scattering law. By inspection $S(\mathbf{Q},\omega)$ is the Fourier transform of $G(\mathbf{r},t)$ in space and time $(dr \, dt)$. The intermediate scattering function $I(\mathbf{Q},t)$ is the Fourier transform of $G(\mathbf{r},t)$ in space and $S(\mathbf{Q},\omega)$ is the Fourier transform of $I(\mathbf{Q},t)$ in time.

There are similar definitions for the self intermediate, $I_s(\mathbf{Q},t)$, self pair-correlation, $G_s(\mathbf{r},t)$, and incoherent scattering functions, $S_i(\mathbf{Q},\omega)$ (Squires 1996). Inspection of the expressions above shows the functions $S(\mathbf{Q},\omega)$ and $S_i(\mathbf{Q},\omega)$ are related to the cross sections:

$$\left(\frac{d^2\sigma}{d\omega dE'}\right)_{coh} = \frac{\sigma_{coh}}{4\pi} \frac{k'}{k} NS(\mathbf{Q},\omega) \quad ; \quad \left(\frac{d^2\sigma}{d\omega dE'}\right)_{inc} = \frac{\sigma_{inc}}{4\pi} \frac{k'}{k} NS_i(\mathbf{Q},\omega)$$

Simplifications of the expression for coherent neutron scattering

Diffraction. One simplification of Equation (8), Van Hove's expression (Van Hove 1954) for the coherent partial differential cross section

$$\left(\frac{d^2\sigma}{d\Omega dE'}\right)_{coh} = \frac{\sigma_{coh}}{4\pi} \frac{k'}{k} \frac{1}{2\pi\hbar} \sum_{j,j'} \int_{-\infty}^{\infty} \left\langle \exp\{-i\mathbf{Q}\cdot\mathbf{r}_{j'}(0)\} \exp\{i\mathbf{Q}\cdot\mathbf{r}_j(t)\} \right\rangle \times \exp(-i\omega t) \, dt$$

is the scattering law for diffraction, which is mainly an elastic process ($\omega = 0$). Neutron detectors integrate over the energies of scattered neutrons and so rather than measuring the partial differential cross-section we integrate over energy (or ω). The result, for a crystal containing a single isotope, is (Pynn 1990)

$$I(\mathbf{Q}) = b_{coh}^2 \sum_{j,j'} \left\langle \exp(-i\mathbf{Q}\cdot(\mathbf{r}_{j'} - \mathbf{r}_j)) \right\rangle$$

where the atomic positions $\mathbf{r}_j$ and $\mathbf{r}_{j'}$ are evaluated at the same instant. If the nuclei were truly stationary, the thermodynamic averaging brackets,<....>, could be removed from this equation because $\mathbf{r}_j$ and $\mathbf{r}_{j'}$ would be constant. The nuclei in a real solid oscillate, and the position of the atom $\mathbf{r}$, changes with time. It is convenient to describe this oscillation in terms of the deviation from an average position, $\mathbf{r}(t) = \mathbf{u}(t) + <r>$, where $\mathbf{u}$ is the deviation $<r>$ is the time-averaged position of the nucleus about its equilibrium positions. When this is taken into account, the thermodynamic average introduces another factor, called the Debye-Waller factor (DWF), and the expression for $I(\mathbf{Q})$ then becomes

$$I(\mathbf{Q}) = b_{coh}^2 \sum_{j,j'} \exp[-i\mathbf{Q}\cdot(\mathbf{r}_{j'} - \mathbf{r}_j)] \exp\left(-\frac{1}{2}Q^2 \left\langle u^2 \right\rangle\right) \equiv S(\mathbf{Q})$$

where $\langle u^2 \rangle$ is the average of the square of the displacement of an atom from its equilibrium position and diffracted intensity is now also called $S(\mathbf{Q})$, the structure factor. This equation is the basis of any crystallographic analysis of neutron-diffraction data. The effect of lattice vibrations is to diminish the intensity of scattering by the factor $\exp(-\frac{1}{2}\mathbf{Q}^2\langle u^2 \rangle)$ but does not broaden Bragg scattering, described below.

Greater simplification in the definition of $S(\mathbf{Q})$ arises for the case of a perfectly periodic lattice. Ignoring the DWF for the moment, we represent a monatomic infinite 1D crystal by a set of equally spaced delta functions.

$$\rho(x) = \sum_{n=-\infty}^{\infty} \delta(x - na)$$

where a is the unit cell repeat distance. The Fourier transform of ρ will be

$$F(x^*) = \frac{1}{a} \sum_{h=-\infty}^{\infty} \delta\left(x^* - \frac{h}{a}\right) = \frac{1}{a} \sum_{h=-\infty}^{\infty} \delta(ax^* - h)$$

where δ is the Dirac delta function again, n is an integer and $a^* = 2\pi/a$. The Fourier transform of a 1D lattice with period a is a 1D lattice with period $1/a$ represented by delta functions with weight $1/a$. For a 3D lattice, scattering only occurs at discrete set of wave-vectors, $\mathbf{K}$, forming the *reciprocal lattice*

$$\mathbf{K} = h a^* + k b^* + l c^* \tag{13}$$

where a^*, b^* and c^* are the reciprocal lattice vectors and h, k and l are integers. The Bragg condition for allowed scattering now restricts $\mathbf{Q} = \mathbf{K}$. For more than one atom in the unit cell we can denote

$$\left\langle \mathbf{r}_{j'} \right\rangle - \left\langle \mathbf{r}_j \right\rangle = \mathbf{r}_k + \mathbf{r}_n - \mathbf{r}_m$$

where $\mathbf{r}_k$ specifies the separation between unit cells, and $\mathbf{r}_n$ and $\mathbf{r}_m$ are the position vectors of the nth and m^{th} atoms within the unit cell. Then using this equation in combination with Equation (5), which relates the scattering cross-section to the square of the scattering amplitude, for coherent elastic scattering in this case.

$$\frac{d\sigma_{coh}}{d\Omega} = \frac{\delta(\mathbf{Q} - \mathbf{K})}{n_c} \sum_{n,m} b_n b_m \exp(-i\mathbf{Q} \cdot (\mathbf{r}_m - \mathbf{r}_n)) \tag{14}$$

$$= \frac{\delta(\mathbf{Q} - \mathbf{K})}{n_c} \sum_{n,m} b_n b_m \exp(-i\mathbf{K}_{hkl} \cdot (\mathbf{r}_m - \mathbf{r}_n))$$

$$= \frac{\delta(\mathbf{Q} - \mathbf{K})}{n_c} \left| \sum_{n,m} b_n b_m \exp(-i\mathbf{K}_{hkl} \cdot \mathbf{r}_n) \right|^2$$

where the sum over n is now taken only over all the atoms in the unit cell, n_c. The term

$$F_{hkl} = \sum_n b_n \exp(-i\mathbf{K}_{hkl} \cdot \mathbf{r}_n) \tag{15}$$

is the well-known crystallographic structure factor that gives the intensity of the Bragg peak located at $\mathbf{Q} = \mathbf{K}_{hkl}$ where h, k and l are the integers defining the particular reciprocal lattice vector in Equation (13). The intensity of the Bragg peak, Equation (14), is related to the positions of the nuclei, $\mathbf{r}_n$, within the unit cell. The enormous simplification in the problem that comes about when the sample is crystalline is appreciated by comparing Equations (14) and (15) with the general equations for the cross section. In the general case the sums run over all the atoms in the sample; in case of crystalline materials, the sum runs only over the atoms in the unit cell. Intermediate cases occur, as in the atomic arrangements of nano-materials where the periodicity might be imperfect. These situations are treated explicitly in the following chapters.

SUMMARY OF FIELDS OF APPLICATION

In the chapters that follow, principles and techniques of neutron scattering and absorption are discussed. It is useful at this stage to summarize comparisons and distinctions between neutron, and X-ray scattering and absorption (Bacon 1975) the techniques most used in the characterization of earth materials (Table 3). Neutrons contribute significant information in: 1) structural investigations of solids, where the aim is to discover the positions of light elements, particularly hydrogen; 2) investigation of magnetic materials; 3) problems of order-disorder where distinction is made between elements of neighboring atomic number, especially those of low Z where application of anomalous X-ray scattering is experimentally difficult; 4) structural investigation of liquids and glasses where the isotropic scattering of neutrons, which does not fall off with angle, means the measurements are easily made to high values of $\sin\theta/\lambda$; 5) The relative transparency of materials to neutrons means it is easier to find containers for non-ambient studies at low T, high T and high PT. A comparison of X-ray and neutron properties and their applicability is summarized in Table 3.

The chief impediment to the wider use of neutron scattering and absorption techniques (Table 3) is the relative brightness attainable at neutron sources and the restrictions this limitation places on the smallest sample that can be studied at neutron sources. By way of comparison the brightness, in terms of photons/neutrons scattered $s^{-1}\,m^{-2}$ sterradian^{-1} for a typical neutron source is 10^{15} while this figure is 10^{16} for a rotating anode X-ray source in the laboratory, and 10^{33} for a typical undulator source at a 3rd generation synchrotron such as the Advanced Photon Source, respectively. While the incident neutron flux will improve with the advent of advanced spallation sources (Vogel and Priesmeyer 2006, this volume) the gains over the next decade might amount to an order of magnitude. The tremendous advantages of X-rays sources, in terms of brightness, continues to grow with the commissioning of ever more powerful (Bentson et al. 2003; Shen et al. 2003) machines that promise gains of 3-9 orders of magnitude. These gains in X-ray flux have allowed (Table 3) selected experiments in magnetic and inelastic scattering to be performed with X-ray sources (Gibbs et al. 1985, 1988, 1999; Gibbs 1993; Islam et al. 1998) that were once the sole domain of neutron sources. For example, while an X-ray incident on an electron is scattered by both the electron's charge and its magnetic moment, the charge (Thomson) scattering is the dominant mechanism. However magnetic X-ray scattering was observed from single crystals of antiferromagnetic NiO (de Bergev and Brunel 1972) using a fixed-target X-ray tube. The intensities of the magnetic reflections were about 10^{-8} smaller than the Bragg peaks. Despite the fact that the magnetic X-ray scattering cross section is substantially smaller than the Thomson cross section for charge measurable signals are obtained with the intense synchrotron radiation beams available from wiggler and undulator sources, especially when single crystals are used and the beam is tuned to favorable edges (Gibbs 1993; Gibbs et al. 1988). Much the same can be said for inelastic X-ray scattering—the extraordinary intensity and brightness available at modern X-ray sources allows for the observation of signals resulting from very small scattering cross sections.

Despite inroads of X-ray-based techniques into traditional strongholds of neutron scattering, the neutron has unique properties (Tables 1-3) that make routine the study of condensed matter with light elements, magnetic structure of powders, and other applications discussed in this book. Apart from these applications, the dependence of the neutron scattering length on isotope, the existence of null scattering alloys and the ability to tune average scattering lengths using isotopes of the same element (Breger et al. 2005) are properties with no equivalent in X-ray scattering or absorption.

Table 3. Comparative properties of X-rays and neutrons for diffraction studies*

Property	X-rays	Neutrons
Wavelength	Characteristic line spectra (CuK_α) or wavelength band at synchrotron storage rings	Wavelength band either separated from Maxwellian spectrum by monochromator at reactor sources or analyzed in terms of time-of-flight at spallation sources (Chapter 2)
Energy, for λ = 1Å	10^{18} h	10^{13} h. Same order as energy quantum of crystal vibrations
General nature of scattering	Electronic charge (Thomson) scattering	Nuclear (+ moment scattering – see next point)
	Form factor dependence on $(\sin\theta)/\lambda$	Scattering is isotropic, no angular dependence; irregular variation with Z; dependent on structure of nucleus and only determined empirically by experiment
	Angular dependence polarization factor; regular increase in scattering amplitude with Z and calculable from known electronic configurations	
	No differences amongst isotopes	Amplitude is different for different isotopes and depends also on nuclear spin, giving nuclear and spin incoherence
	Phase change of 180° on scattering	Phase change of 180° for most nuclei but H, ^{7}Li, Ti, V, Mn, ^{62}Ni give zero phase change
Magnetic scattering	Magnetic X-ray scattering cross section is substantially smaller than the Thomson cross section for charge scattering; typically $<10^{-5}$.	Additional scattering by atoms with magnetic moments—diffuse scattering by paramagnetic materials and coherent diffraction peaks for ferro- and antiferromagnetic materials—comparable to nuclear scattering at low angle.
		Amplitude of scattering falls off with $(\sin\theta)/\lambda$
		Amplitude calculable from magnetic moment; different for ions with different spin quantum numbers, e.g., Fe^{2+}, Fe^{3+}
Absorption coefficient	Large, true absorption being much greater than scattering, increasing with Z, $\mu \sim 10^2 - 10^3$	Absorption usually very small and less than scattering, $\mu \sim 10^{-1}$. Exceptions, e.g., B, Cd and rare earths. Varies with isotopes
Thermal effects	Reduction of coherent scattering by Debye exponential factor	
Inelastic scattering	Diffuse streaks observed	
	Small change in λ	Appreciable change in λ; frequency wave number relation can be found for lattice vibrations and magnetic spin waves
Absolute intensity measurement	Difficult	Straightforward, particularly for powder methods
	Interpretation depends on precise knowledge of atomic scattering factor curves	

* Adapted from Bacon (1975).

Neutron powder diffraction

The most widely used and productive technique in terms of users served at reactor and spallation neutron sources (Vogel and Priesmeyer 2006, this volume) is neutron diffraction, and neutron powder diffraction in particular. This is a straightforward entrée for those who have used powder X-ray techniques in the home laboratory. In principle, the experimental set up required to perform powder neutron and powder X-ray diffraction are similar, and software, such as Rietveld refinement programs (Knorr and Depmeier 2006, this volume) are designed to analyze either X-ray or neutron data or to use them simultaneously to analyze a structural model. The designs of instruments constructed on both steady-state reactor neutron sources and synchrotrons have been very similar to those of conventional laboratory diffractometers (Fig. 6a-c): 1) a source of radiation collimated by a set of slits, and perhaps with an incident beam monochromator; 2) a powdered sample 3) optional diffracted beam slits and/or monochromators; 4) a detector. For both neutron and the X-ray cases the positions of the "Bragg reflections" are easily calculated in terms of angle and wavelength (energy) using Bragg's equation:

$$\lambda = 2d \sin\theta \tag{16}$$

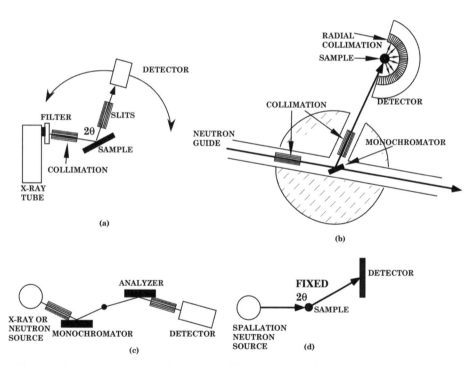

Figure 6. Typical geometries for (a) a laboratory X-ray source with incident and diffracted beams in the vertical plane and for (b) a neutron powder diffractometer at a reactor source. For X-ray scattering the characteristic radiation is chosen by inserting a filter (e.g., Ni filter for CuK$_\alpha$ radiation) while at a steady state (reactor) source a single crystal monochromator selects the desired, narrow, wavelength range from the Maxwellian distribution of energies emerging from the moderated beam (see text above). After the sample, a radial collimator improves signal to noise discrimination for the scattering from the sample detected in a position sensitive detector. If inelastic scattering is to be analyzed then the triple axis (monochromator-sample-analyzer) geometry shown schematically in (c) is adopted at steady state sources with single crystals selecting the energy of the incident and scattered neutrons. (d) At spallation (pulsed) sources the energies of the neutrons scattered at fixed 2θ are determined according to their time of flight along the flight path relative to the origin of the neutron pulse.

For pulsed neutron sources (both accelerator driven, such as ISIS, SNS, IPNS, LANSCE in the USA, and the pulsed reactor source JINR), white beam techniques using time-of-flight methods have been developed (Fig. 6d).

Conventional powder diffraction experiments utilizing monochromatic beams of radiation of known wavelength, λ, are standard on laboratory or synchrotron X-ray sources, or at steady state neutron reactor sources. The scattered radiation is measured as a function of scattering angle, 2θ, either by step scanning a small detector (Fig. 6a), by using a large position sensitive detector (PSD) covering a range of scattering angles (Fig 5b) or perhaps by using an integrating detector such as an imaging plate or charge coupled device. In those cases where it is important to define the energy of the scattered radiation, in inelastic scattering studies for example, an analyzer crystal (Fig. 6c) might be used. This triple axis geometry (monochromator-sample-analyzer) forms the basis for the workhorse instrument for inelastic studies at reactor sources (Shirane et al. 2002).

In time-of flight neutron diffraction (Fig. 6d) at spallation neutron sources the sample is irradiated with a pulsed "white" beam, produced at a repetition rate of say 50 or 60 Hz. The beam contains neutrons of different energies (wavelengths). The neutrons travel a known distance, L, from the source to the sample to the detector, which is positioned at a fixed angle 2λ (Fig. 6d), and their arrival times at the detector is recorded. The neutron's wavelength is related to its momentum through the de Broglie relationship ($\lambda = \mathbf{h}/mv$) and so we can calculate the d-spacing. By combining Bragg's Law (Eqn. 16) with the de Broglie relationship, and substituting values for Planck's constant and the neutron mass (Table 1), we obtain the expressions:

$$t = 505.56\, L\, \sin\theta\, d \qquad (17)$$

or

$$d = 1.978 \times 10^{-3}\, t\, (L\, \sin\theta)^{-1} \qquad (18)$$

where time-of-flight t, is measured in micro-seconds, flight path, L, in meters and d-spacing in Ångstrom units.

A fundamental feature of the time-of-flight technique is its ability to measure a complete diffraction pattern using a single, fixed detector (Fig. 5c). In practice, in order to reduce counting times time-of-flight diffractometers typically contain large numbers of detectors, arranged into a series of different banks. The powder diffractometers capable of achieving the highest count rates, or obtaining data from the smallest samples, typically have large numbers of detectors covering as large an area as is consistent with sample access. A good example of such a diffractometer is the ISIS instrument GEM described at *http://www.isis.rl.ac.uk/ disordered/gem/gem_home.htm*.

The advantages of neutron diffraction in providing unique information which is complementary to X-ray diffraction is illustrated in Figure 7, where calculated X-ray and neutron powder diffraction patterns for ice-Ih (Goto et al. 1990) are compared. Although incoherent scattering from hydrogen will dominate the neutron diffraction pattern of H_2O ice-Ih, the contrast provided with the X-ray or neutron experiment is large. A solution to the well known problem with incoherent contributions to the neutron powder diffraction pattern, minimized for the single crystal neutron diffraction technique because of the inherently high signal to-noise discrimination, is the use of deuterated samples. Hydrogen and deuterium are chemically very similar (Jeffrey 1997). The use of deuterated powder samples for the study of hydrogen bonding in minerals is now a routine part of the characterization of such materials.

Geometry of elastic and inelastic neutron scattering experiments

The laws of energy and momentum conservation governing all diffraction and inelastic neutron scattering (INS) experiments are well known

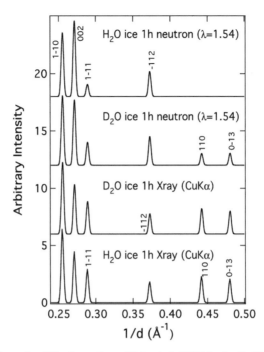

Figure 7. Calculated powder diffraction patterns (Goto et al. 1990) for ice-Ih (*P6₃/mmc*; *a* = 4.523, *c* = 7.367 Å) in deuterated and hydrogenated forms. Only the coherent elastic scattering is included; the incoherent scattering from hydrogen would add considerable background. Note the dramatic changes in calculated peak heights between D_2O and H_2O ice-Ih in the neutron case and between the neutron and X-ray diffraction patterns of H_2O ice-Ih; note also the lack of discrimination between D_2O and H_2O in the X-ray case. Patterns were calculated using CrystalDiffrac© software (*http://www.crystalmaker.com*).

$$\mathbf{Q} = \mathbf{k}' - \mathbf{k} \qquad momentum\ conservation \qquad (19)$$

$$|\mathbf{Q}| = \mathbf{k}^2 + \mathbf{k}'^2 - 2\mathbf{k}\mathbf{k}'\cos\theta_s \qquad (20)$$

$$\hbar\omega = E - E' \qquad energy\ conservation$$

Recall (Table 1) that in the equations above the prime refers to the final state, that the wave-vector magnitude $k = 2\pi/\lambda$, where λ is the neutron wavelength, and that the momentum transferred to the crystal is $h\mathbf{Q}/2\pi = \hbar\mathbf{Q}$. The angle between the incident and final beams is $2\theta_s$ and the energy transferred to the sample is $\hbar\omega$. Again referring to Table 1, the dispersion relation for the neutron is:

$$E = \frac{1}{2}mv^2 = \frac{\hbar^2\mathbf{k}^2}{2m} \qquad \text{or} \qquad E\ [\text{meV}] = 2.072\ \mathbf{k}^2[\text{Å}^{-2}]$$

and the energy conservation law can be written as

$$\hbar\omega = \frac{\hbar^2}{2m}\left(\mathbf{k}^2 - \mathbf{k}'^2\right) \qquad (21)$$

In any scattering experiment we measure the properties of the incident and scattered (') beams and infer the energy and momentum transferred to the sample via Equations 19 and 21.

The geometry of inelastic and elastic scattering is best described in terms of the reciprocal lattice, extending the treatment of the elastic case to that of inelastic scattering. For elastic and inelastic scattering Ewald's construction for a 2D reciprocal lattice is illustrated in Figures 7a and 7b, respectively. The 3D reciprocal lattice, of course, can be observed if we irradiate the real (single crystal) lattice with a beam of collimated neutrons, or other short wavelength radiation. In crystallography the reciprocal lattice (RL) is the set of all vectors $\mathbf{G}$ such that $\exp(2\pi i \mathbf{G} \cdot \mathbf{R}) = 1$ for all real space lattice points. Reciprocal space is also referred to as Fourier space, k-space, or momentum space (Fig. 8). From the crystallographic definition, the RL is constructed by taking the planes (hkl) in the real Bravais lattice (BL) and drawing a vector of magnitude $1/d_{hkl}$ normal to those planes (Fig. 8a); the points at the ends of these vectors define the RL. An example of this construction is given in Figure 8a.

The familiar Bragg equation, which describes the geometric condition for diffraction, can be derived by considering the Ewald construction in 2D (Fig. 8a). From the origin O of the RL a point C is defined by drawing the vector $\mathbf{k}/2\pi$, of magnitude $1/\lambda$ parallel to the direction of the incident neutrons. From C we draw a vector of magnitude $1/\lambda$ in the direction of the scattered neutrons, to terminate at the point P. Only when P happens to lie at a RL point (hkl) will a Bragg reflection take place, since it follows from Figure 8(a) that

$$OP = 2\frac{1}{\lambda}\sin\theta_B = \frac{1}{d_{hkl}}$$

so that

$$\lambda = 2d_{hkl}\sin\theta_B$$

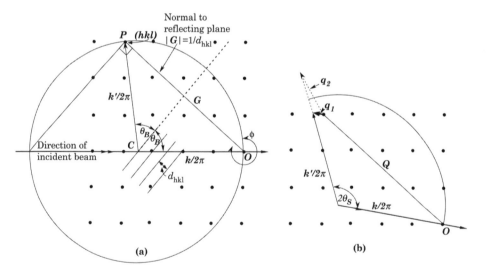

(a) **(b)**

Figure 8. Two dimensional (2D) reciprocal lattice (RL) representation of Ewald's construction for (a) elastic, and (b) inelastic scattering. In each case $\mathbf{k}$ and $\mathbf{k}'$ are wave-vectors of incident and scattered neutrons, respectively. $\mathbf{G}$ is a reciprocal lattice vector, of magnitude $1/d_{hkl}$; $\mathbf{Q}$ is the general vector $\mathbf{k}' - \mathbf{k}$. C is the center of a circle of radius $1/\lambda$, O is the origin of the RL and P is an arbitrary RL point on the circle. The planes shown near C spaced d_{hkl} apart in the real Bravais lattice, are represented by P in RL. Different RL points can be brought to intercept the Ewald circle by rotating the sample (ϕ). In (b) $\mathbf{q}_1$ is the wave-vector of the phonon given up to the crystal vibrations and neutron energy loss ($|\mathbf{k}'| < |\mathbf{k}|$). In the alternative case, indicated by the dotted lines, $\mathbf{q}_2$ is the wave-vector of an annihilated phonon and neutron energy gain ($|\mathbf{k}'| > |\mathbf{k}|$).

which means that the Bragg equation is satisfied for the *hkl* planes which correspond to *P* and a reflection will be produced.

In the special case satisfying the Bragg condition illustrated in Figure 8(a) there is no change in the neutron energy and $|\mathbf{k}| = |\mathbf{k}'|$ and

$$\mathbf{Q} = \mathbf{G} = \mathbf{k}' - \mathbf{k}$$

and from Equation (20) and Figure 8a

$$|\mathbf{Q}| = |\mathbf{G}| = 2\mathbf{k}\sin\theta_B$$

In a diffraction experiment (Ross and Hoffman 2006, this volume) the magnitude of **G** is controlled by adjusting the angle $2\theta_B$ between **k** and **k'**. The orientation of **G** within the RL is set by rotating the sample. For example, with **k** fixed (fixed monochromatic source) any point in the reciprocal space can be measured by an appropriate choice of **k'**, $2\theta_B$, and the orientation ϕ of the sample (Fig. 8a).

The situation for inelastic scattering is a little more complicated. In this case $|\mathbf{k}| \neq |\mathbf{k}'|$ since a difference is needed to transfer energy to the sample. For an inelastic experiment using the triple-axis instrument (Fig. 6c) one again holds the **k** constant while varying **k'**. For a single crystal sample (Fig. 8b) it is convenient to reference the momentum transfer to the nearest reciprocal vector

$$\mathbf{Q} = \mathbf{G} + \mathbf{q}$$

The measured spectrum can be interpreted straightforwardly (Shirane et al. 2002) if **Q** is held constant while the energy transfer is varied. Figure 8b illustrates two cases. In the first case, vector $\mathbf{q}_1$ with $|\mathbf{k}| > |\mathbf{k}'|$ and $\hbar\omega > 0$ represents the creation of a phonon involving transfer of energy from the neutron to the sample; this is equivalent Stokes scattering in optical spectroscopy. In the second case ($\mathbf{q}_2$) $|\mathbf{k}| > |\mathbf{k}'|$ and $\hbar\omega > 0$ so that the sample gives up a quantum of energy to the neutron beam and a phonon is annihilated and we have neutron energy gain. This is anti-Stokes scattering.

In order to keep **Q**, and thus **q**, constant while varying **k'**, the scattering angle $2\theta_s$ must change as well as the relative orientation of the crystal (ϕ). The science enabled by INS experiments and the various instruments used to carry out the measurements are presented by Loong (2006, this volume).

CONCLUSION

This brief introduction into the properties of neutrons, and the geometric and mathematical aspects of some important techniques, provides the basis for some of the more important applications of neutron scattering in earth and materials science. While examples in this chapter emphasize structural and elastic properties and highlight crystallography, the following chapters illustrate a broad range of applications far beyond crystal structures, including kinetics of chemical reactions, microstructural details and mechanical properties such as residual stresses.

ACKNOWLEDGMENTS

Partial financial support was provided by the U.S. National Science Foundation through its CHE (0221934) EAR (0510501) and DMR (0452444) programs and by the U.S. Department of Energy (DE-FG02-03ER46085).

REFERENCES

Bacon GE (1962) Neutron Diffraction. Oxford University Press

Bacon GE (1975) Neutron Diffraction. Clarendon Press

Bentson L, Bolton P, Bong E, Emma P, Galayda J, Hastings J, Krejcik P, Rago C, Rifkin J, Spencer CM (2003) FEL research and development at the SLAC sub-picosecond photon source, SPPS. Nucl Instrum Methods Phys Res A 507:205-209

Breger J, Dupre N, Chupas PJ, Lee PL, Proffen T, Parise JB, Grey CP (2005) Short- and long-range order in the positive electrode material, Li(Ni,Mn)$_{0.5}$O$_2$: A joint X-ray and neutron diffraction, pair distribution function analysis and NMR study. J Am Chem Soc 127:7529-7537

Cowley JM (1975) Diffraction Physics. North-Holland

de Bergev F, Brunel M (1972) Observation of magnetic superlattice peaks by X-Ray-diffraction on an antiferromagnetic NiO crystal. Phys Lett A 39:141-146

Dianoux A-J, Lander G (2002) Neutron Data Booklet. Institute Laue-Langevin, Grenoble

Flook RJ, Freeman HC, Scudder ML (1977) X-ray and neutron-diffraction study of aqua(L-glutamato)cadmium(II) hydrate. Acta Crystallogr B33:801-809

Giacovazzo C (1992) Fundamentals of Crystallography. Oxford Science Publications

Gibbs D (1993) X-ray magnetic scattering - new developments. J Appl Phys 73:6883-6883

Gibbs D, Harshman DR, Isaacs ED, McWhan DB, Mills D, Vettier C (1988) polarization and resonance properties of magnetic-X-Ray scattering in holmium. Phys Rev Lett 61:1241-1244

Gibbs D, Hill JP, Vettier C (1999) Recent advances in X-ray magnetic scattering. Phys Status Solidi B 215: 667-678

Gibbs D, Moncton DE, Damico KL, Bohr J, Grier BH (1985) Magnetic-X-ray scattering studies of holmium using synchrotron radiation. Phys Rev Lett 55:234-237

Goto A, Hondoh T, Mae SJ (1990) The electron-density distribution in ice Ih determined by single-crystal X-ray-diffractometry. J Chem Phys 93:1412-1417

Hill JP, Kao CC, Caliebe WAC, Gibbs D, Hastings JB (1996) Inelastic X-ray scattering study of solid and liquid Li and Na. Phys Rev Lett 77:3665-3668

Islam Z, Detlefs C, Goldman AI, Bud'ko SL, Canfiield PC, Hill JP, Gibbs D, Vogt T, Zheludev A (1998) Neutron diffraction and X-ray resonant exchange-scattering studies of the zero-field magnetic structures of TbNi$_2$Ge$_2$. Phys Rev B 58:8522-8533

Jeffrey GA (1997) An Introduction to Hydrogen Bonding. Oxford University Press

Knorr K, Depmeier W (2006) Application of neutron powder-diffraction to mineral structures. Rev Mineral Geochem 63:99-111

Loong C-K (2006) Inelastic scattering and applications. Rev Mineral Geochem 63:233-254

Materlik G, Sparks CJ, Fischer K (1994) Resonant Anomalous X-ray Scattering. Elsevier

Proffen T (2006) Analysis of disordered materials using total scattering and the atomic pair distribution function. Rev Mineral Geochem 63:255-274

Pynn R (1990) The Mathematical Foundations of Neutron Scattering. *http://www.fas.org/sgp/othergov/doe/lanl/pubs/00326652.pdf*

Ross NL, Hoffman C (2006) Single-crystal neutron diffraction: present and future applications. Rev Mineral Geochem 63:59-80

Sears VF (1995) Scattering lengths for neutrons. *In*: International Tables for Crystallography, Volume C. Wilson AJC (ed) Kluwer, p 383-391

Shen Q, Bilderback DH, Finkelstein KD, Bazarov IV, Gruner SM (2003) Coherent X-ray imaging and microscopy opportunities with a diffraction-limited Energy Recovery Linac (ERL) synchrotron source. J Physique IV 104:21-26

Shirane G, Shapiro SM, Tranquada JM (2002) Neutron Scattering with a Triple-Axis Spectrometer. Cambridge University Press

Singh AK, Ramasesh S (1968a) Use of neutron anomalous scattering in crystal structure analysis. I. Non-centrosymmetric structures. Acta Crystallogr B24:35-39

Singh AK, Ramasesh S (1968b) Use of neutron anomalous scattering in crystal-structure analysis. 2. Centrosymmetric structures. Acta Crystallogr B24:1701-1704

Squires GL (1996) Introduction to the Theory of Thermal Neutron Scattering. Dover Press

Stone HJ, Tucker MG, Meducin FM, Dove MT, Redfern SAT, Le Godec Y, Marshall WG (2005) Temperature measurement in a Paris-Edinburgh cell by neutron resonance spectroscopy. J Appl Phys 98:064905

Van Hove L (1954) Correlations in space and time and born approximation scattering in systems of interacting particles. Phys Rev 95:249-262

Vogel SC, Priesmeyer H-G (2006) Neutron production, neutron facilities and neutron instrumentation. Rev Mineral Geochem 63:27-58

Von Dreele RB (2006) Neutron Rietveld refinement. Rev Mineral Geochem 63:81-98

Waller I, Hartree DR (1929) Intensity of total scattering X-rays. Proc R Soc London Ser A 124:119-142

Winkler B (2006) Applications of neutron radiography and neutron tomography. Rev Mineral Geochem 63: 459-471

Reviews in Mineralogy & Geochemistry
Vol. 63, pp. 27-57, 2006
Copyright © Mineralogical Society of America

2

Neutron Production, Neutron Facilities and Neutron Instrumentation

Sven C. Vogel

Los Alamos Neutron Science Center
Los Alamos National Laboratory
Los Alamos, New Mexico, 87545, U.S.A
e-mail: sven@lanl.gov

Hans-Georg Priesmeyer

Geesthacht Neutron Scattering Facility
GKSS Research Center
21502 Geesthacht, Germany
e-mail: hans-georg.priesmeyer@gkss.de

NEUTRON GENERATION

The main natural source of free neutrons is secondary radiation from *cosmic radiation* (creation of particles by interactions of cosmic radiation particles with atoms of the earth's atmosphere). The main sources of neutrons generated artificially are nuclear reactors. Free neutrons have an average lifetime of about 888 seconds (e.g., Mampe et al. 1989a,b: 887.6 ± 3 s) and decay according to

$$n \rightarrow p + e^- + \bar{v} \qquad (1)$$

into a proton p, an electron e^- and an anti-neutrino $\bar{v}$. Hence, free neutrons must be produced shortly before their use. Besides reactors, where neutrons from nuclear fission reactions after moderation induce further nuclear fissions in a chain reaction, neutrons for research may be generated by a process called spallation. Neutron sources based on these two concepts as well as some special cases of neutron generation are described in more detail below. Figure 1 shows the historical evolution of the performance of neutron sources. In all cases it should be noted that the neutron flux (number of neutrons per unit area and unit time) at the sample is much lower than the peak flux provided by the source. For instance, the spallation neutron source at LANSCE (Los Alamos Neutron Science Center, Los Alamos, U.S.A.) produces 10^{16} neutrons cm^{-2} s^{-1}, but at the HIPPO instrument, with a moderator to sample distance of 9 m, only a flux of 2.4×10^7 neutrons cm^{-2} s^{-1} is available for neutron energies in the "thermal" (< ~0.4 eV) range (see below) suitable for diffraction. Other facilities provide thermal neutron fluxes of similar magnitude. From these low intensities, as compared to the number of atoms in a cubic centimeter of solid material, it is apparent that the radiation damage from thermal neutrons for most materials is negligible, even for days of exposure.

Reactor sources

Nuclear reactors have been, and continue to be, an important and reliable source of neutrons as a research tool with wide application in different scientific fields. Most research reactors allow user access for neutron scattering experiments, with instrumentation that has greatly evolved since the early days of this technique (Shull 1995). As expensive scientific facilities, they tend to be multi-purpose (nuclear physics, nuclear engineering, diffraction, etc.), and many reactors have been operating for more than 30 years.

1529-6466/06/0063-0002$05.00 DOI: 10.2138/rmg.2006.63.2

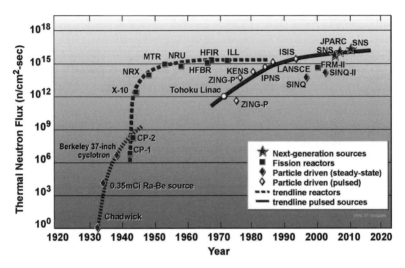

Figure 1. Historical evolution of the performance (peak thermal neutron flux) of neutron sources over time. The Spallation Neutron Source (SNS) has two data points, representing the initial power level and that after the upgrade (Mason et al. 2003).

Nuclear reactors operate on the physical basis of the *fission chain reaction*. In heavy nuclei, such as uranium, the number of neutrons exceeds the number of protons by about 40%. Figure 2 shows a schematic of the fission process. When a neutron is absorbed, for instance, by a ^{235}U nucleus (Fig. 2A), a large amount of energy is transferred into the system. The excited ^{236}U compound nucleus begins to oscillate (Fig. 2B) with increasing amplitude (Fig. 2C), until the repelling Coulomb force drives it into separation (Fig. 2D). Each fission liberates 2 or 3 free neutrons, so that a chain reaction becomes possible. One of these neutrons is definitely required to maintain the chain reaction and, thus, trigger the release of the next generation of neutrons. The number of neutrons produced in a nuclear reactor scales with its thermal energy production: each fissioning nucleus produces energy of about 200 MeV. Thus, approximately 3×10^{16} fission events per second occur per 1 MW reactor power, resulting in about 7.5×10^{16} neutrons per second per MW.

In a research reactor, as many neutrons as possible should be available for the experiments. Careful design of the reactor core and sophisticated neutron optical devices are needed to

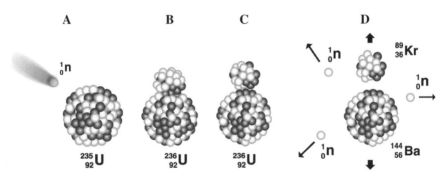

Figure 2. The fission process.

maximize the flux of neutrons at the sample positions of the specific experiments. This requires a very high power density in the reactor core and, therefore, special design features like a compact core and reflectors to reduce neutron loss from the core. Numerous designs of research reactors were developed in the early days of the technology; of these designs the *pool* and *tank type* reactors became the most common ones (Bauer and Thamm 1991). The core of such reactors is a group of fuel elements in a large pool of light or heavy water (^1H_2O or $^2H_2O = D_2O$, respectively). Movable rods containing neutron-absorbing materials like boron or cadmium are distributed among the fuel elements in order to control the fission process by absorbing neutrons and, hence, slowing down the chain reaction. Water moderates (decelerates) the neutrons and cools the reactor, while graphite or beryllium are generally used as reflector materials, reflecting neutrons back into the reactor core. Beam tubes to extract the neutrons from the core region are inserted in the wall of the pool or tank. Figure 3 shows a schematic of the FRM-2 research reactor in Munich, Germany, with the fuel element and control rod in the center of the reactor core and the horizontal beam tubes leading to the instruments. The operation of this modern reactor started in 2004. For safety purposes, the FRM-2 reactor has two independent mechanisms for controlled shutdown of the reactor under all circumstances. The control rod, made of highly neutron absorbing material, is located in the hollow fuel element and controls the reactor power via the amount of control rod area exposed to neutron emitting uranium. It can be decoupled from its driving device to fall down and completely shield the fuel element, hence, stopping the chain reaction. The second, independent shutdown mode are five highly neutron absorbing hafnium shut down rods in the moderator tank of the reactor core. During normal operation, these rods are fully withdrawn. When activated, they drop into positions close to the core and end the chain reaction by absorbing neutrons. Four out of the five rods are sufficient to keep the reactor subcritical, i.e., prevent the chain reaction. Note that both shutdown systems rely on gravity, i.e., they are operational without electrical power. Also note that the neutron beam tubes do not provide a direct sight to the reactor core, markedly reducing the fast neutron and gamma radiation backgrounds (FRM-2 2004). The reactor fuel elements are typically rods of urania (UO_2) of several cm in diameter. Each fuel rod is embedded in a steel or Zircaloy tube (cladding).

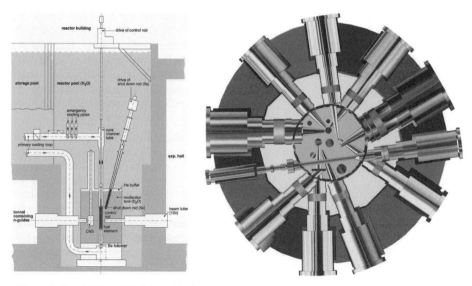

Figure 3. Simplified vertical (left) and horizontal (right) sections of the research reactor FRM-2 at Munich, Germany. The internal diameter of the reactor pool is approximately 5 m (FRM-2 2004).

Neutrons within a reactor have a broad energy distribution, which extends from below 0.001 eV to more than 10 MeV and can roughly be divided into three subgroups:

- Neutrons with energies above 500 keV are called "*fast neutrons.*" They originate from the fission process and their spectral distribution may be approximated by

$$N(E) \propto e^{-E} \sinh \sqrt{2E} \quad (E \text{ in MeV}) \tag{2}$$

- The "*epithermal* range" of neutron energies extends from approximately 200 meV up to 500 keV. In this range, the neutron spectrum is primarily determined by the slowing down process by elastic collisions of the neutrons with moderator nuclei, resulting in a spectral distribution

$$N(E) \propto \frac{1}{E} \tag{3}$$

- Below 200 meV the spectral distribution of the neutrons looks like one of a gas in equilibrium with the thermal motion of the moderator nuclei, resulting in a *Maxwellian distribution*

$$N(E) \propto \frac{E}{k_B T} \cdot e^{-E/k_B T} \tag{4}$$

Neutrons with this energy spectrum are called "*thermal neutrons.*" The average neutron energy emitted from a water moderator at room temperature is 25 meV, corresponding to a neutron wavelength of 0.18 nm.

In general, the absolute intensity of each of these three components depends on the reactor type, the reactor power and the location within the reactor core, with a maximum near the center of the active zone of the core. Some applications, such as small-angle scattering, require neutron energies even below the thermal region. Such neutrons are termed "*cold neutrons.*" They emerge from a so-called *cold source*, which, in fact, is not a source that creates new neutrons, but rather a hydrogenous low-temperature moderator. The neutron wavelength is thereby changed to 0.5 nm and longer.

Since neutrons have become a unique tool to study the structure and dynamics of materials at the atomic level, neutron scattering plays a dominant role among the experimental methods utilized at research reactors. Being electrically neutral, neutrons, unlike charged particles, are not scattered by the electron clouds surrounding the atoms, and thus may penetrate through considerable thickness of material layers. This allows the investigation of bulk samples under differing environmental conditions such as high pressure, high or low temperature, or strong magnetic fields (see below). Figure 4 illustrates the large number of instruments available at the Institut Laue-Langevin (ILL) in Grenoble, France, the world's largest research reactor.

Scattering and absorption of neutrons are competing nuclear processes, whose fractional ratios differ from isotope to isotope. This feature is often used in research reactors to produce radioisotopes. Since almost all materials irradiated by neutrons become radioactive to a certain extent, safe handling of specimens after an experiment is to be assured by health physics measurements. Since the gamma radiation emitted by the nuclei in the sample after neutron capture is characteristic for every nuclide, it is possible to trace elements in the range of parts per billion (ppb). This has led to the installation of equipment into research reactors that allows for so-called neutron activation analysis. At most research reactors, facilities are set up to do neutron radiography (see below and Winkler 2006, this volume) in two or three spatial dimensions. This method may be considered to be complementary to X-ray imaging, because the systematic dependence of transmission contrast prevailing for X-rays does not exist in the case of neutrons. Hence, light elements (especially hydrogenous materials) can be identified and visualized, even if they are embedded within heavy structures.

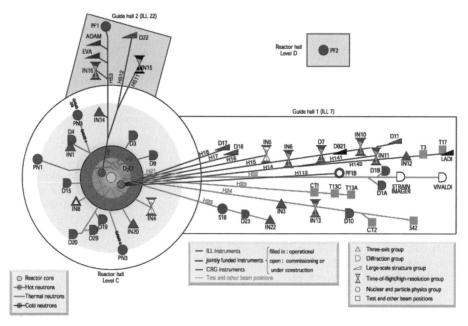

Figure 4. Floor plan of the experimental halls at the Institut Laue-Langevin (ILL), the world's largest research reactor (ILL 2004). ILL (2004) Insititut Laue-Langevin website: *http://www.ill.fr*

The International Atomic Energy Agency keeps track of the international development of research reactor history, which is documented in their research reactor database (*http://www.iaea.org/worldatom/rrdb*).

Spallation sources

For spallation, hydrogen ions with no electron (proton, H^+) or two electrons (H^-) are produced by an ion source and accelerated in vacuum to energies of typically 800 MeV (corresponding to about 84% of the speed of light) and directed to a heavy-metal target (e.g., uranium at the ISIS spallation source in Chilton, U.K., or tungsten at the LANSCE spallation source in Los Alamos, U.S.A.). In a linear accelerator (linac), the ions are accelerated in stages by a standing electric wave (i.e., the points of zero field, the nodes, do not change their location) in a RF (radio frequency) cavity. So-called drift tubes shield the ions from the field when the direction of the electric field would decelerate the ions. In each gap between drift tubes, the ions gain energy and become faster, hence, the length of the drift tubes and gaps has to increase with the ion energy. When the direction of the field has reversed, ions of the opposite charge are accelerated in the gaps, increasing the efficiency of the system. Magnetic fields are used to focus and bunch the ions at the end of each drift tube. The acceleration process for a linear accelerator of this type is illustrated in Figure 5 with H^+ ions being accelerated in the gaps between the drift tubes by the positive half-wave. At this point, the H^- ions travel in the drift tubes and are shielded from the field (H^- ions are not shown in Figure 5). 2.49 ns later (half the period of the 201 MHz frequency wave), the field will have the opposite sign and the H^- ions will be accelerated in the gaps between the drift tubes. At LANSCE, the second stage of the linear accelerator consists of 165 gaps and accelerates the ions to 40% of the speed of light. Another type of accelerator is the synchrotron. At spallation sources both types may be used in combination. Stripper foils are used to remove the electrons from the H^- ions.

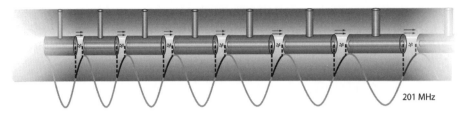

Figure 5. Schematic of a section of the LANSCE accelerator showing the acceleration of protons. The length of the drift tubes and the gaps between the drift tubes are not to scale.

At LANSCE, the positive hydrogen ions, stripped of their electrons, are then injected into a 30-m-diameter Proton Storage Ring (PSR). The PSR converts a 625-μs pulse of negative hydrogen ions into a 125 ns intense burst of protons, which is guided to the spallation target. Proton currents at the brightest spallation sources currently available are 125 μA at LANSCE and 200 μA at ISIS. Figure 6 shows a schematic of the LANSCE target. LANSCE has a split target, consisting of upper and lower targets with the proton beam hitting both targets vertically. Both targets are tungsten cylinders with 10 cm diameter and 7.25 cm and 27 cm length for the upper and lower target, respectively. Even though a so-called slab moderator in a single target piece configuration generates about a factor of two higher thermal flux than a so-called flux-trap moderator, which views the gap between the two targets in split target goemetry, this gain is degraded by the fact that a slab moderator in single target configuration produces about a factor of 20 more high energy neutrons, which comesfrom viewing the target directly. Because of that, the signal to noise ratio favors the flux-trap moderator with the split target configuration. The target-moderator-reflector-shielding (TMRS) module is located in the center of a 3.7 m thick laminated iron/concrete biological shield.

The proton bursts are released from the storage ring to the target 20 (LANSCE) or 50 (ISIS) times per second. When the target is hit by the protons, target nuclei are spalled ("smashed") into many small particles (contrary to fission where basically two nuclei of roughly the same

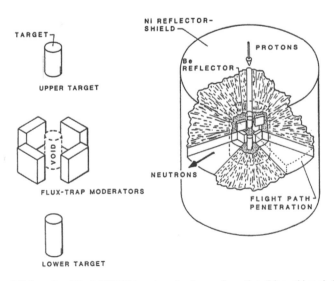

Figure 6. Schematic of the LANSCE target-moderator configuration (Lisowski et al. 1990).

mass and few neutrons are generated) (Windsor 1981, Section 2.2.). Figure 7 shows a schematic of the spallation process: When a pulse of energetic protons (bright grey) hits a tungsten target, each proton that collides with a tungsten nucleus as in (a) causes the nucleus to release ~20 neutrons (dark grey) with different energies, represented by different grey-scales in (b). Moderated neutrons traveling in a particular direction move along a beam line (c). The resulting pulse of neutrons is very short, but as the pulse travels down the beam line, with the higher energy neutrons traveling faster, the pulse stretches out in space, and the arrival times of the neutrons serve to identify their energies and wavelengths (time-of-flight method, as discussed below).

Neutrons generated by spallation in the targets drive the moderator-reflector configuration (Fig. 8). Beryllium has a large neutron scattering cross-section with negligible absorption and is therefore used as a neutron reflector. Moderators slow down the spallation neutrons and are at LANSCE 13 by 13 cm^2 in area and 2.5 cm thick. Three of the LANSCE moderators are ambient temperature water moderators, the fourth is liquid parahydrogen at 20 K. The modular design with target, moderators and reflectors in one unit (Fig. 8, left) allows for changing the entire target/moderator system within 3 weeks, schematically shown in the right part of Figure 8. The proton beam enters the TMRS module from the top, guided by the 90° bending magnets.

As can be seen in Figure 9, in particular the LANSCE accelerator serves as a versatile tool for various scientific applications besides neutron generation for scattering, ranging from proton radiography (King et al. 1999) to nuclear physics and medicine (Lisowski 2005).

The sources at ISIS and LANSCE produce neutron fluxes on the order of 10^{16} neutrons cm^{-2} s^{-1}. Spallation neutron sources have, compared to reactor sources, the advantage that

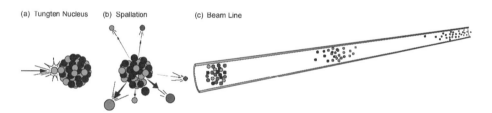

Figure 7. The spallation process (Hurd and Schaefer 2006).

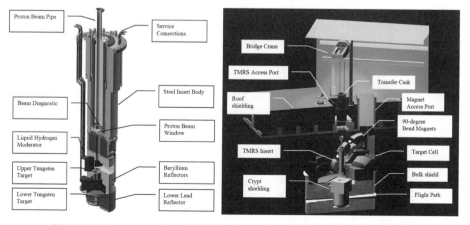

Figure 8. Cutaway view of the LANSCE target-moderator-reflector-shielding (TMRS) module (left) and exchange of the TMRS module (right, Donahue et al. 1999).

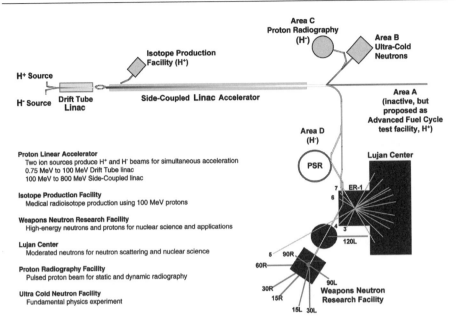

Figure 9. Layout of the LANSCE user facility (Lisowski 2005).

the proton current is virtually unlimited and, hence, the neutron flux can be, in principle, increased without a physical limit. At reactor sources, this is not possible due to reactor safety. Furthermore, spallation neutron sources are, due to the time structure of the neutron flux, ideal for time-of-flight (TOF) measurements. The projected Spallation Neutron Source (SNS) at Oak Ridge National Laboratory, U.S.A., (Fig. 10) is designed for proton currents of 2 mA at 1 GeV proton energy on a liquid mercury target at a repetition rate of 60 Hz, resulting in a peak neutron flux on the order of 10^{17} neutrons cm^{-2} s^{-1} (Mansur et al. 2001; Mason et al. 2002, 2005). The spallation neutron source KENS (Ikeda 2002) under construction at Tokai, Japan, is designed for a 333 μA current of 3 GeV protons, also with a liquid mercury target and generating a peak neutron flux on the order of 10^{17} neutrons cm^{-2} s^{-1}. The U.S. and Japanese spallation neutron sources are scheduled to become available in 2007. The ESS (European Spallation Source, Richter 2002) was designed to a similar flux (ESS 2003), but the project was canceled in 2003 by European authorities.

Other neutron sources

Another way to generate neutron bursts are pulsed reactors: Two neutron reflectors are rotated with fissile material in-between, such that a short chain reaction is induced when both reflectors are in appropriate orientation, generating a neutron pulse. The research reactor IBR-2 at Dubna, Russia, operates according to this principle and generates neutron pulses of 320 μs half-width at 5 Hz with a peak neutron flux of 5×10^{15} neutrons cm^{-2} s^{-1} (JINR 2005). The Swiss spallation neutron source SINQ generates a continuous current of spallation neutrons from a 590 MeV continuous proton beam running at 1.8 mA (i.e., without a storage ring to accumulate the protons to intense bursts), resembling a medium flux research reactor (PSI 2002). The time averaged flux emitted from the moderator is 1.6×10^{13} neutrons cm^{-2} s^{-1}.

Compact, portable neutron sources utilize spontaneous fission, α-emitters or nuclear fusion for neutron generation. Spontaneous fission occurs for instance in the isotope ^{252}Cf,

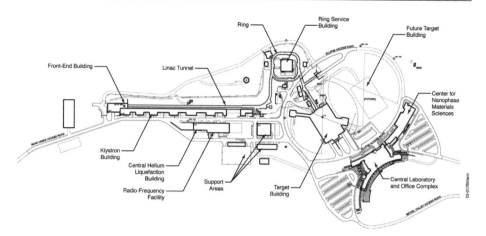

Figure 10. Layout of the Spallation Neutron Source (SNS) in Oak Ridge, Tennessee, U.S.A. (courtesy of Oak Ridge National Laboratory).

which has to be produced from uranium or other transuranic elements by successive neutron capture processes in a nuclear reactor. One milligram of ^{252}Cf produces 2.3×10^9 neutrons s^{-1}. A fresh commercial ^{252}Cf source typically produces of the order of 10^8 neutrons s^{-1}, but ^{252}Cf has a half-life of 2.6 years, resulting in a source intensity decrease to half of its initial intensity over 2.6 years. Neutron generation using α-emitters (Pu, Am, Ra) is based on mixed powders of an isotope decaying by emission of α-particles and a light element, typically Beryllium. The emitted α-particle may generate a neutron based on the nuclear reaction

$$^9\text{Be} + \alpha \rightarrow {}^{12}\text{C} + \text{n} \tag{5}$$

Such sources also produce neutron intensities of 10^8 neutrons s^{-1}. These two types of neutron generators are called isotopic neutron sources and have the disadvantage that they cannot be pulsed or turned off, requiring bulky shielding at all times. Their energy spectrum is broad and the intensities are too low for scattering applications, but they may be used for element identification using neutron activation analysis, simple radiography applications or for detector calibration, development and testing.

Tabletop neutron generators based on nuclear fusion have evolved from large, expensive instruments to compact and affordable devices. These sources utilize deuterium (^{2}H, D) or tritium (^{3}H, T) nuclear fusion reactions to produce neutrons:

$$\text{D} + \text{T} \rightarrow {}^4\text{He} + \text{n} \tag{6}$$

$$\text{D} + \text{D} \rightarrow {}^3\text{He} + \text{n} \tag{7}$$

For the nuclear fusion process, deuterium ions are produced from a metal-hydride based deuterium source, accelerated and steered into a deuterium or tritium target, also made out of a metal hydride containing deuterium or tritium. Since these sources are accelerator based, the neutrons can be produced in pulses and the source can be switched off. Typical neutron intensities are of the order of 10^9 neutrons s^{-1}, but neutron generators with intensities up to 10^{13} neutrons s^{-1} are under development (Reijonen et al. 2005), making such generators potentially interesting for neutron scattering applications in the future. Portable neutron generators are also useful for applications in the field, for instance for the determination of soil humidity (Pouzo et al. 2003) or the detection of drugs and explosives (Womble et al. 2001).

BEAM CONDITIONING

Neutron moderation

Both at steady-state (reactor sources) and pulsed sources, the generated neutrons initially have energies far too high to be suitable to measure properties relevant to solid state physics. Particle kinetic energy E and wavelength λ are related by de Broglie's law:

$$E = \frac{p^2}{2m} = \frac{\left(\hbar k\right)^2}{2m} = \frac{h^2}{2m\lambda^2} \tag{8}$$

where p is the particle's momentum, m its mass, $\hbar$ is Planck's constant h over 2π, and $k = 2\pi/\lambda$ is the wave-vector of the particle. For measurements of the static distribution of the atoms in a solid (e.g., the crystal structure in crystalline or the average atomic distances in amorphous solids), the wavelength must be of the order of the interatomic distance, equivalent to energies of several tens of meV in the case of neutrons (for X-rays, this energy is in the range of several tens of keV). Measuring the dynamic distribution of atoms (i.e., measurements involving phonons) requires neutron energies of the same order of magnitude. Therefore, the neutron energy must be reduced from the order of 2 MeV and 100 MeV for reactor and spallation neutrons, respectively, to the order of 10 meV. The kinetic energy of neutrons can only be changed by collisions with nuclei as neutrons have no charge. By guiding the neutron beam through appropriate materials like water, hydrogen or methane, the neutrons lose energy in collisions with atoms and molecules. This process is called moderation. In thermal equilibrium with the moderator medium, the neutron energies are of the order of magnitude of thermal vibrations of the moderator atoms, and thus the neutrons may gain as well as lose energy. The mean energy of the neutron spectrum after the moderation process is given by

$$E = \tfrac{3}{2} k_B T \tag{9}$$

(k_B is Boltzmann's constant, T the moderator temperature), resulting in an intensity maximum slightly lower than this energy, since the so-called Maxwellian energy distribution of neutrons leaving the moderator is slightly asymmetric. Neutrons with energies corresponding to room temperature are called thermal neutrons. As such, the moderation process in this case is called thermalization. As Bacon (1955, sect. 1.2) states, it is just a 'fortunate circumstance' that these wavelengths and energies are those energies desired for investigations of atomic arrangements and phonon energies in solids, making the moderation process relatively convenient. It is remarkable that a water layer of only 2.5 cm is sufficient to decrease the neutron energy by 10 orders of magnitude. Figure 11 shows examples of neutron flux emerging from moderators at

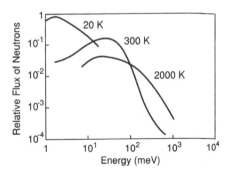
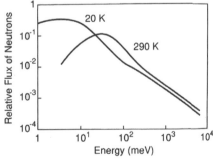

Figure 11. Relative neutron flux as a function of energy for the high-flux reactor at the ILL for different moderator temperatures (*left*). Similar spectra from the spallation source LANSCE (*right*) (Pynn 1990).

different temperatures at the reactor of the ILL and the LANSCE spallation source. The 'hot' source at ILL, with a moderator temperature of 2000 K, consists of radiation heated graphite and is the only one of its kind in the world.

The velocities of thermal neutrons are on the order of 1000 m/s. Thus, relativistic effects can be neglected. The moderation process introduces a broadening of the initially sharp neutron pulse at a spallation source (e.g., ~270 ns initial neutron pulse width at LANSCE, determined by the pulse width of the proton pulse), which can be, for example, approximated for the water moderator used at the HIPPO instrument at LANSCE by

$$\Delta t = \frac{7.1}{\sqrt{E}} \approx 25\lambda \qquad (10)$$

where Δt is the halfwidth in μs of a neutron pulse of wavelength λ (given in Å), and energy E (given in eV). Therefore, the pulse widths are, for instance, 7, 32, and 225 μs for neutron energies of 1 ($\lambda = 0.3°$Å), 0.05 (1.3 Å), and 0.001 eV (9 Å), respectively. Converted to d-spacing resolution $\Delta d/d$, the above figures result in an instrumental contribution to reflection broadening from the moderator of 1.7×10^{-3}. With a 25 mm thick water moderator, both high intensity and high resolution (i.e., small instrumental broadening) moderators can be built by placing a gadolinium foil (e.g., 0.38 mm thick for LANSCE water moderators) either 8 mm or 18 mm away from the spallation target within the water (Windsor 1981, section 3.4). Gadolinium is a strong neutron absorber and the absorption cross-section is proportional to the neutron velocity, hence thermalized, slow neutrons are more likely to be absorbed than still un-moderated fast neutrons. The neutron pulse width or instrument resolution is determined by the distance the already thermalized neutrons travel in the moderator before leaving the moderator towards the sample. Therefore, in the first case the relatively small amount of neutrons already thermalized in the first 8 mm within the moderator is very likely absorbed by the gadolinium while the un-moderated, fast neutrons pass the foil to become moderated in the larger downstream section of the moderator. This results in a relatively high flux at the cost of a relatively large instrumental contribution of the moderator to the width of Bragg-reflections (i.e., poorer resolution). In the second case the gadolinium absorbs the large amount of neutrons thermalized in the larger section towards the source, while the smaller number of neutrons moderated in the 7 mm thick layer towards the sample provides a smaller contribution to the instrumental resolution, resulting in sharper reflection peaks at the cost of a lower neutron flux.

The neutron intensity in an experiment using polychromatic neutrons varies as a function of neutron energy (or wavelength, see Fig. 11). The spectrum of the incident neutrons for a given instrument is required, for instance, to normalize diffraction spectra to allow the comparison of peak intensities. This spectrum can be measured using a "zero scatterer," such as vanadium. A zero scatterer does not produce a coherent diffraction pattern, i.e., no Bragg diffraction peaks occur. The incoherent scattering from the material results in a detected spectrum proportional to the incident neutron intensity. After moderation, the neutron spectrum at a spallation source is an overlay of $1/E$ (Eqn. 3) and Maxwellian (Eqn. 4) distributions, Figure 11. The effect of the normalization on the diffraction pattern of silicon is shown in Figure 12, where the raw diffraction data measured for a silicon powder sample (Fig. 12a) is divided by the incident intensity (Fig. 12b) to produce a normalized diffraction pattern that is used for subsequent analysis (Fig. 12c).

Choppers and filters

Neutron beams emerging from a moderator may require some conditioning for a given application. *Beam choppers* are devices with either a neutron absorbing or highly scattering rotating disk containing one or more openings, or a two-armed rotor (chopper blade) made of neutron absorbing material. In the first case, a constant, typically polychromatic beam is broken up into short pulses, allowing measurement of the energy of a neutron by its time-of-flight for the known distance between sources and the detector. Choppers of the second kind

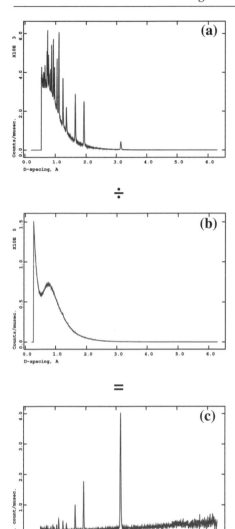

Figure 12. (a) A diffraction pattern measured at a pulsed spallation source (the example shows silicon powder) needs to be normalized by the incident intensity (b) to correct for the wavelength-dependence of the incident intensity (c). The conversion from time-of-flight to d-spacing is performed for the 90° detector bank of the HIPPO instrument to illustrate the neutron intensity available over the d-spacing range of this detector bank.

are called T_0-choppers and used to block high energy neutrons and gamma radiation originating from the spallation process at a spallation neutron source, which would cause severe background or damage to the sample otherwise. They also prevent fast neutrons from arriving at a detector together with slow neutrons from the previous pulse. Figure 13 illustrates this so-called frame overlap: The neutrons in each moderated pulse at a pulsed source begin their flight to the detector at essentially the same time (within a small fraction of a millisecond). However, because they all have different energies (velocities), as time passes they spread out, reaching the same distance from the moderator at different times after departure. That situation is depicted graphically in Figure 13 in plots of distance traveled versus time for neutrons with various energies. Each "explosion" on the time axis represents a pulse of neutrons emerging from the moderator. Note that fast neutrons from, say, the second pulse can reach a detector 12 m distant at the same time as slow neutrons from the first pulse. To prevent such "frame overlap", which interferes with time-of-flight measurements, a spectrometer can be equipped with a frame-overlap chopper. Because of the two rotor blades, the chopper is rotated at half the repetition rate of the neutron pulses (for instance, 10 times per second at LANSCE with a 20 Hz neutron pulse frequency, Fig. 14) and at a certain phase relative to the proton pulse, it blocks neutrons with energies corresponding to wavelengths less than 16 Å. That mode of chopper operation is represented in Figure 13 by a thick horizontal line at a distance of 6.25 m. Changing the phase of the chopper by 90° (moving the thick horizontal line to the left or right a distance equal to its length) blocks neutrons with energies corresponding to the wavelengths between 16 and 32 Å. Figure 14 shows a combination of a T_0-chopper and a frame overlap chopper. The former blocks by scattering and absorption the burst of high-energy neutrons and gamma rays (light grey in Fig. 14) that precedes the pulse of moderated neutrons; the latter absorbs lower-energy neutrons that would cause frame overlap as they spread out along the flight path because of their different energies (Fig. 13). In Figure 14, by $t = 10$ μs the T_0-chopper, which rotates 360° every 50 ms, has eliminated the flash of γ radiation from the spallation and scattered neutrons with energies down to 0.91 MeV.

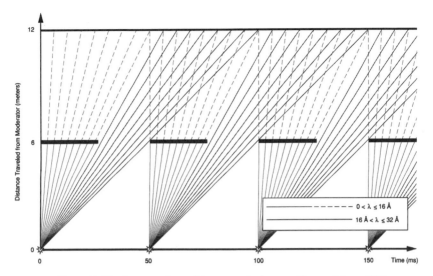

Figure 13. The frame overlap problem at a pulsed source (Hyer and Pynn 1990).

By $t = 15$ ms the T_0-chopper has removed neutrons down to 300 meV and rotated out of the beam. Energetic thermal neutrons (around a wavelength of 1 Å, light to dark grey in Fig. 14) have reached the frame-overlap chopper, which rotates 180° every 50 ms, and are removed. The remaining neutrons in the pulse have spread out along the flight path. By $t = 40$ ms the frame-overlap chopper has absorbed neutrons down to 0.3 meV (dark grey) and rotated out of the beam, allowing very slow neutrons (black to white) to proceed on to the sample. By changing the phase of the frame-overlap chopper, faster neutrons, rather than slower neutrons, can be allowed to reach the sample. Obviously, such choppers have to be carefully synchronized with the neutron source frequency.

A third type of chopper allows extraction of neutrons within a defined energy band from a "white" (polychromatic) neutron beam as illustrated in Figure 15. A neutron (dot) traveling from left to right can, if it has the proper energy (velocity), pass through the center of a curved slit in a rotating disk. Shown in Figure 15 are four snapshots of the neutron making its passage through one of several slits in an energy defining chopper. To an observer moving with the neutron, the slit appears to open up in front of the neutron. Neutrons that do not have the proper velocities strike the walls of the slits and are absorbed. A chopper of this type can be used to select a narrow band of neutron energies from either the incident or the scattered beam. Different neutron energies with a rather broad spectral resolution can be selected by changing the speed of the chopper rotation. Hence, these choppers are also termed "selectors."

Some applications require neutrons to have wavelengths greater than a certain critical wavelength. This can be achieved using a *neutron filter*: in a polycrystalline material, all crystal orientations will be present and all available lattice planes will reflect neutrons away from the incident beam direction. For a sufficiently thick sample, this results in a complete attenuation of all neutrons of wavelengths fulfilling Bragg's law. Only neutrons of wavelengths greater than $2d_{max}$ are transmitted through the sample and, thus, the material acts as a neutron filter.

Neutron transport and neutron optics

Neutrons are neutral particles and hence cannot be steered by electromagnetic fields like protons or electrons. The neutron flux is reduced proportional to the square of the distance L

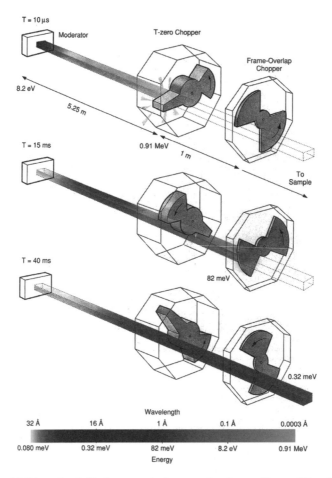

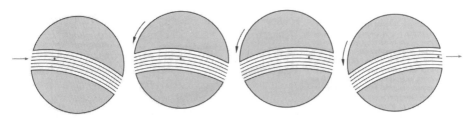

Figure 14. Schematic of a T_0-chopper and a frame-overlap chopper (Hyer and Pynn 1990).

Figure 15. Schematic of an energy defining chopper that can be used as a monochromator (Hyer and Pynn 1990).

from the source since it is proportional to the solid angle Ω covered by the beam spot size A

$$\Omega = \frac{A}{L^2} \tag{11}$$

For example, at the HIPPO beam line at LANSCE, the initial flight path length is 9 m and a typical beam spot size is 10 mm diameter, resulting in a covered solid angle of $\Omega = 79$ mm^2/ 81 m$^2 \approx 10^{-6}$ sr (steradian, this solid angle corresponds to $\sim 8 \times 10^{-6}\%$ of 4π). Additional losses of the neutron intensity caused by absorption and scattering in air can be avoided by using evacuated beam or drift tubes. Without the evacuation, the beam intensity is reduced by $\sim 5\%$ per meter in dry air. The losses can be further reduced by neutron guides or super-mirrors, which utilize the effect of total reflection, well-known from the optics of light.

Along total-reflecting surfaces, like thin Ni (or even more efficiently ^{58}Ni) layers deposited on glass, neutrons can be guided from the source to the instrument with almost no loss. It has already been shown by Fermi and Marshall (1947), that if the *coherent scattering length*, b_{coh}, is positive, a refractive index, n, for neutron waves can be derived:

$$n = 1 - \frac{\lambda^2}{2\pi} \cdot N \cdot b_{coh} \tag{12}$$

where N is the number of identical nuclei of scattering length b_{coh} per cm^3. Thus, most materials are neutron-optically less dense ($n < 1$) than air or vacuum. Therefore, neutrons incident on the surface of such media are totally reflected almost without loss as long as their angle of incidence is below the critical angle, ε, defined by

$$\varepsilon = \lambda \sqrt{\frac{N \cdot b_{coh}}{\pi}} = \lambda_c \cdot \sqrt{\frac{N}{\pi}} \sqrt[4]{\frac{\sigma_{tot}}{4\pi}} \tag{13}$$

where σ_{tot} is the total coherent scattering cross section. For a given angle of incidence $\alpha \leq \varepsilon$, total reflection can only occur if the neutron wavelength is greater than the critical wavelength λ_c. Typical critical wavelengths are ~ 1 Å, meaning that reflection is most efficient for slow neutrons, which have low intensities at reactor and spallation sources. A neutron guide can be given a slight curvature, so that only slow neutrons, which fulfill the condition of total reflection, will follow the bent guide, while fast neutron and gamma ray backgrounds will end up in the absorber material surrounding the guide. The principle of total reflection can also be utilized for polycapillary focusing optics, where neutrons are guided through bend channels of diameters on the order of micrometers. Such polycapillary focusing optics typically consist of a fused bundle of tapered polycapillary glass fibers. Figure 16 shows a schematic of such a device, converging a uniform parallel beam into a point by multiple glancing-angle deflections from the inside walls of the channels (upper part of Fig. 16). Beam spot sizes of ~ 500 µm with a flux gain of the order of 100 compared to the divergent beam on the same area are achievable with this technique (Xiao et al. 1994), which also can be used to bend cold neutron beams by up to $20°$ (Chen et al. 1992). Neutron lenses are also feasible, with converging neutron lenses being concave rather than convex as for light since the refractive index is smaller than 1. With neutron lenses, flux gains of the order of 10 are possible (Eskildsen et al. 1998).

Since neutrons have a spin and a magnetic moment, μ_n, an interesting phenomenon arises from the reflection on magnetized mirrors. In this case the index of refraction depends also on the relative orientation of the direction of magnetization and the neutron spin:

$$n_\pm = 1 - \frac{\lambda^2}{2\pi} \cdot N \cdot \mp \frac{\mu_n \cdot B}{2 \cdot E_n} \tag{14}$$

where $\mp$ refers to the orientation of the magnetic field B with respect to the neutron spin, and

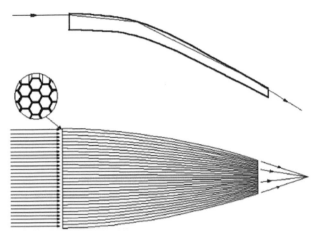

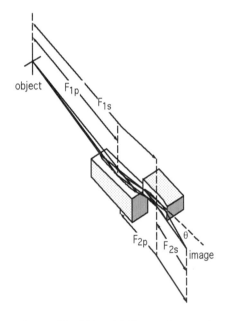

Figure 16. Schematic of a polycapillary monolithic focusing optic. The inset illustrates the individual hexagonal close-packed channels (Gibson et al. 2002).

E_n is the neutron energy. As can be seen from this equation, neutrons with the same energy or wavelength, but opposite spin orientations have different critical angles. As such, magnetized mirrors can provide a method to polarize neutrons, if suitable materials are chosen.

Neutron super mirrors may also be used to focus neutron beams into small cross sections (Goncharenko et al. 1997, 2003). High intensity in small beam spots through focusing are of particular interest for high pressure experiments, small single crystals, or position sensitive strain scanning. The Kirkpatrick-Baez setup, originally developed for synchrotron radiation, was successfully demonstrated for neutrons by Ice and coworkers (Ice et al. 2005) who achieved a neutron spot-size of 100 μm with a flux gain of 37 compared to the divergent beam of the same cross section. In the Kirkpatrick-Baez mirror setup, beam divergences in the horizontal and vertical beam planes are focused sequentially by crossed mirrors that each focus in their plane of reflection (Fig. 17, see also Fig. 13 in Parise 2006, this volume). It can be expected that this concept will become more routinely available

Figure 17. Kirkpatrick-Baez mirror pair. Elliptical mirrors focus from one focus to another in the plane of reflection. The rays act almost independently in the vertical and horizontal focusing planes (Ice et al. 2005).

at neutron sources in the future. As all techniques relying on total reflection, neutron super-mirror based neutron optics work best for neutrons with wavelengths greater than ~1 Å.

The actual beam spot size for scattering experiments is typically defined through movable slits ("jaws"), pin-holes or cylinders made out of neutron absorbing materials. A pair of slits, a distance L apart, provides a rudimentary collimator with a defined beam divergence of

$$\alpha_{max} = \frac{a_1 + a_2}{L} \tag{15}$$

where a_1 and a_2 are the widths or diameters of the slits or pinholes. Elements that have a large absorption cross-section for thermal neutrons in their natural isotope mix are boron, cadmium and gadolinium. Boron-carbide and boron-nitride are materials commonly used for collimators, whereas cadmium metal and gadolinium-oxide based paint are used for experiment-specific masking of sample environments. Fast neutrons still penetrate these materials, hence, if background due to fast neutrons needs to be reduced, T_0 choppers as described above are the preferred method. Similar to X-ray instruments, secondary collimation is used to reduce background originating from air scatter or interaction of the primary beam with the sample environment. Such collimators are typically comprised of mylar blades coated with gadolinium oxide paint. To avoid blind spots and achieve homogeneous transmission through the collimator, radial collimators are generally oscillated or continuously rotated (Wright et al. 1981). The collimators may also be used in a stationary mode to define a diffraction volume buried inside the volume of a larger sample to allow spatially resolved measurements, for instance to measure gradients in phase composition or strain as a function of depth (Withers et al. 2000). Achievable volumes at current sources for this technique are of the order of cubic millimeters, but can be expected to get smaller with neutron focusing improvements and more powerful neutron sources. Figure 18 shows the principle layout of a neutron strain scanner at a reactor source, utilizing neutron guides, collimators, and apertures.

NEUTRON INSTRUMENTS

Diffraction with constant wavelength methods

The neutron wavelengths or energy can be determined before and after interacting with the sample by scattering neutrons off a single crystal or a polycrystal with highly oriented grains of known lattice parameters. In a *constant wavelength* (CW) experiment, utilizing Bragg's law

$$\lambda = 2d \sin\theta \tag{16}$$

the wavelength can be selected from the Bragg-angle θ and the known lattice spacing d of the reflecting set of lattice planes (monochromator). This process is used at reactor sources and

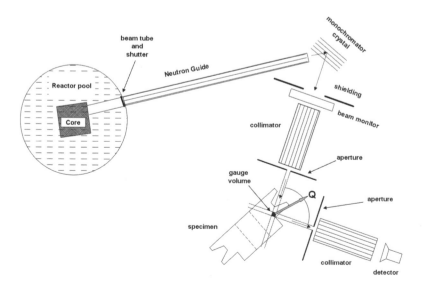

Figure 18. Layout of a neutron strain scanner at a reactor source.

requires large beam windows in sample environments, large collimators and large position sensitive detectors or is time consuming if many different angles have to be probed by a moving detector in order to get an intensity versus diffraction angle ($I(\theta)$) histogram. The principle is very similar to a constant wavelength experiment with X-rays.

Diffraction experiments with monochromatic radiation can be done on single crystals, powders or polycrystals and there is a large variety of instruments. Some use point detectors and reciprocal space is scanned by rotating the sample on a goniometer as well as varying the Bragg angle. Other instruments use 2D position sensitive detectors and a large 2θ range can be recorded simultaneously. Instruments such as D1B, D2B (Fig. 18 in Redfern 2006, this volume) and D20, all at ILL, are of this type and are particularly useful to analyze time-dependent processes such as chemical reactions as function of temperature or to investigate textures (Bunge et al. 1982). The D20 instrument is described in detail in Kuhs and Hansen (2006, this volume; see their Figs. 1 and 2). There are diffractometers with 2D position sensitive detectors, such as D19 at ILL with a detector aperture of 4° by 64° and a *d*-spacing range of 0.5 to 100 Å, which is mainly used for single crystal studies of small samples with large unit cells (Detti et al. 2004). Another single crystal diffractometer is the Very-Intense, Vertical-Axis Laue DIffractometer VIVALDI, also at the ILL (Fig. 1 in Ross and Hoffman 2006, this volume). It utilizes a polychromatic neutron beam and a neutron-sensitive image plate detector to record Laue patterns of stationary single crystals. VIVALDI has been applied to problems such as crystal structure solution of large organic crystals, magnetic structures or high pressure research (McIntyre et al. 2005).

Diffraction with Time-of-Flight (TOF) methods

The second method to determine diffraction patterns uses pulsed neutrons and determines the time, t, it takes for a neutron to travel a given distance, L. This can be achieved by generating neutron pulses, either by chopping a steady neutron current from a reactor or generating the neutrons in pulses. The latter is found at most spallation sources (with the PSI in Switzerland being the exception) and hence the TOF-method is the method of choice at such facilities. By equating the quantum mechanical momentum (de Broglie-relation) and the classical mechanics momentum, one derives

$$p_{QM} = \hbar k = mv = p_{CM} \leftrightarrow \frac{h}{\lambda} = m\frac{L}{t} \leftrightarrow \lambda = \frac{ht}{mL} \tag{17}$$

With Bragg's law, this becomes

$$\lambda = 2d\sin\theta = \frac{ht}{mL} \leftrightarrow d = \frac{h}{\underbrace{2mL\sin\theta}_{\text{const}}} \cdot t \tag{18}$$

Hence, using the time-of-flight technique, the d-spacing is directly proportional to the measured time-of-flight, with all other parameters being constant. Compared to a constant wavelength experiment, the most important difference is that any given detector records the full diffraction pattern $I(t)$ (energy-dispersive) with each neutron pulse for its unique scattering vector such as shown in Figure 12. To improve counting statistics, multiple pulses are summed up and multiple detectors are typically integrated after appropriate corrections for differences in L and θ. The constant Bragg angle allows for smaller beam windows and stationary detectors, simplifying design of sample environments.

Also at pulsed sources there are instruments specialized for single crystals such as SCD at IPNS and LANSCE with 2D position sensitive detectors (Fig. 3 in Ross and Hoffman 2006, this volume) and the instrument TOPAZ under construction at SNS (Fig. 11 in Ross and Hoffman 2006, this volume). Most powder diffractometers have multiple detectors for more efficient counting. Examples are GEM and POLARIS (Fig. 10 in Redfern 2006, this volume) and

ENGIN-X (Fig. 3c in Daymond 2006, this volume) at ISIS, GPPD at IPNS, SKAT at Dubna (Fig. 9 in Wenk 2006, this volume) and NPDF (Fig. 4 in Proffen 2006, this volume), HIPD, SMARTS and HIPPO at LANSCE. Depending on the application, some of these instruments are optimized for resolution (such as residual strain measurements), intensity (e.g., to record time-dependent processes), angular range (such as small angle neutron scattering, SANS; see Radlinski 2006, this volume) or texture (with detectors in different positions recording differently oriented crystals (Wenk 2006, this volume). The layout of the time-of-flight powder diffractometer HIPPO (which stands for High Pressure - Preferred Orientation) at LANSCE (Wenk et al. 2003; Vogel et al. 2004) is shown in Figure 19. It is unique in having a total of 1360 ^{3}He detector tubes distributed over 50 detector panels arranged on 5 banks at different Bragg angles of 140°, 90°, 40°, 20°, and 10°. The large sample chamber can accommodate a wide range of ancillary equipment such as an automatic sample changer, a load frame, various high pressure cells, cryostats, furnaces, as well as user supplied equipment.

It is worth noting that all lattice planes reflecting into a given detector in a TOF experiment share the same normal direction. This fact can be used for instance in an *in situ* deformation experiment to probe responses of various lattice planes *hkl* (in different grains) in the same direction relative to the applied load (Daymond 2006, this volume). In a constant wavelength experiment each probed lattice plane has a different direction relative to the applied load unless the sample is rotated appropriately.

d-spacing resolution

For constant wavelength as well as pulsed neutron beams, the instrument resolution can be derived from Bragg's law by calculating the propagated uncertainty:

$$d = \frac{\lambda}{2\sin\theta} \Rightarrow \frac{\Delta d}{d} = \sqrt{\left(\frac{\Delta\lambda}{\lambda}\right)^2 + \left(\Delta\theta\cot\theta\right)^2} \tag{19}$$

This means that the sharpest peaks appear in back-scattering geometry ($2\theta > 90°$). For a TOF experiment, the wavelength is proportional to the TOF, and hence $\Delta d/d$ has a contribution from

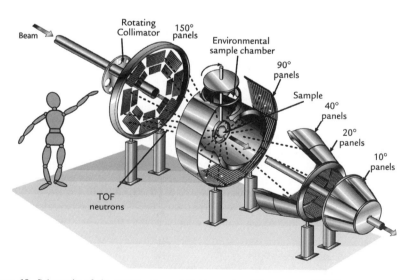

Figure 19. Schematic of the neutron time-of-flight powder diffractometer HIPPO at LANSCE. The detector tubes are arranged on 50 detector panels on 5 rings. The distance between the 150° and the 10° banks is ~3 m (Wenk et al. 2003).

$\Delta t/t$. The time uncertainty in a pulsed experiment is typically determined by the moderation time, which is wavelength dependent, and the bin width of the detector electronics. As shown in Equation (17), the wavelength in a pulsed neutron beam experiment is also proportional to $1/L$, resulting in a contribution of $\Delta L/L$ to the instrument resolution. The main contributions to the flight-path uncertainty are moderator thickness and divergence. This means that peaks get sharper with increasing flight-path length, resulting in a trade-off between resolution and flux, which decreases with $1/L^2$, in the design of an instrument (Johnson and Daymond 2002). Figure 20 shows a section of the diffraction pattern of CaF_2 measured on three detector banks with different diffraction angles on the HIPPO (9 m flight-path from moderator to sample) and SMARTS (31 m flight-path from moderator to sample) instruments at LANSCE. The variation in resolution with the diffraction angle (HIPPO detectors) and flight path length (HIPPO 90° and SMARTS 90° detectors) is evident in the peak broadening. For comparison, the same material was measured with a conventional laboratory X-ray source (Rigaku Ultima III, 2 hours data collection, divergent beam of 0.6° between source and detector, 0.6 mm detector slit, graphite monochromator in front of detector). The resolution of the X-ray powder diffractometer is sufficient to resolve the K_α/K_β splitting of the peaks, a much better resolution than even the relatively good resolution of the SMARTS neutron diffractometer. Also, the differences in the structure factors for the two types of radiation are evident, resulting in some peaks becoming almost extinct with X-rays.

Nuclear Resonance Spectroscopy

The time-of-flight method not only allows resolving the wavelength of scattered neutrons, but also the $I(E)$ spectrum of the transmitted neutrons. For a transmission experiment, the detector is located in the direct beam behind the sample ($2\theta = 0°$)[1]. This facilitates the utilization of nuclear resonance spectroscopy for temperature measurements: Certain isotopes have sharp and well defined absorption resonances, i.e., only neutrons (typically epithermal) of a well defined energy are absorbed. A change in temperature increases the thermal motion of the nuclei and causes a Doppler broadening of the resonance spectrum. By fitting appropriate resonance profiles to the measured data, the broadening can be accurately determined and, after calibration, the sample temperature can be measured to ±10 K without thermocouples or optical access to the sample for pyrometry. Thin foils of resonant material in the hot zone of a furnace or temperature/pressure cell allow temperature measurements without major additional diffraction peaks (Stone et al. 2005a,b). Figure 21 shows an example of nuclear resonance spectra of hafnium at three different temperatures. Note that the resonance width does not change with pressure, hence, the observed change in width is due to the change in temperature only.

Inelastic scattering

The energy of thermal neutrons is comparable to the energy of lattice phonons and molecular vibrations, resulting in *inelastic neutron scattering* during which a neutron can gain or lose energy due to interaction with a phonon or vibrational modes of a molecule (Loong 2006, this volume). The triple axis spectrometer (TAS), found at most research reactors, is the most versatile instrument for single crystal spectroscopic measurements. Figure 22 shows a schematic of the cold neutron three-axis spectrometer IN12 at ILL as an example of this type of instrument. Similar instruments are IN14 at ILL and T2 at LLB. Utilizing Bragg diffraction, a monochromator defines the wave vector (i.e., direction and energy/wavelength) of the incident neutrons, k_i. After interaction with the sample, Bragg reflection by an analyzer crystal

---

[1] The term "transmission" is also used in "transmission diffraction," describing a diffraction geometry where the diffracted beam exits the sample volume on the opposite side of the incident beam. In this case, neutrons which were scattered by the sample are measured, whereas in a true transmission experiment those neutrons are measured which did not interact with the sample.

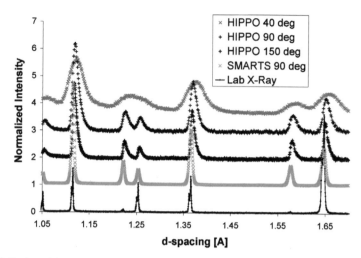

Figure 20. Section of the diffraction pattern of CaF_2 measured on the HIPPO (different detector banks) and SMARTS neutron diffractometers at LANSCE and with a conventional X-ray source.

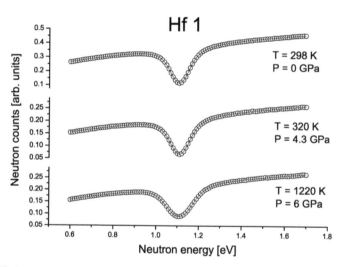

Figure 21. Nuclear resonance spectra of hafnium as a function of temperature (pressure changes do not affect the resonance shape, courtesy of B. Winkler).

determines the wave vector after scattering, k_f. From these two, the momentum transfer

$$Q = k_i - k_f \tag{20}$$

and the energy transfer, ω, from neutron to sample or vice versa can be determined. Phonon dispersion curves can be measured using this technique, even with samples at non ambient conditions in a furnace, cryostat, or pressure cell. Inelastic neutron scattering is particularly powerful in combination with molecular dynamics simulations (Trouw 1992; Johnson et al. 2000). Instruments to measure phonon dispersion curves also exist at pulsed sources, where choppers are used to select the wavelength of the incident neutrons. Vibrational spectra are obtained by neutron energy loss during scattering from the sample as incoming neutrons excite

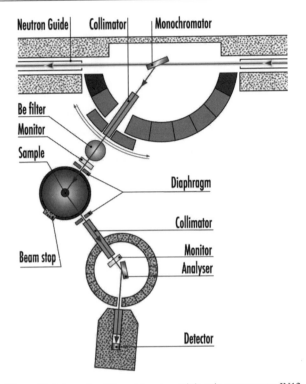

Figure 22. Schematic of the cold neutron triple axis spectrometer IN12
at the ILL (Schmidt and Raymond 2005).

molecular vibrations by giving up energy (Kearley 1995; Mitchell et al. 2005). Typical results from such experiments are interatomic force constants and estimates of bond strengths, in particular for bonds involving hydrogen. Instruments for this type of measurement exist both at pulsed sources and reactor sources, e.g., QENS at IPNS (Fig. 6 in Loong 2006, this volume).

Neutron radiography

The first study using neutron radiography was made by Kallmann and Kuhn in the 1930's in Germany (Kallmann 1948), but only after the pioneering work of Harold Berger (Berger 1965), did neutron radiography become widely accepted as a two-dimensional imaging method for use in non-destructive testing. Today most research reactors are equipped with radiographic facilities (Winkler 2006, this volume). Radiographic imaging methods utilize the attenuation properties of materials (cross sections and thicknesses) to produce a shadow image of the object under investigation (Fig. 23). X-ray and neutron imaging are based upon the same physical principle in that respect. However, since the cross sections of interaction differ considerably, neutron radiography is complimentary to, rather than competitive with, X-ray imaging. Neutron attenuation is governed by the total neutron cross section, which varies irregularly from isotope to isotope, while X-ray attenuation depends on the electron density of the material and therefore increases monotonically with increasing atomic number. Thus, thermal neutrons can visualize hydrogenous material or neutron-absorbing material, even if it is enclosed in metal shields (e.g., lubrication oil in a car engine or the hydrogen bearing rose inside a lead container in Fig. 23). Another field where neutron radiography is the unique method of choice is the quality control imaging of used nuclear fuel elements, which are highly radioactive.

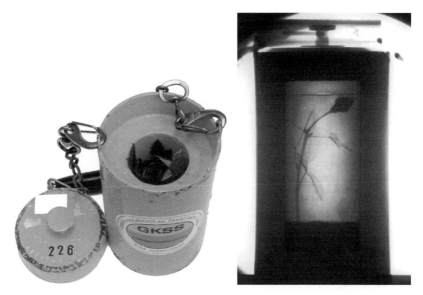

Figure 23. A rose in a lead container used for transporting radioactive materials. The picture to the left shows the setup as a conventional photograph. To the right is a neutron radiography image. The neutrons readily penetrate the lead container and the hydrogen in the rose provides sufficient contrast to see even the leaves of the flower.

The radiography setup is very simple: neutrons emitted by a source are collimated and hit the investigated object, producing a shadow image either on a converter foil or a dynamic neutron detector. The distance from the collimator to the imaging plane L and the pinhole diameter D of the collimator define the so-called collimation ratio L/D, which is an important parameter, since it determines the spatial resolution of the radiograph. Converter foils are sheets of either dysprosium or gadolinium, which by readily absorbing the transmitted neutrons map the image as their excited nuclei emit electrons. The foils are processed using ordinary photographic film. Dynamic detectors enable the investigation of live events, like the distribution of the cooling fluid in a refrigerator or measurement of the rate of descent of a gadolinium sphere in molten rock (Bayon et al. 2001; Winkler et al. 2002). Modern developments of cooled CCD cameras and further development of multi-wire proportional chambers (MPC's) have opened the field of three-dimensional neutron tomography (Schillinger 1996). Using this technique, accurate images for example of metallic mechanical structures and plastic parts within them, or micro-fissures in materials with water intrusion, can be made visible in a non-destructive way. The rose in a lead container in Figure 23 is an example of such an image, showing even the structure of the leaves of the flower. With neutron energies chosen near the Bragg edges of different constituents of poly-crystalline material, phase-specific radiography has been made. Spatial resolutions to date are of the order of 10 μm.

Sample environments

The low attenuation of neutrons by most materials enables the design of special sample environments to measure the response of samples to temperature (typically ~2000 K to a few mK), pressure (vacuum to ~20 GPa), magnetic field (up to 10 T), uni-axial stress (up to ~2 GPa) or combinations of these. Most neutron facilities have beamlines available which provide such sample environments. As an example, Figure 24 shows the SMARTS load-frame/furnace combination, allowing *in situ* deformation of a sample in the neutron beam uni-axially in the horizontal plane (up to 250 kN) while the sample is at a temperature up to 1500 °C (Bourke et

Figure 24. The load-frame/furnace combination of the SMARTS
instrument at LANSCE (Bourke et al. 2002).

al. 2002). The changes in peak positions, peak intensities and peak widths provide insight into deformation mechanisms, in particular in combination with modeling approaches (Clausen et al. 1998). Figure 25 gives an example of a recent high pressure cell for neutron diffraction experiments, the ZAP cell. This cell uses a pair of single crystal SiC anvils, from 5 to 100 carats, to apply pressure on a small sample volume (3~30 mm³) between the anvils. The cell is pre-loaded under high hydraulic loading forces (up to 100 tons) and the hydrostatic pressure is locked into the inner cell. The cell can be removed from the press and then easily transported to experimental setups, such as a neutron diffractometer, where the same sample can be studied under identical pressure-temperature (*P-T*) conditions. The optically transparent windows of SiC anvils are particularly useful for the measurement of vibrational spectra (Raman and IR) on the same sample under identical *P-T* conditions. The straightforward anvil-sample-anvil setting allows applications of acoustic transmission and ultrasonic interferometry techniques for elasticity measurements at high pressures. High pressure neutron scattering techniques are discussed in detail in Parise (2006, this volume). Compared to X-rays, neutrons have the disadvantage for high pressure research that larger sample volumes are required, resulting in larger devices and lower maximum pressures achievable with pressure cells designed for neutron scattering.

Neutrons also allow for *in situ* observation of crystallographic changes under special conditions and environments, such as order-disorder phase transformations (Redfern 2006, this volume), or the change in crystal structure as a function of charge state in a commercial battery (Rodriguez et al. 2004) or during the reduction of $NiAl_2O_4$ to Nickel and Al_2O_3 triggered by a reducing atmosphere (Üstündag et al. 2000). Studies of clathrates (Kuhs et al. 1997; Lokshin et al. 2004; Kuhs and Hansen 2006, this volume) are specific to neutron diffraction since hydrogen is virtually invisible to synchrotron radiation and the required moderate pressures of few kilobars at low temperatures are not achievable with high pressure setups using diamond anvil cells at synchrotron facilities. Figure 26 shows the low temperature pressure cell developed by Lokshin and Zhao (2005).

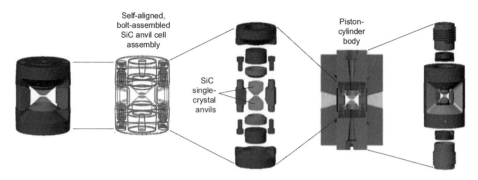

Figure 25. The ZAP cell, a SiC Cell assembly (Migliori et al. 2006).

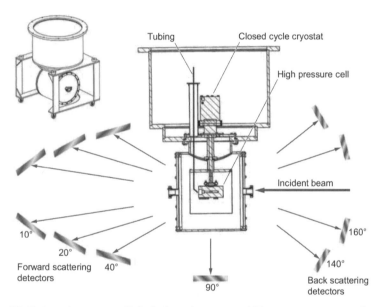

Figure 26. Hydrostatic pressure cell for hydrostatic pressure and low temperature research on HIPPO (Lokshin and Zhao 2005). The outer diameter of the top flange is 0.83 m.

In general, specimen size for neutron experiments is larger (mm^3 to cm^3) than for comparable X-ray or synchrotron measurements (μm^3 to mm^3). This results in negligible artifacts from the surface or sample preparation, and averages the derived information over larger numbers of grains than is possible with X-rays.

NEUTRON DETECTION

To record neutron intensity spectra, neutrons must be converted to charged particles which in turn can be detected electronically. Great care has to be taken to discriminate neutrons from other types of radiation, in particular gamma rays typically present at neutron sources or from cosmic radiation. Due to the low count rates compared to even laboratory X-ray sources, neutron detectors have to be insensitive to gamma radiation. For thermal neutrons, typically scintillation counters or gas counters are employed. Scintillation counters consist of a scintillator and a photomultiplier (Fig. 27, left). In the scintillator, the neutron absorption according to one of the exothermic reactions

$$^6\text{Li} + \text{n} \rightarrow {}^3\text{H} + {}^4\text{He} + 4.78 \text{ MeV} \tag{21}$$

$$^{10}\text{B} + \text{n} \rightarrow {}^7\text{Li} + {}^4\text{He} + 2.79 \text{ MeV} \quad (6\%)$$

$$^{10}\text{B} + \text{n} \rightarrow {}^7\text{Li}^* + {}^4\text{He} + 2.31 \text{ MeV} \quad (94\%) \tag{22}$$
$$\downarrow$$

$$^7\text{Li} + 480 \text{keV}$$

(where $^7\text{Li}^*$ designates an excited state of the ^7Li nucleus) is followed by the emission of an energetic charged particles which excites atoms of the working medium, causing fluorescent radiation[2]. Some of the emitted light quanta reach the photocathode of the photomultiplier and initiate an electron avalanche which creates a measurable electric charge pulse at the anode. The pulses can be counted, resulting in the desired neutron intensity versus time or diffraction angle histograms. As background radiation, e.g., cosmic radiation or gamma radiation emitted from the neutron source, also causes scintillation, neutrons and other radiation have to be discriminated electronically. This can be performed using the pulse heights and/or the different pulse shapes of the radiation types. Special scintillating glasses, containing lithium enriched in ^6Li, such as NE905 or GS20, are commercially available. They can be either glued directly on a photomultiplier tube or cut into small cylinders or cubes of a few mm^2 surface area to form pixelated detectors. In this case, the scintillation light is transmitted through fiber optic light guides to photomultiplier tubes for conversion into electronic signals. Scintillation detectors are the prevalent technology at ISIS.

In a gas counter, helium enriched with ^3He, or BF_3 enriched with ^{10}B is placed in a metal cylinder with a thin wire anode (Fig. 27, right). Between the anode and cylinder, a high voltage is applied. Again the neutron is absorbed and according to the above reactions for boron or

$$^3\text{He} + \text{n} \rightarrow {}^3\text{H} + {}^1\text{H} + 0.765 \text{ MeV} \tag{23}$$

fast charged particles are generated. The charged particles are accelerated in the electric field and produce more charged particles by collisions with the gas molecules, generating an avalanche which is converted to an electric pulse at the anode or cathode. The gas counter must be run in the "proportional counter" mode, allowing electronic discrimination of neutrons from gamma ray background by the pulse heights. After discrimination, the pulses may be counted the same way as those of scintillation counters. This type of detectors can be modified to show the location at which a particle arrived, making it a position sensitive detector (PSD). Such neutron detector tubes are available commercially and, for instance, are the standard detectors at IPNS and LANSCE.

In addition to the classical tube setup, so-called microstrip neutron detectors have been developed (Oed 1988): rather than having a potential between a wire in the center of a tube and the tube wall, a grid of strips of alternating potentials is applied to an insulating substrate by means of photolithography. In Figure 28, the potential of the thin conductor strip is positive relative to the two adjacent strips. A charged particle produced in the gas volume above the substrate and reaching this electric field will be accelerated and produce an avalanche amplification. A repulsive electric field is applied from the back of the detector to prevent positive gas ions from accumulating on the insulating substrate surface and degrading the electric field between the strips. This allows production of large position sensitive neutron detectors without dead areas between elements as is the case for detector tubes. The "banana"

---

[2] The isotope ^{10}B has a natural abundance of about 20%, ^6Li of 7.5%.

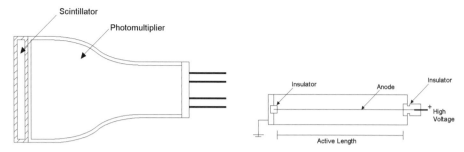

Figure 27. Scintillator/photomultiplier (left) and gaseous proportional counter (right).

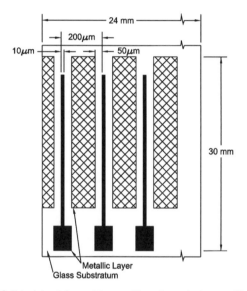

Figure 28. Principle of the position sensitive microstrip detector (Oed 1988).

detectors of the D1B and D20 instruments at ILL are such devices and shown in Kuhs and Hansen (2006, this volume) (Clergeau et al. 2001). Gas counters have, in general, a longer dead time after a detection event compared to scintillation detectors, hence, for instruments with a high event rate at the detector, scintillation detectors might be the only choice. For very fast neutron detection, gas scintillation detectors on the basis of reaction (23) have been developed and are also commercially available.

So-called converter foils (a few microns thickness) use nuclear reactions such as:

$$^{155}Gd + n \rightarrow Gd^* \quad \text{and} \quad ^{157}Gd + n \rightarrow Gd^* \tag{24}$$

to generate a gadolinium nucleus in an excited state which decays by emission of electrons and/or γ radiation. This radiation can then be detected and integrated using electron or photon detectors such as X-ray or photographic films, CCD cameras, or image plates. Such detectors do not allow discrimination of thermal neutrons against other types of radiation and thus are primarily used for neutron imaging techniques such as neutron radiography or beam alignment studies. The textbook by Knoll (2000) covers most aspects of neutron detection.

HEALTH PHYSICS

Working with nuclear radiation requires precautions to be taken to minimize the exposure of persons involved in an experiment. While facilities in general have issued site-specific rules and codes-of-practice, it is important that the experimenter is familiar with a few facts which help to fulfill the needs of self-control. The radiation types which have to be considered primarily are neutrons, gamma rays and, to a lesser extent, electrons. The biological impact on living organisms depends on the type and energy of the radiation involved: neutrons have a biological effectiveness which is an order of magnitude higher than that of gamma rays and electrons. Because of their absent electric charge, neutrons are readily absorbed by the nuclei of most isotopes, which get into an excited state and release this energy by the emission of electrons and gamma rays. This is why specimens which have spent some time in the neutron beam may have become radioactive and must be monitored before they are removed from an experiment. The intensity of the radiation will decay with time, but the half-lives are very different, depending on isotopes. For instance, 10 g of silver, chlorine, chromium, or potassium exposed for 24 h to the beam on the HIPPO instrument at LANSCE, require about 20 days, 0.1 days, 42 days, and 1.1 days, respectively, to decay to 74 Bq/g (74 disintegrations per second per gram), the limit accepted for shipping material as "non-radioactive." This is the reason why some specimens must stay shielded for a certain time, sometimes for years, in order for the radiation level to drop below the accepted limit. In any case, it is advisable to estimate the expected radiations—types and energies—during the planning phase of an experiment. This will help to comply with the commonly adopted ALARA (as-low-as-reasonably-achievable) principle, which every experimenter should have in mind in order to minimize his or her radiation exposure.

There are a number of ways to shield against nuclear radiation, depending on the type. Neutrons are best shielded by first decreasing their energy, e.g., through scattering in hydrogeneous materials or nuclear inelastic scattering ($(n,n'\gamma)$ reactions), and then capturing them in ^{6}Li or ^{10}B. Gamma rays interact with electrons and, thus, heavy elements like lead can efficiently be utilized to shield against electromagnetic radiation.

NEUTRONS OR X-RAYS?

Although neutrons have clear advantages over X-rays when studying hydrogen, magnetic structures, phonons, or molecular vibrations, in general, neutrons and X-rays are complementary research tools. Beam time applications and experiments at neutron and synchrotron radiation facilities cost both time and money and many months may transpire between proposal and experiment. As such, care has to be taken that the right tool is used in order to tackle a specific problem. Averaging over sample volumes of a cubic centimeter to cover large numbers of grains for even coarse grained materials might be required, for instance, in re-crystallized metals or rocks to get sufficient grain statistics. In such cases, neutron diffraction with its typically large beam spot sizes and the deep penetration through most materials is the probe of choice. For crystal structure determination, it might be beneficial to use neutrons for studies of compounds consisting of elements of similar atomic numbers or light elements, in particular hydrogen. For more complex structures, the location of certain atoms might be distinguishable by neutrons only, while other atomic species might have good contrast with X-rays and poor contrast with neutrons. In such cases, a combined Rietveld refinement of datasets from both sources may be the best solution (Williams et al. 1988). Isotopic substitution, e.g., D vs. H, allows studies of a specific bond using crystal structure solution and vibrational spectroscopy methods by selectively replacing an isotope in a molecule and investigating the difference in scattering signal.

FURTHER READING

Naturally, a complex topic such as neutron generation and instrumentation cannot be completely covered on a few pages. The textbooks by Bacon (1955), Windsor (1981), Dobrzynski and Blinowski (1994), Byrnes (1996), and Squires (1997), provide further details and in-depth material of the topics covered in this section. The introductory booklets by Pynn (1990) (freely available at *http://www.mrl.ucsb.edu/~pynn/*) and Dianoux and Lander (2003) may serve as entry level literature sufficient for the occasional user of neutron diffraction. A list with links to neutron sources around the world and their user programs can be found at *http://www.ncnr.nist.gov/nsources.html* or *http://idb.neutron-eu.net/facilities.php*.

ACKNOWLEDGMENTS

We wish to thank B. Clausen, J. J. Wall and H.-R. Wenk for valuable comments on this chapter. N. T. Callaway and S. J. Veenis provided Figure 5 and J. J. Wall provided Figures 27 and 28. L. L. Daemen contributed the X-ray diffraction data of CaF_2 in Figure 20. The International Union of Crystallography granted permission to reproduce Figure 16 for this chapter.

REFERENCES

Bacon GE (1955) Neutron Diffraction. Clarendon Press
Bauer GS, Thamm G (1991) Reactors and neutron scattering instruments in western Europe – an update on continuous neutron sources. Physica B 174:476-490
Bayon G, Winkler B, Kahle A, Hennion B, Boutrouille P (2001) Application of dynamic neutron imaging in the earth sciences to determine viscosities and densities of silicate melts. Nondestructive Testing and Evaluation 16:287-296
Berger H (1965) Neutron Radiography - Methods, Capabilities and Applications. Elsevier
Bourke MAM, Dunand DC, Üstündag E (2002) SMARTS – a spectrometer for strain measurement in engineering materials. Appl Phys A 74:S1707-S1709
Bunge HJ, Wenk HR, Pannetier J (1982) Neutron diffraction texture analysis using a 2D position sensitive detector. Textures and Microstructures 5:153-170
Byrnes J (1996) Neutrons, Nuclei and Matter: An Exploration of the Physics of Slow Neutrons. CRC Press
Chen H, Downing RG, Mildner DFR, Gibson WM, Kumakhov MA, Ponomarev IY, Gubarev MV (1992) Guiding and focusing neutron beams using capillary optics. Nature 357:391-393
Clausen B, Lorentzen T, Leffers T (1998) Self-consistent modeling of the plastic deformation of fcc polycrystals and its implications for diffraction measurements of internal stresses. Acta Mater 46:3087-3098
Clergeau JF, Convert P, Feltin D, Fischer HE, Guerard B, Hansen T, Manzin G, Oed A, Palleau P (2001) Operation of sealed microstrip gas chambers at the ILL. Nucl Instrum Methods Phys Res A 471:60–68
Daymond MR (2006) Internal stresses in deformed crystalline aggregates. Rev Mineral Geochem 63:427-458
Detti S, Forsyth VT, Roulet R, Ros R, Tassan A, Schenk, KJ (2004) Comparative neutron and X-ray study of [PPN][HIr₄(CO)₉(μ-Ph₂PCH₂PPh₂)]. Z Kristallogr 219:47-53
Dianoux AJ, Lander G (2003) Neutron Data Booklet. Old City Publishing
Dobrzynski L, Blinwoski K (1994) Neutrons and Solid State Physics. Ellis Horwood Limited
Donahue JB, Baker GD, Bultman NK, Brun TO, Ferguson PD, Macek RJ, Njegomir MM, Plum MA, Roberts JE, Russell GE, Sommer WF, Tuzel WM (1999) The Lujan Center target upgrade. Transactions of the American Nuclear Society, Accelerator Applications, Long Beach, CA, November 14-18, 1999
Eskildsen MR, Gammel PL, Isaacs ED, Detlefs C, Mortensen K, Bishop DJ (1998) Compound refractive optics for the imaging and focusing of low-energy neutrons. Nature 391:563-566
ESS (2003) European Spallation Source website: *http://www.fz-juelich.de/ess/*
Fermi E, Marshall L (1947) Interference phenomena of slow neutrons. Phys Rev 71:666-677
FRM-2 (2004) Research Reactor Munich II website: *http://www.frm2.tum.de*
Gibson WM, Schultz AJ, Chen-Mayer HH, Mildner DFR, Gnäupel-Herold T, Miller ME, Prask HJ, Vitt R, Youngman R, Carpenter JM (2002) Polycapillary focusing optic for small sample neutron crystallography. J Appl Cryst 35:677-838
Goncharenko IN, Mirebeau I, Mignot JM, Goukasov A (2003) Neutron diffraction on microsamples under high pressures. Neutron News 14:21-25

Goncharenko IN, Mirebeau I, Molina P, Böni P (1997) Focusing neutrons to study small samples. Physica B 234-236:1047-1049

Hurd AJ, Schaefer DW (2006) Introduction to materials and bioscience neutron-scattering research. Los Alamos Science 30:146-151

Hyer DK, Pynn R (1990) LANSCE – a facility for users. Los Alamos Science 19:46-63

Ice GE, Hubbard CR, Larson BC, Pang JWL, Budai JD, Spooner S, Vogel SC (2005) Kirkpatrick–Baez microfocusing optics for thermal neutrons. Nucl Instrum Methods Phys Res A 539:312–320

Ikeda S (2002) Japanese spallation neutron source. Appl Phys A 74:S15-S17

JINR (2005) JINR website: *http://nfdfn.jinr.dubna.su/ibr-2/index.html*

Johnson MR, Kearley GJ, Eckert J (2000) Condensed phase structure and dynamics: A combined neutron scattering and numerical modeling approach. Chem Phys 261:1-274

Johnson MW, Daymond MR (2002) An optimum design for a time-of-flight neutron diffractometer for measuring engineering stresses. J Appl Cryst 35:49-57

Kallmann H (1948) Neutron radiography. Research 1:254-260

Kearley GJ (1995) A review of the analysis of molecular vibrations using INS. Nucl Instrum Methods Phys Res A 354:53-58

King NSP, Ables E, Adams K, Alrick KR, Amann JF, Balzar S, Barnes Jr. PD, Crow ML, Cushing SB, Eddleman JC et al. (1999) An 800-MeV proton radiography facility for dynamic experiments. Nucl Instrum Methods Phys Res A 424:84-91

Knoll GF (2000) Radiation Detection and Measurement. Chapter 14 &15 on Neutron Detection, John Wiley

Kuhs WF, Hansen TC (2006) Time-resolved neutron diffraction studies with emphasis on water ices and gas hydrates. Rev Mineral Geochem 63: 171-204

Kuhs WF, Chazallon B, Radaelli PG, Pauer F (1997) Cage occupancy and compressibility of deuterated N_2-clathrate hydrate by neutron diffraction. J Incl Phen Molec Recog Chem 29:65–77

Lisowski PW (2005) Basic and applied science research at the Los Alamos Neutron Science Center. AIP Conference Proceedings 769:712-717

Lisowski PW, Bowman CD, Russell GJ, Wender SA (1990) The Los Alamos National Laboratory spallation neutron sources. Nucl Sci Eng 106:208-218

Lokshin KA, Zhao Y (2005) Advanced setup for high-pressure and low-temperature neutron diffraction at hydrostatic conditions. Rev Sci Instrum 76:1-4

Lokshin KA, Zhao Y, He D, Mao WL, Mao HK, Hemley RJ, Lobanov MV, Greenblatt M (2004) Structure and dynamics of hydrogen molecules in the novel clathrate hydrate by high pressure neutron diffraction. Phys Rev Lett 93:125503

Loong C-K (2006) Inelastic scattering and applications. Rev Mineral Geochem 63:233-254

Mampe W, Ageron P, Bates JC, Pendlebury JM, Steyerl A (1989a) Neutron lifetime from a liquid walled bottle. Nucl Instrum Methods Phys Res A 284:111-115

Mampe W, Ageron P, Bates JC, Pendlebury JM, Steyerl A (1989b) Neutron lifetime measured with stored ultracold neutrons. Phys Rev Lett 63:593-596

Mansur LK, Gabriel TA, Haines JR, Lousteau DC (2001) R&D for the Spallation Neutron Source mercury target. J Nucl Mater 296:1-16

Mason TE, Abernathy D, Ankner J, Ekkebus A, Granroth G, Hagen M, Herwig K, Hoffmann C, Horak C, Klose F, Miller S, Neuefeind J, Tulk C, Wang XL (2005) The Spallation Neutron Source: A powerful tool for materials research. AIP Conference Proceedings 773:21-25

Mason TE, Arai M, Clausen KN (2003) Next-generation neutron sources. MRS Bull 28:923-928

Mason TE, Crawford RK, Bunick GJ, Ekkebus AE, Belanger D (2002) The spallation neutron source is taking shape. Appl Phys A 74:S11-S14

McIntyre GJ, Melesi L, Guthrie M, Tulk CA, Xu J, Parise JB (2005) One picture says it all - High-pressure cells for neutron Laue diffraction on VIVALDI. J Physics: Condens Matter 17:S3017-S3024

Migliori A, Hurd AJ, Zhao Y, Pantea C (2006) Filling the gap in plutonium properties - studies at intermediate temperatures and pressures. Los Alamos Science 30:86-89

Mitchell PCH, Parker SF, Ramirez-Cuesta AJ, Tomkinson J (2005) Vibrational Spectroscopy with Neutrons. World Scientific Publishing

Oed A (1988) Position-sensitive detector with microstrip anode for electron multiplication with gases. Nucl Instrum Methods Phys Res A 263:351-359

Parise JB (2006) High pressure studies. Rev Mineral Geochem 63:205-231

Pouzo J, Milanese M, Moroso R (2003) Portable neutron probe for soil humidity measurements. AIP Conference Proceedings 669:277-280

Proffen T (2006) Analysis of disordered materials using total scattering and the atomic pair distribution function. Rev Mineral Geochem 63:255-274

PSI (2002) SINQ website: *http://asq.web.psi.ch/ASQ/facilities/SINQSYSTEMS.html*

Pynn R (1990) Neutron scattering – a primer. Los Alamos Science 19:1-31

Radlinski AP (2006) Small-angle neutron scattering and the microstructure of rocks. Rev Mineral Geochem 63:363-397

Redfern SAT (2006) Neutron powder diffraction studies of order-disorder phase transitions and kinetics. Rev Mineral Geochem 63:145-170

Reijonen J, Gicquel F, Hahto SK, King M, Lou TP, Leung KN (2005) D–D neutron generator development at LBNL. Appl Radiat Isot 63:757–763

Richter D (2002) The European spallation source. Appl Phys A 74:S18-S22

Rodriguez MA, Ingersoll D, Vogel SC, Williams DJ (2004) Simultaneous in situ neutron diffraction studies of the anode and cathode in a lithium-ion cell. Electrochem Solid-State Lett 7:A8-A10

Ross NL, Hoffman C (2006) Single-crystal neutron diffraction: present and future applications. Rev Mineral Geochem 63:59-80

Schillinger B (1996) 3D Computer tomography with thermal neutrons at FRM Garching. J Neutron Res 4: 57-63

Schmidt W, Raymond S (2005) CRG - cold neutron three-axis spectrometer IN12. ILL yellow book, *http: //www.ill.fr/YellowBook/IN12/*

Shull CG (1995) Early development of neutron scattering. Rev Mod Phys 67:753-757

Squires GL (1997) Introduction to the Theory of Thermal Neutron Scattering. Dover Publications

Stone HJ, Tucker MG, Le Godec Y, Méducin FM, Cope ER, Hayward SA, Ferlat GPJ, Marshall WG, Manolopoulos S, Redfern SAT, Dove MT (2005a) Remote determination of sample temperature by neutron resonance spectroscopy. Nucl Instrum Methods Phys Res A 547:601-615

Stone HJ, Tucker MG, Méducin FM, Dove MT, Redfern SAT, Le Godec Y, Marshall WG (2005b) Temperature measurement in a Paris-Edinburgh cell by neutron resonance spectroscopy. J Appl Phys 98:64905-1-10

Trouw FR (1992) Molecular dynamics simulation and inelastic neutron scattering. Spectrochim Acta A 48: 455-476

Üstündag E, Clausen B, Bourke MAM (2000) Neutron diffraction study of the reduction of $NiAl_2O_4$. Appl Phys Lett 76:694-696

Vogel SC, Hartig C, Lutterotti L, Von Dreele RB, Wenk HR, Williams DJ (2004) Texture measurements using the new neutron diffractometer HIPPO and their analysis using the Rietveld method. Powder Diffraction 19:64-68

Wenk H-R (2006) Neutron diffraction texture analysis. Rev Mineral Geochem 63:399-426

Wenk HR, Lutterotti L, Vogel S (2003) Texture analysis with the new HIPPO TOF diffractometer. Nucl Instrum Methods Phys Res A 515:575-588

Williams A, Kwei GK, Von Dreele RB, Larson AC, Raistrick ID, Bish DL (1988). Joint X-ray and neutron refinement of the structure of superconducting $YBa_2Cu_3O_{7-x}$: precision structure, anisotropic thermal parameters, strain, and cation disorder. Phys Rev B 37:7960-7962

Windsor CG (1981) Pulsed Neutron Scattering. Taylor & Francis

Winkler B (2006) Applications of neutron radiography and neutron tomography. Rev Mineral Geochem 63: 459-471

Winkler B, Knorr K, Kahle A, Von Tobel P, Lehmann E, Hennion B, Bayon G (2002) Neutron imaging and neutron tomography as non-destructive tools to study bulk-rock samples. Eur J Mineral 14:349-354

Withers PJ, Johnson MW, Wright JS (2000) Neutron strain scanning using a radially collimated diffracted beam. Physica B 292:273-285

Womble PC, Campbell C, Vourvopoulos G, Paschal J, Gacsi Z, Hui S (2001) Detection of explosives with the PELAN system. AIP Conference Proceedings 576:1069-1072

Wright AF, Berneron M, Heathman SP (1981) Radial collimator system for reducing background noise during neutron diffraction with area detectors. Nucl Instrum Methods 180:655-658

Xiao QF, Chen H, Sharov VA, Mildner DFR, Downing RG, Gao N, Gibson DM (1994) Neutron focusing optic for submillimeter materials analysis. Rev Sci Instrum 65:3399-3402

Reviews in Mineralogy & Geochemistry
Vol. 63, pp. 59-80, 2006
Copyright © Mineralogical Society of America

3

Single-crystal Neutron Diffraction: Present and Future Applications

Nancy L. Ross

Crystallography Laboratory, Department of Geosciences
Virginia Polytechnic Institute and State University
Blacksburg, Virginia, 24061, U.S.A.
e-mail: nross@vt.edu

Christina Hoffmann

Oak Ridge National Laboratory/SNS
One Bethel Valley Road
P.O. Box 2008, MS6474
Oak Ridge, Tennessee, 37831-6474, U.S.A.
e-mail: hoffmanncm@ornl.gov

INTRODUCTION

Single-crystal neutron diffraction provides a powerful complementary probe to X-ray diffraction for the characterization of earth materials. The ability of neutron diffraction to determine the position of the atomic nucleus rather than electron density is key to its use in structural studies. A typical neutron beam spectrum contains neutrons in the energy ranges of epithermal, hot, thermal and cold. The relation between the neutron kinetic energy and the wavelength is:

$$E = \frac{1}{2}mv^2 = \left(\frac{\hbar^2 2\pi^2}{m}\right)\frac{1}{\lambda^2} \quad \text{(de-Broglie)} \tag{1}$$

where m is the neutron mass and v is the velocity. This can be transformed in (1) via de Broglie using the Plancks constant, $\hbar$, and the wavelength, λ. This equation is frequently used in neutron diffraction to convert energies (E) into wavelengths that directly describe the useful experimental range:

$$E\,[\text{meV}] = 81.796\lambda^{-2}\,[\text{Å}^{-2}] \tag{2}$$

Conventional ranges for neutron energies and corresponding wavelengths for neutron scattering experiments are:

i.	"epithermal neutrons	$E > 500$ meV	$\lambda \leq 0.5$ Å
ii.	"hot" neutrons	$E = 100\text{-}500$ meV	$\lambda = 0.5\text{-}1$ Å
iii.	"thermal" neutrons	$E = 10\text{-}100$ meV	$\lambda = 1\text{-}3$ Å
iv.	"cold" neutrons	$E = 0.1\text{-}10$ meV	$\lambda = 3\text{-}30$ Å

Cold neutrons are used for macromolecular crystallography (3-10 Å) and hot neutrons are typically used for high resolution measurements, magnetic form factor studies, and amorphous materials. Single-crystal neutron diffraction typically uses thermal to hot neutrons for small and medium-size unit cells < 25 Å per unit cell basal vector (approx. 0.5-7 Å).

This chapter presents a summary of some of the applications to which single-crystal neutron diffraction has been put in the study of earth and related materials, together with an account of

1529-6466/06/0063-0003$05.00 DOI: 10.2138/rmg.2006.63.3

the techniques and instrumentation used for these experiments. Recent developments in instrumentation, and the new scientific applications that have resulted from these, are described along with a forward look to some of the exciting developments currently taking place in this field.

REACTOR SOURCES AND SINGLE-CRYSTAL INSTRUMENTATION

As described in Vogel and Priesmeyer (2006, this volume), production of neutrons requires either a steady state (usually reactor) source or a pulsed (usually spallation) source. In the fission process in a reactor source, slow neutrons interact with metastable ^{235}U nuclei. The excited nucleus decays in a cascade of fission products. In the average 2.5 neutrons are produced by the fission of one ^{235}U nucleus. These neutrons possess high energies of about 2 MeV and are not suitable for inducing further fission processes. Moderators slow the fast neutrons down to meV energies which are used to sustain the chain reaction in a nuclear reactor. Some of the moderated neutrons escape from the core region through neutron beam tubes as free neutrons for scientific use. Neutrons are produced at a constant rate without an implicit time structure. A reactor is a continuous source and it is impossible to determine the specific energy of the produced neutrons. Therefore either mechanical choppers that are phased to only permit a narrow wavelength band or reflecting crystal monochromators are used as energy selection devices. Examples of steady state sources with instruments for single-crystal neutron diffraction are summarized in Table 1.

Four-circle diffractometers

On a steady state neutron source, four-circle diffractometer techniques (Arndt and Willis 1966) have traditionally been used for single-crystal diffraction, with a monochromatic beam and a single detector. Rotations of the crystal (and detector) are used to allow measurement of each reflection sequentially. It is necessary to set up a scan in order that the reflection passes entirely through the Ewald sphere, allowing its intensity to be recorded. There is also the potential for increasing the region of reciprocal space accessed in a single measurement by using an area detector. If an area detector is combined with a monochromatic incident beam it is still necessary to scan the crystal and possibly also the detector to observe the diffracted intensity. Alternatively using an area detector with a broad band (white) beam allows Laue or quasi-Laue diffraction, with a stationary crystal and detector. In this case the Ewald sphere itself is finite in extent, effectively a shell with inner and outer radii of $1/\lambda_{max}$ and $1/\lambda_{min}$ and no scanning is required to allow intensity measurements.

Quasi-Laue diffractometers

A particularly exciting development in single-crystal neutron diffraction has been the implementation of Laue methods of data collection from single-crystal samples. LADI, the quasi Laue diffractometer at the ILL (Wilkinson and Lehmann 1991), incorporates recently developed neutron image plate detectors wrapped around a cylinder surrounding the sample to give more than 2π solid angle coverage. LADI operates with a relatively narrow wavelength range (hence the name "quasi-Laue") in order to reduce the problems of reflection overlap and to prevent the accumulation of a large background in the Laue pattern. This instrument has already demonstrated its potential to revolutionize neutron crystallography, especially for proteins, with dramatically reduced data collection times (Helliwell 1997; Niimura et al. 1997). The LADI concept also has applications in the study of earth materials and in particular, magnetism. For magnetic applications, the longer wavelength option, as implemented on LADI is probably optimal, while the chemical version is ideally situated on a thermal neutron beam. Figure 1 shows a schematic diagram of the new thermal LADI instrument VIVALDI (Very-Intense, Vertical-Axis Laue Diffractometer) installed at the ILL. The main difference with LADI is that it has a vertical rather than horizontal geometry which makes the provision of sample environment more straightforward. VIVALDI allows rapid preliminary

Table 1. Examples of instruments for single-crystal neutron diffraction at steady state sources.

Neutron Facility	Instrument	Applications
Institut Laue Langevin (ILL)	D3: Polarized hot neutron diffractometer	Polarized neutron diffraction
	D9: Hot neutron 4-circle diffractometer	Nuclear density maps; phase transitions
	D10: Thermal neutron 4-circle & three-axis diffractometer	Structures including modulated structures, quasielastic scattering and diffuse scattering.
	D15: Thermal neutron normal-beam diffractometer	Magnetic and *P-T* phase diagrams, magnetic and nuclear structures
	D19: 4-circle diffractometer with 2D PSD	Structures with large unit cells; phase transitions, diffuses scattering
	LADI Laue diffractometer	Studies of small protein systems at medium or high resolution
	VIVALDI thermal beam Laue diffractometer	Magnetism, charge (nuclear) density waves, high-*P* studies and structural phase transitions
Berlin Neutron Scattering Center (BENSC)	E2: Flat-Cone- and Powder Diffractometer	Bragg and superstructure reflections in 3D; diffuse scattering
	E4: 2-Axis Diffractometer	Magnetic structure determination under various *P-T* conditions
	E5: four-circle diffractometer	Crystal structures
	E6: Focusing diffractometer	Magnetic structures
Laboratoire Léon Brillouin (LLB)	5C2: Hot neutron four-circle diffractometer	Crystal and magnetic structures
	6T2: Thermal Neutron Four-Circle Diffractometer with Lifting Counter	Structural studies of large unit cells; ase transitions; magnetic studies at *P* and *T*

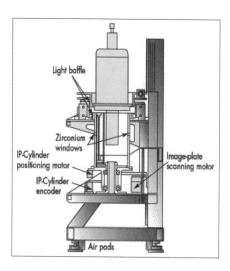

Figure 1. A schematic diagram of the Very-Intense, Vertical-Axis Laue Diffractometer (VIVALDI) at the Institut Laue Longevin (ILL). The main difference between VIVALDI and the Laue diffractometer, LADI, at the ILL is that the detector cylinder is vertical to allow free access for different sample holders such as cryostats, furnaces, magnets or pressure cells.

investigation of new materials, even when only small single-crystals are available, and covers fields of interest such as magnetism, charge (nuclear) density waves, high-pressure studies and structural phase transitions. The detector is also suitable for some types of diffuse scattering experiments on a monochromatic beam.

While structure factor data collected on a monochromatic steady state source at present yields the ultimate in accuracy for neutron single-crystal structure determination, the high time-averaged neutron flux makes Laue diffraction at a steady-state source also extremely powerful. Wilson (2005) summarized the benefits of single-crystal diffraction instrumentation on a steady state source:

- Methods developed worldwide on four-circle X-ray diffractometers can be directly applied to monochromatic neutron instrument. Control software for the diffractometers and step-scanning methods for intensity extraction can be directly transferred to neutron sources.

- The constant wavelength nature of the data collection eliminates the need for wavelength-dependent corrections, and the steady state of the source also removes the need for correcting data for the incident flux profile leading to more straightforward error analysis.

- The time-averaged flux at current high flux steady state sources is substantially higher than at present day pulsed sources, allowing better counting statistics to be obtained in the same time, and allowing the study of smaller crystals or larger unit cells, particularly with the Laue technique (see below).

These factors lead to more accurate structure factors, better internal agreement and ultimately to lower crystallographic R factors and somewhat more precise atomic parameters. Constant wavelength single-crystal diffraction is the method of choice if ultimate precision is required in an individual structure determination, and the high flux reactor sources are also currently favored for larger unit cells or smaller crystals.

PULSED NEUTRON SOURCES AND
SINGLE-CRYSTAL INSTRUMENTATION

The characteristics of a pulsed spallation source are very different from those of a reactor neutron source. In the spallation process, protons are produced by an H discharge process and rapidly accelerate in a linear accelerator (LINAC). The protons are collected in an accumulator or storage ring and bunched into short high energy proton pulses. One pulse hits a heavy metal target like uranium, tungsten and lead or mercury at a time at a fixed frequency of 10, 20, 50 or 60 Hz. Through the impact excited metal nuclei boil off up to 20 highly energetic neutrons. These high energy neutrons are moderated/slowed in a moderator. Here neutrons collide with cooled, H/D rich phases, i.e., of liquid D_2O, H_2O, H_2, D_2, or solid CH_4, to name a few. The emission spectrum of the moderator depends on the moderator temperature and follows a Maxwell-Boltzmann distribution as not all neutron spend equal time in the moderator and some escape earlier with higher energy whereas others stay longer and carry less energy. As neutrons are produced in pulsed mode their kinetic energy and wavelength is known. The distribution maximum is dependent on the moderator temperature. Examples of pulsed neutron sources that have instruments for single-crystal neutron diffraction are the single-crystal neutron diffractometer, SXD, at ISIS (*http://www.isis.rl.ac.uk/crystallography/sxd*), SCD at IPNS (*http://www.pns.anl.gov/instruments/scd*) and SCD at LANSCE (*http://www.lansce.lanl.gov/ lujan/instruments/SCD/index.html*), and the Laue method single-crystal diffractometer, FOX, at the neutron science laboratory, KENS (*http://neutronwww.kek.jp/kens_e/spectrometer/fox.html*).

Time-of-flight Laue diffraction

The single-crystal time-of-flight (TOF) Laue technique can be traced back fifty years ago to a paper by Lowde (1956) that set the stage for a series of technical developments leading to the first collection and analysis of single-crystal data using a position-sensitive detector (PSD) at a pulsed spallation neutron source some 26 years later (e.g., Schultz et al. 1982). For neutrons produced at a pulsed source using accelerator-based methods, the production time of the neutrons can be precisely defined as the moment when the proton beam hits the target. Providing the flight path is known, it is possible to determine the wavelength of each of each neutron arriving at the detector by recording its arrival time. The velocities of thermal neutrons are extremely convenient for this method. Each neutron detected is thus time-stamped, giving a direct determination of its velocity and hence its energy and wavelength. The use of white beams, sorted using the TOF technique, allows fixed scattering geometries to be adopted that greatly simplify the use of complex and extreme sample environments.

The TOF Laue diffraction technique at pulsed sources exploits the capability to access large volumes of reciprocal space in a single measurement. This is due to combining the wavelength-sorting inherent in the TOF technique with large area PSDs. TOF Laue diffraction thus samples a large three dimensional volume of reciprocal space in a single measurement with a stationary crystal and detector. Figure 2 depicts the Ewald spheres for the minimum wavelength (λ_{min} at t_{min}) and the maximum wavelength (λ_{max} at t_{max}) during each pulse of neutrons. With a stationary crystal and with a stationary *point* detector at $2\theta = 90°$, four Bragg reflections along a linear reciprocal lattice vector can be measured "simultaneously" since they will satisfy the Bragg condition $\lambda_n = 2d_n\sin\theta$ at different times-of-flight, where n in Figure 2 varies from 2 to 5. With a linear PSD spanning the range from $2\theta_{min}$ to $2\theta_{max}$, all of the Bragg peaks within the shaded area could be measured. Going one step further, a two-dimensional area PSD will sample a solid volume of reciprocal space which may contain hundreds of Bragg peaks. This feature of the TOF Laue technique with a large area PSD is particularly useful in the measurement of numerous high Q data.

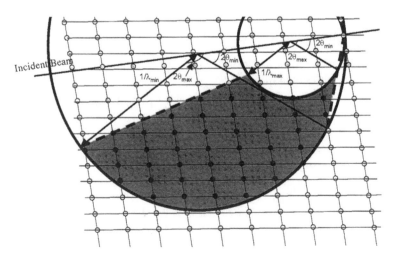

Figure 2. The geometry of (time-of-flight) Laue diffraction: the two Ewald spheres in reciprocal space are defined by the minimum and maximum wavelengths used and have radii of $1/\lambda_{min}$ and $1/\lambda_{max}$, respectively. The $2\theta_{min}$ and $2\theta_{max}$ values are defined by the angles subtended by the area detector. In this projection, the shaded region shows all regions of reciprocal space observed in a single measurement with stationary crystal and detector (modified from Wilson 2005).

The characteristics of the structure factor data collected on a pulsed source (described in detail below) have certain advantages for structural refinements (e.g., Wilson 2005):

- The collection of many Bragg reflections simultaneously in the detector allows the accurate determination of crystal cell and orientation from a single data frame (collected in one fixed crystal/detector geometry).

- The white nature of the incident beam allows straightforward measurement of reflections at different wavelength and is therefore useful in studies of wavelength-dependent effects such as extinction and absorption.

- The collection of data to very high $\sin\theta/\lambda$ values allows more precision in parameters that depend on very high resolution (e.g., charge density, anharmonicity effects), enabling examination of subtle structural features.

- The nature of the Laue method allows greater possibilities for the rapid collection of data sets by removing the need to measure each reflection individually.

TOF single-crystal diffraction is therefore ideal for surveying reciprocal space, rapid determination of large numbers of reflections and following structural changes using a subset of reflections. Such instrumentation developments offer substantial advantages to the earth sciences community, as both the data-collection time and the volume of sample required for the experiments are significantly reduced. As a consequence, a large number of natural and high-pressure phases only available as small single-crystals may be studied with neutron techniques. Another important feature of the technique is that not just the Bragg peaks are measured, but all of reciprocal space within a solid volume is sampled. This is advantageous for searching and measuring satellite and superlattice reflections and diffuse scattering. Structural and magnetic phase transitions associated with varying temperature or pressure can be observed and characterized without an *a priori* knowledge of where in reciprocal space the new scattering will appear, as for example in the study of the phase transitions in proustite, Ag_3AsS_3 (Nelmes et al. 1984). Such studies of the diffraction pattern as a function of time, temperature, or pressure will be of great aid in future experimental investigations of phase transformations and reaction kinetics.

SAMPLE REQUIREMENTS AND DATA COLLECTION PROCEDURES

In general neutron diffraction experiments require a much larger single-crystals than X-ray diffraction. Many samples will only form small single-crystals in the 0.001-0.1 mm^3 range, which are ideal for X-ray diffraction but generally too small for neutron studies. The requirement for rather large single-crystals is one of the main drawbacks of single-crystal neutron diffraction compared with X-ray methods. With the relatively low flux of neutron sources and the rather weak scattering of most materials, usually crystals of several mm^3 are required to allow collection of a good data set in a reasonable data collection time with the present-day neutron sources.

The first step of a typical single-crystal neutron data collection is to mount a crystal on the instrument, check its quality and orient it. It is advisable to discuss this aspect of the experiment with the beamline scientists well in advance of the experiment. For room and low temperature experiments, it is often most convenient to glue the samples directly onto aluminum pins using rapid setting Araldite. As much of the glue as possible should be masked by cadmium to reduce hydrogen scattering. It is important to allow the glue sufficient time to set (typically one hour) and not to jolt the sample on mounting into the apparatus. Samples for a furnace should be wired, or glued using high temperature cement, onto a stainless steel or similar sample pin. Gadolinium masks must be used if shielding is required in the furnace—low-melting cadmium must never be used.

Since the basic method of data collection using a time-of-flight Laue diffractometer is substantially different from standard four-circle methods used at steady state sources, the following example describes a typical experiment on a single-crystal diffractometer, SCD, at the IPNS (Schultz 1993; Schultz et al. 2006). Some manuals and user guides for operating the instrument and analyzing data are available online at: *http://www.pns.anl.gov/instruments/ scd/subscd/scd.shtml*. As shown in Figure 3, the SCD consists of Huber χ and ϕ circles with full 360° rotation about ϕ and rotation about χ from 90° to 180°. Neutrons travel from the upper right in Figure 3 and pass into the vacuum chamber where the crystal is mounted on a Displex cold stage. There are two 15 × 15 cm² detectors. One is at a 2θ angle of 75° and at a

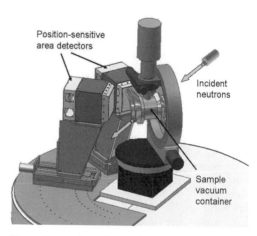

Figure 3. A schematic drawing of the single-crystal diffractometer, SCD, at the IPNS. There are two 15 × 15 cm² detectors: one is centered at a 2θ angle of 75° and at a distance of 23 cm from the sample and the second detector is centered at a 2θ angle of 120° and at a distance of 17 cm from the sample.

distance of 23 cm from the sample. The second detector is centered at a 2θ angle of 120° and at a distance of 17 cm from the sample. The ω angle, the detector angle, the sample-to-detector distance, and the sample-to-moderator distance are fixed. The detectors are position-sensitive scintillation detectors with a positional resolution of ~1.5 mm. Each detected neutron is characterized by three histogram coordinates x,y,t representing horizontal and vertical detector positions and time-of-flight, respectively. The time channel widths Δt are varied such that the time (wavelength) resolution $\Delta t/t$ is constant for all TOFs (wavelengths). The histogram coordinates can be mapped into *hkl* reciprocal lattice coordinates and plotted in various ways when searching for superlattice and satellite reflections.

A search program is used to locate and store the positions of peaks in a reflection file and an auto-indexing program (Jacobson 1986) is then used to obtain a unit cell and the crystal orientation. The design and operation of environmental equipment (temperature, pressure) is easier since alignment requirements are less stringent and, during actual data collection, the sample is stationary. Based on the symmetry and orientation of the crystal, the χ and ϕ angles of 10 to 35 histograms are selected in order to cover a unique region of reciprocal space. Bragg intensities are obtained by integrating procedures such as fitting a box around the peak (Wilkinson and Schultz 1989) or by use of an ellipsoidal contour technique which maximizes the $I/\sigma(I)$ ratio (Wilkinson et al. 1988). The integrated intensities I_{hkl} are reduced to structure factor amplitudes $|F_{hkl}|$ based on the Laue formula (Buras 1965):

$$I_{hkl} = k \ \tau(\lambda) \ \phi(\lambda) \ \varepsilon(\lambda,\mathbf{r}) \ A(\lambda) \ y(\lambda) \ |F_{hkl}|^2 \ \lambda^4 \ / \sin^2\theta \qquad (3)$$

where k is a scale factor and $\tau(\lambda)$ is the deadtime loss. The incident flux spectrum $\phi(\lambda)$ is obtained by measuring the incoherent scattering from a vanadium sample. The detector efficiency $\varepsilon(\lambda,\mathbf{r})$ is calculated as a function of wavelength λ and position $\mathbf{r}$ on the detector for each Bragg peak. Sample absorption $A(\lambda)$ includes the wavelength dependence of the linear absorption coefficients. Because of the large wavelength dependence of extinction, equivalent data are not averaged and the extinction correction $y(\lambda)$ is evaluated during the least-squares refinement of the structure. The set of structure factor amplitudes which are obtained at this point can be used to refine a structure using conventional least-squares methods. The procedure outlined above is similar to that encountered at other pulsed neutron sources. For comparison, the reader is encouraged to peruse the comprehensive user's manual for the single-crystal diffractometer, SXD, at ISIS (Keen and Wilson 1996) which is available online at: *http://www.isis.rl.ac.uk/crystallography/SXD/index.htm.*

APPLICATIONS OF SINGLE-CRYSTAL
NEUTRON DIFFRACTION STUDIES

Single-crystal neutron diffraction is the method of choice for many crystallographic experiments. The unique properties of neutrons summarized by Parise (2006, this volume) provide single-crystal neutron diffraction with the following advantages compared with other methods: (1) the location of hydrogen and light elements in the presence of heavy elements in structures; (2) the use of the neutron-scattering contrast to study the site partitioning of iso-electronic or near-iso-electronic atoms; (3) accurate measurements of the atomic displacement parameters that provide insight into properties such as zero-point vibrational motion, phase transitions, diffusion profiles and ionic conductivity; (4) diffuse scattering; (5) the study of the magnetic ordering and magnetic moments in crystal structures under ambient and non-ambient conditions; and (6) the charge density distribution in crystals. Some recent examples of these studies with application to minerals follow.

Location of light atoms in minerals

One of the main problems that can be solved using neutron diffraction is the determination of the position of H atoms (as D) and water molecules within a crystal structure. Knowledge of the position of the H atoms defines H bonding patterns and plays a fundamental role in the mechanistic understanding of H bonding and reactivity in minerals. A few examples with relevance in the earth sciences are given below and the reader is referred to Artioli (2002) for more examples.

Single-crystal neutron diffraction has been used to for locating light atoms (alkali cations) and molecules (H_2O, OH, CO_2) disordered in the channels of hydrous minerals, such beryl and cordierite (Cohen et al. 1977; Artioli et al. 1993, 1995a). The water molecules within such structures are weakly bonded and show a high degree of mobility. In structures such as these, the position of the proton must be located amidst a region of positive nuclear density related to the oxygen atoms and, sometimes, other cations or molecules located in the same region. Gatta et al. (2005) recently studied an alkali-poor beryl with single-crystal neutron diffraction. A high-quality, inclusion-free crystal with dimensions of $1.1 \times 1.3 \times 2.0$ mm^3 was used for the experiment. The single-crystal neutron diffraction experiment was performed at room temperature with a Huber four-circle diffractometer at the DIDO reactor at the Forschungszentrum Juelich, Germany. The unit-cell parameters determined from the neutron measurement are: $a = b = 9.2099(35)$ Å and $c = 9.1894(18)$ Å. A total of 1534 reflections were recorded with $-8 \leq h \leq 9$, $-9 \leq k \leq 8$ and $-9 \leq l \leq 9$ (maximum $2\theta = 100.78°$), of which 191 were unique. An initial structural refinement was carried out with isotropic displacement parameters in space group *P6/mcc* using the SHELXL-97 package (Sheldrick 1997), starting from the atomic coordinates of Artioli et al. (1993) and only considering the framework atoms shown in Figure 4. Correction for secondary isotropic extinction was applied using a fixed weighting scheme $(1/\sigma(F_0)^2)$. A convergence was rapidly achieved after the first cycles of refinement and the variance-covariance matrix did not show any strong correlation between the refined parameters (scale factor, atomic position of the framework atoms and their isotropic displacement parameters). The structural refinement conducted with only the framework sites produced two intense residual peaks in the final difference-Fourier map of the nuclear density: one positive positive (~ +2.9 fm/Å^3) at 0,0,1/4 (Fig. 5a) and at one negative (~ −0.6 fm/Å^3) at

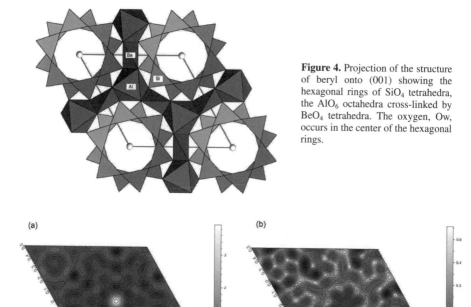

Figure 4. Projection of the structure of beryl onto (001) showing the hexagonal rings of SiO$_4$ tetrahedra, the AlO$_6$ octahedra cross-linked by BeO$_4$ tetrahedra. The oxygen, Ow, occurs in the center of the hexagonal rings.

Figure 5. Results from Gatta et al. (2005) showing the difference-Fourier maps of the nuclear density of beryl projected onto (001) at (a) $z = 0.25$ and (b) $z = 0.332$ after the first cycles of isotropic refinement with a channel-free structure. At $z = 0.25$ a residual peak of about $+2.9$ fm/Å^3 is seen lying on the six-fold axis, whereas at $z = 0.332$ there are six negative residual peaks (with a star shaped distribution) of about 0.6 fm/Å^3 around the six-fold axes. The orientation is the same as that in Figure 4.

0.03,−0.07,0.33 (Fig. 5b). Therefore, a further refinement was performed assigning the oxygen scattering length to the position with the positive residual peak and hydrogen scattering length to the site with the negative residual peak. The final agreement index (R_1) was 0.037 for 34 refined parameters and 160 unique reflections. The framework site positions were found to be in excellent agreement with those reported by Artioli et al. (1993) for an alkali-poor beryl. Analysis of the nuclear density Fourier maps shown in Figure 5 indicates that the (water) oxygen is located along the six-fold axis at the $2a$ site whereas the (water) protons are at the $6b$ site. Thus the topological configuration of water molecules appears to be very complex: for every oxygen site, the H atoms are distributed in 6×2 equivalent positions, above and below the oxygen site. The geometrical configuration of the water molecule appears to be different from that found in alkali-rich beryl (Artioli et al. 1993). It is worth noting that complementary methods such as inelastic neutron scattering (see Loong 2006, this volume) can be used to study the dynamics of the H$_2$O molecules in the channels of the structure (e.g., Winkler 1999).

Natural aluminosilicate zeolites provide another example of complex hydrous compounds in which AlO$_4$ and SiO$_4$ forming tetrahedral frameworks with channels and cavities containing cations and water molecules. The cations are necessary for framework charge compensation and are generally partially hydrated, having in their coordination sphere both framework oxygens and water molecules that are H-bonded to the framework or to other water molecules

The water molecules not linked to the cations usually fill all the remaining accessible space in the structure forming a complex network of hydrogen bonds. High-quality neutron-diffraction data is required to unravel the chemical complexity of the hydrogen bonds present in zeolites. A number of single-crystal neutron studies have been performed on large zeolite single-crystals both in the natural (Artioli et al. 1984, 1985, 1986, 1989) and cation-exchanged forms (Meneghinello et al. 2000). In yugawaralite, $CaAl_2Si_6O_{16} \cdot 4(H_2O)$, a low site-occupancy water molecule was found in the cavity that is not bonded to the Ca cations, but it is only hydrogen bonded to other water molecules and to framework oxygen atoms (Kvick et al. 1986). The presence of water molecules is the cause of the static site disorder observed in two of the fully occupied water molecules bonded to the Ca cations. Artioli et al. (2001) also found that some of the cation-bonded water molecules are expelled from the cavities before the weakly bonded molecules upon thermal activation.

In other mineralogical examples, the behavior of H atoms and the evolution of the H bonds have also been studied as a function of temperature in a number of hydrous minerals including muscovite (Liang et al. 1998), vesuvianite (Pavese et al. 1998; Lager et al. 1999), and gypsum (Schofield et al. 2000). Friedrich et al. (2001) used single-crystal neutron diffraction to determine accurate H positions of a natural F-bearing chondrodite from the Tilley Foster mine (Brewster, New York), an F-bearing titanian clinohumite from Kukh-i-Lal (Tadjikistan) and F-free titanian hydroxylclinohumite from Val Malenco (Italy). The goal of this study was to determine the structural environment of H in chemically diverse (OH/F ratio) natural humites at both ambient and lower temperatures. The humite group minerals have been proposed as possible candidates for the storage of water in the Earth's upper mantle as Ti-rich clinohumite and chondrodite have been found in rocks derived from the upper mantle, primarily as inclusions in kimberlites (McGetchin et al. 1970; Aoki et al. 1976; Smith 1979). Friedrich et al. (2001) collected data on a four-circle diffractometer at the High Flux Isotope Reactor of the Oak Ridge National Laboratory. Structures were refined in space group $P2_1/b$ from single-crystal neutron diffraction data and accurate H atom positions were determined at 295 K, 100 K, and 20 K. For each structure, only one H position of approximately 50% occupancy was observed, confirming a disordered H model. A few, very weak, symmetry-forbidden reflections were observed in the clinohumites at both ambient and lower temperatures. No temperature dependence is indicated and the intensity of the reflections are sample dependent. It appears that the real structure is made up of $P2_1$ and Pb domains so that violations are due to ordering of both H and Ti.

Site partitioning of iso-electronic or near-iso-electronic atoms

A common problem in mineral structures is the determination of the partitioning of cations between two or more symmetry-independent crystallographic sites. Cation partitioning has a direct bearing in many geological studies because the partitioning process of the cations is the result of the complex history of the mineral. Modeling of the thermodynamic and/or kinetic aspects of the partitioning process can provide important information on the *T-P*-time history of the rock under study (e.g., Carpenter et al. 1994; Redfern et al. 1997, 2000).

In many of the rock-forming silicates, Si and Al occupy tetrahedral sites in the structures. It is extremely difficult to determine the relative site occupancies of Si and Al with X-ray diffraction because the atoms are nearly isoelectronic. Most X-ray studies of the Si-Al site distribution in silicates therefore rely on the indirect evidence of the mean Si-O and Al-O tetrahedral distances. However, Si and Al site occupancies can be directly refined by using single-crystal neutron-diffraction data because of the neutron-scattering contrast between Si and Al. As an example, Artioli et al. (1984) investigated the structure of a single-crystal natrolite, $Na_2Al_2Si_3O_{10} \cdot 2H_2O$, at 20 K. The small equivalent isotropic temperature factors observed at 20 K are consistent with the lack of substitutional disorder and the presence of only weak thermal disorder. Complete Al_2Si_3 ordering on the tetrahedral sites (Fig. 6) was demonstrated by the refinement of scattering lengths. A similar study showed that nearly perfect tetrahedral order exists in Amelia albite (Smith et al. 1986) and long-range Si/Al tetrahedral order was confirmed in the zeolite, bikitaite, from Bikita (Ståhl et al. 1989).

Neutron diffraction can also be used to study the cation partitioning between different sites *in situ* under non-ambient conditions. It is well known, for example, that cation diffusion at high temperature is too fast to be frozen by rapid quenching, so that *in situ* studies at high temperature are required to determine the equilibrium configuration. As discussed by Redfern (2006, this volume), neutron-diffraction techniques are very useful for experiments using furnaces and pressure cells, and therefore high-quality data can be collected *in situ* at various temperatures and pressures. As an example, single-crystal

Figure 6. View of the structure of natrolite, $Na_2Al_2Si_3O_{10} \cdot 2H_2O$, down [001] showing the framework with fully ordered AlO_4 (light gray) and SiO_4 tetrahedra (black).

neutron data confirmed Mg-Fe ordering at high temperature in natural olivines (Artioli et al. 1995b; Rinaldi et al. 2000). The detection of the cation partitioning in the case of natural olivines is a difficult task because of the low iron content (the fayalite content in mantle or meteoritic samples is about 12% Fa) and because a severe parameter correlation between site-occupancy factors and the atomic displacement parameters. The single-crystal neutron studies were performed *in situ* using different crystals at both pulsed and steady-state neutron sources and proved beyond any experimental uncertainty that there is an anomalous segregation effect of the Fe cations into the octahedral *M*2 site above 900 °C. A similar partitioning process has also been observed by using powder neutron techniques in olivine samples, showing a more favorable neutron-scattering contrast, such as Fe-Mn and Mg-Mn (Redfern et al. 1997) and synthetic Mg-Fe olivine with iron content of 50% Fa (Redfern et al. 2000).

Atomic displacement parameters

The atomic displacement parameters (ADPs) are an important piece of information resulting from the structure refinements of single-crystal neutron diffraction data. Atomic displacement parameters measure the mean-square displacement amplitude of an atom about its equilibrium position in a crystal. The value of the mean-square atomic displacement can be due to the vibration of the atom or to static disorder. Obtaining quantitative ADP data using X-rays is, in general, more difficult than with neutrons. Atomic-displacement parameters resulting from a structural refinement are biased by several physical factors (i.e., extinction model, anisotropic primary absorption, thermal diffuse scattering, multiple scattering, etc.) that may not have been properly modeled during data analysis. Displacement parameters derived from neutron nuclear scattering lack the complications due to electron-density deformations and do not suffer from the fall-off in scattering power as a function of scattering angle. Therefore single-crystal neutron diffraction is the preferred technique to measure reliable ADPs.

Kunz et al. (2006) investigated the same sample of F-bearing chondrodite characterized by Friedrich et al. (2001) by single-crystal neutron diffraction in order to determine if the strengthening of the hydrogen bond in chondrodite, as suggested by spectroscopic measurements, could be detected by diffraction methods. They collected single-crystal neutron diffraction data at 500, 700, and 900 K on instrument D9 at the ILL at a wavelength of 0.8397(3) Å. The full mean square displacement matrix Σ of the O–H pair was determined from the temperature dependence of the ADPs, enabling a proper correction of the O–H bond for thermal vibration without assumptions about the correlation of O and H movements. The results show that the perpendicular O–H motions in chondrodite are intermediate between the riding and the independent motion models. The corrected O–H bond lengths do not change with temperature whereas the corrected H···F distances show an increase of ~0.02 Å with temperature, as do the Mg–O distances. This result shows that spectroscopic observations on the strength of the covalent O–H bond cannot be interpreted unambiguously in terms of a corresponding behavior of the associated hydrogen bond.

Tribaudino et al. (2003) studied spodumene, $LiAlSi_2O_6$, at 54 K using single-crystal neutron diffraction. X-ray refinements suggested that the atomic displacement parameter for Li in spodumene is significantly larger than that at $M2$ in other pyroxenes (Cameron et al. 1973). The possible presence of site splitting was suggested by Knight and Schofield (2000). Tribaudino et al. (2003) distinguished dynamic from static disorder by an analysis of the evolution of the displacement parameters with temperature. As previously shown (e.g., Benna et al. 1990; Pavese et al. 1995; Pilati et al. 1996; Prencipe et al. 2000), a positive intercept for a linear extrapolation of the high-temperature data at 0 K can provide an indication of the presence of positional disorder. As shown in Figure 7, an extrapolation of the B_{eq} (equivalent to isotropic displacement parameters) from data for spodumene as a function of temperature shows no significant residual for any atom. The B_{eq} are significant (about 60% of the room-temperature data) even at 54 K, suggesting that there is zero-point motion.

The zero-point motion has also been analyzed in another clinopyroxene, diopside ($CaMgSi_2O_6$). Prencipe et al. (2000) reported the first single-crystal neutron diffraction study of diopside at 10 K. The refinement at low temperature was of interest to evaluate the extent of zero-point motion in diopside and to compare with theoretical predictions provided by lattice dynamics (Pilati et al. 1996). The results from Prencipe et al.'s (2000) study indicated the presence of a significant zero-point motion, which ranges from 25 to 35% of the room-T value (apart from the $M2$ cation, which has a higher value, 45%), in reasonable agreement with the predictions by Pilati et al. (1996). Similar zero-point contributions have been reported in albite (Smith et al. 1986) and pyrope (Pavese et al. 1995).

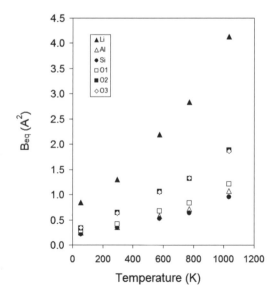

Figure 7. Equivalent isotropic displacement parameters, $B_{eq} = 8\pi^2 U_{eq}$, for spodumene as a function of temperature: data at 54 and 298 K are from Tribaudino et al. (2005); data at 573, 733 and 1033 K are taken from Cameron et al. (1973).

Analysis of atomic displacement parameters can also help explain other phenomena such as superionic conductivity. Many oxides and halides exhibit the perovskite structure, ABX_3. The alkaline earth fluorides are well known to exhibit fast ion conduction over a range of a few hundred degrees, up to their melting points. Indeed, it has been suggested that the high pressure phase of $MgSiO_3$, which has a distorted perovskite structure, might be a fast ion conductor under the pressures and temperatures of the Earth's lower mantle. Demetriou et al. (2005) used single-crystal neutron diffraction to investigate the nature of the high conductivity observed in the cubic perovskite $KCaF_3$ (space group *Pm3m*). It was apparent from previous powder neutron diffraction studies that there was a need for single-crystal neutron measurements using short wavelengths in order to obtain high resolution data down to low *d*-spacings, thus permitting differentiation between the point defect formation and the large dynamic effects observed at high temperature in superionic systems. The fact that fluorine has a higher scattering length than either Ca or K is an additional feature that favored a neutron study. Single-crystal neutron diffraction data were collected at 673 K and in the superionic region at 973 K. The modeling of the harmonic thermal parameters of the F yield an oblate ellipsoid shape for the vibrational envelope whereas K and Ca remain approximately spherical (Fig. 8a). The anharmonic model yields a significantly improved fit and, similar to the harmonic model, the F ions show a large degree of anisotropy in their vibrational envelopes and appear to vibrate towards a saddle point, at the centre of a cube face, between two neighboring K ions. On the basis of the F vibrations, Demetriou et al. (2005) proposed a vacancy migration model that involves the F ion jumping to a next-nearest neighbor site towards the face centre saddle point between two K ions (Fig. 8b). The results also indicate that there is no significant occupation of interstitial sites in $KCaF_3$ and it is likely that the high conductivity is due to a small population of highly mobile vacancies.

Zirconia, ZrO_2, doped with cations such as Ca^{2+}, Y^{3+}, Mg^{2+}, Sc^{3+} or anions (e.g., N^{3-}), has attracted considerable interest over several decades due to its favorable material properties and interesting order–disorder phenomena. Pure ZrO_2 is monoclinic at room temperature (space group $P2_1/c$) and transforms into a tetragonal ($P4_2/nmc$) and a cubic ($Fm3m$) structure at higher temperatures. Depending on the type and amount of doping as well as on temperature various short- and long-range order phenomena occur. All structures can be derived from the parent cubic fluorite type structure. Here the cations occupy the corners and face centers of the cubic unit cell, while the anions are located in the tetrahedral interstices forming a primitive cubic

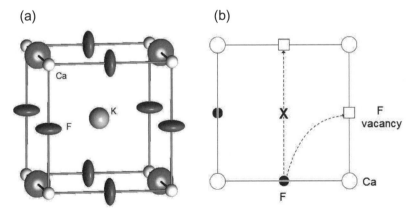

Figure 8. (a) Thermal ellipsoid plot of $KCaF_3$ perovskite based on a harmonic model at 700 °C (Demetriou et al. 2005). While the K and Ca atoms appear to show spherical vibrations, the F atoms exhibit a flattened sphere, with large thermal vectors in the direction of the unit cell face centers. (b) Projection onto (010) showing the proposed vacancy mechanism for F migration in $KCaF_3$ where "X" is a saddle point.

lattice with half the lattice constant. The doping with lower valence cations and/or higher valence anions leads to vacancies on the anion lattice generated for charge compensation. These are responsible for the short- and long-range order phenomena mentioned as well as the enhanced anionic conductivity that therefore proceeds via a vacancy mechanism. Kaiser-Bishoff et al. (2005) carried out single-crystal neutron diffraction experiments on three Y- and N-doped ZrO_2 samples at high temperatures using the instrument 5C2 of the LLB and D9 of the ILL to shed light on the anion diffusion processes in these materials. Neutron (instead of X-ray) diffraction was essential for these investigations because of the relatively larger scattering power of the lighter anions and the general higher reliability for the determination of Debye-Waller factors. The analyses involved the calculation of the probability density function which is derived by a Fourier transformation of the Debye-Waller factor. Contributions to the probability density function arise from the thermal vibrations of the atoms and the static displacements around the defects. These two contributions are not always easy to disentangle and the reader is referred to Boysen (2003) for more details. A major conclusion of this study is that diffusion jumps take place directly to vacant nearest-neighbor anion sites through the edges of the surrounding cation tetrahedra along <100> directions. Activation enthalpies of migration for O (1.09 eV) and N (1.99 eV) are in good agreement with experimental values obtained from tracer diffusion measurements (Kilo et al. 2003, 2004). The diffusion process is facilitated by local short range order and anharmonic thermal vibrations, suggesting that phonons have to be taken into account in the description of the diffusion process.

The detailed analysis of the temperature behavior of the thermal components of the atomic-displacement parameters also yields important information about phase transitions and insight into unusual phenomena occurring at phase transitions. Harris et al. (1996), for example, confirmed that lattice melting occurs in ferroelastic Na_2CO_3 using a combination of single-crystal neutron diffraction and inelastic neutron scattering. Lattice melting is an unusual effect that corresponds to a continuous loss of long-range order at a ferroelastic phase transition. This involves a continuous divergence of the mean-squared atomic displacements at the transition, so that the crystalline long range order is destroyed and Bragg scattering is replaced by diffuse scattering (described below). The ferroelastic phase transition in Na_2CO_3 occurs at ~760 K and involves a change in symmetry from $P6_3/mmc$ (Fig. 9a) to $C2/m$ (Fig. 9b) upon cooling (Swainson et al. 1995). The crystal structure of the hexagonal phase of Na_2CO_3 contains one-dimensional chains of NaO_6 face-sharing octahedra lying parallel to c. These chains are linked to each other laterally by carbonate groups that lie parallel to the (001) planes. Upon cooling, the restoring force for shear motions of the chains against each other (with the carbonate groups acting as "hinges") becomes smaller and smaller, and the shear fluctuations become larger in amplitude, until eventually the critical point, T_c, is reached. Below this temperature, the shear fluctuations freeze in to produce the static monoclinic strain. The effect of lattice melting in Na_2CO_3 is evident from the significant diffuse scattering that appears perpendicular to c (but not along c) at the phase transition. The data gathered by Harris et al. (1996) are consistent with a special form of lattice melting where the long-range order appears to be destroyed in a two-dimensional sense, but is preserved in the third dimension.

Diffuse scattering

Diffuse scattering is the weak background scattering that occurs in the diffraction patterns of crystalline materials from the simplest, e.g., NaCl, to the most complex, e.g., proteins. While the strong sharp diffraction peaks in the pattern are used by conventional crystallography to deduce the average repetitive arrangements of atoms in crystals, diffuse scattering contains information about the deviations from the average. Quite often it is just these deviations from the average, or types of disorder, rather than the average structure itself, that give materials their unique or novel properties. Traditionally, X-ray diffraction is used for surveying measurements of reciprocal space. The recent return to the use of 2D detectors in

(a)   **(b)**

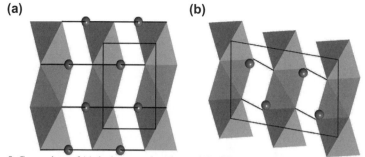

Figure 9. Comparison of (a) the hexagonal α-phase of Na_2CO_3 and (b) the monoclinic β-phase of Na_2CO_3 viewed down [010] with outlines of unit cells shown. The $Na(1)O_6$ octahedra are shown with $Na(2)$ atoms represented as shaded spheres and the planes of the carbonate groups represented by short, black lines. The shear strain and the coupled hinged rotations of the carbonate groups are clearly shown in this projection.

diffraction experiments has meant that many scientists are seeing diffuse scattering for the first time. For almost a generation diffuse scattering has been overlooked entirely due to the use of single point detectors. Enormous improvements in measuring power (sources and detectors) and computing power have made and will continue to make the collection and analysis of diffuse scattering data more commonplace.

Keen et al. (1994) demonstrated that it was feasible to measure good quality diffuse scattering from single-crystals at pulsed neutron sources. The diffuse scattering of a 1 cm³ single-crystal of α-AgI was measured on the SXD diffractometer at ISIS. The structure of the α-AgI exists between 420 K and the melting point $(T_m = 825$ K). The iodide sublattice in α-AgI possesses a body-centered cubic arrangement, with the smaller Ag ions located in a disordered manner in the anion interstices (space group *Im3m*). In this study, the single-crystal of α-AgI was grown from the melt *in situ* on the diffractometer. Weak rings of diffuse scattering were observed together with strong scattering around some Bragg peaks. The results indicate significant correlations between the motion of the Ag ions and the vibrations of the I ions.

Keen et al. (1998) further demonstrated that single-crystal TOF neutron diffraction using large area position-sensitive detectors gives extremely rapid measurement of large volumes of reciprocal space to high Q and described how more subtle, localized diffuse features can be measured using neutron TOF instrumentation with highly collimated detectors. Welberry et al. (2001), for example, performed experiments on the molecular crystal benzil, $C_{14}D_{10}O_2$, using SXD at ISIS and showed that it was possible to access a large fraction of the total three-dimensional reciprocal space out to a Q value of 15 Å⁻¹ using only four individual exposures and by making use of the $\bar{3}m$ Laue symmetry of the crystal. Features were seen in the diffuse neutron scattering data of benzil that could not be observed in the X-ray patterns (Fig. 10a). A model previously derived from analysis of X-ray data observed over a limited range of Q was used to calculate neutron patterns over the full Q range. Comparison with the neutron data showed that while the model gives a good description of the form of the diffuse patterns, the magnitudes of the atomic displacements are underestimated by a factor of approximately 2.25. Neutrons of lower energies allow the inelastic nature of the scattering to be manifested. In particular, a splitting of the diffuse lines which form a hexagon of scattering (Fig. 10b) is attributed to scattering in which phonon energies are added to or subtracted from the incident neutron energy. The phonon mode largely responsible for the diffuse hexagon was postulated to be a relatively dispersion-less vibration involving the coupling of an internal molecular mode with more distant neighbors via a network of hydrogen bonds. The form of this diffuse scattering in benzil is very similar to that reported for quartz (Tucker et al. 2000, 2001), for which similar low-lying relatively dispersion-less modes are known to exist.

(a)   (b)

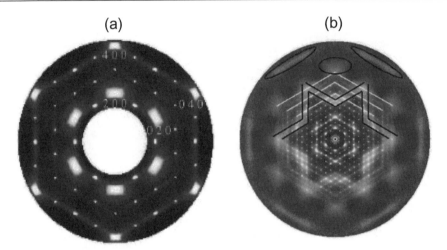

Figure 10. Comparison of (a) the diffuse X-ray scattering for the $hk0$ reciprocal section of benzil and (b) single-crystal neutron scattering data for the $hk0$ section. The dark lines in (b) are drawn to emphasize broad regions of diffuse intensity and the white lines to emphasize narrow lines of scattering. The neutron data extend to very high Q, far beyond the region where Bragg peaks are observed (Welberry et al. 2001).

Magnetic structural studies

Neutron diffraction is the best probe for measuring microscopic ordering of magnetic moments, and can be used to determine magnetic structures of many magnetic elements. Iron, for example, is the most abundant element by weight in the Earth and the magnetic structures and transitions of iron-bearing minerals present in the Earth's crust, upper mantle, transition zone and lower mantle are of paramount importance to elucidate their physical properties and behavior. Harrison (2006, this volume) describes examples of magnetic studies in Fe-bearing minerals including a polarized single-crystal neutron diffraction study of nanoscale hematite, Fe_2O_3, exsolution in ilmenite, $FeTiO_3$. We describe below a series of neutron experiments of a member of the pyrochlore family, $Tb_2Ti_2O_7$, including results from a high-pressure single-crystal neutron diffraction study from which the magnetic phase diagram of $Tb_2Ti_2O_7$ was determined.

Pyrochlore compounds, with chemical composition $A_2B_2O_7$, crystallizes in a face-centered-cubic structure (space group $Fd3m$) in which the A site is occupied by a trivalent rare-earth ion with eightfold coordination with oxygen and the B site is occupied by a tetravalent transition-metal ion with sixfold coordination with oxygen. The sublattice of each of the two metal ions form an interpenetrating network of corner-sharing tetrahedra. Many experimental studies have been carried out on pyrochlore and related systems and many have revealed classic spin-glass behavior at low temperatures (e.g., Gardner et al. 2001; Mirebeau et al. 2002). $Tb_2Ti_2O_7$ belongs to a family of rare-earth titanate pyrochlores, $R_2Ti_2O_7$, whose magnetic rare-earth ion, Tb^{3+}, resides on a network of corner-sharing tetrahedra. Such a local geometry is known to give rise to geometrical frustration in the presence of antiferromagnetic interactions. Neutron scattering experiments on polycrystalline and single-crystals of $Tb_2Ti_2O_7$ (e.g., Gardner et al. 1999, 2001) showed that the system displays a cooperative paramagnetic or spin liquid state at low temperatures, with neither long-range Néel order nor spin glass ordering at temperatures as low as 0.07 K, despite developing short-range antiferromagnetic correlations as high as 50 K. Recent high-pressure neutron powder studies further demonstrated that it is possible to induce long range magnetic order in $Tb_2Ti_2O_7$ by applying a quasi-hydrostatic pressure (Mirebeau et al. 2002; Mirebeau and Goncharenko 2004). A complex anti-ferromagnetic structure was found, coexisting with the spin liquid phase down to 1.4 K. While these experiments opened new possibilities about the complex energy balance that controls the stability of these systems, they also raised several crucial questions about the exact magnetic structure, the role of pressure, and the true ground state of $Tb_2Ti_2O_7$.

To address these questions, Mirebeau and Goncharenko (2005) studied $Tb_2Ti_2O_7$ by means of single-crystal neutron diffraction at high pressures (up to 2.8 GPa), high magnetic fields (up to 7 T) and at low temperatures (down to 0.14 K). A large single-crystal was grown and plates from 1 to 0.15 mm thickness and 1 mm^2 surface were cut perpendicular to the principal axes [111], [100] or [110] of the cubic cell. The sample was inserted in a Kurchatov–LLB pressure cell. By choosing the appropriate transmitting medium, they were able to impose either a uniaxial stress (no transmitting medium) or a hydrostatic pressure (liquid ethanol–methanol transmitting medium), or a combination of the two (NaCl solid transmitting medium). The measurements were performed on the single-crystal diffractometer 6T2 of the LLB with a lifting arm with an incident neutron wavelength of 2.34 Å. In the single-crystal experiments, the magnitude and orientation of the stress can be controlled with respect to the single-crystal which is a critical for inducing magnetic order. The single-crystal data provide insight into the mechanism which suppresses the spin liquid state in $Tb_2Ti_2O_7$. Mirebeau and Goncharenko (2005) found that by applying a combination of isotropic ($P_i = 2.4$ GPa) and anisotropic pressure ($P_u = 0.3$ GPa) along the [011] axis of a single-crystal of $Tb_2Ti_2O_7$, they could determine the magnetic phase diagram for these pressures (Fig. 11). They observe only one smeared transition towards the paramagnetic phase, where all magnetic peaks disappear. The transition temperature strongly increases with the applied field, by more than one order of magnitude between 0 and 4 T, reaching 23 K at 4 T. Applying a magnetic field under pressure induces a second transition towards a canted ferromagnet (Fig. 11). Thus while the mechanism that relieves the very strong frustration of this compound has been now identified, the stability of the spin liquid state in $Tb_2Ti_2O_7$ remains a mystery.

Charge density distributions

Neutron diffraction complements X-ray diffraction in terms of obtaining full descriptions of structures. Since neutrons are scattered by the atomic nuclei and X-rays by the surrounding

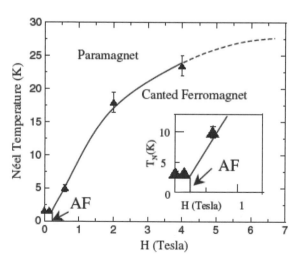

Figure 11. The magnetic phase diagram of $Tb_2Ti_2O_7$ as determined by Mirebeau and Goncharenko (2005) in the pressure induced ordered state $P_i = 2.4$ GPa and $P_u = 0.3$ GPa. P_u and H are oriented along the [011] axis.

electrons, use of both techniques allows a complete picture to be built up of the atoms. This is of particular relevance in studies of bonding density, as the neutron diffraction provides accurate descriptions of the nuclear "anchor points" of the atoms, from which the electron density can be better modeled and understood. Single-crystal neutron diffraction has therefore been used in conjunction with single-crystal X-ray diffraction to determine the charge density distributions in crystals (e.g., Artioli 2002). X-ray diffraction yields information on the time- and volume-averaged electron-charge distribution in the crystal whereas neutron diffraction yields information on the position of the atomic nuclei and their thermal motion. Diffraction data obtained by both techniques on the same compound can be analyzed in order to define the differences between the distribution of the electrons and the nuclei in the crystal. Once the positional and the thermal atomic parameters obtained by the neutron structure refinement are fixed, then the residual charge density visible in the maps obtained from refinement of the neutron model against the X-ray data can be determined and information about the distortion of the electron distribution of the atoms involved in chemical bonds can be derived.

As an example, McIntyre et al. (1990) determined the electron density of tetragonal nickel sulfate hexadeuterate $NiSO_4 \cdot 6D_2O$ (space group $P4_32_12$) at room temperature from combined X-ray and neutron diffraction data. The deformation electron density of the compound was determined by multipole refinement against single-crystal X-ray intensity data with the H(D) positional and displacement parameters fixed to values determined by refinement against single-crystal neutron data. They found that the experimental deformation density around Ni is in good agreement with that expected from simple ligand field theory for an ideally octahedral $[Ni(D_2O)_6]^{2+}$ complex while the individual densities of the water molecules show clear polarization of the lone-pair densities according to the coordination of the water molecules.

There have been several recent studies comparing theoretical and experimental calculations of the charge-density distribution in minerals looking in particular at the distribution of the valence electrons or the electrons confined in lone pairs. Gibbs et al. (2005), for example, compared theoretical electron density distributions and bond critical point properties with those derived from high energy single-crystal synchrotron X-ray diffraction data and from high resolution X-ray diffraction data for coesite, stishovite, forsterite, fayalite, cuprite, coesite and senarmonite. The overall agreement between the bond critical point properties generated with computational quantum methods and experimental methods provides a basis for improving our understanding of the crystal chemistry and bonded interactions of earth materials. Indeed, theoretical bond critical point properties generated with computational quantum methods now rival the accuracy of those determined experimentally and provide a powerful and efficient method for evaluating electron density distributions and the bonded interactions for a wide range of earth materials.

LOOKING TO THE FUTURE

Some of the common applications of single-crystal neutron diffraction have been reviewed with examples derived from earth and related materials. Historically, the main drawback of using neutron diffraction in mineralogical applications has been the requirement for relatively large crystals (e.g., several cubic millimeters in volume), the long data acquisition times owing to the available neutron flux and the limited access to neutron sources. As the next generation sources such as the SNS and J-PARC come on line, it will be possible to work with crystals approaching the size commonly used in X-ray studies and thus to greatly expand the range of materials that are open to investigation. In addition, by utilizing the enhanced flux of the next generation spallation sources along with the increased detector coverage and improvements in focusing optics, data collection times will be reduced from days to hours. The new single-crystal instrument, Topaz, under development for SNS will measure a large portion of reciprocal space simultaneously, employing an array of highly pixilated two-dimensional position and time sensitive detectors to utilize the time-of-flight Laue technique (Fig. 12). The instrument is being optimized for high-throughput on samples with moderate-size unit cells, up to approximately 50 Å on edge. Provision is also intended for the measurement of diffuse scattering and for the production of polarized beams to facilitate the study of magnetic structures.

Neutron scattering will continue to provide an excellent choice for *in situ* studies under non- ambient conditions because of the low absorption coefficient for neutrons of materials suitable for the construction of conditioning chambers (furnaces, cryostats, pressure cells). In addition, the increased flux from next generation neutron sources will increase the pressure range for single-crystal studies. Nelmes et al. (2003) recognized the increasing need for high pressure single-crystal neutron diffraction studies to deal with issues of structural complexity that power diffraction could not resolve. The availability of large gem moissanite (silicon carbide) for use as anvils has made development of single-crystal techniques feasible with >1 mm^3 sample volumes (Xu et al. 2000). Mao et al. (2004) reported preliminary results from an experiment in which mixtures of deuterated water and deuterium gas were loaded and grown into single-crystals in a panoramic cell. With further advances in the growth of 30 and 300 carat diamond anvils, the 30-300 GPa ultrahigh pressure field will be brought into the main stream neutron investigations. This will be carried out using a combination of high-pressure

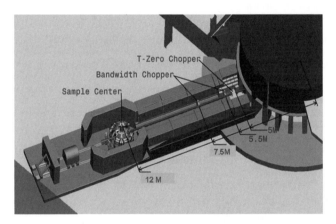

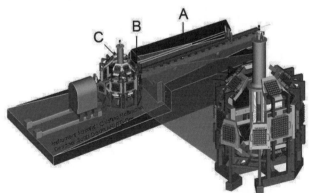

Figure 12. Upper diagram shows a schematic layout of the single-crystal diffractometer, Topaz, under construction at the Spallation Neutron Source (SNS) in Oak Ridge, TN. In the bottom diagram, neutrons travel from right to left down a short flight path (A), through focusing optics (B), to a versatile sample chamber (C) which is enlarged to show the large, real space detector coverage.

synthetic diamond plus single-crystal CVD diamond which Yan et al. (2002) have shown can be produced at high growth rates in the laboratory and has already been shown to reach pressures in excess of 200 GPa. The combination of increased flux and larger diamonds will push capabilities for "mainstream" neutron studies and reconnaissance work into regimes now becoming available at synchrotron sources.

ACKNOWLEDGMENTS

We are grateful to Diego Gatta and Art Schultz for providing figures and access to manuscripts in press. NLR acknowledges support from COMPRES, the Consortium for Materials Properties Research in Earth Sciences under NSF Cooperative Agreement EAR 01-35554.

REFERENCES

Aoki K, Fujino K, Akaogi M (1976) Titanochondrodite and titanoclinohumite derived from the upper mantle in the Buell Park kimberlite, Arizona, U.S.A. Contrib Mineral Petrol 56:243-253
Arndt UW, Willis BTM (1966) Single-crystal Diffractometry. Cambridge University Press
Artioli G (2002) Single-crystal neutron diffraction. Eur J Mineral 14:233-239
Artioli G, Rinaldi R, Kvick Å, Smith, JV (1986) Neutron diffraction structure refinement of the zeolite gismondine at 15 K. Zeolites 6:361-366
Artioli G, Rinaldi R, Ståhl K, Zanazzi PF (1993) Structure refinement of natural beryls by single-crystal neutron and X-ray diffraction. Am Mineral 78:762-768
Artioli G, Rinaldi R, Zanazzi PF, Wilson CC (1995a) Single-crystal pulsed neutron diffraction of a highly hydrous beryl. Acta Crystallogr B51:733-737
Artioli G, Rinaldi R, Zanazzi PF, Wilson CC (1995b) High temperature Fe/Mg cation partitioning in olivine: in situ single-crystal neutron diffraction study. Am Mineral 80:197-200
Artioli G, Smith JV, Kvick Å. (1985) Multiple hydrogen positions in the zeolite brewsterite, $(Sr_{0.95},Ba_{0.05})Al_2Si_6O_{16}\cdot 5H_2O$. Acta Crystallogr C41:492-497
Artioli G, Smith JV, Kvick Å. (1989) Single-crystal neutron diffraction study of partially dehydrated laumontite at 15 K. Zeolites 9:377-391
Artioli G, Smith, JV, Kvick Å (1984) Neutron diffraction study of natrolite, $Na_2Al_2Si_3O_{10}\cdot 2H_2O$, at 20 K. Acta Crystallogr C40:1658-1662
Artioli G, Ståhl K, Cruciani G, Gualtieri A, Hanson JC (2001) In situ dehydration of yugawaralite. Am Mineral 86:185-192
Benna P, Tribaudino M, Zanini G, Bruno E (1990) The crystal structure of $Ca_{0.8}Mg_{1.2}Si_2O_6$ clinopyroxene $(Di_{80}En_{20})$ at T = −130°, 25°, 400° and 700°C. Z Kristallogr 192:183-199
Boysen H (2003) The determination of anharmonic probability densities from static and dynamic disorder by neutron powder diffraction. Z Kristallogr 218:123-131
Buras B, Mikke K, Lebech B, Leciejewicz J (1965) The time-of-flight method for investigations of Single-Crystal Structures . Physica Status Solidi 11:567-573
Cameron M, Sueno S, Prewitt CT, Papike JJ (1973) High-temperature crystal chemistry of acmite, diopside, hedenbergite, jadeite, spodumene, and ureyite. Am Mineral 58:594-618
Carpenter MA, Salje EKH (1994) Thermodynamics of non-convergent cation ordering in minerals: II Spinels and orthopyroxene solid solution. Am Mineral 79:770-776
Cohen JP, Ross FK, Gibbs GV (1977) An X-ray and neutron diffraction study of hydrous low cordierite. Am Mineral 62:67-78
Demetriou DZ, Catlow CRA, Chadwick A, McIntyre GJ, Abrahams I. (2005) The anion disorder in the perovskite fluoride $KCaF_3$. Solid State Ionics 176:1571-1575
Friedrich A, Lager GA, Kunz M, Chakoumakos BC, Smyth JR, Schultz AJ (2001) Temperature-dependent single-crystal neutron diffraction study of natural chondrodite and clinohumites. Am Mineral 86:981-989
Gardner JS, Dunsiger SR, Gaulin BD, Gingras MJP, Greedan JE, Kiefl RF, Lumsden MD, MacFarlane WA, Raju NP, Sonier JE, Swainson I, Tun Z (1999) Cooperative paramagnetism in the geometrically frustrated pyrochlore antiferromagnet $Tb_2Ti_2O_7$. Phys Rev Lett 82:1012-1015
Gardner JS, Gaulin BD, Berlinsky AJ, Waldron P, Dunsiger SR, Raju NP, Greedan JE (2001) Neutron scattering studies of the cooperative paramagnet pyrochlore $Tb_2Ti_2O_7$. Phys Rev B 64:224416 (1-9)
Gatta GD, Nestola F, Bromiley GD, Mattauch S (2005) The real topological configuration of the extra-framework content in alkali-poor beryl: A multi-methodological study. Am Mineral 91:29-34

Gibbs GV, Cox DF, Ross NL, Crawford TD, Downs RT, Burt JB (2005) Comparison of the electron localization function and deformation electron density maps for selected earth materials. J Phys Chem A 109:10022-10027

Harris M J, Dove MT, Godfrey KW (1996) A single-crystal neutron scattering study of lattice melting in ferroelastic Na_2CO_3. J Phys Condens Matt 8:7073–7084

Harrison RJ (2006) Neutron diffraction of magnetic materials. Rev Mineral Geochem 63:113-143

Helliwell JR (1997) Neutron Laue diffraction does it faster. Nature Struct Biol 4:874-876

Jacobson RA (1986) An orientation-matrix approach to Laue indexing. J Appl Crystallogr 19:283-286

Kaiser-Bischoff I, Boysen H, Scherf C, Hansen T (2005) Anion diffusion in Y- and N-doped ZrO_2. Phys Chem Chem Phys 7:2061-2067

Keen DA, Harris MJ, David WIF (1998) Neutron time-of-flight measurements of diffuse scattering. Physica B 241-243:201-203

Keen DA, Nield VM, McGreevy RL (1994) Diffuse neutron scattering from an *in situ* grown α-AgI single-crystal. J Appl Crystallogr 27:393-398

Kilo M, Argirusis C, Borchardt G, Jackson RA (2003) Oxygen diffusion in yttria stabilised zirconia - experimental results and molecular dynamics calculations. Phys Chem Chem Phys 5:2219-2224

Kilo M, Taylor MA, Argirusis C, Borchardt G, Lerch M, Kaitasov O, Lesage B (2004) Nitrogen diffusion in nitrogen-doped yttria stabilized zirconia. Phys Chem Chem Phys 6:3645-3649

Knight KS, Schofield PF (2000) Structural basis for the negative eigenvalue in the thermal expansion tensor of spodumene. ISIS Experimental Report 10782

Kunz M, Lager G, Burgi HB, Fernandez-Diaz MT (2006) High-temperature single-crystal neutron diffraction study of natural chondrodite. Phys Chem Minerals 33:17-27

Kvick Å, Artioli G, Smith JV (1986) Neutron diffraction study of the zeolite yugawaralite at 13 K. Z Kristallogr 174:265-281

Lager GA, Xie QY, Ross FK, Rossman GR, Armbruster T, Rotella FJ, Schultz AJ (1999) Hydrogen-atom positions in *P4/nnc* vesuvianite. Can Mineral 37:763-768

Liang JJ, Hawthorne FC, Swainson IP (1998) Triclinic muscovite: X-ray diffraction, neutron diffraction and photoacoustic FTIR. Can Mineral 36:1017-1027

Loong C-K (2006) Inelastic scattering and applications. Rev Mineral Geochem 63:233-254

Lowde RD (1956) A new rationale of structure-factor measurement in neutron-diffraction analysis. Acta Crystallogr 9:151-155

Mao HK, Nelmes RJ (2004) Crystallography of high pressure hydrogen hydrate. ISIS Experimental Report, Rutherford Appleton Laboratory, 14361

McGetchin TR, Silver LT, Chodos AA (1970) Titanoclinohumite: A possible mineralogical site for water in the upper mantle. J Geophys Res 75:255-259

McIntyre GJ, Ptasiewicz-Bak H, Olovsson I (1990): Bonding deformation and superposition effects in the electron density of tetragonal nickel sulfate hexadeuterate $NiSO_4 \cdot 6D_2O$. Acta Crystallogr B46:27-39

Meneghinello E, Alberti A, Cruciani G, Sacerdoti M, McIntyre G, Ciambelli P, Rapacciuolo MT (2000) Single-crystal neutron diffraction study of the natural zeolite barrerite in its ND_4-exchanged form. Eur J Mineral 12:1123-1129

Mirebeau I, Goncharenko IN (2004) Spin liquid and spin ice under high pressure: a neutron study of $R_2Ti_2O_7$ (R = Tb, Ho). J Phys Condens Matt 16:S653-S663

Mirebeau I, Goncharenko IN (2005) $Tb_2Ti_2O_7$: a "spin liquid" single-crystal studied under high pressure and high magnetic field. J Phys Condens Matt 17:S771-S782

Mirebeau I, Goncharenko IN, Cadavez-Peres P, Bramwell ST, Gingras MJP, Gardner JS (2002) Pressure-induced crystallization of a spin liquid. Nature 420:54-57

Mirebeau I, Goncharenko IN, Dhalenne G, Revcolevschi A (2004): Pressure and field induced magnetic order in the spin liquid $Tb_2Ti_2O_7$ as studied by single-crystal neutron diffraction. Phys Rev Lett 93:187204

Nelmes RJ, Howard CJ, Ryan TW, David WIF, Schultz AJ (1984) A neutron and X-ray diffraction study of the phase transitions in proustite (Ag_3AsS_3) between 35 K and room temperature. J Phys C Solid State Phys 17:L861-L865

Nelmes RJ, Loveday JS, Klotz S, Hamel G, Strässle T (2003) Single-crystal neutron diffraction at high pressure. ISIS Experimental Report, Rutherford Appleton Laboratory, 14007

Niimura N, Minezaki Y, Nonaka T, Castagna JC, Cipriani F, Hoghoj P, Lehmann MS, Wilkinson C (1997): Neutron Laue diffractometry with an imaging plate provides an effective data collection regime for neutron protein crystallography. Nature Struct Biol 4:909-914

Parise JB (2006) Introduction to neutron properties and applications. Rev Mineral Geochem 63:1-25

Pavese A, Artioli G, Prencipe M (1995) X-ray single-crystal diffraction study of pyrope in the temperature range 30–973 K. Am Mineral 80:457-464

Pavese A, Prencipe M, Tribaudino M, Aagaard SS (1998) X-ray and neutron single-crystal study of *P4/n* vesuvianite. Can Mineral 36:1029-1037

Pilati T, Demartin F, Gramaccioli CM (1996) Lattice-dynamical evaluation of atomic displacement parameters of minerals and its implications: the example of diopside. Am Mineral 81:811-821

Prencipe M, Tribaudino M, Pavese A, Hoser A, Reehuis M (2000): Single-crystal neutron-diffraction investigation of diopside at 10 K. Can Mineral 38:183-189

Redfern SAT (2006) Neutron powder diffraction studies of order-disorder phase transitions and kinetics. Rev Mineral Geochem 63:145-170

Redfern SAT, Artioli G, Rinaldi R, Henderson CMB, Knight KS, Wood BJ (2000) Octahedral cation ordering in olivine at high temperature. II: an *in situ* neutron powder diffraction study on synthetic MgFeSiO₄ (Fa50). Phys Chem Minerals 27:630-637

Redfern SAT, Henderson CMB, Knight KS, Wood BJ (1997) High temperature order-disorder in $(Fe_{0.5}Mn_{0.5})_2SiO_4$ and $(Mg_{0.5}Mn_{0.5})_2SiO_4$ olivines: an *in situ* neutron diffraction study. Eur J Mineral 9:287-300

Rinaldi R, Artioli G, Wilson CC, McIntyre G (2000) Octahedral cation ordering in olivine at high temperature. I: *in situ* neutron single-crystal diffraction studies on natural mantle olivines (Fa12 and Fa10). Phys Chem Minerals 27:623-629

Schofield PF, Wilson CC, Knight KS, Stretton IC (2000) Temperature related structural variation of the hydrous components in gypsum. Z Kristallogr 215:707-710

Schultz AJ (1993) Single-crystal time-of-flight neutron diffraction. Trans Am Crystallogr Assoc 29:29-41

Schultz AJ, De Lurgio PM, Hammonds JP, Mikkelson DJ, Mikkelson RJ, Miller ME, Naday I, Peterson PF, Porter RR, Worlton TG (2006) The upgraded IPNS single crystal diffractometer. Physica B (in press)

Schultz AJ, Teller RG, Peterson SW, Williams JM (1982) Collection and analysis of single-crystal time-of-flight neutron diffraction data. *In*: AIP Conference Proceedings #89: Neutron Scattering – 1981. Faber J Jr (Ed) American Institute of Physics, p 35-41

Sheldrick GM (1997) SHELX-97. Programs for crystal structure determination and refinement. University of Goettingen, Germany (*http://shelx.uni-ac.gwdg.de/SHELX/*)

Smith D (1979) Hydrous minerals and carbonates in peridotite inclusions from the Green knobs and Buell Park kimberlite diatremes on the Colorado Plateau. *In*: The Mantle Sample: Inclusions in Kimberlites and Other Volcanics Boyd FR, Meyer HOA (eds) American Geophysical Union, p 345–356

Smith JV, Artioli G, Kvick Å. (1986): Low albite, $NaAlSi_3O_8$: neutron diffraction study of crystal structure at 13 K. Am Mineral 71:727-733

Ståhl K, Kvick Å, Ghose S (1989) One-dimensional water chain in the zeolite bikitaite: neutron diffraction study at 13 and 295 K. Zeolites 9:303-311

Swainson IP, Dove MT, Harris MJ (1995) Neutron powder diffraction study of the ferroelastic phase transition and lattice melting in sodium carbonate, Na_2CO_3. J Phys Condens Matt 7:4395-4417

Tribaudino M, Nestola F, Prencipe M, Rundlof H (2003) A single-crystal neutron-diffraction investigation of spodumene at 54 K. Can Mineral 41:521-527

Tucker MG, Dove MT, Keen DA (2000) Simultaneous analysis of changes in long-range and short-range structural order at the displacive phase transition in quartz. J Phys Condens Matt 12:L723-L730

Tucker MG, Keen DA, Dove MT (2001) A detailed structural characterization of quartz on heating through the alpha-beta phase transition. Mineral Mag 65:489-507

Vogel SC, Priesmeyer H-G (2006) Neutron production, neutron facilities and neutron instrumentation. Rev Mineral Geochem 63:27-58

Welberry TR, Goossens DJ, Edwards AJ, David WIF (2001) Diffuse X-ray scattering from benzil, $C_{14}H_{10}O_2$: analysis *via* automatic refinement of a Monte Carlo model. Acta Crystallogr A57:101-109

Wilkinson C, Khamis HW, Stansfield RFD, McIntyre GJ (1988) Integration of single-crystal reflections using area multidetectors. J Appl Crystallogr 21:471-478

Wilkinson C, Lehmann MS (1991) Quasi-Laue neutron diffractometer. Nucl Instrum Methods Phys Res A310: 411-415

Wilkinson C, Schultz AJ (1989) Integration of Bragg reflections with an Anger camera area detector. J Appl Crystallogr 22:110-114

Wilson CC (2005) Neutron single-crystal diffraction: techniques and applications in molecular systems. Z Kristallogr 220:385-398

Winkler B (1999) Introduction to the application of neutron spectroscopy in the Earth Sciences. *In*: Microscopic Properties and Processes in Minerals. Wright K, Catlow R (eds) Kluwer Academic Publishers, p 93-144

Xu JA, Mao HK (2000) Moissanite: A window for high-pressure experiments. Science 290:783-785

Yan C, Vohra YK, Hemley RJ, Mao HK (2002) Very high growth rate chemical vapor deposition of single-crystal diamond. Proc Nat Acad Sci USA 99:12523-12525

Reviews in Mineralogy & Geochemistry
Vol. 63, pp. 81-98, 2006
Copyright © Mineralogical Society of America

4

Neutron Rietveld Refinement

Robert B. Von Dreele

IPNS/APS
Argonne National Laboratory
Argonne, Illinois, 60439, U.S.A.
e-mail: vondreele@anl.gov

INTRODUCTION

A polycrystalline powder can be represented in reciprocal space as a set of nested spherical shells positioned with their centers at the origin (Warren 1990) (Fig. 1). These shells each arise from a reciprocal lattice point from the myriad (e.g. ~10^9 mm^{-3} for 1μm crystallites) of small crystals, ideally with random orientation, in the sample. These shells each have some thickness or broadening from both instrumental effects and the characteristics of the crystalline grains themselves. Their magnitude is related to the crystalline structure factors (in this instance for neutron scattering) as well as the symmetry driven overlaps (i.e., reflection multiplicities). An experimentally measured powder diffraction pattern is a scan through this suite of shells which by its nature is a smooth curve comprising a sequence of peaks resting upon a slowly varying background.

The techniques for obtaining this data using neutron scattering are discussed elsewhere in this volume (e.g., Vogel and Priesmeyer 2006, this volume).

Early data analysis attempted to extract values of the individual structure factors from peak envelopes and then apply standard single crystal methods to obtain structural information. This approach was severely limited because the relatively broad peaks in a neutron powder pattern resulted in substantial reflection overlap and the number of usable structure factors that could be

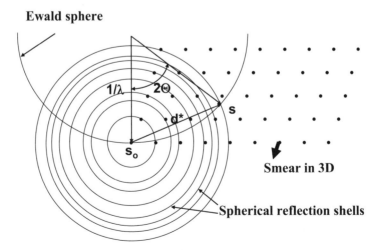

Figure 1. Reciprocal space construction for a powder diffraction experiment. The myriad reciprocal lattice points for the crystallites combine to form nested spherical shells centered at the reciprocal space origin.

1529-6466/06/0063-0004$05.00   DOI: 10.2138/rmg.2006.63.4

obtained in this way was very small. Consequently, only very simple crystal structures could be examined by this method. To overcome this limitation, H.M. Rietveld (1967, 1969) realized that a neutron powder diffraction pattern is a smooth curve comprised of Gaussian peaks on top of a smooth background and that the best way of extracting the maximum information from it was to write a mathematical expression to represent the observed intensity at every step in this pattern

$$Y_c = Y_b + \sum Y_\mathbf{h} \tag{1}$$

This expression has both a contribution from the background (Y_b) and each of the Bragg reflections ($Y_\mathbf{h}$; $\mathbf{h} = hkl$) which are in the vicinity of the powder pattern step (Fig. 2). Each of these components is represented by a mathematical model which embodies both the crystalline and noncrystalline features of a powder diffraction experiment.

The adjustable parameters for this model are refined by a least-squares minimization of the weighted differences between the observed and calculated intensities. This approach to the analysis of powder patterns has been so successful (Cheetham and Taylor 1977; Hewat 1985; Bish and Post 1989; Young 1993) that it has lead to a renaissance in powder diffraction and this technique of treating powder diffraction data is now known as "Rietveld refinement."

RIETVELD THEORY

The current and more complete description of the powder pattern model includes the possibility of multiple phases in the sample and the possibility that the incident intensity is not the same for all data points. Thus, the observed intensities are normalized for neutron time-of-flight data (TOF) by

$$Y_o = \frac{Y_o{}'}{WY_i} \tag{2}$$

where Y'_o is the number of counts observed in a channel of width W, Y_i is the incident intensity for that channel. The sum in Equation (1) is over those reflections from all phases in the

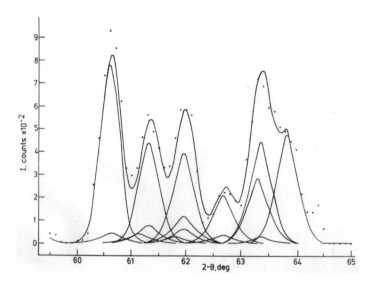

Figure 2. Portion of a constant wavelength (CW) neutron powder diffraction pattern showing the contributions to the calculated pattern from 15 reflections.

sample that are sufficiently close to the profile point to make a significant contribution. Then the function

$$M = \sum w \left(Y_o - Y_c \right)^2 \tag{3}$$

is minimized by least-squares, where ideally the weight, w, is computed from the variances in both Y_o and Y_i. The quality of the least squares refinement is indicated by some residual functions,

$$R_p = \frac{\sum |Y_o - Y_c|}{\sum Y_o} \tag{4}$$

and

$$R_{wp} = \sqrt{\frac{M}{\sum w Y_o^2}} \tag{5}$$

The reduced χ^2 or "goodness of fit" is defined from the minimization function as

$$\chi^2 = \frac{M}{N_{obs} - N_{var}} \tag{6}$$

and the "expected R_{wp}" from

$$R_{wp(exp)} = \frac{R_{wp}}{\sqrt{\chi^2}} \tag{7}$$

In a least squares analysis, the minimum in Equation (3) must be found iteratively because, as we will see below, the model functions are nonlinear. A set of derivatives of the powder pattern profile intensity with respect to all of the model parameters, p_j, is formed. These may be modified by various constraints, U_{ij}, which reduce their number to the set of refined variables, v_i, via

$$\frac{\partial Y_c}{\partial v_i} = U_{ij} \frac{\partial Y_c}{\partial p_j} \tag{8}$$

The normal equations are then formed as a sum over all the observed powder profile points as

$$\sum w (Y_o - Y_c)^2 \frac{\partial Y_c}{\partial v_i} \tag{9}$$

Restraints, which act as additional observations, are represented as additional terms in this summation. After expansion as a Taylor series and ignoring high order terms, the set of observational equations becomes in matrix form

$$\mathbf{Ax} = \mathbf{y} \tag{10}$$

where

$$y_i = \sum w (Y_o - Y_c)^2 \frac{\partial Y_c}{\partial v_i} \quad x_i = \Delta v_i \quad a_{ij} = \sum w \frac{\partial Y_c}{\partial v_i} \frac{\partial Y_c}{\partial v_j} \tag{11}$$

This matrix equation is solved for the desired variable shifts by

$$\mathbf{A^{-1}Ax} = \mathbf{A^{-1}y} \; \mathbf{x} = \mathbf{A^{-1}y} = \mathbf{By} \tag{12}$$

and the parameter shifts are found by a reapplication of the constraints.

$$\Delta p_i = \Delta v_j U_i. \tag{13}$$

The inverse matrix, **B**, is normalized by the reduced χ^2 to give the variance-covariance matrix. The square roots of the diagonal elements of this normalized matrix are the estimated errors in the values of the shifts and thus those for the parameters themselves. These error estimates are based solely on the statistical errors in the original powder diffraction pattern intensities and can not accommodate the possible discrepancies arising from systematic flaws in the model. Consequently, the models described next attempt to provide a close correspondence to the scattering process that gives rise to features in an observed powder diffraction pattern.

Bragg intensity contribution

The contributed intensity, Y_h, from a Bragg peak to a particular profile intensity point will depend on several factors. Obviously the value of the structure factor and the amount of that particular phase will determine the contribution; the structure factors for neutron powder diffraction can arise from the atom positions within the crystal structure (cf. Parise 2006, this volume) or an ordering of magnetic ions (cf. Harrison 2006, this volume). In addition the peak shape and width in relation to its position will have an effect. The intensity is also affected by extinction and absorption as well as some geometric factors. Thus,

$$Y_h = SF_h^2 H(T - T_h) K_h \tag{14}$$

where S is the scale factor for the particular phase, F_h is the structure factor for a particular reflection, $H(T-T_h)$ is the value of the profile peak shape function for that reflection at the position, T, which is displaced from its expected position, T_h, and K_h is the product of the various geometric and other intensity correction factors for that reflection. Each of these contributions will be discussed in turn.

Systematic effects on intensity

The intensity correction factors, K_h, consist of those factors which are dependent on the sample, the instrument geometry, and the type of radiation used,

$$K_h = \frac{E_h A_h O_h m_h L}{V} \tag{15}$$

where E_h is an extinction correction, A_h is an absorption correction, O_h is the preferred orientation correction, m_h is the reflection multiplicity, L is the angle-dependent Lorentz correction, and V is the unit cell volume for the phase. Some of these will be discussed in turn.

Extinction in powders. Extinction in powders is a primary extinction effect within the crystal grains and can be calculated according to a formalism developed by Sabine (1985) and Sabine et al. (1988). It is only of importance for TOF neutron data because extinction is strongly dependent on wavelength and not on scattering angle, and if it is not allowed for, the refined values of the atomic temperature factors can be seriously in error. From the Darwin energy transfer equations Sabine (1988), by following the formalisms of Zachariasen (1945, 1967) and Hamilton (1957), developed intensity expressions for both the symmetric Laue and Bragg cases of diffraction by an infinite plane parallel plate. The extinction correction E_h for a small crystal is a combination of Bragg and Laue components

$$E_h = E_B \sin^2 \Theta + E_L \cos^2 \Theta \tag{16}$$

where for the Bragg component

$$E_B = \frac{1}{\sqrt{1+x}} \tag{17}$$

and for the Laue component

$$E_L = 1 - \frac{x}{2} + \frac{x^2}{4} - \frac{5x^3}{48}... \quad \text{for } x < 1 \tag{18}$$

or

$$E_L = \sqrt{\frac{2}{\pi x}} \left[1 - \frac{1}{8x} - \frac{3}{128x^2}... \right] \text{for } x > 1 \tag{19}$$

where

$$x = E_x \left(\frac{\lambda F_h}{V} \right)^2 \tag{20}$$

The units for these expressions are such that E_x is in μm^2 and is a direct measure of the mosaic block size in the powder sample. Note that this is not necessarily the same as either the crystallite size or the powder particle size as crystallites may have more than one mosaic block and powder particles may be composed of multiple crystallites. Sabine et al. (1988) demonstrated this by examining the extinction effects for neutron powder diffraction by hot pressed MgO samples characterized by electron microscopy. The extinction effects in some of these samples decreased the observed intensity in low order reflections by nearly 70%. The Sabine model allowed correction of these intensities and gave temperature factors which were independent of grain size and matched both theoretical models and other experimental results. Moreover, the refined extinction coefficients correlated very well with the measured particle size distribution in these samples.

Powder absorption factor. The absorption, A_h, for a cylindrical sample is calculated for powder data according to an empirical formula (Rouse et al. 1970; Hewat 1979). It is assumed that the linear absorption of all components in the sample varies with λ and is indistinguishable from multiple scattering effects within the sample,

$$A_h = \exp(-T_1 A_B \lambda - T_2 A_B^2 \lambda^2) \tag{21}$$

where

$$T_1 = 1.7133 - 0.368 \sin^2 \Theta \tag{22}$$

and

$$T_2 = -0.0927 - 0.3750 \sin^2 \Theta \tag{23}$$

For a fixed wavelength, this expression is indistinguishable from thermal motion effects and hence the coefficient can not be determined independently of atomic thermal motion parameters and it is only effective for the very low absorptions ($\mu R > 1$) typically encountered in neutron powder diffraction. Therefore, it is only of importance for the analysis of TOF neutron data. A more complex function was developed by Lobanov and Alte da Veiga (1998) which can be used for higher absorptions ($\mu R > 10$) and is available in some Rietveld refinement programs (e.g. GSAS; Larson and Von Dreele 2004).

Preferred orientation of powders. Many powders are made up of crystallites which have a strongly preferred cleavage so that they are not isotropic in shape. Consequently when the powder is packed into a diffraction mount the crystallites can take up a preferred orientation. For example platy crystals in a typical Bragg-Brentano X-ray mount tend to lie with the plate normals perpendicular to the sample surface. Preferred orientation is rarely a significant problem for neutron diffraction because the sample volumes are usually quite large (1-5 cm³). Thus, Rietveld (1969) utilized a simple Gaussian description for the preferred orientation,

$$O_h = \exp(-G\alpha^2) \tag{24}$$

where α is the acute angle between the plate normal and the diffraction vector for the reflection, **h**. This function is only useful when the preferred orientation is small as would be the case for neutron diffraction on loose powders. A number of modifications for this basic function had been proposed since then that have met with only moderate success in describing the effects of preferred orientation. After a review of the available preferred orientation models, Dollase (1986) selected a special case from the more general description by March (1932) and developed what he regarded as the best preferred orientation correction, O_h. For this model, the crystallites are assumed to be effectively either rod or disk shaped. When the powder is packed into a diffraction sample holder, the crystallite axes may acquire a preferred orientation that is approximated by a cylindrically symmetric ellipsoid. In the usual diffraction geometry for neutron powder diffraction, the unique axis of this distribution is normal to the diffraction plane while for Bragg-Bretano X-ray geometry this axis is coincident with the diffraction vector. Integration about this distribution at the scattering angle for each reflection in either description gives the same very simple form for the correction,

$$O_h = \sum \left(R_o^2 \cos^2 \alpha_h + \frac{\sin^2 \alpha_h}{R_o} \right)^{-3/2} \tag{25}$$

where α_h is the angle between the preferred orientation direction and the reflection vector **h**. The sum is over the reflections equivalent to **h** so this expression is effectively a correction to the ideally random sample reflection multiplicity, m. The one refinable coefficient, R_o, gives the effective sample compression or extension due to preferred orientation and is the axis ratio for the ellipsoid. If there is no preferred orientation then the distribution is spherical and $R_o = 1.0$ and thus $O_h = 1.0$.

Samples that consist of a polycrystalline mass and not a loose powder may also have been subject to a process that may have deformed the material and induced a texture to the crystallite orientations which will change the Bragg reflection intensities. For example, a metal plate that was manufactured in a rolling mill frequently displays the effects of rolling texture on the diffraction intensities. A complete description of this texture (or preferred orientation) is formulated as a probability for finding a particular crystallite orientation within the sample; this is the orientation distribution function (ODF). For an ideally random powder the ODF is the same everywhere (ODF $\equiv 1$) while for a textured sample the ODF will have positive values both less and greater than unity. This ODF can be used to formulate a correction to the Bragg intensities *via* a 4-dimensional surface (general axis equation) that depends on the both the direction in reciprocal space and the direction in sample coordinates (Bunge 1982; Von Dreele 1997)

$$O(\phi,\beta,\psi,\gamma) = 1 + \sum_{L=2}^{N_L} \frac{4\pi}{2L+1} \sum_{m=-L}^{L} \sum_{n=-L}^{L} C_L^{mn} k_L^m(\phi,\beta) k_L^n(\psi,\gamma) \tag{26}$$

In a diffraction experiment the crystal reflection coordinates (ϕ,β) are determined by the reflection index (**h**) while the sample coordinates (ψ,γ) are determined by the orientation of the sample on the diffractometer. This formulation assumes that the probability surface is smooth and can be described by a sum of N_L spherical harmonic terms, k_L^m and k_L^n, that depend on **h** and sample orientation, respectively, to some maximum harmonic order, N_L. The coefficients, C_L^{mn}, then determine the strength and details of the texture. It should be noted that only the even order, $L = 2n$, terms in these harmonic sums affect the intensity of Bragg reflections; the odd order terms in the ODF are invisible to diffraction.

These even order harmonic coefficients can be determined in a Rietveld refinement with a suite of neutron TOF powder diffraction patterns obtained for a selection of sample orientations. By selecting a particular reflection, **h**, the general axis equation can be used to calculate a pole figure for that reflection (Von Dreele 1997). For example, Figure 3 shows pole

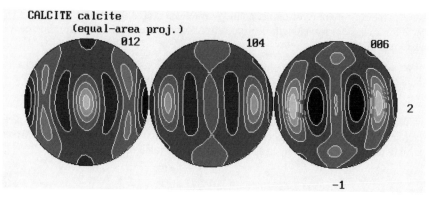

Figure 3. Pole figures computed via the general axis equation (Eqn. 21) and harmonic coefficients obtained in a Rietveld refinement with 52 neutron TOF patterns from an experimentally deformed calcite limestone.

figures determined in this way for an experimentally deformed limestone sample (calcite) that was used in a round robin study (Wenk 1991). Details of methods for determination of the ODF from neutron diffraction data are discussed by Wenk (2006, this volume).

While the general axis equation in Equation (26) can be used to describe the effect of texture on diffraction intensities in the most general case; most powder diffraction experiments are performed to simplify the problem. Rotation of the sample about the cylinder axis for a typical neutron Debye-Scherrer experiment will simplify Equation (26) *via* symmetry to

$$O(\phi,\beta,\gamma) = 1 + \sum_{L=2}^{N_L} \frac{4\pi}{2L+1} \sum_{m=-L}^{L} C_L^{m0} \, k_L^m(\phi,\beta) k_L^0(\gamma) \tag{27}$$

with substantially fewer coefficients than the more general case. It must be noted that spinning the sample does not remove the effect of preferred orientation; it only simplifies the form of the correction.

Other angle dependent corrections. The only other angle dependent correction for neutron powder diffraction data is the Lorentz factor. For TOF neutron data there is an additional factor for the variation of scattered intensity with wavelength or

$$L = d^4 \sin\Theta \tag{28}$$

while for constant wavelength neutrons

$$L = \frac{1}{2\sin 2\Theta \cos\Theta} \tag{29}$$

There is no polarization effect for neutron diffraction.

Profile functions for CW and TOF neutron powder diffraction

The contribution a given reflection makes to the total profile intensity depends on the shape function for that reflection profile, its width coefficients and the displacement of the peak from the profile position. The locations of the peak are usually given in microseconds of TOF or in degrees 2Θ. Discussion of these values is given first followed by details of some of the peak shape functions presently in use.

Reflection positions in powder patterns. The reflection position in a constant-wavelength experiment is obtained from Bragg's Law,

$$T = \arcsin(\lambda/2d) + Z \tag{30}$$

where Z is the zero-point error on the counter arm position and the result is in degrees. Both λ and Z for a constant wavelength neutron diffractometer are obtained from refinement of a standard material.

For a neutron TOF powder diffractometer the relationship between the d-value for a particular powder line and its TOF is empirically given by the simple quadratic

$$T = C\,d + A\,d^2 + Z \tag{31}$$

The three parameters C, A and Z are characteristic of a given counter bank on a TOF powder diffractometer. C may be calculated with good precision from the flight paths, diffraction angle, and counter tube length by use of the de Broglie equation,

$$C = 252.816 \times 2\sin\Theta\left(L_1 + \sqrt{L_2^2 + \frac{L_3^2}{16}}\right) \tag{32}$$

where Θ is the Bragg angle, L_1 is the primary flight path, L_2 is sample to detector center distance and L_3 is the height of the detector; all distances are in meters. The units of C are then μsec/Å. A is a small second-order correction and Z depends on the various electronic delays in the counting system. Precise values for constants C, A and Z must be obtained by fitting to a powder diffraction pattern of a standard material.

CW profile functions. A CW neutron diffractometer (Fig. 4) is constructed by allowing the "white" beam from a reactor to fall on a large monochromating crystal (Cu or Ge) selecting a narrow wavelength band for the diffraction experiment. The divergence of the beam is controlled by placing two or three sets of Soller slits in the beam paths between the source and the detector (Hewat 1975; Hewat and Bailey 1976).

The combination of slit divergences, mosaic spread of the monochromator and scattering angles results in a virtually pure Gaussian diffraction line profile (Rietveld 1967).

$$G(\Delta T - \tau)\frac{1}{\sqrt{2\pi\sigma^2}}\exp\left\{-\frac{(\Delta T - \tau)^2}{2\sigma^2}\right\} \tag{33}$$

Using a modified form of the Caglioti et al. (1968) function for the variation of FWHM with

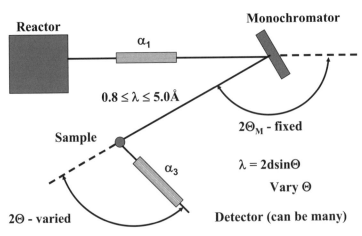

Figure 4. A schematic view of a constant wavelength neutron powder diffractometer. The overall geometry is very similar to a Debye-Scherrer X-ray camera.

scattering angle gives

$$\sigma^2 = U \tan^2 \Theta + V \tan \Theta + W + \frac{P}{\cos \Theta} \qquad (34)$$

Modern CW neutron powder diffractometers give a line profile that is much narrower than could be obtained from early instruments. Thus, more of the characteristic line shape arises from effects from within the sample and gives significant deviations from a simple Gaussian line shape. To accommodate the experimental line shape as well as that obtained from modern X-ray diffractometers, several modifications of the Gaussian function have been made as well as the introduction of other broadening functions (Young et al. 1977; Young and Wiles 1982).

To accommodate both the instrumental and sample contributions to the line shape, the most successful function for CW data of either kind employs a pseudo-Voigt, $F(\Delta T)$, described by Thompson et al. (1987) and is the one installed in a comprehensive Rietveld refinement package GSAS (Larson and Von Dreele 2004),

The pseudo-Voigt is a linear combination of a Gaussian and Lorentzian as

$$F(\Delta T) = \eta L(\Delta T, \Gamma) + (1 - \eta) G(\Delta T, \Gamma) \qquad (35)$$

where the Lorentzian function is

$$L(\Delta T) = \frac{\gamma}{2p} \left[\frac{1}{(\gamma/2)^2 + \Delta T^2} \right] \qquad (36)$$

and the Gaussian expression is given above (Eqn. 33). In the Thompson et al. (1987) formalism the mixing factor, η, is represented by empirical functions so that the result is a close approximation to the true Voigt (a convolution of Gaussian and Lorentzian functions).

As seen in Figure 5 the fit of this function to the peak shapes observed in neutron powder diffraction is excellent.

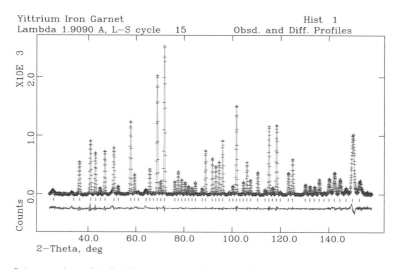

Figure 5. A comparison of the fit of the constant wavelength profile function to neutron powder diffraction data from a synthetic yttrium iron/aluminium garnet obtained on the D1A diffractometer at the Institut Laue-Langevin, Grenoble, France at room temperature. This garnet is cubic, space group *Ia3d*; *a* = 12.19Å. There is a partial disordering of the Fe and Al on the tetrahedral and octahedral sites.

Interpretation of CW profile coefficients. The coefficients from a CW powder profile function can be interpreted to give both strain and crystallite size information. More detailed interpretation of these factors is discussed by Daymond (2006, this volume).

The strain in a lattice can be visualized as a defect-induced distribution of unit cell dimensions about the average lattice parameters. In the reciprocal space (Fig. 6) associated with a sample with isotropic strain, there is a broadening of each point which is proportional to the distance of the point from the origin, i.e.,

$$\frac{\Delta d^*}{d^*} = \text{constant} \tag{37}$$

Then the strain broadening in real space is related to 2Θ broadening by

$$\frac{\Delta d}{d} = \Delta 2\Theta \cot\Theta = \text{constant} \tag{38}$$

or

$$\Delta 2\Theta = \frac{\Delta d}{d} \tan\Theta \tag{39}$$

In this expression $\Delta 2\Theta$ is in radians. Examination of the Caglioti expression for the Gaussian broadening (Eqn. 34) indicates that the first term contains a strain broadening component. This is a variance and the instrument contribution can be subtracted. This variance must be converted to radians to yield strain, thus

$$S = \frac{\pi}{180}\sqrt{8\ln 2\left(U - U_i\right)} \cdot 100\% \tag{40}$$

Alternatively, the strain term is the one that varies with $\tan\Theta$ in the Lorentzian component of a CW peak shape obtained from

$$\gamma = \frac{X}{\cos\Theta} + Y\tan\Theta \tag{41}$$

Any instrumental or spectral contribution must first be subtracted to yield the strain component. This is in degrees and is already a full width at half maximum so the strain is

$$S = \frac{\pi}{180}\left(Y - Y_i\right) \cdot 100\% \tag{42}$$

For small crystallites the assumption that the lattice is infinite no longer holds so that the reciprocal lattice points, **h**, are not δ-functions but are all smeared out uniformly depending on the average crystallite size. Thus, all the points are the same size independent of the distance from the origin (Fig. 7) and

$$\Delta d^* = \text{constant} \tag{43}$$

The reciprocal of this quantity is the average crystallite size. In real space the broadening is

$$\frac{\Delta d}{d^2} = \frac{\Delta\Theta \cot\Theta}{d} = \text{constant} \tag{44}$$

From Bragg's law and $\Delta 2\Theta = 2\Delta\Theta$ then

$$\frac{\Delta d}{d^2} = \frac{\Delta 2\Theta \cot\Theta \sin\Theta}{\lambda} \tag{45}$$

and the broadening is

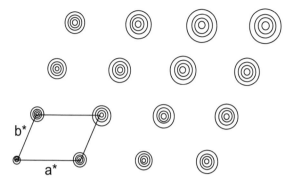

Figure 6. Illustration of the strain broadening of the reflection maxima in reciprocal space.

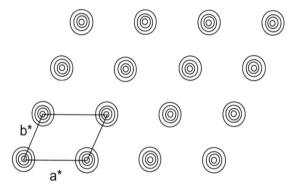

Figure 7. Illustration of the particle size broadening of the reflection maxima in reciprocal space.

$$\Delta 2\Theta = \frac{\lambda \Delta d}{d^2 \cos \Theta} \tag{46}$$

The first term in the expression for the Lorentzian broadening is of this form where

$$X = \frac{\Delta d}{d^2} \tag{47}$$

The crystallite size can be obtained by rearrangement of this expression and converting from degrees to radians by

$$p = \frac{180 K \lambda}{\pi X} \tag{48}$$

which includes the Scherrer constant, K. The units are Å.

The corresponding term in the Gaussian expression is the fourth one in Equation (34). Converting from degrees to radians gives the expression

$$p = \frac{180 \lambda}{\pi \sqrt{8 P \ln 2}} \tag{49}$$

and again the units are Å.

TOF profile functions. A time-of-flight neutron powder diffractometer (Fig. 8) consists of a sample placed at a certain distance from a pulsed "white" beam source with detectors arrayed about it at fixed scattering angles (Buras and Holas 1968; Jorgenson and Rotella 1982).

In the case of a spallation neutron source (Carpenter et al. 1975) the diffraction line profile is strongly asymmetric and results from a convolution of the neutron source pulse and the Gaussian broadening from the sizes of the source, sample and detectors. The best known profile function for spallation TOF neutron data is the empirical convolution function of Jorgensen et al. (1978) and Von Dreele et al. (1982),

$$H(\Delta T) = N[\exp(u)\,\mathrm{erfc}(y) + \exp(v)\mathrm{erfc}(z)] \tag{50}$$

where ΔT is the difference in TOF between the reflection position, T_h, and the profile point, T; the terms N, u, v, y and z are dependent on the profile coefficients. The function erfc is the complementary error function. This profile function is the result of convoluting two back-to-back exponentials with a Gaussian.

$$H(\Delta T) = \int G(\Delta T - \tau)P(\tau)d\tau \tag{51}$$

where

$$P(\tau) = 2N\exp(\alpha\tau) \text{ for } \tau < 0 \tag{52}$$

and

$$P(\tau) = 2N\exp(-\beta\tau) \text{ for } \tau > 0 \tag{53}$$

for the two exponentials; α and β are the rise and decay coefficients for the exponentials and are largely characteristic of the specific moderator viewed by the instrument. The Gaussian function is the same as used in the CW function (Eqn. 33).

The Gaussian variance is the coefficient σ^2 and is largely characteristic of the instrument design (scattering angle and flight path) and the sample. These functions when convoluted give the profile function shown above. The normalization factor, N, is

$$N = \frac{\alpha\beta}{2(\alpha+\beta)} \tag{54}$$

The coefficients u, v, y and z are

$$u = \frac{\alpha}{2}(\alpha\sigma^2 + 2\Delta T) \tag{55}$$

$$v = \frac{\beta}{2}(\beta\sigma^2 - 2\Delta T) \tag{56}$$

$$y = \frac{(\alpha\sigma^2 + \Delta T)}{\sqrt{2\sigma^2}} \tag{57}$$

$$z = \frac{(\beta\sigma^2 - \Delta T)}{\sqrt{2\sigma^2}} \tag{58}$$

Each of the three coefficients α, β and σ^2 show a specific empirical dependence on the reflection d-value,

$$\alpha = \alpha_o + \alpha_1/d \tag{59}$$

$$\beta = \beta_o + \beta_1/d^4 \tag{60}$$

$$\sigma^2 = \sigma_o^2 + \sigma_1^2 d^2 + \sigma_2^2 d^4 \tag{61}$$

As seen in Figure 9 this function gives a good fit to a TOF neutron powder diffraction profile.

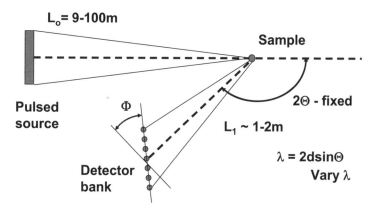

Figure 8. A schematic view of a time-of-flight neutron powder diffractometer.

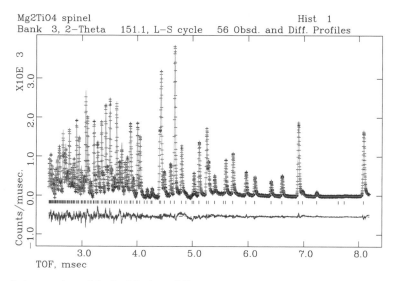

Figure 9. A comparison of the fit of the time-of-flight neutron profile function to typical neutron time-of-flight powder diffraction data obtained from a synthetic magnesium titanate inverse spinel (Weschler and Von Dreele 1989).

Interpretation of TOF profile coefficients. The profile coefficients from a TOF neutron powder pattern Rietveld refinement can also give information about the microtexture of the sample. This discussion will describe how this information can be extracted from the coefficients.

Again starting from the reciprocal space picture of the effects of microstrain (Fig. 6), the peak broadening in real space (the regime of a TOF experiment) from strain is

$$\frac{\Delta d}{d} = \text{constant} \tag{62}$$

Thus, examination of the function for the Gaussian component of the peak shape from a TOF pattern (Eqn. 59) implies that the second term contains an isotropic contribution from strain broadening. The other major contribution to σ_1^2 is from the instrument; because it is expressed as a variance, σ_{1i}^2, it can simply be subtracted. The remaining sample dependent contribution is then converted to strain (S), a dimensionless value which is frequently expressed as percent strain or fractional strain as full width at half maximum (FWHM).

$$S = \frac{1}{C}\sqrt{8\ln 2(\sigma_1^2 - \sigma_{1i}^2)} \cdot 100\% \tag{63}$$

where C is the diffractometer constant from Equation (31). The term $\sqrt{8\ln 2}$ is the ratio between FWHM and the standard deviation, σ, of a Gaussian distribution.

For crystallite size broadening, we start from the reciprocal space picture (Fig. 7) where all the points are smeared equally or

$$\Delta d^* = \text{constant} \tag{64}$$

The reciprocal of this quantity is the average particle size. In real space (for TOF) the broadening is

$$\frac{\Delta d}{d^2} = \text{constant} \tag{65}$$

From the functional form for the Gaussian broadening of a TOF peak (Eqn. 61), the particle size affects the third term (σ_2^2) in the expression. This term generally has no instrument contribution and is used directly to calculate the crystallite size (p) by

$$p = \frac{CK}{\sqrt{8\ln 2\sigma_2^2}} \tag{66}$$

where C is the diffractometer constant, K is the Scherrer constant and the units for p are Å.

Background intensity contribution

In the original program written by Rietveld, the background in a powder pattern was visually estimated from relatively clear portions and subtracted by linear interpolation from the observed intensity. These modified data were then fitted by the program to a refined structural model. Recent modifications to the method now include a description of the background with a set of refinable coefficients. There are several different functions in use, one of the more successful is the entirely empirical cosine Fourier series which including a leading constant term,

$$I_b = B_o + \sum_{j=1}^{n} B_j \cos\left(P(j-1)\right) \tag{67}$$

In the case of CW data, P is in degrees 2Θ and is the detector position for the step. For TOF data the times are scaled by $180/T_{\text{MAX}}$, where T_{MAX} is the maximum TOF in the pattern.

In some cases the sample or container (e.g., fused silica) will show an oscillating background which arises from a glass-like contribution to the scattering. This can be represented by the major features of a radial distribution function (RDF). The RDF for an amorphous phase has peaks which correspond to the interatomic distances (Fig. 10).

Thus this background function can model diffuse background from an amorphous phase in the sample or the background from a silica sample holder. The function is

$$I_b = \sum_{i=1}^{n} \frac{A_i \sin(QR_i)}{QR_i} \exp(-2U_iQ^2) \tag{68}$$

where

$$Q = 2\pi/d \tag{69}$$

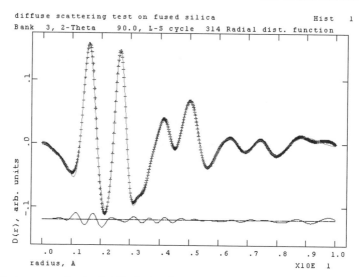

diffuse scattering test on fused silica   Hist 1
Bank 3, 2-Theta   90.0, L-S cycle 314 Radial dist. function

Figure 10. An unnormalized radial distribution function for fused silica as measured by time-of-flight neutron scattering. The peaks correspond to various Si-O, O-O and Si-Si distances in the glass.

The three coefficients for each term (A_i, R_i and U_i) give a scattering power, interatomic distance and thermal motion for atom pairs within the glass, respectively. It should be noted that the RDF may contain local structure information not easily found by Rietveld refinement techniques (see Proffen 2006, this volume, for details).

A third method for fitting the background is a modification of the original method used by Rietveld (1969). The user selects a set of specific points across the powder pattern and the background is then fitted in the least squares refinement by adjusting the background intensity at these points calculating the background contribution for all other points by linear interpolation. This method is somewhat less satisfying that the previous three because the selection of the interpolation points can be subjective.

The background is normally quite flat. There is usually only a slight curvature arising from the thermal diffuse scattering from the sample. Larger background contributions will be fitted by these functions given sufficient terms, but short period fluctuations not associated with an amorphous phase cannot be fit very well. Examples of background problems include frame overlap peaks in a TOF pattern, miscellaneous peaks from sources outside the experiment and unaccounted extra phases.

RIETVELD REFINEMENT PRACTICE

Software

There are some 25 Rietveld refinement software packages available. Many of these are freely available for download (see *http://www.ccp14.ac.uk/* for details) while others are offered commercially. Of the former, the ones in most common use include the General Structure Analysis System (GSAS) (Larson and Von Dreele 2004) and its graphical interface EXPGUI (Toby 2001), FULLPROF (Rodriguez-Carvajal 1990) and MAUD (Lutterotti et al. 1999). By and large the commercial Rietveld programs are designed for specific laboratory X-ray powder diffraction instruments and form part of their data acquisition/analysis system and frequently are not applicable to neutron powder diffraction.

Practicalities

The quality of the result of a Rietveld refinement of powder diffraction data will depend on many factors; these have been reviewed by McCusker et al. (1999). Although much of the discussion in that work focused on analysis of data collected on standard laboratory X-ray powder diffractometers, many of the recommendations and conclusions can be applied to neutron powder diffraction practice. In particular, a "best practice" is to ensure that there are an adequate number of powder profile steps in the pattern; 5-10 steps across the expected FWHM of any Bragg peak is recommended. There are also recommendations with respect to counting times and d-spacing coverage. Most useful is a survey of common systematic errors and their effect on the detailed fit between the observed profile and the one calculated *via* a model with a specific error. Careful examination of a graphical display of the observed and model calculation of the powder diffraction profile and their difference curve will provide a most useful guide for further progress in these refinements. It should be noted that Rietveld refinement is not always a numerically stable procedure and a wide variety of techniques are used to provide stability including damping or providing additional information in the form of constraints or restraints. Finally, the residuals given above in Equations (4-7) give only a statistical summary of the quality of the refinement and are no substitute for a graphical display of the observed, calculated and difference profiles from the final refinement. They also can not be used in the absence of a full stereochemical evaluation of the result. It should make "chemical sense" and give reasonable values for interatomic distances (including non-bonded ones) and angles.

SUMMARY

The Rietveld refinement method has evolved considerably since the original introduction (Rietveld 1967, 1969). It has now been extended to both X-ray and neutron time-of-flight powder diffraction data by development of mathematical descriptions for a wide variety of the commonly observed systematic effects on both the powder pattern intensity and shape. The modern version of Rietveld refinement is thus a powerful tool for crystal structure analysis.

ACKNOWLEDGMENT

Support was provided by the U.S. Department of Energy, Office of Science, Basic Energy Sciences to IPNS and APS under Contract No. W-31-109-ENG-38.

REFERENCES

Bish DL, Post JE (eds) (1989) Modern Powder Diffraction. Reviews in Mineralogy 20. Mineralogical Society of America
Buras B, Holas A (1968) Intensity and resolution in neutron time-of-flight powder diffractometry. Nukleonika 13:591-619
Cagliot-i G, Paoletti A, Ricci FP (1958) Choice of collimators for a crystal spectrometer for neutron diffraction. Nucl Inst 3:223-228
Carpenter JM, Mueller MH, Beyerlein RA, Worlton TG, Jorgenson JD, Brun TO, Sköld K, Pelizzari CA, Peterson SW, Watanabe N, Kimura M, Gunning JE (1975) Neutron diffraction measurements on powder samples using the ZING-P pulsed neutron source at Argonne. Proc Neut Diff Conf, Petten Netherlands, 5-6 August, RCN Report, No. 234:192-208
Cheetham AK, Taylor JC (1977) Profile analysis of powder neutron diffraction data: its scope, limitations, and applications in solid state chemistry. J Solid State Chem 21:253-275
Daymond MR (2006) Internal stresses in deformed crystalline aggregates. Rev Mineral Geochem 63:427-458
Dollase WA (1986) Correction of intensities for preferred orientation in powder diffractometry: application of the March model. J Appl Cryst 19:267-272
Hamilton WC (1957) The effect of crystal shape and setting on secondary extinction. Acta Crystallogr 10: 620-34

Harrison RJ (2006) Neutron diffraction of magnetic materials. Rev Mineral Geochem 63:113-143

Hewat AW (1975) Design for a conventional high-resolution neutron powder diffractometer. Nuclear Inst Methods 127:361-370

Hewat AW (1979) Absorption corrections for neutron diffraction. Acta Crystallogr A35:248-250.

Hewat AW (1985) High-resolution neutron and synchrotron powder diffraction. Chemica Scripta 26A:119-130

Hewat AW, Bailey I (1976) D1a, a high resolution neutron powder diffractometer with a bank of mylar collimators. Nucl Instrum Methods Phys Res 137:463-471

Howard CJ (1982) The approximation of asymmetric neutron powder diffraction peaks by sums of Gaussians. J Appl Crystallogr 15:615-620

Jorgensen JD, Johnson DH, Mueller MH, Worlton JG, Von Dreele RB (1978) Profile analysis of pulsed-source neutron powder diffraction data. Proc Conf Diffraction Profile Analysis, Cracow, 14-15 Aug., 20-22

Jorgensen JD, Rotella FJ (1982) High-resolution time-of-flight powder diffractometer at the ZING-P' pulsed neutron source. J Appl Cryst 15:27-34

Klug HP, Alexander LE (1974) X-ray Diffraction Procedures for Polycrystalline and Amorphous Materials. Wiley-Interscience

Larson AC, Von Dreele RB (2004) GSAS - General Structure Analysis System. Los Alamos National Laboratory Report No. LA-UR-86-748. *http://www.ccp14.ac.uk/solution/gsas/gsas_with_expgui_install.html* (with EXPGUI)

Lobanov NN, Alte da Veiga L (1988) Analytic absorption factors for cylinders to an accuracy of 0.5%. 6th European Powder Diffraction Conference, Abstract P12-16, Aug. 22-25, 1998 and N.N. Lobanov, private communication

Lutterotti L, Matthies S, Wenk H-R (1999). MAUD: a friendly Java program for materials analysis using diffraction. Int Union Crystallogr Comm Powder Diffraction Newsletter 21:14-15. *http://www.ing.unitn.it/~maud/*

March A (1932) Mathematische theorie der regelung nach der korngestalt bei affiner deformation. Z Kristallogr 81:285-297

McCusker LB, Von Dreele RB, Cox DE, Louer D, Scardi P (1999) Rietveld refinement guidelines. J Appl Crystallogr 32:36-50

Parise JB (2006) Introduction to neutron properties and applications. Rev Mineral Geochem 63:1-25

Proffen T (2006) Analysis of disordered materials using total scattering and the atomic pair distribution function. Rev Mineral Geochem 63:255-274

Rietveld HM (1967) Line profiles of neutron powder-diffraction peaks for structure refinement. Acta Crystallogr 22:151-2

Rietveld HM (1969) A profile refinement method for nuclear and magnetic structures. J Appl Crystallogr 2:65-71

Rodriguez-Carvajal J (1990) FULLPROF: a program for Rietveld refinement and pattern matching analysis. Abstracts of the Satellite Meeting on Powder Diffraction of the XV Congress of the IUCr, p. 127, Toulouse, France. *http://www-llb.cea.fr/fullweb/fp2k/fp2k.htm*

Rouse KD, Cooper MJ, York EJ, Chakera A (1970) Absorption corrections for neutron diffraction. Acta Crystallogr A26:682-691

Sabine TM (1985) Extinction in polycrystalline materials. Aust J Phys 38:507-18

Sabine TM (1988) A reconciliation of extinction theories. Acta Crystallogr A44:368-373

Sabine TM, Von Dreele RB, Jorgensen J-E (1988) Extinction in time-of-flight neutron powder diffractometry. Acta Crystallogr A44:374-379

Thompson P, Cox DE, Hastings JB (1987) Rietveld refinement of Debye-Scherrer synchrotron X-ray data from Al_2O_3. J Appl Crystallogr 20:79-83

Toby BH (2001) EXPGUI, a graphical user interface for GSAS. J Appl Crystallogr 34:210-221. *http://www.ccp14.ac.uk/solution/gsas/gsas_with_expgui_install.html* (with GSAS)

Vogel SC, Priesmeyer H-G (2006) Neutron production, neutron facilities and neutron instrumentation. Rev Mineral Geochem 63:27-58

Von Dreele RB, Jorgensen JD, Windsor CG (1982) Rietveld refinement with spallation neutron powder diffraction data. J Appl Crystallogr 15:581-589

Warren BE (1990) X-ray Diffraction. Dover

Wenk H-R (2006) Neutron diffraction texture analysis. Rev Mineral Geochem 63:399-426

Wenk H-R (1991) Standard project for pole figure determination by neutron diffraction. J Appl Crystallogr 24:920-927

Weschler BA, Von Dreele RB (1989) Structure refinements of Mg_2TiO_4, $MgTiO_3$ and $MgTi_2O_5$ by time-of-flight neutron powder diffraction. Acta Crystallogr B45:542-549

Wiles DB, Young RA (1981) A new computer program for Rietveld analysis of X-ray powder diffraction patterns. J Appl Crystallogr 14:149-151

Young RA (ed) (1993) The Rietveld Method. IUCr Monograph on Crystallography 5. Oxford University Press
Young RA, Mackie PE, Von Dreele RB (1977) Application of the pattern-fitting structure-refinement method to X-ray powder diffractometer patterns. J Appl Crystallogr 10:262-269
Young RA, Wiles DB (1982) Profile shape functions in Rietveld refinements. J Appl Crystallogr 15:430-438
Zacharaisen WH (1945) Theory of X-ray Diffraction in Crystals. Wiley
Zacharaisen WH (1967) A general theory of X-ray diffraction in crystals. Acta Crystallogr 23:558-64

Reviews in Mineralogy & Geochemistry
Vol. 63, pp. 99-111, 2006
Copyright © Mineralogical Society of America

5

Application of Neutron Powder-Diffraction to Mineral Structures

Karsten Knorr and Wulf Depmeier

Christian-Albrechts-Universität zu Kiel
Institut für Geowissenschaften, Mineralogie / Kristallographie
D 24098 Kiel, Germany
e-mail: knorr@min.uni-kiel.de

INTRODUCTION

The application of neutrons to study the crystal structures of powdered minerals is of growing popularity among earth scientists. This has just recently been demonstrated at a workshop on Neutrons at the Frontiers of Earth Sciences and Environments (NESE) (Rinaldi and Schober 2006).

There are a number of physical reasons to study minerals with neutrons. They are related to the fundamental properties of neutrons regarding their interaction with matter (Parise 2006a, this volume). The major points are the isotope specific scattering lengths and absorption cross-sections of the neutron, being independent of the number of electrons. The scattering power of many elements is of similar magnitude (see Table 1 for some elements of interest for minerals). As a consequence neutrons are well suited to localize light elements (also in the direct neighborhood of heavy elements) and to distinguish between ions or atoms having the same or a very similar number of electrons. A large advantage of neutrons over X-rays is the independence of the scattering power of the scattering angle. Consequently, structural data from neutron experiments are in principle of higher precision.

The focus of this Chapter is on powder diffraction. There are several answers to the question why to study powders. The material may not be available as a single crystal or cannot be grown

Table 1. Coherent (b_c) and incoherent (b_i) neutron scattering lengths of selected elements, as well as the corresponding absorption cross sections (σ_a) (adapted from Sears VF in Wilson 1992, Table 4.4.4.1).

Element	b_c / fm	b_i / fm	σ_a / barn
^{1}H	−3.74	25.22	0.33
^{2}H = D	6.67	4.03	0.00
Li	−1.90		70.5
Be	7.79	0.20	0.007
Al	3.45	0.27	0.23
Si	4.15		0.17
Ti	−3.44		6.09
V	−0.382		5.08
Mn	−3.73	1.79	13.3
Fe	9.54		2.56
Pb	9.40		0.17
U	8.42		7.57

1529-6466/06/0063-0005$05.00

DOI: 10.2138/rmg.2006.63.5

to the appropriate size. Since the interaction potential of neutrons with matter is small, single crystals should be in the order of about 1 mm³ (Ross and Hoffman 2006, this volume). Many natural samples are multi phased or structurally and chemically heterogeneous. In order to measure a representative sample the use of a large volume is beneficial. It ensures appropriate powder statistics and also reduces preferred orientation of the crystallites. Another reason for studying powders is the investigation of phase transitions that destroy single crystals.

The number of questions that can be addressed with neutron powder diffraction is large. They range from quantitative phase analysis, the determination of cell parameters and their dependence on pressure, temperature and concentration, to the refinement or even determination of crystal structures. The method of choice for evaluating the measured data is in many cases the Rietveld-method (e.g., Rietveld 1969), i.e., the evaluation of the full powder pattern based on the fit of a model function that calculates the intensities at each step in a diffraction pattern as the sum over all overlapping diffraction peaks of all phases in the sample on the basis of their crystal structures (Von Dreele 2006, this volume).

The low absorption of neutrons by several metals makes them an ideal probe for studying materials in pressure cells, furnaces, magnets etc. Consequently, powder diffraction is applied in investigations at high and low temperature (e.g., Kuhs and Hansen 2006, this volume), order/disorder and phase transitions (Redfern 2006, this volume), or at high-pressure (Parise 2006b, this volume). Since neutrons have a magnetic moment that interacts with the spin structure of minerals, neutron powder diffraction can be used to determine magnetic structures (Harrison 2006, this volume). Examples from such applications are therefore not considered in depth here.

This review is intended to give an overview over typical mineral specific structural investigations. It is mainly based on a search of the Inorganic Crystal Structure Database (ICSD) for minerals that were investigated utilizing powder neutron-diffraction. Naturally, the examples given below are far from being complete. We grouped the examples into three sections, namely weak X-ray scatterer, neighboring elements and miscellaneous, the latter including strong X-ray absorber and representative sampling.

WEAK X-RAY SCATTERER

Many minerals contain elements that are weak X-ray scatterers such as hydrogen or lithium. Hydrogen mostly occurs in the form of hydroxyl groups or water. Minerals containing ammonia are rare. The study of hydrogen in minerals is of particular interest to earth sciences since it has strong influence on physical properties and phase transitions. Some minerals are microporous and may host, in addition to water, organic molecules that usually also contain hydrogen. These compounds are not only of fundamental interest but also have a major technological potential.

It is of great advantage to study deuterated samples. The scattering power of the different elements is isotope specific. Selected scattering lengths are given in Table 1. It shows that deuterium, $^2H = D$, has a positive and protons, i.e., ^1H, a negative scattering length. Therefore, if hydrogen in the form of protons is to be located it will show up with negative values in scattering length distribution maps, such as Fourier maps. In contrast, deuterium has maxima in the scattering length distribution maps. The difficulty of measuring hydrogen is the incoherent scattering that results in high background and consequently, gives a poor signal over noise ratio.

Hydrogen

The ICSD contains about 190 entries of deuterated minerals that were studied by neutron powder diffraction and about the same number is reported for non deuterated minerals, i.e., samples containing protons, 18 entries deal with partial deuteration. This Chapter reports on a selection of these minerals, not including ice clathrates for which the reader is referred to Kuhs and Hansen (2006, this volume).

Hydroxides. The precise determination of hydrogen positions in several hydroxides was successful on fully or partially deuterated samples of e.g., γ-AlO(OH,D) boehmite (Christensen et al. 1982), FeO(OH,D) (Szytula et al. 1968; Christensen et al. 1982) CrOOH, grimaldite (Christensen et al. 1963, 1977), guyanaite (Christensen et al. 1976) and heterogenite CoOOH (Delaplane et al. 1969).

In brucite, $Mg(OD)_2$, the low temperature structure and dynamics were studied by Chakoumakos et al. (1997), while Parise et al. (1994) investigated the high pressure behavior up to 9.3 GPa. They found a marked anisotropy in the compression behavior of that layer like material with the *c*-axis being twice as compressible as the *a*-axis. Furthermore, it was found, that the length of the O-D bonds remains virtually unaffected by the compression up to 9 GPa. Instead, a threefold split-position was found for D and its pressure evolution could be determined (Fig. 1; see also Parise 2006b, this volume).

Other examples of more complex minerals containing OD groups studied by neutron powder diffraction include hydroxylapatite (Leventouri et al. 2001, 2003), burtite, $CaSn(OH)_6$ (Cohen-Addad 1968; Basciano et al. 1998), hydrogarnets (e.g., katoite, $Ca_3Al_2(SiO_4)_{3-x}(OH)_{4x}$ with $x = 1.5$-3) (Lager et al. 1987, 1989; Lager and von Dreele 1996) and hydroxyl-clinohumite [$Mg_7Si_4O_{14}·2Mg(OH)O$] (Berry and James 2001). The protons of the hydroxyl groups in hydroxyl-clinohumite were found to be disordered over two crystallographic positions in the monoclinic space group $P2_1/b$, each with occupancy of approximately 0.5.

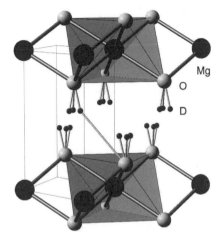

Figure 1. Crystal structure of brucite, $Mg(OD)_2$, redrawn after Parise et al. (1994). For pressures up to 9 GPa the length of the hydrogen bond D-O (indicated in the graph) shortens by more than 15%. The 3-fold deuterium split-site also changes with pressure, such that the O-D-O angle increases from 148 to 156°.

Friedrich et al. (2002) studied the high-pressure crystal structure of a natural F-bearing chondrodite $Mg_{4.62}Fe_{0.28}Mn_{0.014}Ti_{0.023}(Si_{1.01}O_4)_2F_{1.16}OH_{0.84}$ and the structure of the deuterated synthetic analogue $Mg_5(SiO_4)_2F_{1.10}OD_{0.90}$. From the analysis of the lattice parameters they found that the substitution of OH by F makes the chondrodite structure less compressible. The axial compressibilities are anisotropic. The general compression mechanism was determined from the X-ray structural investigations in a diamond-anvil cell. The greatest changes with pressure occur in octahedra, which are located at flexion points of octahedral chains. Neutron powder data were collected at pressures up to 7 GPa in a Paris-Edinburgh cell. These data allowed investigating changes in the O-D...O/F hydrogen bond geometry. It was found that the OD vector changes its orientation. Furthermore, the static disorder associated with the hydrogen bond acceptor/donor site is reduced by increasing pressure.

Water in minerals. Gypsum $CaSO_4·(D_2O)_2$ is a prominent example of a water bearing mineral. Schofield et al. (1996) investigated the crystal structure and the thermal expansion behavior of gypsum between 4 and 300 K. The expansion behavior was found to be strongly anisotropic. Along the *b*-axis of the monoclinic lattice the expansion is largest and this can be related to changes of a particular hydrogen bond.

Apart from several zeolites that may contain water, ikaite $Ca(CO_3)(D_2O)_6$ is another water containing mineral that has been studied by neutron powder diffraction (Hesse et al. 1970; Swainson and Hammond 2003). Lawsonite, $CaAl_2Si_2O_7(OH)_2(H_2O)$ (Meyer et al. 2001) and palygorskite $(Mg,Al)_2Si_4O_{10}(OH)\cdot4(H_2O)$ (Giustetto and Chiari 2004) are examples of minerals containing both water and hydroxyl groups.

Ammonia. There are some minerals known that contain ammonia ions. The positions of the protons could be determined from deuterated samples of e.g., niahite $(ND_4)Mn(PO_4)$ (D_2O) (Carling et al. 1995), nitrammite ND_4NO_3 (Ahtee et al. 1983a,b; Lucas et al. 1979, 1980), and oxammite $(ND_4)_2(CO_2)_2(D_2O)$ (Taylor and Sabine 1972), all by neutron powder diffraction. Several authors report on the determination of fully deuterated ammonia in zeolite-Rho(Fischer et al. 1989; Corbin et al. 1990; Gilles et al. 2004)

Organics in zeolites. The use of deuterated organic molecules allowed investigating the molecule positions in the open framework structures of zeolites. This is of particular technological relevance, since it facilitates the understanding of catalytically active sites in zeolites. Prominent examples are the cases of benzene (Fitch et al. 1986), xylene (Czjzek et al. 1991a) or aniline (Czjzek et al. 1991b) in zeolite–Y. Another example is the localization of mono- and tri-methylamine in zeolite-Rho (Weidenthaler et al. 1997), or the determination of the positions of water in Ba exchanged zeolite-X (Pichon et al. 1999).

Lithium

Lithium is difficult to study by X-ray methods since its scattering power is low. The ICSD lists 48 mineral entries belonging to 17 structures that were studied by neutron powder diffraction. This is notable insofar as Li is a strong neutron absorber (Table 1).

Examples of Li bearing minerals studied by neutrons are e.g., acmite $LiFe(Si_2O_6)$ an inosilicate with single width unbranched chains (Lottermoser et al. 1998), or Li-hollandite $Li_{2.72}Mn_{6.64}O_{16}$, where Li is substituted for Ba and Mn in hollandite, $Ba(Mn^{4+},Mn^{2+})_8O_{16}$, (Botkovitz et al. 1994). Lithium is also found in various concentrations in the anatase structure (Wagemaker et al. 2003). It can be substituted in the chromite structure, e.g., as $LiFe_3Cr_2O_8$ (Dargel et al. 1972) or $Li(CrTO_4)$ (Arillo et al. 1996). Furthermore, Li occurs in LiTi-spinel superconductors (Dalton et al. 1994) or LiMn-spinel (Fong et al. 1994) which may also contain Ni (Gryffroy et al. 1991). Ibarra Palos et al. (2001) reported on the occurrence of Li at the tetrahedral A and the octahedral B sites of $(Li,H)(Mn,Li)O_4$ spinel.

Microporous lithosilicates. The incorporation of Li in zeolite like structures is of potential technological interest since these materials are discussed as ion-conductors as well as having large ion exchange potential. The nesosilicate eucryptite $Li(AlSiO_4)$ also described as a zeolite, corresponds to the stuffed cristobalite type. The thermally induced transformation from γ-eucryptite to pseudo-eucryptite was studied by Norby (1990) by neutron powder diffraction and Sartbaeva et al. (2004) investigated the cooperative motion of Li in β-eucryptite.

Li is also found in several other zeolite topologies such as the sodalite-type (Weller and Wong 1989; Brenchley and Weller 1994; Mead and Weller 1995), analcime (Seryotkin et al. 2001) and chabazite (Smith et al. 2000), and in the topology of the synthetic zeolites A (Norby et al. 1986), X (Feuerstein and Lobo 1998), Y (Forano et al. 1989), LSX (Plevert et al. 1997), RUB-29 (Park et al. 2000), or RUB-31 (Park et al. 2002). In particular the RUBs (named after the Ruhr-Universität-Bochum) attracted some interest as novel microporous lithosilicates. Number 29, for example, exhibits an extraordinary high ratio of Li to Si (1:4) compared to other open framework structures synthesized so far. The crystal structure of RUB-29 was obtained from synchrotron single-crystal diffraction intensities by direct methods and Fourier calculations. To confirm Li-sites and to find the extra-framework (non-tetrahedrally coordinated) Li^+ cations, Rietveld analysis using neutron powder diffraction data was performed. Note that using neutrons did not only allow obtaining the framework structures but also the positions of D_2O molecules in several of the above-mentioned examples.

Beryllium

Structural studies on minerals phases containing beryllium are rare. The ICSD reports just the cases of a beryllophosphate having the faujasite topology with the chemical formula $Mg_{0.2}Na_{0.6}(BePO_4)$ (Nenoff et al. 1992) and $Rb_{24}(Be_{24}As_{24}O_{96})(D_2 O)_{3.2}$ (Parise et al. 1992), having zeolite-Rho topology.

NEIGHBORING ELEMENTS

Several elements commonly occurring in minerals are difficult to distinguish with X-rays, since they have the same or very similar number of electrons and hence a very similar scattering power. Using neutrons, ions as Ti^{4+}, Ca^{2+}, K^+ and Cl^-, or Na^+, Mg^{2+}, Al^{3+}, and Si^{4+}, or Fe^{2+} and Mn^{2+} can easily be identified. This is of particular importance in Earth Sciences since that allows the direct determination of site occupancies and therefore the study of order/disorder processes in minerals (Redfern 2006, this volume).

Si/Al distribution

Where does our insight into the various Al/Si distribution patterns come from? Normally, knowledge about the atomistic structure of a given crystalline material is gained by analyzing the diffraction patterns. The most available and economical radiation is certainly X-rays from laboratory sources. However, because the scattering power for X-rays of an atom is proportional to its number of electrons, there is less than 10% difference between isoelectronic ions Al^{3+} and Si^{4+}. In addition, they are relatively weak X-ray scatterers and, hence, their contribution to the total scattered intensity is low. Thus, the determination of the Al/Si-distribution would be very difficult, were it not for the marked difference in the interatomic distances, which can be measured with high precision. Al-O and Si-O bond lengths amount to 1.74 and 1.61 Å, respectively; substitution of Al for Si, or vice versa, on the same tetrahedral (*T*) position results in intermediate mean values. The question arises of how the average bond length changes as a function of the molar ratio. Early researchers proposed an essentially linear relationship between mean bond length and *T* occupancy (e.g., Smith 1954; Smith and Bailey 1963; Jones 1968; Ribbe and Gibbs 1969). Being by and large correct, the rule is limited since individual bond lengths are also influenced by other factors, introducing uncertainties in the determination of the Al/Si ratio from bond length considerations. It is thus highly desirable to have means of directly determining the Al/Si ratio.

The method of choice is neutron diffraction. The difference in scattering power for Al and Si amounts to almost 20% (Fig. 2).

Feldspars, aluminosilicates. About 60% of the Earth's crust consists of minerals belonging to the feldspar family. As an example we consider a particular member of this family, albite $NaAlSi_3O_8$. It is found in a high and a low temperature form, which differ with respect to their symmetry. This difference is the result of a different distribution of the Al^{3+} and Si^{4+} among a set of TO_4-tetrahedra, which make up the characteristic framework of feldspars. The TO_4-tetrahedra may contain variable proportions of Al^{3+} and Si^{4+} at the same *T* site. In high albite the distribution is disordered, whereas in low albite Al and Si are ordered. Similar relationships are found for potassium feldspars, with Si/Al-disordered sanidine and ordered microcline (Barth 1934). The transitions from an ordered into a disordered state and vice versa are complex temperature-, pressure-, composition- and time-dependent processes. In plagioclase the processes are even more complicated because of variable Al/Si stoichiometry. Many other aluminosilicates, natural or synthetic, show variable ratios and distribution patterns of Si and Al, with often significant influence on the properties.

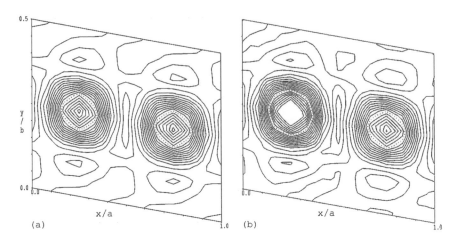

Figure 2. Fouriermaps calculated for a hypothetic structure (space group $P\bar{1}$, atoms at 0.25, 0.25, 0.5 and 0.75, 0.25, 0.5). Sections through the neutron scattering density at $z = 0.5$ are shown for (a) complete disorder of Si and Al, and (b) full order of Si and Al. Panel (b) clearly demonstrates the larger maximum in the scattering length distribution at 0.25, 0.25, 0.5 that is associated with the larger scattering length of Si compared to Al.

Many papers are dedicated to Al-Si order studies in a large variety of aluminosilicates and related compounds. We just mention a few of them as examples. Harlow and Brown (1980) studied the structure of low albite by using both X-rays and neutrons. They could confirm that the results with respect to the positional and thermal atomic parameters are essentially identical. The neutron data allowed the direct determination of the Al/Si occupancy. The result was in accordance with the assignments based on bond length considerations, i.e., Al and Si are fully ordered. These preliminary results were confirmed and improved by a refinement with neutron data taken at the very low temperature of 13 K (Smith et al. 1986). Mc Mullan (1996) went in the opposite direction and investigated the effect of high temperature on the structure of sodalite. They found that Al and Si remain fully ordered up to 1200 K. The distribution is ordered because the natural sample used most probably grew at much lower temperatures and for kinetic reasons the Al/Si distribution did not change in the heating experiment. By contrast a sodalite sample used by Tarling (1988) was synthesized at high temperatures and Si/Al are fully disordered and this disorder prevails upon quenching. ^{29}Si MAS NMR could confirm the disorder determined by neutron diffraction. The observed disorder in this Al:Si = 1 aluminosilicate necessarily requires direct linkage between AlO_4-tetrahedra. According to Loewenstein's rule such linkages should be avoided (Loewenstein 1954). There is recent evidence which suggest that the restrictions set by Loewenstein's rule can be considerably relaxed (Peters 2005; Peters et al. 2005, 2006a,b).

Of the many neutron studies on zeolites we only mention as prominent examples the early work on zeolite-A (Cheetham et al. 1982; Cheetham and Eddy 1983), the combined synchrotron and neutron work on zeolite LSX (Vitale et al. 1995), the combination of neutrons and computational crystallography to determine $CFCl_3$ in zeolite NaY (Mellot-Draznieks et al. 2003), or studies on zeolite Y (Hriljac et al. 1993) and X (Vitale et al. 1997; Zhu et al. 1999). A recent high pressure study is on natrolite (Colligan et al. 2005; Fig. 3). Full order of Al and Si could be established and the effect of superhydration under the application of high pressure explained.

Biotite. The structure of the layer silicate Biotite is a typical example for a complex crystal structure refined from neutron data. The chemical formula of the natural sample from

Bancroft (Ontario) studied by Chon et al. (2003) is $K_2(Mg_{3.15}Fe_{2.59}Ti_{0.17}Mn_{0.09})(Si_{5.98}Al_{1.92}Ti_{0.10})O_{20}((OH)_{1.47}F_{1.98})$. It crystallizes in the monoclinic space group $C2/m$. The bulk chemical composition was obtained from X-ray fluorescence and thermal gravimetric analysis and the site occupancies were initially set according to the chemical analysis with Si, Al and Ti sharing the $8j$ site, Mg, Fe, Ti and Mn on $2c$ and $4h$, as well as O and F on $4i$ (Fig. 4). No sign of cation ordering was obtained for Mg and Fe, while refinements of the OH/F site occupancies were unstable. The sample was heated to 900 °C in vacuum and in air. For both processes a different thermal expansion behavior was found that is related to different structural changes. The change of the unit cell in vacuum is dominated by a strong expansion in c-direction. It results from the expansion of the K coordination sphere along that direction. Samples heated in air show a decrease of the magnitude of the unit cell above 400 °C due to dehydroxylation and the oxidation of iron.

Fe/Mn in olivines and spinel

A substitution frequently occurring in minerals is that of iron and manganese. It is not only of structural interest but also might have great implications on the magnetic behavior of minerals.

Cation ordering in minerals belonging to the olivine group has been studied on fayalite–tephroite solid solutions $(Fe_xMn_{1-x})_2SiO_4$ (Henderson et al. 1996; Redfern et al. 1997, 1998). Fe and Mn share the $2a$ and the $4c$ sites in space group $Pbnm$.

Hausmannite belongs to the spinel group and is a mixed-valence compound with the chemical formula Mn_3O_4 (or better $Mn^{2+}Mn^{3+}_2O_4$). The structure may contain Fe^{3+}-ions on the octahedral $8d$ site of space group $I4_1/amd$ that changes the magnetic properties (Baron et al. 1998). Because of the large difference in the scattering lengths of Fe and Mn (see Table 1) neutron powder diffraction experiments are ideally suited to study the influence of the substitution on the crystal structure and the magnetic properties. The unit-cell parameters of hausmannite illustrate a decreasing Jahn-Teller distortion

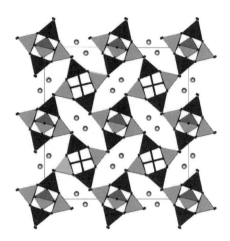

Figure 3: Projected structure of natrolite, redrawn after Colligan et al. (2005). Light tetrahedra represent SiO_4-units and dark tetrahedra AlO_4-units that are fully ordered. Grey spheres indicate sodium atoms and small black spheres stand for oxygen. Water molecules are omitted for the sake of simplicity

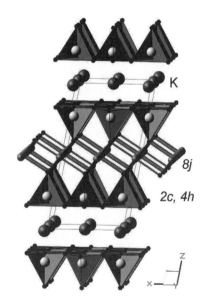

Figure 4: View of the biotite structure along the monoclinic *y*-axis (space group $C2/m$), redrawn after Chon et al. (2003). Tetrahedra indicate the environment of the $2c$- and $4h$-sites filled by Si, Al and Ti (light spheres). Small black spheres represent oxygen atoms. The $8j$-site hosts Mg, Fe, Ti and Mn atoms. OH and F are not drawn for clarity.

with increasing Fe content, whereas the Curie temperature was found to increase with increasing Fe content.

Other spinel-type compounds investigated with focus on the cation distribution and its influence on magnetic properties are jacobsite $MnFe_2O_4$, franklinite $(Mn,Zn)Fe_2O_4$ (Koenig and Chol 1968) and chromian jacobsite $(Mn,Fe)(MnFeCr)_2O_4$ (Pickart and Nathans 1959). Pseudobrookite $(Fe^{3+}, Fe^{2+})_2(Ti,Fe^{2+})O_5$ is a further example of a mineral with mixed valence of Fe. In addition, the structure may contain manganese. The distribution of the cations in kennedyite $Ti_2(Fe_{.33}Ti_{.52}Mn_{.05})O_5$ was studied by Teller et al. (1990). The CaMn-bicarbonate kutnohortite may also have Fe. This was examined by Farkas et al. (1988) for a sample having the chemical composition $(Ca_{0.86}Mn_{0.14})$ $(Ca_{0.14}Mn_{0.50}Fe_{0.13}Mg_{0.23})$ $(CO_3)_2$.

MISCELLANEOUS

Strong absorption

A major problem in X-ray diffraction is the determination of the positions of light elements such as hydrogen if the compound also contains heavy elements such as e.g., lead or uranium. Since the interaction of neutrons with atoms does not depend on the number of electrons the refinement of structures with light and very heavy elements is a very powerful application for neutrons. However, since the resolution of many neutron powder diffractometers is poor—compared to synchrotron instruments—a combination of both methods may be necessary.

Here we take uranyl minerals as representatives for compounds showing strong X-ray absorption. The very rare mineral metaankoelite $K(UO_2)(PO_4)(D_2O)_3$ was investigated by Fitch and Cole (1991). The crystal structure of this deuterated potassium uranyl phosphate trihydrate has been simultaneously refined from high-resolution powder neutron and synchrotron X-ray diffraction data. The structure is tetragonal ($a = 6.99379(2)$ Å and $c = 17.78397(7)$ Å) and isostructural with the analogous arsenate, the mineral abernathyite. A phase transition is observed at about 270 K. Other examples of uranium minerals studied with neutrons are uramphite $(ND_4)(UO_2)(PO_4)(D_2O)_3$ (Fitch and Fender 1983) and troegerite $D(UO_2)(AsO_4)(D_2O)_4$ (Fitch et al. 1983). They all crystallize in the tetragonal space group $P4/ncc$. D_2O is disordered in metaankoelite at ambient temperature. Below 270 K the symmetry of metaankoelite is reduced and the water molecules are ordered (Cole et al. 1993). The use of neutrons was essential for solving and refining the crystal structure of the low temperature phase. A neutron powder pattern showed no visible peak splitting or systematically absent reflections. However, the structure refinement in lower symmetric tetragonal space groups failed. In high-resolution synchrotron data, broadening of several peaks and small splitting was found, being indicative for an orthorhombic metric. The lattice parameters at 150 K are $a = 6.9919(1)$ Å, $b = 6.9738(1)$ Å, and $c = 17.6632(1)$ Å. Systematic absences of reflections suggest space group $Pccn$. However, the reevaluation of the neutron data with the orthorhombic cell, obtained from the synchrotron pattern, showed the presence of an intense peak violating the c-glide plane perpendicular to the a-direction. The final structure was refined simultaneously from the synchrotron and neutron data in space group $P2_1cn$, optimizing 61 positional and thermal parameters, and the deuterium content in addition to global and profile parameters.

Representative sampling

Since the interaction potential of neutrons with matter is small, compared to X-rays or electrons, large samples have to be studied. This is of advantage for the study of natural samples or products from industrial processes since it improves representative sampling. This is of particular importance for quantitative phase analysis as it has e.g., been performed on iron ores (Böhm et al. 2005). Low grade carbonaceous iron ore from the Austrian Erzberg is thermally treated to turn it into a magnetic oxide phase. A valuable iron ore concentration is achieved by magnetic separation. Since the chemical analysis and the X-ray investigation were not conclusive, a neutron study was performed. From the structure refinement of the multi phase samples with the Rietveld method it was found that the raw material contained sideroplesite and ankerite with different proportions of Mg, Ca and Mn in addition to Fe. Furthermore, wuestite (FeO), hematite (Fe_2O_3), magnetite (Fe_3O_4) and complex ferrites with different cation ratios may be present. The result of the quantitative phase analysis strongly depends on the number of cations being present in the minerals. Therefore, a combination of multi phase Rietveld refinement, density measurements and magnetic structure refinement was applied. A typical example for the results of a Rietveld quantitative phase analysis of processed material is: sideroplesite 2.80% (0.27), magnetite 54.17% (0.49), wuestite 39.59% (0.40), and ankerite 3.44% (0.21).

SUMMARY AND OUTLOOK

We have presented a broad range of examples demonstrating the power of neutron powder diffraction for tackling problems in mineral structures related to weak X-ray scatterering, neighboring elements in the periodic table, strong X-ray absorption or representative sampling.

So much for the advantages of neutron diffraction. Are there any caveats? One of the most important drawbacks of neutrons is certainly their limited availability. Another limitation has been the comparably low intensity of neutron sources. This requires large sample quantities that are often not available. The results of neutron or X-ray diffraction experiments are to some degree averages over space and time and only global information is obtained. Therefore, if local information is required, complementary methods such as spectroscopy or electron microscopy must be used for data interpretation. With the advent of new powerful synchrotron facilities anomalous X-ray scattering may become a seriously competing method for solving problems like Al/Si occupancy. However, this will not mean that neutrons powder diffraction will become obsolete as demonstrated by other applications described in this volume.

ACKNOWLEDGMENT

We like to thank A. Böhm (Montanuniversität Leoben, Austria) for helpful discussions and providing unpublished data.

REFERENCES

Ahtee M, Smolander KJ, Lucas BW, Hewat AW (1983a) Low-temperature behaviour of ammonium nitrate by neutron diffraction. Acta Crystallogr B 39:685-687

Ahtee M, Smolander KJ, Lucas BW, Hewat AW (1983b) The structure of the low-temperature phase V of ammonium nitrate, ND_4NO_3. Acta Crystallogr C 39:651-655

Arillo MA, Lopez ML, Fernandez MT, Veiga ML, Pico C (1996) Preparation and magnetic properties of $LiCr_{1-x}Al_xTiO_4$ (0<x<0.4). J Solid State Chem 125:211-215

Baron V, Gutzmer J, Rundloef H, Tellgren R (1998) The influence of iron substitution on the magnetic properties of hausmannite, $Mn^{(2+)}(Mn,Fe)_2^{(3+)}O_4$. Am Mineral 83:786-793

Barth TFW (1934) Polymorphic phenomena and crystal structure. Am J Sci 27:273-286

Basciano LC, Peterson RC, Roeder PL, Swainson I (1998) Description of schoenfliesite, $MgSn(OH)_6$ and roxbyite, $Cu_{1.72}S$, from a 1375 BC shipwreck, and Rietveld neutron-diffraction refinement of synthetic schoenfliesite, wickmanite, $MnSn(OH)_6$ and burtite, $CaSn(OH)_6$. Can Mineral 36:1203-1210

Berry AJ, James M (2001) Refinement of hydrogen positions in synthetic hydroxyl-clinohumite by powder neutron diffraction. Am Mineral 86:181-184

Böhm A, Böhm M, Cheptiakov D, Kogelbauer A, Steiner HJ (2005) In-situ neutron diffraction studies on thermally treated carbonacious iron ore materials. Geophys Res Abstracts 7:08462

Botkovitz P, Brec R, Deniard P, Tournoux M, Burr G (1994) Electrochemical and neutron diffraction study of a prelithiated hollandite-type Li_xMnO_2 phase. Mol Cryst Liq Cryst 244:233-238

Brenchley ME, Weller MT (1994) Synthesis and structures of $M_8(AlSiO_4)_6$. $(XO_4)_2$, M = Na, Li, K; X = Cl, Mn sodalites. Zeolites 14:682-686

Carling SG, Day P, Visser D (1995) Crystal and magnetic structures of layer transition metal phosphate hydrates. Inorg Chem 34:3917-3927

Chakoumakos BC, Loong CK, Schultz AJ (1997) Low-temperature structure and dynamics of brucite. J Phys Chem 101:9458-9462

Cheetham AK, Eddy MM (1983) Neutron diffraction studies of zeolite-A and synthetic faujasite. ACS Sym Ser 218:132-142

Cheetham AK, Eddy MM, Jefferson DA, Thomas JM (1982) A study of Si,Al ordering in thallium zeolite-A by powder neutron diffraction. Nature 299:24-26

Chon C-M, Kim Shin A, Moon H-S (2003) Crystal structures of biotite at high temperatures and of heat-treated biotite using neutron powder diffraction. Clay Clay Mineral 51:519-528

Christensen AN, Hamilton WC, Ibers JA (1963) Structures of $HCrO_2$ and $DCrO_2$. Acta Crystallogr 16:1209-1212

Christensen AN, Hansen P, Lehmann MS (1977) Isotope effects in the bonds of alpha-CrOOH and alpha-CrOOD. J Solid State Chem 21:325-329

Christensen AN, Lehmann MS, Convert P (1982) Deuteration of crystalline hydroxides. Hydrogen bonds of gamma-AlOO(H,D) and gamma-FeOO(H,D). Acta Chem Scand A 36:303-308

Christensen N, Hansen P, Lehmann MS (1976) Isotope effects in the bonds of beta-CrOOH and beta-CrOOD. J Solid State Chem 19:299-304

Cohen-Addad C (1968) Etude structurale des hydroxystannates $CaSn(OH)_6$ et $ZnSn(OH)_6$ par diffraction neutronique, absorption infrarouge et resonance magnetique nucleaire. B Soc Fr Mineral Cr 91:315-324

Cole M, Fitch AN, Prince E (1993) Low-temperature structure of $KUO_2PO_4\cdot3D_2O$ determined from combined synchrotron radiation and neutron powder diffraction measurements. J Mater Chem 3:519-522

Colligan M, Lee YS, Vogt T, Celestian AJ, Parise JB, Marshall WG, Hriljac JA (2005) High pressure neutrondiffraction study of superhydrated natrolite. J Phys Chem B 109:18223-18225

Corbin DR, Abrams L, Jones GA, Eddy MM, Harrison WTA, Stucky GD, Cox DE (1990) Flexibility of the zeolite RHO framework. In situ X-ray and neutron powder structural characterization of divalent cation-exchanged zeolite RHO. J Am Chem Soc 112:4821-4830

Czjzek M, Fuess H, Vogt T (1991a) Structural evidence for pi complexes in catalytically active Y zeolites with o-, m-, and p-xylene. J Phys Chem 95:5255-5261

Czjzek M, Vogt T, Fuess H (1991b) Aniline in Yb, Na-Y: a neutron powder diffraction study. Zeolites 11:832-836

Dalton M, Gameson I, Armstrong AR, Edwards PP (1994) Structure of the $Li_{1+x}Ti_{2-x}O_4$ superconducting system. A neutron diffraction study. Physica C 221:149-156

Dargel L, Kubel W, Olkiewicz K (1972) Influence of thermal treatment on the crystallographic structure of a series of $Li_2O_{(5-2t)}(Fe_2O_3)(Cr_2O_3)_{2t}$ ferrites. Acta Phys Pol A 41:689-700

Delaplane RG, Ibers JA, Ferraro JR, Rush JJ (1969) Diffraction and spectroscopic studies of the cobaltic acid system $HCoO_2$ - $DCoO_2$. J Chem Phys 50:1920-1927

Farkas L, Bolzenius BH, Schaefer W, Will G (1988) The crystal structure of kutnohorite $CaMn(CO_3)_2$. Neues Jb Miner Monat 1988:539-546

Feuerstein M, Lobo RF (1998) Characterization of Li cations in zeolite LiX by solid-state NMR spectroscopy and neutron diffraction. Chem Mater 10:2197-2204

Fischer RX, Baur WH, Shannon RD, Parise JB, Faber J Jr., Prince E (1989) New different forms of ammonium loaded and partially de-ammoniated zeolite rho studied by neutron powder diffraction. Acta Crystallogr C 45:983-989

Fitch AN, Bernard L, Howe AT, Wright AF, Fender BEF (1983) The room-temperature structure of $DUO_2AsO_4\cdot4D_2O$ by powder neutron diffraction. Acta Crystallogr C 39:159-162

Fitch AN, Cole M (1991) The structure of $KUO_2PO_4\cdot3D_2O$ refined from neutron and synchrotron-radiation powder diffraction data. Mater Res Bull 26:407-414

Fitch AN, Fender BEF (1983) The structure of deuterated ammonium uranyl phosphate Trihydrate, $ND_4UO_2PO_4(D_2O)_3$ by Powder Neutron Diffraction. Acta Crystallogr C 39:162-166

Fitch AN, Jobic H, Renouprez A (1986) Localization of benzene in sodium-Y zeolite by powder neutron diffraction. J Phys Chem 90:1311-1318

Fong C, Kennedy BJ, Elcombe MM (1994) A powder neutron diffraction study of lambda and gamma manganese dioxide and of $LiMn_2O_4$. Z Kristallogr 209:941-945

Forano C, Slade RCT, Krogh-Andersen E, Krogh-Andersen IG, Prince E (1989) Neutron diffraction determination of full structures of anhydrous Li-X and Li-Y zeolites. J Solid State Chem 82:95-102

Friedrich A, Lager GA, Ulmer P, Kunz M, Marshall WG (2002) High-pressure single-crystal X-ray and powder neutron study of F,(OH)/(OD)-chondrite: compressibility, structure and hydrogen bonding. Am Mineral 87:931-939

Gilles F, Blin JL, Mellot-Draznieks C, Cheetham AK, Su BL (2004) Neutron diffraction evidence of double interaction between NaY zeolite and ammonia and migration of $Na^{(+)}$ ions upon ND_3 adsorption. Chem Phys Lett 390:236-239

Giustetto R, Chiari G (2004) Crystal structure refinement of palygorskite from neutron powder diffraction. Eur J Miner 16:521-532

Gryffroy D, Vandenberghe RE, Legrand E (1991) A neutron diffraction study of some spinel compounds containing octahedral Ni and Mn at a 1:3 ratio. Mater Sci For 79:785-790

Harlow GE, Brown GE (1980) Low albite: an X-ray and neutron diffraction study. Am Mineral 65:986-995

Harrison RJ (2006) Neutron diffraction of magnetic materials. Rev Mineral Geochem 63:113-143

Henderson CMB, Knight KS, Redfern SAT, Wood BJ (1996) High-temperature study of octahedral cation exchange in olivine by neutron powder diffraction. Science 271:1713-1715

Hesse KF, Küppers H, Suess E (1970) Refinement of the structure of ikaite, $CaCO_3 \cdot 6(H_2O)$ Z Kristallogr 163: 227-231

Hriljac JA, Eddy MM, Cheetham AK, Donohue JA, Ray GJ (1993) Powder neutron diffraction and ^{29}Si MAS NMR studies of siliceous zeolite-Y. J Solid State Chem 106:66-72

Ibarra Palos A, Anne M, Strobel P (2001) Topotactic reactions, structural studies and lithium intercalation in cation-deficient spinels with formula close to $Li_2Mn_4O_9$. J Solid State Chem 160:108-117

Jones JB (1968) Al-O and Si-O tetrahedral distances in aluminosilicate framework structures. Acta Crystallogr B 24:355-358

Koenig U, Chol G (1968) Röntgenbeugungs- und Neutronenbeugungsuntersuchungen an Ferriten der Reihe $Mn_xZn_{1-x}Fe_2O_4$. J Appl Crystallogr 1:124-126

Kuhs WF, Hansen TC (2006) Time-resolved neutron diffraction studies with emphasis on water ices and gas hydrates. Rev Mineral Geochem 63: 171-204

Lager GA, Armbruster T, Faber J, Jr. (1987) Neutron and X-ray diffraction study of hydrogarnet $Ca_3Al_2(O_4H_4)_3$. Am Mineral 72:756-765

Lager GA, Armbruster T, Rotella FJ, Rossman GR (1989) OH substitution in garnets: X-ray and neutron diffraction, infrared, and geometric-modeling studies. Am Mineral 74:840-851

Lager GA, von Dreele RB (1996) Neutron powder diffraction study of hydrogarnet to 9.0 GPa. Am Mineral 81:1097-1104

Leventouri T, Bunaciu CE, Perdikatsis V (2003) Neutron powder diffraction studies of silicon-substituted hydroxyapatite. Biomaterials 24:4205-4211

Leventouri T, Chakoumakos BC, Papanearchou N, Perdikatsis B (2001) Comparison of crystal structure parameters of natural and synthetic apatites from neutron powder diffraction. J Mater Res 16:2600-2606

Loewenstein W (1954) The distribution of aluminum in the tetrahedra of silicates and aluminates. Am Mineral 39:92-98

Lottermoser W, Redhammer GJ, Forcher K, Amthauer G, Paulus W, Andre G, Treutmann W (1998) Single crystal Moessbauer and neutron powder diffraction measurements on the synthetic clinopyroxene Li-acmite $LiFeSi_2O_6$. Z Kristallogr 213:101-107

Lucas BW, Ahtee M, Hewat AW (1979) The crystal structure of phase II ammonium nitrate. Acta Crystallogr B 35:1038-1041

Lucas BW, Ahtee M, Hewat AW (1980) The structure of phase III ammonium nitrate. Acta Crystallogr B 36: 2005-2008

Mc Mullan RK, Ghose S, Naga N, Schomaker V (1996) Sodalite, $Na_4Si_3Al_3O_{12}Cl$: Structure and ionic mobility at high temperatures by neutron diffraction. Acta Crystallogr B 52:616-627

Mead PJ, Weller MT (1995) Synthesis, structure, and characterization of halate sodalites: $M_8(AlSiO_4)_6$ $(XO_3)_x(OH)_{2-x}$; M = Na, Li, or K; X = Cl, Br, or I. Zeolites 15:561-568

Mellot-Draznieks C, Rodriguez-Carvajal J, Cox DE, Cheetham AK (2003) Adsorption of chlorofluorocarbons in nanoporous solids: a combined powder neutron diffraction and computational study of $CFCl_3$ in NaY zeolite. Phys Chem Chem Phys 5:1882-1887

Meyer HW, Marion S, Sondergeld P, Carpenter MA, Knight KS, Redfern SAT, Dove MT (2001) Displacive components of the low-temperature phase transitions in lawsonite. Am Mineral 86:566-577

Nenoff TM, Harrison WTA, Gier TE, Nicol JM, Stucky GD (1992) Structural characterization of a dehydrated magnesium/sodium beryllophosphate faujasite phase. Zeolites 12:770-775

Norby P (1990) Thermal transformation of zeolite Li-A(BW). The crystal structure of gamma-eucryptite, a polymorph of $LiAlSiO_4$. Zeolites 10:193-199

Norby P, Norlund Christensen A, Krogh Andersen IG (1986) Hydrothermal preparation of zeolite Li-A(BW), $LiAlSiO_4 \cdot H_2O$, and structure determination from powder diffraction data by direct methods. Acta Chem Scand A 40:500-506

Parise JB (2006a) Introduction to neutron properties and applications. Rev Mineral Geochem 63:1-25

Parise JB (2006b) High pressure studies. Rev Mineral Geochem 63:205-231

Parise JB, Corbin DR, Gier TE, Harlow RL, Abrams L, von Dreele RB (1992) Flexibility of the RHO framework: a comparison of the Rb-exchanged zeolite and the novel composition $Rb_{24}Be_{24}As_{24}O_{96} \cdot 3.2($ $H_2O)$. Zeolites 12:360-368

Parise JB, Leinenweber K, Weidner DJ, Tan K, von Dreele RB (1994) Pressure-induced H bonding: neutron diffraction study of brucite, $Mg(OD)_2$, to 9.3 GPa. Am Mineral 79:193-196

Park SH, Gies H, Toby BH, Parise JB (2002) Characterization of a new microporous lithozincosilicate with ANA topology. Chem Mater 14:3187-3196

Park SH, Parise JB, Gies H, Liu HM, Grey CP, Toby BH (2000) A new porous lithosilicate with a high ionic conductivity and ion-exchange capacity. J Am Chem Soc 122:11023-11024

Peters L, Knorr K, Depmeier W (2006a) Coupled substitution in the melilite type structure from gehlenite to $(Ln_xCa_{2-x})Al[Al_{1+x}Si_{1-x}O_7]$ with $0<x<1$ and Ln: La, Eu, Er. Z Anorg Allg Chem 632:301-306

Peters L, Knorr K, Fechtelkord M, Appel P, Depmeier W (2006b) Structural variations in the solid solution series of sodalite-type $|(Eu_xCa_{2-x})_4(OH)_8|[(Al_{2+x}Si_{1-x}O_{24}]$-SOD with $0<x<1$, determined by X-ray powder diffraction and ^{27}Al MAS NMR spectroscopy. Z Kristallogr 221:643-648

Peters L, Knorr K, Knapp M, Depmeier W (2005) Thermal expansion of gehlenite, $Ca_2Al[AlSiO_7]$, and the related aluminates $(LnCa)Al[AlSiO_7]$, Ln: Tb, Sm. Phys Chem Mineral 32:460-465

Peters L. (2005) Gekoppelte Substitution im Melilith- und Sodalith-Strukturtyp, PhD thesis. University of Kiel

Pichon C, Methivier A, Simonot-Grange MH, Baerlocher C (1999) Location of water and xylene molecules adsorbed on prehydrated zeolite BaX. A low-temperature neutron powder diffraction study. J Phys Chem B 103:10197-10203

Pickart SJ, Nathans R (1959) Neutron study of the crystal and magnetic structure of $MnFe_{1-t}Cr_tO_4$. Phys Rev 116:317-322

Plevert J, di Renzo F, Fajula F, Chiari G (1997) Structure of dehydrated zeolite Li-LSX by neutron diffraction: evidence for a low-temperature orthorhombic faujasite. J Phys Chem 101:10340-10346

Redfern SAT (2006) Neutron powder diffraction studies of order-disorder phase transitions and kinetics. Rev Mineral Geochem 63:145-170

Redfern SAT, Henderson CMB, Knight KS, Wood BJ (1997) High-temperature order-disorder in $(Fe_{0.5}Mn_{0.5})_2SiO_4$ and $(Mg_{0.5}Mn_{0.5})_2SiO_4$ olivines: an in situ neutron diffraction study. Eur J Miner 9: 287-300

Redfern SAT, Knight KS, Henderson CMB (1998) Fe-Mn cation ordering on fayalite-tephroite $(Fe_xMn_{1-x})_2SiO_4$ olivines: a neutron diffraction study. Mineral Mag 62:607-615

Ribbe PH, Gibbs GV (1969) Statistical analysis of mean Al/Si-bond distances and the aluminium content of tetrahedra in feldspars. Am Mineral 54:85-94

Rietveld H (1969) A profile refinement method for nuclear and magnetic structures. J Appl Crystallogr 2: 65-71

Rinaldi R, Schober H (2006) Neutrons at the frontiers of Earth Sciences and environments. Neutron News 17: 3-4

Ross NL, Hoffman C (2006) Single-crystal neutron diffraction: present and future applications. Rev Mineral Geochem 63:59-80

Sartbaeva A, Redfern SAT, Lee WT (2004) A neutron diffraction and Rietveld analysis of cooperative Li motion in beta-eucryptite. J Phys Cond Mat 16:5267-5278

Schofield PF, Knight KS, Stretton IC (1996) Thermal expansion of gypsum investigated by neutron powder diffraction. Am Mineral 81:847-851

Seryotkin YV, Bakakin VV, Bazhan IS (2001) Crystal structure of lithium-sodium $(Li_{0.7}Na_{0.3})$-analcime. Zh Strukt Khim 42:607-611

Smith JV (1954) A review of Al-O and Si-O distances. Acta Crystallogr 7:479-481

Smith JV, Artioli G, Kvick Å (1986) Low albite, $NaAlSi_3O_8$: Neutron diffraction study of crystal structure at 13 K. Am Mineral 71:727-733

Smith JV, Bailey SW (1963) Second review of Al-O and Si-O tetrahedral distances. Acta Crystallogr 16:801-811

Smith LJ, Eckert H, Cheetham AK (2000) Site preferences in the mixed cation zeolite, Li, Na-chabazite: a combined solid-state NMR and neutron diffraction study. J Am Chem Soc 122:1700-1708

Swainson IP, Hammond RP (2003) Hydrogen Bonding in ikaite, $CaCO_3 \cdot 6(H_2O)$. Mineral Mag 67:555-562

Szytula A, Burewicz A, Dimitrijevic Z, Krasnicki S, Rzany H, Todorovic J, Wanic A, Wolski W (1968) Neutron Diffraction studies of alpha-FeOOH. Phys Stat Sol 26:429-434

Tarling SE, Barnes P, Klinowski J (1988) The structure and Si, Al distribution of the ultramarin. Acta Crystallogr B 44:128-135

Taylor JC, Sabine TM (1972) Isotope and bonding effects in ammonium oxalate monohydrate, determined by the combined use of neutron and X-ray diffraction analyses. Acta Crystallogr B 28:3340-3351

Teller RG, Antonio MR, Grau AE, Gueguin M, Kostiner E (1990) Structural analysis of metastable pseudobrookite ferrous titanium oxides with neutron diffraction and Moessbauer spectroscopy. J Solid State Chem 88:334-350

Vitale G, Bull LM, Morris RE, Cheetham AK, Toby BH, Coe CG, MacDougall JE (1995) Combined neutron and X-ray powder diffraction study of zeolite Ca LSX and a $^{(2)}$H NMR study of its complex with benzenes. J Phys Chem 99:16087-16092

Vitale G, Mellot CF, Bull LM, Cheetham AK (1997) Neutron diffraction and computational study of zeolite NaX: influence of SIII' cations on its complex with benzene. J Phys Chem 101:4559-4564

Von Dreele RB (2006) Neutron Rietveld refinement. Rev Mineral Geochem 63:81-98

Wagemaker M, Kearley GJ, van Well AA, Mutka H, Mulder FM (2003) Multiple Li positions inside oxygen octahedra in lithiated TiO_2 anatase. J Am Chem Soc 125:840-848

Weidenthaler C, Fischer RX, Abrams L, Hewat AW (1997) Zeolite Rho loaded with methylamines. I. Monomethylamine loadings. Acta Crystallogr B 53:429-439

Weller MT, Wong G (1989) Characterization of novel sodalites by neutron diffraction and solid state NMR. Solid State Ionics 32:430-435

Wilson AJC (Ed.) (1992) International Tables for Crystallography: Mathematical, Physical and Chemical Tables. Kluwer Academic Publishers

Zhu L, Seff K, Olson DH, Cohen BJ, von Dreele RB (1999) Hydronium ions in zeolites. 1. Structures of partially and fully dehydrated Na,(H_3O)-X by X-ray and neutron diffraction. J Phys Chem B 103:10365-10372

Reviews in Mineralogy & Geochemistry
Vol. 63, pp. 113-143, 2006
Copyright © Mineralogical Society of America

6

Neutron Diffraction of Magnetic Materials

Richard J. Harrison

Department of Earth Sciences
University of Cambridge
Downing Street
Cambridge CB2 3EQ, United Kingdom
e-mail: rjh40@esc.cam.ac.uk

INTRODUCTION

Neutron diffraction is a powerful tool for studying magnetic materials. Neutrons have a magnetic moment, and are scattered by the magnetic moments of atoms in a sample. The cross section for magnetic scattering is sensitive to the relative orientation of the neutron magnetic moment, the atomic magnetic moment, and the scattering vector. This allows magnetic structures to be determined from the intensities of magnetic diffraction peaks, in much the same way that crystal structures are determined from the intensities of nuclear diffraction peaks (Rodríguez-Carvajal 1993). Small-angle neutron scattering and polarized-neutron reflectometry can yield magnetic information over a range length scales, and can be applied to the study of magnetic nanoparticles, spin glasses, and magnetic multilayers (Arai et al. 1985a,b; Arai and Ishikawa 1985; Mangin et al. 1993; Ott et al. 2004). Inelastic magnetic scattering can be used to probe a range of magnetic excitations, providing quantitative information about the magnetic exchange forces between neighboring spins (Brockhouse 1957; Alperin et al. 1967; Samuelsen 1969; Samuelsen and Shirane 1970; Hansen et al. 1997, 2000; Lefmann et al. 1999, 2001; Klausen et al. 2003, 2004).

The earliest neutron diffraction study of a magnetic material was performed by Shull and Smart (1949), who provided the first experimental proof of the existence of antiferromagnetic ordering in MnO. Since then, neutron diffraction has been used to study magnetic structures of increasing complexity, from simple commensurate structures with collinear spins to complex incommensurate structures with non-collinear spins. This review is aimed at researchers requiring an introduction to the use of neutron diffraction to determine the magnetic structure of a material. The essential equations required to interpret neutron scattering from magnetic materials will be presented, along with practical information on how to approach the problem of solving the magnetic structure of a material using neutron diffraction, illustrated with appropriate examples. A description of the additional magnetic information that can be obtained through the use of polarized neutrons will be given. The main focus of the chapter is the determination of magnetic structures using coherent elastic magnetic scattering, and only brief reference will be made to inelastic magnetic scattering. The theory of neutron scattering from magnetic materials is described in full detail in a number of excellent reference works. For a detailed technical treatment the reader is referred to Bacon (1975), Squires (1978), Lovesey (1984), and Shirane et al. (2002).

MAGNETIC NEUTRON SCATTERING THEORY

The magnetic moment of the neutron

Magnetic scattering results from the dipole-dipole interaction between the magnetic

   DOI: 10.2138/rmg.2006.63.6

moment of the neutron and the magnetic moment of an atom. A neutron has a spin of 1/2 and generates a magnetic moment of $\gamma = -1.913\ \mu_N$, where μ_N is the nuclear magneton (1 μ_N = 5.05×10^{-27} Am^2). This moment is around a factor of 1000 smaller than that of an electron.

The magnetic moment of an atom

The magnetic moment of an atom arises either from its nuclear spin or from the presence of unpaired electron spins and unquenched orbital angular momentum. The nuclear spins remain disordered at all temperatures and contribute only to the incoherent scattering cross section. Since we are primarily concerned with the analysis of coherent magnetic scattering from ordered magnetic materials, we will only consider the electronic contributions (both spin and orbital) to the atomic magnetic moment.

The magnetic moment of a paramagnetic atom is defined in terms of its total spin vector, **S**, and its total orbital angular momentum vector, **L**. The spin contribution to the paramagnetic moment is (Crangle 1977):

$$\mu_S = 2\sqrt{S(S+1)}\mu_B \tag{1}$$

where μ_B is the Bohr magneton (1 μ_B = 9.27×10^{-24} Am^2). The orbital contribution to the paramagnetic moment is:

$$\mu_L = \sqrt{L(L+1)}\mu_B \tag{2}$$

The total angular momentum of the atom is defined by the quantum number **J**, which can take any of the values $(L+S)$, $(L+S-1)$, ... , $(L-S-1)$, $(L-S)$. Spin-orbit coupling yields a preferred value of $J = (L-S)$ for electron shells that are less than half full and $J = (L+S)$ for electron shells that are more than half full. The total paramagnetic moment of the atom is then:

$$\mu_J = g\sqrt{J(J+1)}\mu_B \tag{3}$$

where g is the Landé splitting factor:

$$g = 1 + \frac{J(J+1)+S(S+1)-L(L+1)}{2J(J+1)} \tag{4}$$

Note that $g = 2$ for a pure spin moment $(L = 0)$ and $g = 1$ for a pure orbital moment $(S = 0)$. For most transition elements the orbital contribution is "quenched" to zero (or close to zero) by the electrostatic interaction of the spin-bearing valence electrons with the surrounding ligand field (Kittel 1976). For the rare-earth elements, however, both the spin and orbital contributions must be taken into account. This has consequences for both the size of the magnetic moment and the magnetic form factor (see section "Magnetic form factors").

In a material with ordered magnetic moments, the projection, M_J, of J in a given direction of quantization takes one of $2J + 1$ states between $-J$ and $+J$, yielding a magnetic moment:

$$\mu = g M_J \mu_B \tag{5}$$

The maximum moment observed in an ordered magnetic material is, therefore, $gJ\mu_B$.

Amplitude of magnetic scattering

The amplitude of magnetic neutron scattering was first determined by Halpern and Johnson (1939):

$$\alpha_M(\mathbf{Q}) = p f(\mathbf{Q})\mu_\perp \tag{6}$$

The constant p is defined as:

$$p = \frac{\gamma}{2}\left(\frac{e^2}{mc^2}\right) = 0.2696 \times 10^{-12} \, \text{cm} \tag{7}$$

where γ is the magnetic moment of the neutron in nuclear magnetons, e is the electron charge, m is the electron mass, and c is the speed of light. The term in parentheses is equivalent to the classical electron radius ($r_e = 0.282 \times 10^{-12}$ cm). f is the magnetic form factor (see section "Magnetic form factors"). Q is the scattering vector (Parise 2006, this volume; Redfern 2006, this volume). $\mu_\perp$ is the component of the atomic magnetic moment (in Bohr magnetons) perpendicular to Q:

$$\mu_\perp = \hat{Q} \times \mu \times \hat{Q} \tag{8}$$

where $\hat{Q}$ is a unit vector parallel to Q (Fig. 1). Neutron scattering is sensitive only to $\mu_\perp$ (a consequence of the Fourier transform of Maxwell's equation $\text{div}\mathbf{B} = 0$). Hence, the magnetic contribution to a Bragg diffraction peak will be exactly zero if the moments are normal to the diffracting planes. This is one of the key factors that allows magnetic structures to be determined from neutron diffraction.

A comparison of magnetic and nuclear scattering amplitudes for a number of transition elements is given in Table 1 (Brown 1975). The forward-scattering values of the magnetic scattering amplitude can be readily calculated from Equation (6), given that $f(Q) = 1$ for $Q = 0$ (see section "Magnetic form factors"). For example, an Fe^{3+} ion with a spin-only J of 5/2 has a maximum magnetic moment of 5 μ_B (Eqn. 5), yielding a magnetic scattering amplitude of $5*0.2696 \times 10^{-12}$ cm $= 1.35 \times 10^{-12}$ cm. It is clear from Table 1 that the amplitude of magnetic scattering is generally of the same order of magnitude as the nuclear scattering at small Q. It is also clear from Table 1 that, unlike the nuclear scattering amplitude, the magnetic scattering amplitude varies significantly with the valence state of the atom.

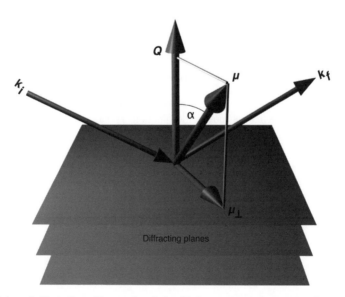

Figure 1. Schematic illustration of the angular relationship between scattering vector and magnetic interaction vector for magnetic neutron scattering. The incident and scattered neutron beams are described by the wave vectors k_i and k_f. The scattering vector, Q, is defined as $Q = k_f - k_i$. For Bragg diffraction, Q points normal to the set of diffracting planes. The magnetic moment of the atom is defined by the vector μ. The scattering amplitude is determined by the component of μ perpendicular to Q, labeled $\mu_\perp$ (Eqns. 6 and 8).

Table 1. Comparison of nuclear and magnetic scattering
amplitudes for selected elements (after Brown 1975).

Atom or ion	Nuclear scattering amplitude b (10^{-12} cm)	Effective spin quantum number S	Magnetic scattering amplitude p (10^{-12} cm)	
			$\theta = 0$	$\sin\theta/\lambda = 0.25\ \text{Å}^{-1}$
Cr^{2+}	0.35	2	1.08	0.45
Mn^{2+}	−0.37	5/2	1.35	0.57
Fe (metal)	0.96	1.11	0.6	0.35
Fe^{2+}	0.96	2	1.08	0.45
Fe^{3+}	0.96	5/2	1.35	0.57
Co (metal)	0.28	0.87	0.47	0.27
Co^{2+}	0.28	2.2	1.21	0.51
Ni (metal)	1.03	0.3	0.16	0.1
Ni^{2+}	1.03	1.0	0.54	0.23

Magnetic form factors

Neutrons are scattered by the magnetization density of an atom. Since the magnetic moment originates from the electrons, interference between neutrons scattered from different parts of the electron cloud causes the amplitude of magnetic scattering to decrease with increasing Q. This decrease is described by the magnetic form factor $f(\mathbf{Q})$ (Fig. 2). Unlike the nuclear contribution, which remains constant as a function of Q, the magnetic contribution to a neutron diffraction pattern is restricted to small Q.

$f(\mathbf{Q})$ is defined as the Fourier transform of the normalized unpaired spin density (Shirane et al. 2002):

$$f(\mathbf{Q}) = \int \rho_S(\mathbf{r}) e^{i\mathbf{Q}\cdot\mathbf{r}} d\mathbf{r} \qquad (9)$$

For a spin-only moment, Equation (9) can be expressed in the form:

$$f(\mathbf{Q}) = \int_0^\infty r^2 j_0(Qr) |\Phi(r)|^2 dr \equiv \langle j_0 \rangle \qquad (10)$$

where $j_n(Qr)$ is a spherical Bessel function of order n and $\Phi(r)$ is the radial wavefunction of the unpaired spin. For practical purposes, the function $\langle j_0 \rangle$ can be parameterized in the form (Brown 1995):

$$\langle j_0(s) \rangle = Ae^{-as^2} + Be^{-bs^2} + Ce^{-cs^2} + D \qquad (11)$$

where $s = \sin(\theta)/\lambda$ (in units of Å^{-1}) and A, a, etc., are constants. For transition elements with a small amount of unquenched orbital angular momentum, higher order terms are needed to describe $f(\mathbf{Q})$:

$$f(\mathbf{Q}) = \langle j_0 \rangle + \left(\frac{g-2}{2}\right)\langle j_2 \rangle \qquad (12)$$

where $\langle j_2 \rangle$ is defined in a similar way to Equation (10) and can be parameterized in the form (Brown 1995):

$$\langle j_2(s) \rangle = (Ae^{-as^2} + Be^{-bs^2} + Ce^{-cs^2} + D)s^2 \qquad (13)$$

For rare-earth elements with large orbital contributions to the magnetic moment, the form factor is given by (Shirane et al. 2002):

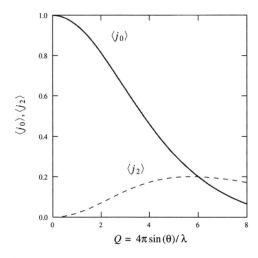

Figure 2. $<j_0>$ and $<j_2>$ components of the magnetic form factor for an Fe^{3+} ion as a function of scattering vector, Q (Brown 1995).

$$f(\mathbf{Q}) = \langle j_0 \rangle + \frac{g_L}{g} \langle j_2 \rangle \tag{14}$$

where:

$$g_L = \frac{J(J+1) + L(L+1) - S(S+1)}{2J(J+1)} \tag{15}$$

Values for the coefficients in Equations (11) and (13) are tabulated by Brown (1995) (see also *http://www.ill.fr/dif/ccsl/ffacts/ffachtml.html*). The form factors vary with the valence state of the ion. One should be aware that the default form factors used by Rietveld refinement programs such as GSAS are often for the neutral atom. Ionic values of the form factor coefficients may have to be entered manually.

The magnetic interaction vector

The nuclear contribution to the intensity of a Bragg peak, $I_N(\mathbf{Q})$, is defined by the nuclear structure factor, $F(\mathbf{Q})$ (Parise 2006, this volume):

$$I_N(\mathbf{Q}) = F(\mathbf{Q})F^*(\mathbf{Q}) \tag{16}$$

$$F(\mathbf{Q}) = \sum_{j=1}^{n_c} \bar{b}_j e^{i\mathbf{Q} \cdot \mathbf{r}_j} \tag{17}$$

where $F^*(\mathbf{Q})$ is the complex conjugate of $F(\mathbf{Q})$, $\bar{b}_j$ and $\mathbf{r}_j$ are the scattering length and position of the jth atom in the crystallographic unit cell, respectively, and n_c is the total number of atoms in the unit cell. The choice of scattering vector is restricted to the set of reciprocal lattice vectors, $\mathbf{H}$, where:

$$\mathbf{Q} = 2\pi \mathbf{H} \tag{18}$$

$$\mathbf{H} = h\,\mathbf{a}^* + k\,\mathbf{b}^* + l\,\mathbf{c}^* \tag{19}$$

The magnetic structure factor, $\mathbf{M}(\mathbf{Q})$, is defined as (Rodríguez-Carvajal 1993):

$$\mathbf{M}(\mathbf{Q}) = p \sum_{j=1}^{n_m} f_j(\mathbf{Q})\boldsymbol{\mu}_j e^{i\mathbf{Q} \cdot \mathbf{r}_j} \tag{20}$$

Note that $\mathbf{M(Q)}$ is a vector and that the sum involves only the n_m magnetic atoms in the unit cell, with moments μ_j in Bohr magnetons. The intensity of the Bragg reflection is determined by the magnetic interaction vector, $\mathbf{M_\perp(Q)}$, defined as the component of $\mathbf{M(Q)}$ perpendicular to $\mathbf{Q}$ (cf. Eqn. 8):

$$I_M(\mathbf{Q}) = \mathbf{M_\perp(Q)} \cdot \mathbf{M_\perp^*(Q)} \tag{21}$$

$$\mathbf{M_\perp(Q)} = \hat{\mathbf{Q}} \times \mathbf{M(Q)} \times \hat{\mathbf{Q}} \tag{22}$$

For collinear magnetic structures, the intensity of the magnetic Bragg peak can be written in the form:

$$M_\perp^2(\mathbf{Q}) = \sin^2(\alpha) M(\mathbf{Q})^2 \tag{23}$$

where α is the angle between the magnetic structure factor and the scattering vector (Fig. 1).

In general, the periodicity of the magnetic structure will be different from that of the nuclear structure. In this case, the scattering vector takes the form:

$$\mathbf{Q} = 2\pi (\mathbf{H} + \mathbf{k}) \tag{24}$$

where $\mathbf{k}$ is referred to as the propagation vector (the use of propagation vectors to describe magnetic structures is discussed in detail in the section "Classification and description of magnetic structures"). When the nuclear and magnetic unit cells are the same (i.e., $\mathbf{k} = 0$), the nuclear and magnetic Bragg peaks coincide. The total intensity for unpolarized neutrons is then simply the sum of nuclear and magnetic contributions:

$$I_{tot}(\mathbf{Q}) = F(\mathbf{Q})F^*(\mathbf{Q}) + \mathbf{M_\perp(Q)} \cdot \mathbf{M_\perp^*(Q)} \tag{25}$$

Domain and powder averaging

A single crystal of a magnetic material may contain a number of symmetry-related magnetic domains, each of which will be described by a different magnetic interaction vector. Similarly, in a powder diffraction experiment, the intensity of the (hkl) Bragg peak is obtained by summing contributions from all symmetry-related planes of the form $\{hkl\}$. Domain and powder averaging limits the amount of information that can be obtained about the crystallographic orientation of the magnetic moments.

For example, consider a tetragonal material with moments lying in the (001) plane (Fig. 3). We wish to determine the orientation of the moments within the (001) plane by measuring the intensity of the (110) Bragg peak. If the angle between the moment and the (110) scattering vector is α, then the intensity will be proportional to $\sin^2(\alpha)$ (Eqn. 23). However, the symmetry-related ($1\bar{1}0$) plane also contributes to the intensity of the (110) reflection in a powder diffraction experiment, yielding a total intensity proportional to $\sin^2(\alpha) + \cos^2(\alpha) = 1$. The observed intensity is independent of α, and, therefore, cannot be used to determine the orientation of moments within the (001) plane.

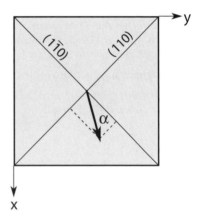

Figure 3. Illustration of the effect of powder averaging on the intensity of magnetic scattering in a tetragonal system. For the orientation of magnetic moments shown, scattering from the symmetry related (110) and ($1\bar{1}0$) planes is proportional to $\sin^2\alpha$ and $\cos^2\alpha$, respectively. The total intensity is, therefore, independent of α, and cannot be used to constrain the orientation of moments within the plane.

For the tetragonal system, it is only possible to constrain the angle between the moments and the z axis, since the symmetry-related (001) and (00$\bar{1}$) planes both have intensities proportional to $\sin^2(\alpha)$. The same restrictions apply to the hexagonal and rhombohedral systems. The problem of averaging is most severe in the cubic system, where an average value of $\sin^2(\alpha)$ = 2/3 is obtained by summing over all symmetry-related planes, irrespective of the orientation of the magnetic moments. Hence, no orientational information can be obtained from a powder diffraction experiment on a cubic material. Only in the orthorhombic, monoclinic, and triclinic systems, can all three direction cosines of a magnetic moment be uniquely determined. In a single crystal, the problems of domain averaging may be overcome by application of a suitably oriented magnetic field. In some cases, the ambiguities can be resolved with the use of single-crystal polarized neutron diffraction (see section "Polarized neutron diffraction").

CLASSIFICATION AND DESCRIPTION OF MAGNETIC STRUCTURES

The solution of magnetic structures using neutron diffraction is intrinsically linked to the formalism of how they are described using combinations of basis vectors and propagation vectors (Rodríguez-Carvajal 2001a,b). Knowledge of this formalism is essential for anyone wishing to use Rietveld refinement programs such as FullProf (Rodriguez-Carvajal 1990, 1993) and GSAS (Larson and Von Dreele 2000) for the analysis of magnetic neutron diffraction patterns. In the following section a brief introduction to the use of basis vectors and propagation vectors to describe simple and complex magnetic structures is presented. Group theory is an essential tool for determining the types of magnetic structures that may be adopted by a given material (Wills 2001a). The availability of free software for performing group theoretical analysis of magnetic structures now permits non-specialists to apply this technique to their problems. A worked example of how group theory can be used to predict the possible magnetic structures of the mineral hematite is provided.

Basis vectors

To describe a magnetic structure we need to specify the magnitude and orientation of the magnetic moment on each atom within the magnetic unit cell. This task becomes difficult when the magnetic unit cell is much larger than, or incommensurate with, the nuclear unit cell. The problem can be simplified using a general approach to the description of magnetic structures based on combinations of basis vectors and propagation vectors (Wills 2001a).

A basis vector, Ψ_j^k, is defined for each magnetic atom, j, within the *nuclear* unit cell. The *magnetic* unit cell is then generated by means of the propagation vector, $\mathbf{k}$. In the general case, the basis vectors are complex, consisting of real and imaginary parts:

$$\Psi_j^k = [u,v,w] + i[u',v',w'] \tag{26}$$

where $[u, v, w]$ and $[u', v', w']$ are vectors referred to the set of *crystallographic axes* (note: these are not necessarily orthogonal!). When a basis vector is real, u, v, and w correspond to the components of magnetic moment parallel to the three crystallographic axes. Complex basis vectors are used to describe helical magnetic structures.

The form of the basis vector for a given j and $\mathbf{k}$ is strictly constrained by the symmetry of the crystal structure, and can be calculated using representational analysis (i.e., the application of group theory to magnetic structures). Free software such as BasIreps (Rodríguez-Carvajal 2001b) and SARAh (Wills 2000) can be used to determine the symmetry-allowed forms of the basis vectors for a given crystal structure and propagation vector (see worked example later). There may be several symmetry-allowed basis vectors for a given atom, in which case Ψ_j^k is expressed as a linear combination of the individual basis vectors, $\Psi_{j,\nu}^k$:

$$\Psi_j^k = \sum_\nu C_{j,\nu}^k \psi_{j,\nu}^k \tag{27}$$

In a magnetic structure refinement, the coefficients, $C_{j,v}^{\mathbf{k}}$, are varied until the best fit is obtained between the observed and calculated diffraction patterns.

The propagation vector

The propagation vector, $\mathbf{k}$, describes how the magnetic moments in the *zero*[th] nuclear unit cell are propagated throughout the rest of the structure (Fig. 4). $\mathbf{k}$ is a vector in reciprocal space:

$$\mathbf{k} = h'\mathbf{a}* + k'\mathbf{b}* + l'\mathbf{c}* \tag{28}$$

where $\mathbf{a}*$, $\mathbf{b}*$, and $\mathbf{c}*$ are the principle reciprocal lattice vectors. The origin of the lth unit cell is specified by the real lattice vector:

$$\mathbf{t} = U\mathbf{a} + V\mathbf{b} + W\mathbf{c} \tag{29}$$

where $\mathbf{a}$, $\mathbf{b}$, and $\mathbf{c}$ are the unit cell parameters. The magnetic moment of atom j in unit cell l is then given by:

$$\boldsymbol{\mu}_{jl} = \boldsymbol{\Psi}_j^{\mathbf{k}} e^{-2\pi i \mathbf{k} \cdot \mathbf{t}} \tag{30}$$

where:

$$\mathbf{k} \cdot \mathbf{t} = h'U + k'V + l'W \tag{31}$$

The concept is illustrated in Figure 4 for the case of a simple antiferromagnet. The *zero*[th] nuclear unit cell (outlined in light grey in the upper left) contains one magnetic atom at its origin. The orientation of its magnetic moment is specified by the basis vector $\boldsymbol{\Psi} = [0\ 1\ 0]$. With

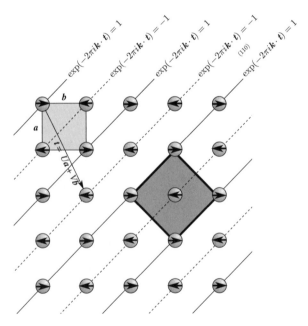

Basis Vector $\boldsymbol{\Psi} = [0\ 1\ 0]$
Propagation vector $\boldsymbol{k} = (1/2\ 1/2\ 0)$

Figure 4. Illustration of the basis vector/propagation vector formalism to describe a simple antiferromagnetic structure.

a propagation vector $\mathbf{k} = (\frac{1}{2} \frac{1}{2} 0)$, the term $\exp(-2\pi \mathbf{k} \cdot \mathbf{t})$ has a value of $+1$ for $U + V = 2n$ and -1 for $U + V = 2n + 1$, causing the moments on alternate (110) planes to point along $[0\ \bar{1}\ 0]$. This has the effect of doubling the periodicity of the magnetic structure in the direction of $\mathbf{k}$. The enlarged magnetic unit cell is outlined by the dark grey square.

Classifications of magnetic structures

Simple magnetic structures. A fundamental requirement of any magnetic structure is that the magnetic moments are real. Rewriting Equation (30) in its expanded form:

$$\mu_{jl} = \Psi_j^k \left[\cos(-2\pi \mathbf{k} \cdot \mathbf{t}) + i\sin(-2\pi \mathbf{k} \cdot \mathbf{t}) \right] \tag{32}$$

it can be seen that the simplest way of satisfying this requirement is to choose real basis vectors and a propagation vector such that the sine term in Equation (32) is zero. If the sine term is zero, the cosine term is either 1 or -1, implying that only the sign, and not the magnitude, of the magnetic moment on a given atom varies from cell to cell. An example of such a structure was given in Figure 4. A range of simple magnetic structures that can be described in this way is illustrated in Figure 5.

Complex magnetic structures. If the basis vectors are complex, or the propagation vector is chosen so that the sine term in Equation (32) is non-zero, the system must be described as the sum of $\mathbf{k}$ and $-\mathbf{k}$ structures in order to obtain real magnetic moments (Wills 2001a):

$$\mu_{jl} = \Psi_j^k \left[\cos(-2\pi \mathbf{k} \cdot \mathbf{t}) + i\sin(-2\pi \mathbf{k} \cdot \mathbf{t}) \right] + \Psi_j^{-k} \left[\cos(2\pi \mathbf{k} \cdot \mathbf{t}) + i\sin(2\pi \mathbf{k} \cdot \mathbf{t}) \right] \tag{33}$$

where:

$$\Psi_j^{-k} = \Psi_j^{k^*} = [u, v, w] - i[u', v', w'] \tag{34}$$

This then yields:

$$\mu_{jl} = 2[u, v, w]\cos(-2\pi \mathbf{k} \cdot \mathbf{t}) - 2[u', v', w']\sin(-2\pi \mathbf{k} \cdot \mathbf{t}) \tag{35}$$

A range of magnetic structures that can be described by Equation (35) is illustrated in Figure 6. If the basis vector is real, the sine term in Equation (35) disappears, and the structure consists of a cosine modulation in the amplitude of the magnetic moments (Fig. 6a). A complex basis vector can lead to helical magnetic structures of the form (Fig. 6b):

$$\mu_{jl} = 2[1,0,0]\cos(-2\pi \mathbf{k} \cdot \mathbf{t}) - 2[0,1,0]\sin(-2\pi \mathbf{k} \cdot \mathbf{t}) \tag{36}$$

whereby the magnetic moment rotates from layer to layer, describing a circle in the x-y plane. If the magnitude of the real and imaginary components of the basis vector are different, the structure becomes an elliptical helix. Addition of a constant component of magnetic moment parallel to the propagation vector leads to conical spiral (Fig. 6c). If the rotation of moments occurs in a plane parallel to the propagation vector, the result is a cycloidal spiral (Fig. 6d). Addition of a constant component of magnetic moment perpendicular to the propagation vector yields a conical cycloid (Fig. 6e).

More complicated magnetic structures can result when a number of differently oriented symmetry-related propagation vectors are involved (so-called multi-k structures). The diffraction patterns of multi-k structures are indistinguishable from those of a single-k structures with multiple domains. This ambiguity can be resolved by applying a magnetic field.

Example: representational analysis of hematite (Fe_2O_3)

To illustrate the use of basis vectors to describe magnetic structures, representational analysis is now used to determine the possible magnetic structures of the mineral hematite (Fe_2O_3). All calculations were performed using the program BasIreps (part of the FullProf

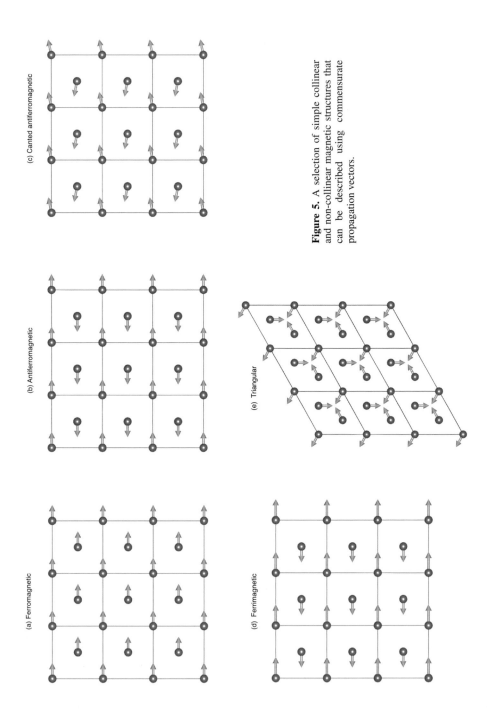

Figure 5. A selection of simple collinear and non-collinear magnetic structures that can be described using commensurate propagation vectors.

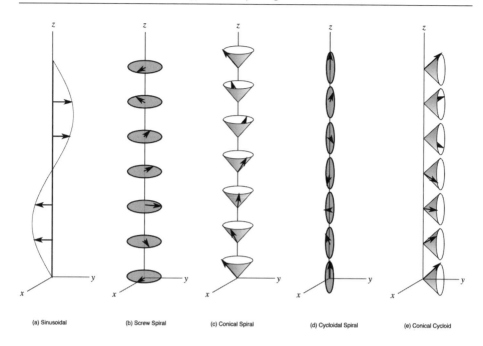

(a) Sinusoidal (b) Screw Spiral (c) Conical Spiral (d) Cycloidal Spiral (e) Conical Cycloid

Figure 6. A selection of complex magnetic structures that can be described using an incommensurate propagation vector (**k** // z).

suite of Rietveld refinement programs; Rodríguez-Carvajal 2001b). The group theory behind the calculations is described by Wills (2001a). Examples of the technique applied to the determination of magnetic structures from neutron diffraction data are given by Champion et al. (2001) and Wills (2001b,c).

Input data for the group theory calculations. To determine the symmetry-allowed forms of the basis vectors for a given material, you need only three pieces of information: i) the crystallographic space group of the paramagnetic phase (i.e., the space group of the material *before* magnetic ordering has taken place); ii) the propagation vector; and iii) the Wyckoff positions of the magnetic sites.

The choice of propagation vector is governed by the periodicity of the magnetic structure (see section "Magnetic neutron diffraction patterns"). Determining the propagation vector from single-crystal diffraction data is relatively straightforward. Programs such as SuperCell (part of the FullProf suite of Rietveld refinement programs; Rodríguez-Carvajal 1990) can help with the more difficult task of determining the propagation vector from powder diffraction data.

The crystallographic space group for hematite is $R\bar{3}c$ (the hexagonal setting will be used throughout). The magnetic unit cell is equal to the nuclear cell, so **k** = 0. Fe atoms occupy the 12c Wycoff site, which generates four Fe atoms per lattice point: Atom 1 at (0, 0, 0.3553), Atom 2 at (0, 0, 0.1447), Atom 3 at (0, 0, 0.6447), and Atom 4 at (0, 0, 0.8553). In addition to the lattice point at the origin of the unit cell, there are two more lattice points given by the centering translations **t** = (2/3, 1/3, 1/3) and **t** = (1/3, 2/3, 2/3), yielding a total 12 Fe atoms per unit cell. Since **k·t** = 0 for both centering translations, the basis vectors for atoms 1-4 are simply repeated for the other two sets of atoms.

The little group G_k. Representational analysis is performed on the subset of space group symmetry operations that leave the propagation vector invariant (i.e., those symmetry operations that map the propagation vector onto itself or onto a propagation vector that is simply translated by a reciprocal lattice vector). This subset is known as the little group, G_k. For $\mathbf{k} = 0$, all symmetry operations map the propagation onto itself, so G_k contains the same number of symmetry operations as the space group (12 for hematite). If the number of elements in G_k is less than in the space group, then the difference is made up by the generation of magnetic domains. For hematite, there are 6 irreducible representations (IRs) of the little group, labeled Γ_1 to Γ_6, four of which are one dimensional and two of which are two dimensional.

The magnetic representation. Operating the symmetry elements of G_k on the 4 atoms of Wycoff site 12c has two effects: the positions of the atoms within the set are interchanged, and the components of spin on each atom are transformed. The combined effect of interchanging atomic positions and transforming spin orientations is described by the magnetic representation, Γ (Wills 2001a). This reducible representation is decomposed into a combination of the 6 irreducible representations of G_k. For hematite, the magnetic representation is given by:

$$\Gamma = 1\Gamma_1^{(1)} + 1\Gamma_2^{(1)} + 1\Gamma_3^{(1)} + 1\Gamma_4^{(1)} + 1\Gamma_5^{(2)} + 1\Gamma_6^{(2)} \tag{37}$$

Each of the IRs appearing in Equation (37) will generate a corresponding set of permitted basis vectors (one set for each dimension of the IR and for each instance of the IR in Γ). For example, the one-dimensional IR Γ_1 appears once in Γ and will be associated with one set of basis vectors. The two-dimensional IR Γ_5 is contained twice in Γ and will be associated with 4 sets basis vectors. The full set of basis vectors is listed in Table 2.

Permitted basis vectors and possible magnetic structures. In general, a continuous magnetic phase transition is driven by a single active IR. The resulting magnetic structure is a linear combination of all the permitted basis vectors within the active IR (Eqn. 27). Γ_1 to Γ_4 are associated with a single set of real basis vectors, each of which generates a different permitted magnetic structure (Fig. 7a-d). For example, the basis vectors for Γ_1 are: Atom 1 [0 0 1];

Table 2. Basis vectors associated with the 6 irreducible representations contained within the magnetic representation of the hematite structure.

IR		Atom 1 (0, 0, 0.3553)	Atom 2 (0, 0, 0.1447)	Atom 3 (0, 0, 0.6447)	Atom 4 (0, 0, 0.8553)
Γ^1		[0 0 1]	[0 0 −1]	[0 0 1]	[0 0 −1]
Γ^2		[0 0 1]	[0 0 −1]	[0 0 −1]	[0 0 1]
Γ^3		[0 0 1]	[0 0 1]	[0 0 1]	[0 0 1]
Γ^4		[0 0 1]	[0 0 1]	[0 0 −1]	[0 0 −1]
Γ^5	v^1	Re: [1.5 0 0]	Re: [0 0 0]	Re: [−1.5 0 0]	Re: [0 0 0]
		Im: [−0.87 −1.73 0]	Im: [0 0 0]	Im: [0.87 1.73 0]	Im: [0 0 0]
	v^2	Re: [0 1.5 0]	Re: [0 0 0]	Re: [0 −1.5 0]	Re: [0 0 0]
		Im: [1.73 0.87 0]	Im: [0 0 0]	Im: [−1.73 −0.87 0]	Im: [0 0 0]
	v^3	Re: [0 0 0]	Re: [0 1.5 0]	Re: [0 0 0]	Re: [0 −1.5 0]
		Im: [0 0 0]	Im: [1.73 0.87 0]	Im: [0 0 0]	Im: [−1.73 −0.87 0]
	v^4	Re: [0 0 0]	Re: [1.5 0 0]	Re: [0 0 0]	Re: [−1.5 0 0]
		Im: [0 0 0]	Im: [−0.87 −1.73 0]	Im: [0 0 0]	Im: [0.87 1.73 0]
Γ^6	v^1	Re: [1.5 0 0]	Re: [0 0 0]	Re: [1.5 0 0]	Re: [0 0 0]
		Im: [−0.87 −1.73 0]	Im: [0 0 0]	Im: [−0.87 −1.73 0]	Im: [0 0 0]
	v^2	Re: [0 1.5 0]	Re: [0 0 0]	Re: [0 1.5 0]	Re: [0 0 0]
		Im: [1.73 0.87 0]	Im: [0 0 0]	Im: [1.73 0.87 0]	Im: [0 0 0]
	v^3	Re: [0 0 0]	Re: [0 1.5 0]	Re: [0 0 0]	Re: [0 1.5 0]
		Im: [0 0 0]	Im: [1.73 0.87 0]	Im: [0 0 0]	Im: [1.73 0.87 0]
	v^4	Re: [0 0 0]	Re: [1.5 0 0]	Re: [0 0 0]	Re: [1.5 0 0]
		Im: [0 0 0]	Im: [−0.87 −1.73 0]	Im: [0 0 0]	Im: [−0.87 −1.73 0]

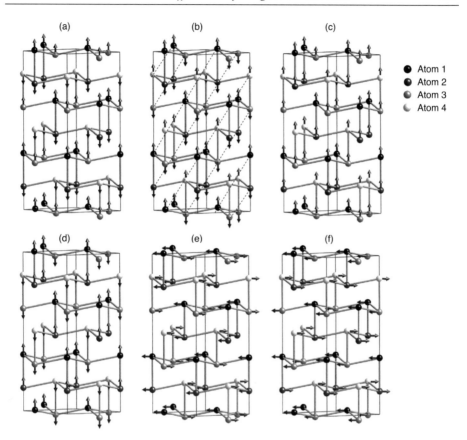

Figure 7. Range of symmetry-permitted magnetic structures for the mineral hematite with **k** = 0. (a) Structure corresponding to IR Γ_1 (Table 2). This is the observed magnetic structure of hematite below the Morin transition. (b) Structure corresponding to IR Γ_2. This is the observed magnetic structure of Cr_2O_3. Dashed lines indicate ferromagnetically coupled layers. (c) Structure corresponding to IR Γ_3. (d) Structure corresponding to IR Γ_4. (e) One possible structure based on a linear combination of basis vectors v_{1-4} of IR Γ_5. Symmetry dictates antiferromagnetic coupling of moments within the (006) layers. (f) One possible structure based on a linear combination of basis vectors v_{1-4} of IR Γ_6. Symmetry dictates ferromagnetic coupling of moments within the (006) layers. The structure shown is the observed magnetic structure of hematite above the Morin transition.

Atom 2 [0 0 −1]; Atom 3 [0 0 1]; Atom 4 [0 0 −1] (Fig. 7a). This corresponds to an antiferromagnetic structure, with magnetic moments pointing along either +z or −z. Examination of Figure 7a shows that the moments are arranged in layers parallel (006), with ferromagnetic ordering with respect to the moments within each layer and antiferromagnetic ordering with respect to the moments in adjacent layers. This corresponds to the observed magnetic structure of hematite at temperatures below 263 K (the Morin transition; Morrish 1994). The magnetic structure corresponding to Γ_2 (Fig. 7b) is also antiferromagnetic with moments parallel to ±z. However, the ordering with respect to moments in the (006) cation layers is now also antiferromagnetic. This corresponds to the observed magnetic structure in the isostructural compound Cr_2O_3. The magnetic structure corresponding to Γ_3 (Fig. 7c) is ferromagnetic, with all moments parallel to +z. The magnetic structure corresponding to Γ_4 (Fig. 7d) is similar to that in Figure 7b, except that there is now ferromagnetic, rather than antiferromagnetic, ordering with respect to the nearest-neighbor moments in adjacent (006) layers.

The basis vectors for Γ_5 and Γ_6 are complex, leading to magnetic structures of the form described by Equation (35). For $\mathbf{k} = 0$, however, the sine term in Equation (35) is zero, and we need only consider the real part of the basis vectors. Using Equation (27), the magnetic moment of each atom is written as as a linear combination of the 4 basis vectors for a given IR. For example, Γ_6 yields magnetic moments of the form:

$$\mu_1 = [u_1, v_1, 0] \tag{38}$$

$$\mu_2 = [u_2, v_2, 0] \tag{39}$$

$$\mu_3 = [u_1, v_1, 0] \tag{40}$$

$$\mu_4 = [u_2, v_2, 0] \tag{41}$$

where u_1, v_1, u_2, and v_2 are independent variables related to the coefficients $C_{j,v}^{\mathbf{k}}$ in Equation (27). Note that $\mu_1 = \mu_3$ and $\mu_2 = \mu_4$. These constraints enforce ferromagnetic coupling with respect to the moments within each (006) layer (Fig. 7f). By comparison, the basis vectors for Γ_5 enforce antiferromagnetic coupling with respect to moments within the (006) layers (Fig. 7e). All spins lie in the x-y plane, and are free to adopt any orientation within that plane. Symmetry does not place any restriction on the relative orientation of the moments in adjacent layers, which may be parallel to each other (ferromagnetic), antiparallel to each other (antiferromagnetic) or somewhere in between (e.g., canted antiferromagnetic). The observed magnetic structure of hematite above the Morin transition is canted antiferromagnetic (canting angle less than $0.1°$), with $u_1 \sim -u_2$ and $v_1 \sim -v_2$ (Fig. 7f). As discussed earlier, it is not possible to determine the direction of magnetic moments within the x-y plane using neutron diffraction due to the problem of domain and/or powder averaging. However it is easy to distinguish between the observed structures above and below the Morin transition by examining the intensity of the (003) Bragg peak (Corliss et al. 1954). The nuclear contribution to the (003) peak is zero (it is systematically absent due to the c-glide). Antiferromagnetic ordering of the type shown in Figure 7f destroys the c-glide, producing a strong magnetic (003) Bragg reflection. Antiferromagnetic ordering of the type shown in Figure 7a also destroys the c-glide, but since the magnetic moments all lie parallel to the (003) scattering vector, the magnetic interaction vector (Eqn. 22) is zero and the peak is disappears (Fig. 8).

Application to magnetic structure refinement from powder diffraction data. The ability to refine magnetic structures based on powder neutron diffraction data is built into many popular Rietveld refinement programs, such as FullProf and GSAS. The starting point for any magnetic structure refinement is a good starting model for the crystal structure of the material of interest. This is entered into the refinement software in the usual way. Methods for entering the magnetic structure differ significantly from program to program. FullProf, for example, uses the basis vector/propagation vector formalism described above to generate the magnetic unit cell from the nuclear unit cell. This is a very flexible approach, as it allows for both commensurate and incommensurate magnetic structures of arbitrary size and complexity to be generated in an efficient manner, while ensuring that the resulting

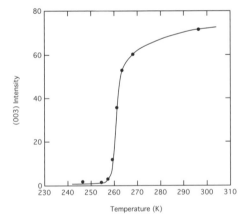

Figure 8. Plot of the intensity of the (003) magnetic reflection of hematite as a function of temperature (Corliss et al. 1954). Above the Morin transition (263 K), magnetic moments lie perpendicular to the (003) scattering vector (Fig. 7f), yielding a large magnetic interaction vector. Below the Morin transition, magnetic moments lie parallel to the (003) scattering vector, and the magnetic interaction vector is zero.

magnetic structure is fully compatible with the symmetry properties of the starting crystal structure. BasIreps produces a formatted output file containing all the basis vector information, suitable for direct pasting into the input file of a FullProf refinement.

Entering magnetic structures into GSAS is more tricky. There are several methods that can be used. The first alternative is to specify the magnetic structure in terms of its magnetic space group (or Shubnikov group). This method can be an efficient way of describing certain magnetic structures. However, there are some structures that cannot be described in this way (e.g., incommensurate structures). Magnetic ordering is driven by an active IR of the little group, G_k, which lowers the symmetry of the system. Without prior knowledge of what that lowered symmetry is, it is difficult to assign the correct Shubnikov group. For example, the structures illustrated in Figures 7e and 7f destroy the rhombohedral symmetry of the hematite structure, and could not be described in terms of one of the three Shubnikov groups associated with the crystallographic space group $R\bar{3}c$ The second alternative is to enter the nuclear and magnetic structures separately. The nuclear structure is entered in the normal way, but is only used to calculate the nuclear contribution to the diffraction pattern. The magnetic structure, containing only the magnetic atoms, is used to calculate the magnetic contribution to the diffraction pattern. It is usually assigned symmetry $P1$, and the direction of the magnetic moments on each atom are specified manually. This approach is unwieldy for large unit cells, and involves setting up a large number of constraints to maintain the desired crystallographic orientations of the moments and to ensure that the lattice parameters and atomic positions of the magnetic phase match those of the nuclear phase. A third alternative is to use an auxiliary program such as SARAh Refine (Wills 2000), that allows the basis vector/propagation vector formalism to be used to describe and refine the magnetic structure (via a simulated annealing algorithm), using GSAS simply as an engine to generate the calculated diffraction patterns.

MAGNETIC NEUTRON DIFFRACTION PATTERNS

Magnetic peaks and the propagation vector

The form of the single-crystal neutron diffraction pattern for a range of magnetic structures is illustrated schematically in Figure 9. Magnetic diffraction peaks occur at positions in reciprocal space that are displaced by $\mathbf{k}$ from the nuclear diffraction peaks (Eqn. 24). For $\mathbf{k} = 0$, the magnetic and nuclear unit cells are of the same size, and the magnetic and nuclear Bragg peaks coincide (Fig. 9a). This is always the case for ferromagnetic materials, but can also apply to antiferromagnetic and ferrimagnetic materials. In some special cases (e.g. hematite), pure magnetic reflections may be observed if there are systematic absences due to glide planes or screw axes that apply to the nuclear structure but not to the magnetic structure. If $\mathbf{k}$ is some commensurate fraction of a reciprocal lattice vector then the magnetic unit cell is a supercell of the nuclear unit cell, and the magnetic Bragg peaks appear as superlattice peaks inbetween the nuclear Bragg peaks (Fig. 9b). This is often the case for antiferromagnetic materials. If $\mathbf{k}$ is an incommensurate fraction of a reciprocal lattice vector, magnetic peaks appear as satellites surrounding the nuclear Bragg peaks (Fig. 9c). Note the presence of both $+\mathbf{k}$ and $-\mathbf{k}$ components (see section "Complex magnetic structures"). Complex magnetic structures involving multiple propagation vectors lead to satellite peaks in symmetry-related positions (Fig. 9d). Such diffraction patterns are equivalent to those of samples with single propagation vectors (e.g., Fig. 9c) containing multiple magnetic domains. The two cases can be distinguished by applying a magnetic field.

Examples

Magnetite (Fe₃O₄). Magnetite at room temperature has the cubic inverse spinel structure, with an Fe^{3+} ion occupying one tetrahedral site and Fe^{2+} and Fe^{3+} ions occupying two octahedral sites per formula unit. Néel (1948) proposed a ferrimagnetic structure for

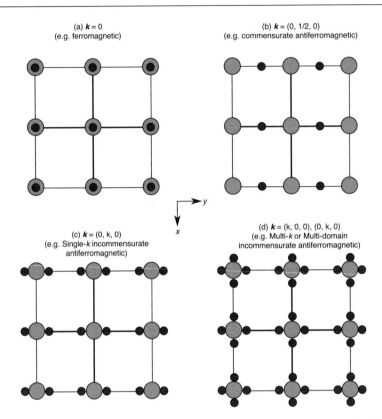

Figure 9. Schematic illustration of the relationship between nuclear (light grey) and magnetic (dark grey) diffraction peaks for a range of different magnetic structures. (a) For **k** = 0 structures, the nuclear and magnetic peaks coincide. (b) If the magnetic cell is a supercell of the nuclear structure, pure magnetic reflections appear as superlattice peaks in between the nuclear reflections. (c) Single-**k** incommensurate structures are characterized by the presence of satellite peaks at positions +**k** and −**k** from each nuclear reflection. (d) Multi-**k** structures and single-**k** structures with multiple domains yield identical diffraction patterns consisting of multiple sets of satellite peaks.

magnetite, with ferromagnetic ordering of moments within tetrahedral and octahedral sublattices, and antiferromagnetic ordering of the tetrahedral sublattice with respect to the octahedral sublattice. Since the tetrahedral and octahedral sites are not symmetry related to each other, this ordering scheme yields a magnetic unit cell that is equal to the nuclear unit cell (**k** = 0). Figure 10a shows a comparison of the powder X-ray and neutron diffraction patterns of magnetite at room temperature (Shull et al. 1951a). Neutron peaks occur in the same positions as the X-ray peaks due to the coincidence of magnetic and nuclear scattering contributions for **k** = 0 (Fig. 9a). The magnetic contribution to the neutron diffraction pattern is evident by the enhanced intensity of the (111) peak. Structure factor calculations indicate that the magnetic contribution to the intensity of this peak is a factor of 30 higher than the nuclear contribution. Figure 10b illustrates the effect of an applied magnetic field on the intensity of the (111) peak. In zero field, the intensity is given by $C^2 + 2/3D^2$, where C and D refer to the nuclear and magnetic structure factors, respectively, and the factor of 2/3 is the average value of $\sin^2(\alpha)$ for a cubic material (Eqn. 23; see section "Domain and powder averaging"). The intensity of the peak reduces to C^2 as the sample is gradually saturated in a direction parallel to the scattering vector. This is a direct consequence of Equation (22), which states that $\mathbf{M}_\perp$ is zero when the

(a)

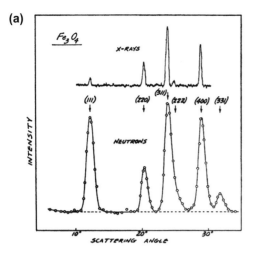

(b)

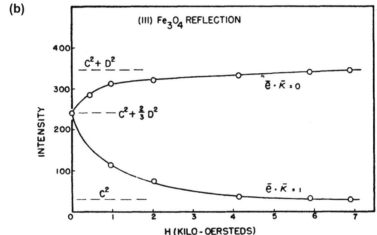

Figure 10. (a) Comparison between X-ray and neutron diffraction patterns of magnetite at room temperature (Shull et al. 1951a). The large difference in the (111) intensities is caused by the magnetic contribution to the neutron diffraction pattern. (b) Plot of the (111) neutron intensity as a function of magnetic field applied either perpendicular (upper curve) or parallel (lower curve) to the scattering vector. C^2 and D^2 refer to the nuclear and magnetic contributions to the peak, respectively. The magnetic contribution is reduced to zero when the sample is saturated parallel to the scattering vector.

magnetic moments lie parallel to **Q**. Conversely, the intensity rises to $C^2 + D^2$ when the sample is saturated in a direction perpendicular to **Q**. Such measurements have been used to study the mechanisms of magnetization reversal in multi-domain and single-domain magnetite powders (Plant 1982). Other studies of magnetite using neutron diffraction include investigations of the crystal and magnetic structure below the Verwey transition (Wright et al. 2000, 2001, 2002; Yang et al. 2004), determination of spin-wave dispersion curves (Brockhouse 1957; Watanabe and Brockhouse 1962; Alperin et al. 1967; McQueeney et al. 2005), studies of magnetite thin films using polarized neutron reflectometry (Morrall et al. 2003), studies of the effect of Zn-doping on the Verwey transition (Kozlowski et al. 1999), and small-angle neutron scattering of magnetite from magnetotactic bacteria (Krueger et al. 1990).

MnO. MnO has the cubic NaCl structure with cell parameter 4.43 Å. Neutron diffraction patterns acquired at temperatures above and below the magnetic ordering temperature are shown in Figure 11a (Shull and Smart 1949). Superlattice reflections appear below the Néel temperature, which can be indexed using a magnetic unit cell that is doubled in size along all three crystallographic axes. The magnetic structure has ferromagnetic ordering with respect to the moments within each (111) plane, and antiferromagnetic ordering with respect to the

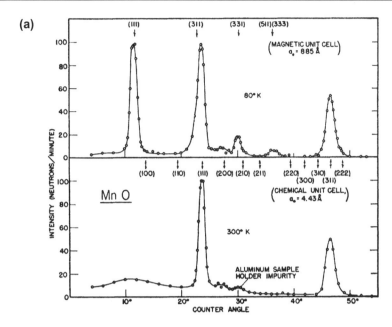

(a)

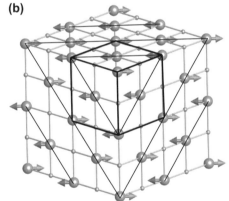

(b)

Figure 11. (a) Comparison of the neutron diffraction pattern of MnO above and below its Néel temperature of 120 K (Shull et al. 1951b). Diffuse scattering around the (½ ½ ½) position of the room temperature nuclear structure becomes a coherent magnetic superlattice peak at low temperature. The low-temperature diffraction pattern can be indexed on a cubic unit cell that is double the size of the nuclear unit cell. (b) Magnetic structure of MnO. There is ferromagnetic coupling with respect to moments within individual (111) layers and antiferromagnetic coupling between moments in adjacent (111) layers, leading to a doubling of the periodicity in the [111] direction. Moments lie parallel to the (111) layers.

moments in adjacent (111) planes. This leads to doubling of the periodicity in the (111) direction, which can be described by a propagation vector of the form $\mathbf{k} = (½, ½, ½)$. Three other symmetry-related $\mathbf{k}$ vectors could have been chosen, each corresponding to a different domain orientation.

Fe_2O_3-Cr_2O_3. Fe_2O_3 and Cr_2O_3 are isostructural, and there is complete solid solution between the two endmembers. Despite the similar size and identical charge of Fe^{3+} and Cr^{3+}, the solid solution displays non-ideal behavior due to the presence of competing magnetic interactions. In Fe_2O_3, all spins lie parallel to the (006) layers. The (006) intralayer superexchange interactions are positive and the interlayer interactions are negative, leading to the magnetic structure shown in Figure 7f (Shull et al. 1951b). In Cr_2O_3, all spins lie perpendicular to the (006) layers and the sign of the dominant intra- and interlayer superexchange interactions are interchanged, leading to the magnetic structure shown in Figure 7b. The different exchange interactions are caused by the different electronic configurations of

Cr^{3+} and Fe^{3+} (Cr^{3+} has a less-than-half-filled $3d$ shell, whereas Fe^{3+} has a more-than-half-filled shell). The incompatibility of the Fe_2O_3 and Cr_2O_3 magnetic structures can be seen from the fact that the Néel temperature of each phase decreases with increasing substitution of the other phase, reaching a minimum at a composition of $20\%Fe_2O_3$-$80\%Cr_2O_3$ (Fig. 12).

Powder neutron diffraction patterns measured at 77 K are shown in Figure 13 for a range of compositions spanning the Néel temperature minimum (Cox et al. 1963). There are three prominent Bragg reflections in this 2θ range: (003), (101), and (012), The (003) and (101) peaks are purely magnetic in origin, whereas the (012) peak contains both magnetic and nuclear contributions. In order to remove the nuclear contribution to the (012) peak, the corresponding room-temperature (paramagnetic) diffraction pattern has been subtracted from each low-temperature diffraction pattern. The different magnetic structures of endmember Fe_2O_3 and Cr_2O_3 can be readily distinguished by their different magnetic structure factors: for Fe_2O_3 the (003) and (101) peaks are strong and the (012) is absent; for Cr_2O_3 the (003) and (101) peaks are absent and the (012) peak is strong. For intermediate compositions, satellite peaks appear on either side of the (012) nuclear peak position, indicating the presence of an incommensurate component to the magnetic structure. For a composition $85\%Cr_2O_3$-15% Fe_2O_3, the satellites surround a small (012) magnetic peak. Cox et al. (1963) demonstrated that the magnetic structure in this case is the sum of two components: a fundamental component equivalent to that of the Cr_2O_3 endmember and an incommensurate component with **k** parallel to z. This yields a conical spiral structure (e.g., Fig. 6c), in which the cone axis is parallel to the propagation vector (Fig. 14a). The spacing of the satellite peaks indicates a spiral periodicity of 80 Å. The cone semi-angle can be determined from the relative intensities of the satellite and fundamental peaks. As the fundamental peak intensity tends to zero, the cone semi-angle

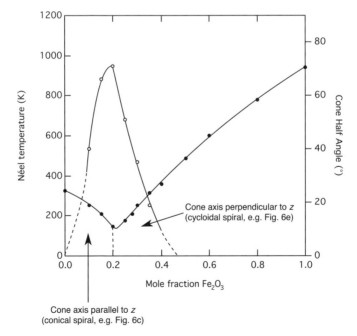

Figure 12. Plot of Néel temperature (solid circles) and cone semi-angle (open circles) as a function of composition in the solid solution Cr_2O_3-Fe_2O_3 (Cox et al. 1963). The magnetic structures of the endmembers are incompatible with each other, leading to non-ideal mixing and the occurrence of incommensurate helical magnetic structures in the solid solution.

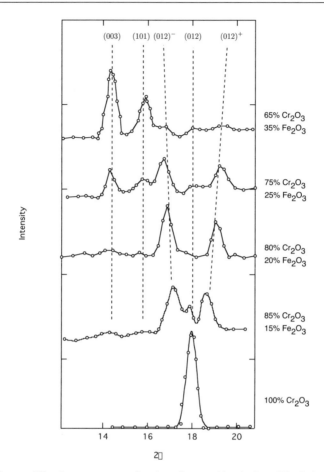

Figure 13. Neutron diffraction patterns as a function of composition in the solid solution Cr_2O_3-Fe_2O_3 (Cox et al. 1963). The diffraction pattern of pure Cr_2O_3 consists of a single (012) peak, which contains both nuclear and magnetic contributions. The nuclear contribution has been removed by subtracting the room temperature paramagnetic diffraction pattern. The diffraction pattern of pure Fe_2O_3 (not shown) consists of (003) and (101) magnetic peaks only. The diffraction patterns of intermediate compositions contain a fundamental contribution from either the Cr_2O_3-like magnetic structure (for $x < 20\%$ Fe_2O_3) or the Fe_2O_3-like magnetic structure (for $x > 20\%$ Fe_2O_3) plus two satellite peaks (012)⁺ and (012)⁻. The relative intensity of the fundamental and satellite components can be used to determine the cone semi-angle of the corresponding conical spiral and conical cycloid structures (Figs. 12 and 14).

tends towards 90° and the structure becomes a screw spiral (Fig. 6b). The variation in the cone semi-angle with composition is shown in Figure 12 (open circles). The rapid decrease in the intensity of the (012) peak with increasing composition corresponds to a rapid increase in the cone semi-angle, which reaches a maximum for a composition of 80%Cr_2O_3-20% Fe_2O_3. For a composition 75%Cr_2O_3-25% Fe_2O_3, the fundamental component switches from (012) to (003) and (101), indicating a switch in magnetic structure from a Cr_2O_3-like conical spiral to an Fe_2O_3-like conical cycloid (e.g., Fig. 6e), with the cone axis now perpendicular to the propagation vector (Fig. 14b). The intensity of the (003) and (101) peaks increases with increasing Fe_2O_3 content. This corresponds to a decrease in the cone semi-angle and a gradual transition to endmember Fe_2O_3 structure.

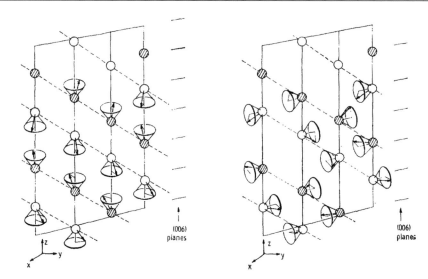

Figure 14. (a) Conical spiral structure adopted in the solid solution Cr_2O_3-Fe_2O_3 for compositions < 20% Fe_2O_3. The cone axes are oriented parallel and antiparallel to the z axis (equivalent to the orientation of moments in the endmember Cr_2O_3 magnetic structure). Dashed lines indicate the planes of ferromagnetically coupled spins (cf. dashed lines in Fig. 7b). (b) Conical cycloid structure adopted in the solid solution Cr_2O_3-Fe_2O_3 for compositions > 20% Fe_2O_3. The cone axes are oriented perpendicular to the z axis (equivalent to the orientation of moments in the endmember Fe_2O_3 magnetic structure).

POLARIZED NEUTRON DIFFRACTION

The main disadvantage of using unpolarized neutrons to study magnetic materials is that it is often difficult to separate out the nuclear and magnetic contributions to the diffraction signal. If measurements are made above and below the Néel temperature, then it may be possible to remove the nuclear contribution by subtracting the diffraction signal of the paramagnetic phase (e.g., Fig. 13). This may cause problems, however, due to the different lattice parameters and Debye-Waller factors for measurements made at different temperatures. A more reliable approach is to use polarized neutrons. By determining the change in the polarization state of the scattered beam relative to the incident beam (polarization analysis), the magnetic and nuclear contributions can be fully separated at any temperature. Polarization analysis can yield information about the magnetic interaction vector that cannot be obtained using unpolarized neutrons. For some complex magnetic structures, polarization analysis provides the only method to determine the magnetic structure unambiguously (Brown 1993).

Theory

The polarization of the incident neutron beam, $\mathbf{P}_i$, is given by summing over the polarization vectors of the individual neutrons, $\mathbf{P}_j$:

$$\mathbf{P}_i = \frac{1}{N}\sum_j \mathbf{P}_j \tag{42}$$

where N is the total number of neutrons in the beam. The magnitude of the polarization is given by:

$$P_i = \frac{N_+ - N_-}{N_+ + N_-} \tag{43}$$

where N_+ is the number of neutrons with spin up and N_- is the number of neutrons with spin down. Hence, $P_i = 0$ for an unpolarized beam ($N_+ = N_-$), and $P_i = 1$ for a fully polarized beam ($N_+ = 1$). The total scattering cross section for a polarized beam is given by (Roessli and Böni 2002):

$$\sigma = FF^* + \mathbf{M}_\perp \cdot \mathbf{M}_\perp^* + P_i(\mathbf{M}_\perp F^* + \mathbf{M}_\perp^* F) + iP_i(\mathbf{M}_\perp^* \times \mathbf{M}_\perp) \tag{44}$$

For an unpolarized neutron beam, the last two terms in Equation (44) are zero and the cross section reduces to the independent sum of the nuclear and magnetic contributions (cf. Eqn. 25). The third term in Equation (44) is referred to as the magnetic-nuclear interference term (Halpern and Johnson 1939), and is non-zero only for reflections that have both a nuclear and a magnetic scattering contribution. The fourth term in Equation (44) is referred to as the chiral term, and is non-zero when $\mathbf{M}_\perp$ is not parallel to $\mathbf{M}^*_\perp$. This is typically the case for helical structures.

The polarization of the scattered beam, $\mathbf{P}_f$, is given by (Roessli and Böni 2002):

$$\mathbf{P}_f\sigma = \mathbf{P}_i FF^* + (-1)\mathbf{P}_i(\mathbf{M}_\perp \cdot \mathbf{M}_\perp^*) + \mathbf{M}_\perp(\mathbf{P}_i \cdot \mathbf{M}_\perp^*) + \mathbf{M}_\perp^*(\mathbf{P}_i \cdot \mathbf{M}_\perp)$$
$$+ \mathbf{M}_\perp F^* + \mathbf{M}_\perp^* F + i(\mathbf{M}_\perp F^* - \mathbf{M}_\perp^* F) \times \mathbf{P}_i + i\mathbf{M}_\perp \times \mathbf{M}_\perp^* \tag{45}$$

The most advanced form of polarization analysis is spherical neutron polarimetry, which allows all three components of the scattered beam polarization to be measured. For many complex magnetic structures, this is the only technique that allows unambiguous determination of the magnetic interaction vector (Brown 1993). Most polarized neutron experiments, however, rely on longitudinal, or uniaxial, polarimetry, where only the vertical component of polarization (i.e., spin up or spin down) is measured (Moon et al. 1969). The neutron beam is polarized in the z direction (see section "Obtaining a polarized neutron beam"). The measured intensity is divided into two channels: the non-spin-flip (nsf) channel, containing those neutrons whose spin is preserved during scattering, and the spin-flip (sf) channel, containing those neutrons whose spin is flipped during scattering. The ratio of the nsf to sf intensity is referred to as the flipping ratio. Whether or not a neutron spin is flipped depends on the relative orientation of $\mathbf{M}_\perp$ and $\mathbf{P}_i$. If $\mathbf{M}_\perp$ is parallel to $\mathbf{P}_i$, precession of the neutron spin around $\mathbf{M}_\perp$ preserves the z component (nsf scattering; Fig. 15a). If $\mathbf{M}_\perp$ is perpendicular to $\mathbf{P}_i$, precession of the neutron spin around $\mathbf{M}_\perp$ reverses the z component (sf scattering; Fig. 15b). The cross section for nsf scattering is, therefore, only sensitive to the z component of $\mathbf{M}_\perp$:

$$\sigma^{++} = \left| F + M_{\perp,z} \right|^2 \tag{46}$$

$$\sigma^{--} = \left| F - M_{\perp,z} \right|^2 \tag{47}$$

where $++$ ($--$) refers to neutrons that are spin up (spin down) in both the incident and scattered beams. Note that the nuclear contribution to the scattering cross section, F, also appears in the nsf channel. The cross section for sf scattering is sensitive to the x and y components of $\mathbf{M}_\perp$:

$$\sigma^{+-} = \left| M_{\perp,x} + iM_{\perp,y} \right|^2 \tag{48}$$

$$\sigma^{-+} = \left| M_{\perp,x} - iM_{\perp,y} \right|^2 \tag{49}$$

where $+-$ ($-+$) refers to neutrons that are spin up (spin down) in the incident beam and spin down (spin up) in the scattered beam. If $\mathbf{P}_i$ is chosen to be parallel to $\mathbf{Q}$ (Fig. 15c) then $M_{\perp,z} = 0$ and the nsf channel contains only nuclear scattering contributions and the sf channel contains only magnetic contributions. This provides an efficient method of separating magnetic and nuclear scattering contributions to a diffraction pattern—one of the main applications of polarization analysis. If $\mathbf{P}_i$ is chosen to be perpendicular to $\mathbf{Q}$ (Fig. 15d) then the ratio of nsf

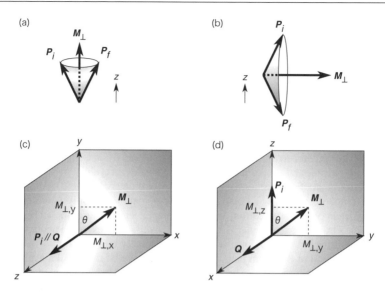

Figure 15. (a) Classical precession of a neutron spin around a subparallel magnetic interaction vector leads to conservation of its z component (i.e., non spin-flip scattering). (b) Classical precession of a neutron spin about a perpendicular magnetic interaction vector leads to reversal of its z component (i.e., spin flip scattering). (c) and (d) Two different geometries for longitudinal polarization analysis. The polarization of the incident beam, $\mathbf{P}_i$, is along z (note the different labeling of axes in c and d). The scattering vector, $\mathbf{Q}$, lies in the horizontal scattering plane. In (c) $\mathbf{P}_i$ is parallel to $\mathbf{Q}$. In this case $\mathbf{M}_\perp$ is perpendicular to $\mathbf{P}_i$, and all magnetic scattering is spin flip. In (d) $\mathbf{P}_i$ is perpendicular to $\mathbf{Q}$. In this case, $\mathbf{M}_\perp$ lies in the x-z plane, and the ratio of spin-flip to non-spin-flip scattering depends on the angle θ.

to sf scattering varies with the orientation of $\mathbf{M}_\perp$ in the y-z plane. For a pure magnetic peak, the nsf and sf intensities will be proportional to $\cos^2\theta$ and $\sin^2\theta$, respectively, allowing the orientation of $\mathbf{M}_\perp$ to be determined.

Experimental details

Obtaining a polarized neutron beam. The simplest method to produce a polarized neutron beam is to diffract the unpolarized beam from a single-crystal of a centrosymmetric ferromagnetic material (Shull et al. 1951a). The crystal is saturated in a direction perpendicular to the scattering vector by applying a suitably large magnetic field. From Equation (44) and (45) with $\mathbf{P}_i = 0$, $F = F^*$, and $\mathbf{M}_\perp = \mathbf{M}^*_\perp$, it follows that:

$$P_f = \frac{2FM_\perp}{\sigma} = \frac{2FM_\perp}{F^2 + M_\perp^2} \tag{50}$$

Hence the diffracted beam can be completely polarized ($P_f = 1$) if a reflection is chosen such that $F = M_\perp$. The first reflection to be used in this way was the (220) reflection of magnetite (Nathans et al. 1959). Those in common usage today include the (111) reflection of Heusler alloy (Cu_2MnAl) and the (200) reflection of $Co_{0.92}Fe_{0.08}$.

An alternative approach relies on the total external reflection of the neutron beam from a magnetized thin film. The critical angle for total external reflection varies with the orientation of the neutron spin relative to the magnetization of the thin film. If the angle of incidence is chosen to be greater than the critical angle for spin-down neutrons but less than the critical angle for spin-up neutrons, then only spin-up neutrons will be reflected. The critical angle can be optimized by the use of multilayer "supermirrors," comprising magnetized thin films separated by non-magnetic spacer layers (Mezei 1976). Common combinations are Co/Ti, Fe/Si, and $Fe_{50}Co_{48}V_2/TiN_x$. Many of these supermirrors have square hysteresis loops with a large remanent magnetization, so that the magnetization state of the mirror can be maintained without the need to apply a constant magnetic field (Stahn 2004). The magnetization state of the mirror can be switched in order to select either spin up or spin down neutrons, removing the need for an additional radio frequency flipper. The use of polarized ^3He gas, which selectively absorbs neutrons of a given spin, is also becoming increasingly popular. For a more detailed review of the techniques used to produce polarized neutron beams, the reader is referred to Anderson et al. (2000).

Typical experimental setup. Longitudinal polarization analysis is most commonly used in conjunction with a triple axis spectrometer, as pioneered by Moon et al. (1969). The experimental setup differs from that of a conventional triple axis spectrometer (Loong 2006, this volume) in several important aspects (Fig. 16). Polarization analysis requires not only a polarized beam but the ability to detect only those neutrons with a given spin. This requires the addition of both a polarizing monochromator and a polarizing analyzer. In order to preserve the polarization of the neutrons as they travel through the instrument, it is necessary to apply a guide magnetic field along the entire length of the beam path. This can be achieved, for example, using a series of closely spaced permanent magnets. The guide field at the sample position is often maintained using Helmholtz coils. Any inhomogeneity in the guide field can lead to depolarization of the beam, and care must be taken that any magnetic field applied to the sample is in the same direction as the guide field. In order to determine the σ^{++}, σ^{--}, σ^{+-}, and σ^{-+} cross sections, it is necessary to be able to flip the polarization of the incident and scattered beams relative to the polarization of the monochromator and analyzer. This can be achieved using radio-frequency spin flippers, which use the precession of the neutron spin around a radio-frequency magnetic field (cf. Fig. 15b) to reverse the spin of any incoming neutron.

Correcting the data for the flipping ratio of the incident beam. No polarizer is 100% efficient. The polarization of the incident beam is affected by many factors, such as the energy of the neutrons, the homogeneity of the guide field, and the presence of sample environment equipment (e.g., cryostats, furnaces, magnets, etc.). To get quantitative data for the nsf and sf

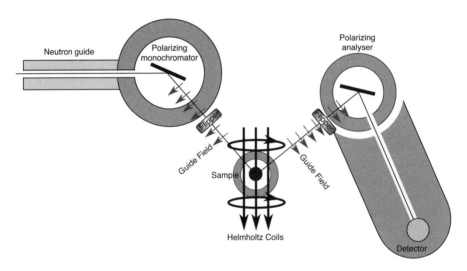

Figure 16. Schematic illustration of a triple axis spectrometer set up for longitudinal polarization analysis (Moon et al. 1969).

scattering intensities, it is necessary to correct for the polarization state of the incident beam. One way to do this is to measure the flipping ratio for a pure nuclear peak. As discussed above, nuclear scattering is entirely nsf. However, since the incident beam contains a fraction of neutrons with the opposite spin, this leads to an apparent sf component to the peak (Fig. 17a). The observed nsf and sf intensities are related to the true nsf and sf cross sections via:

$$\begin{pmatrix} I_{nsf} \\ I_{sf} \end{pmatrix} = \begin{pmatrix} N_+ & N_- \\ N_- & N_+ \end{pmatrix} \begin{pmatrix} \sigma_{nsf} \\ \sigma_{sf} \end{pmatrix} \tag{51}$$

For a pure nuclear peak, $\sigma_{sf} = 0$, so that the observed flipping ratio R is given by:

$$R = \frac{I_{nsf}}{I_{sf}} = \frac{N_+}{N_-} = \frac{1+P_i}{1-P_i} \tag{52}$$

where P_i is given by Equation (43) with $N_+ + N_- = 1$. Rearranging Equation (52) yields:

$$P_i = \frac{R-1}{R+1} \tag{53}$$

Inverting Equation (51) yields the true scattering cross sections:

$$\begin{pmatrix} \sigma_{nsf} \\ \sigma_{sf} \end{pmatrix} = \frac{1}{P_i} \begin{pmatrix} \dfrac{P_i+1}{2} & \dfrac{P_i-1}{2} \\ \dfrac{P_i-1}{2} & \dfrac{P_i+1}{2} \end{pmatrix} \begin{pmatrix} I_{nsf} \\ I_{sf} \end{pmatrix} \tag{54}$$

For example, the ratio of the intensity of nsf and sf contributions to the nuclear peak in Figure 17a is $R = 5.45$, corresponding to polarization of $P_i = 0.69$. Substituting this value into Equation (54) yields the corrected intensities shown in Figure 17b.

Example: nanoscale hematite exsolution in ilmenite

The magnetic properties of the ilmenite-hematite solid solution are profoundly influenced by nanoscale microstructures associated with subsolvus exsolution and cation ordering. Slowly

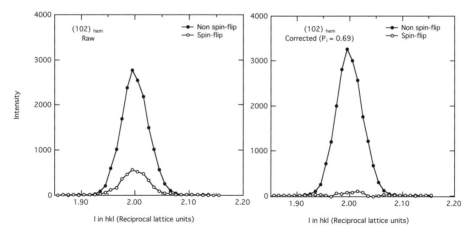

Figure 17. (a) Measurement of the non spin-flip (solid) and spin-flip (open) intensity profiles across the (102) nuclear peak in a natural intergrowth of ilmenite and hematite. The presence of spin-flip intensity for a pure nuclear peak is due to the imperfect polarization of the incident beam ($P_i = 0.69$). (b) The same data after correcting using Equation (54).

cooled rocks containing finely exsolved members of the hematite-ilmenite series have strong and extremely stable magnetic remanence, suggesting an explanation for some magnetic anomalies in the deep crust and on planetary bodies that no longer retain a magnetic field, such as Mars (McEnroe et al. 2001, 2002, 2004a,b,c; Kasama et al. 2004). This remanence has been attributed to a stable ferrimagnetic substructure originating from the coherent interface between nanoscale ilmenite and hematite exsolution lamellae (the so-called "lamellar magnetism hypothesis;" Harrison and Becker 2001; Robinson et al. 2002 and 2004). The characteristic nanoscale exsolution microstructure of a natural sample with bulk composition 84%Fe-TiO$_3$-16%Fe$_2$O$_3$ is shown in Figure 18. Nanoscale precipitates of hematite appear as thin white lines surrounded by dark 'strain shadows' in Figure 18b.

A single crystal extracted from the same sample was investigated using

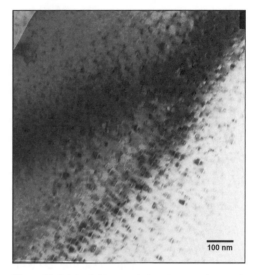

Figure 18. Bright-field transmission electron micrograph of a natural intergrowth of nanometer-scale hematite precipitates (bright lines surrounded by dark shadows) within a matrix of ilmenite. A different crystal from the same sample was used for the neutron diffraction work (Figs. 17 and 19).

longitudinal polarization analysis on the TASP triple-axis spectrometer at the Paul Scherrer Institute, Zürich. The crystallographic orientation of the sample was determined using electron backscattered diffraction (Robinson et al. 2006). The orientation of the natural magnetic remanence (NRM) carried by the sample was determined using a 3-axis, 2G cryogenic magnetometer, and found to lie along a <100> direction (Robinson et al. 2006). The sample was mounted in the geometry shown in Figure 15d, with the crystallographic *c* axis pointing along the $+x$ direction and the NRM pointing along the $-z$ direction. For polarization analysis, the sample was exposed to a guide field of 0.05 T in a direction parallel to the NRM using Helmholtz coils. The flipping ratio of the incident beam was determined by scanning the (102) nuclear peak (Fig. 17).

Scans across the (003) peak were recorded at 280 K and in fields of 0.05 T and 2 T (Fig. 19). Although the resolution of the spectrometer is relatively poor, the use of polarization analysis enables the diffraction signals from ilmenite and hematite to be separated. Ilmenite has the same rhombohedral crystal structure as hematite, except that the Fe and Ti cations are ordered onto alternating (006) layers (Harrison et al. 2000a,b; Harrison and Redfern 2001; Harrison 2006). This cation ordering lowers the symmetry of the structure from $R\bar{3}c$ to $R\bar{3}$. Consequently, there is a strong nuclear contribution to the (003) peak from ilmenite, whereas the nuclear contribution is systematically absent from hematite due to the presence of the *c*-glide plane. Ilmenite is paramagnetic at 280 K, and contributes only to the nsf intensity (closed circles in Fig. 19). Hematite is magnetic at 280 K, and in the chosen experimental geometry (Fig. 15d), can contribute to both the nsf and sf intensities. The ratio of nsf to sf scattering varies with the orientation of spins within the basal plane (i.e., the angle θ in Fig. 15d).

Figure 19a was recorded in the guide field of 0.05 T. The expected positions of the ilmenite and hematite contributions—based on the lattice parameters determined by X-ray diffraction—are indicated by the arrows. There is a broad peak in the nsf intensity that

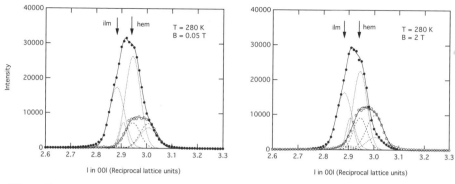

Figure 19. Non spin-flip (closed circles) and spin-flip (open circles) intensity across the (003) peak in a natural intergrowth of ilmenite and hematite. Measurements were made at 280 K and in applied magnetic fields of (a) 0.05 T and (b) 2 T. Dotted and dashed curves show the four fitted Gaussian contributions to the non spin-flip and spin-flip intensities, respectively. Black arrows indicate the expected positions of the dominant ilmenite and hematite reflections, based on lattice parameters determined via X-ray diffraction.

encompasses both the ilmenite and hematite positions. The peak in the sf intensity, however, is centered on the hematite position. Four Gaussian peaks are required to describe each profile: two associated with the ilmenite phase and two associated with the hematite phase. The nsf and sf profiles were fitted simultaneously, such that for each Gaussian contribution to the nsf signal there is a corresponding contribution with the exact same position and width to the sf signal. The presence of multiple contributions from each phase is most likely due to the presence of multiple generations of exsolution lamellae in the sample (Fig. 18; see discussion below). The results of the fitting are shown as the dashed lines in Figure 19 and summarized in Table 3.

As expected, the sf contributions to the two ilmenite peaks are within error of being zero, consistent with the nuclear origin of the (003) ilmenite peak. The two hematite peaks have very similar sf scattering components but very different nsf scattering components. Hematite Peak 2 (see Table 3 for definitions) has nsf and sf scattering components of roughly equal magnitude. This implies that the magnetic interaction vector makes an angle of $\theta = 45°$ to the z axis. A more likely interpretation is that the region of the sample responsible for Hematite Peak 2 is demagnetized. The sample can be magnetized by applying a magnetic field (Nathans et al. 1964), causing a significant change in the magnetic scattering (Fig. 19b). With a saturating field of 2 T applied along the $-z$ direction, the nsf component of Hematite

Table 3. Gaussian peak fitting parameters for the nsf and sf signals in Fig. 19a,b.

Fit parameters*		Ilmenite 1	Ilmenite 2	Hematite 1	Hematite 2
0.05 T	x_0	2.88	2.908	2.946	3.01
	w	0.043	0.02	0.041	0.047
	A_{nsf}	17514	7379.42	26362	5884
	A_{sf}	398	326	7265	7096
2 T	x_0	2.88	2.908	2.946	3.00
	w	0.044	0.02	0.037	0.047
	A_{nsf}	16415	9245	22621	3624
	A_{sf}	1054	1128	9242	9942

*Gaussian is of the form $I = A \exp\left[-\left(\dfrac{x - x_0}{w}\right)^2\right]$

Peak 2 decreases and the sf component increases. The ratio of sf to nsf intensity yields $\theta \sim 60°$. This is intermediate between the values expected for spin canting ($\theta = 90°$) and lamellar magnetism ($\theta = 0°$), suggesting that the net moment is a vector sum of the two contributions (Robinson et al. 2006). Hematite Peak 1 is dominated by a large nsf component, implying that the corresponding region of the sample is strongly magnetized. The signal is not strongly changed by applying a magnetic field, implying that the magnetization is much harder than that associated with Peak 2. The ratio of sf to nsf intensity in a field of 0.05 T yields $\theta \sim 28°$. In a field of 2 T, this angle changes only slightly to $\theta \sim 33°$. This is also intermediate between the values expected for canting and lamellar magnetism, but much closer to the lamellar magnetism case than Peak 2.

One possible interpretation of these observations is that the two hematite peaks correspond to different generations of exsolution lamellae. This sample contains a mixture of micron-scale first-generation hematite lamellae and nanoscale second-generation hematite lamellae. The micron-scale lamellae themselves contain nanoscale exsolution precipitates of ilmenite. Peak 2 behaves in a manner consistent with the nanoscale hematite lamellae, whose magnetization is dominated by a spin-canted moment with a smaller lamellar contribution. Such small lamellae are close to the superparamagnetic limit, and are more likely to be demagnetized initially. The magnetic moments of the nanoscale hematite precipitates will be easily reoriented in a magnetic field, leading to an increase in the sf scattering and a decrease in the nsf scattering, as observed. Peak 1 behaves in a manner consistent with single-domain hematite containing nanoscale ilmenite exsolution lamellae, whose magnetization is dominated by the defect moment associated with spin imbalance at the ilmenite lamellar interfaces. Such regions are less influenced by the application of a magnetic field, since the orientation of magnetization is controlled by the physical placement of the ilmenite within the hematite host (Robinson et al. 2004). If this interpretation is correct, the neutron data support the hypothesis that the NRM in this sample is predominantly carried by the lamellar moment associated with the fine scale ilmenite precipitates within the larger hematite exsolution lamellae (Robinson et al. 2006).

ACKNOWLEDGMENTS

The polarized neutron study of ilmenite-hematite was performed in collaboration with Luise Theil Kuhn, Kim Lefmann, and Bertrand Roessli, with a sample kindly provided by Suzanne McEnroe and Peter Robinson. Equations (51) to (54) were provided by K. Lefmann. The theme of this review was inspired by the workshop "Magnetic Structure Determination from Powder Neutron Diffraction Data" organized by Paolo Radaelli, Andrew Wills, and Juan Rodríguez, held at ISIS 12-14 Dec. 2002. R.J.H. is funded via an NERC Advanced Fellowship (NE/B501339/1 "Mineral magnetism at the nanometer scale").

REFERENCES

Alperin HA, Steinsvoll O, Nathans R, Shirane G (1967) Magnon scattering of polarised neutrons by the diffraction method: Measurements on magnetite. Phys Rev 154:508–514
Anderson IS, Hamelin B, Høghøj P, Courtois P, Humblot H (2000) Novel Trends in Neutron Optics. *In*: Proceedings of the Seventh School on Neutron Scattering. Furrer A (ed) World Scientific, p 44-71
Arai M, Ishikawa Y (1985) A new oxide spin glass system of $(1-x)FeTiO_3–xFe_2O_3$. III. Neutron scattering studies of magnetization processes in a cluster type spin glass of $90FeTiO_3-10Fe_2O_3$. J Phys Soc Japan 54:795–802
Arai M, Ishikawa Y, Saito N, Takei H (1985a) A new oxide spin glass system of $(1-x)FeTiO_3–xFe_2O_3$. II. Neutron scattering studies of a cluster type spin glass of $90FeTiO_3-10Fe_2O_3$. J Phys Soc Japan 54: 781–794
Arai M, Ishkawa Y, Takei H (1985b) A new oxide spin glass system of $(1-x)FeTiO_3–xFe_2O_3$. IV. Neutron scattering studies on a reentrant spin glass of 79 $FeTiO_3-21$ Fe_2O_3 single crystal. J Phys Soc Japan 54: 2279-2286

Bacon GE (1975) Neutron Diffraction. Clarendon Press

Brockhouse BN (1957) Scattering of neutrons by spin waves in magnetite. Phys Rev 106:859–864

Brown PJ (1993) Magnetic structure studied with zero-field polarimetry. Physica B 192:14–24

Brown PJ (1995) *In*: International Tables for Crystallography Volume C. Wilson AJC (ed) Kluver Academic Publishers, Dordrecht, p. 391

Champion JDM, Wills AS, Fennell T, Bramwell ST, Gardner JS, Green MA (2001) Order in the Heisenberg pyrochlore: The magnetic structure of $Gd_2Ti_2O_7$. Phys Rev B 64:140407

Corliss LM, Hastings JM, Goldman JE (1954) Neutron diffraction study of the anisotropy transition in α-Fe_2O_3. Phys Rev 93:893–894

Cox DE, Takei WJ, Shirane G (1963) A magnetic and neutron diffraction study of the Cr_2O_3-Fe_2O_3 system. J Phys Chem Solids 24:405–423

Crangle J (1977) The Magnetic Properties of Solids. Edward Arnold Limited

Halpern O, Johnson MH (1939) On the magnetic scattering of neutrons. Phys Rev 55 898–923

Hansen MF, Bodker F, Morup S, Lefmann K, Clausen KN, Lindgård P-A (1997) Dynamics of magnetic nanoparticles studied by neutron scattering. Phys Rev Lett 79:4910–4913

Hansen MF, Bodker F, Morup S, Lefmann K, Clausen KN, Lindgård P-A (2000) Magnetic dynamics of fine particles studied by inelastic neutron scattering. J Magnetism Magnetic Mater 221:10–25

Harrison RJ (2006) Microstructure and magnetism in the ilmenite-hematite solid solution: a Monte Carlo simulation study. Am Mineral 91:1006-1024

Harrison RJ, Becker U (2001) Magnetic ordering in solid solutions. *In*: Solid solutions in silicate and oxide systems. Geiger C (ed) Eur Mineral Soc Notes Mineral Vol. 3, p 349-383

Harrison RJ, Becker U, Redfern SAT (2000b) Thermodynamics of the R-3 to R-3c phase transition in the ilmenite-hematite solid solution. Am Mineral 85:1694-1705

Harrison RJ, Redfern SAT (2001) Short- and long-range ordering in the ilmentie-hematite solid solution, Phys Chem Minerals 28:399–412

Harrison RJ, Redfern SAT, Smith RI (2000a) In-situ study of the R-3 to R-3c phase transition in the ilmenite-hematite solid solution using time-of-flight neutron powder diffraction. Am Mineral 85:194–205

Kasama T, McEnroe SA, Ozaki N, Kogure T, Putnis A (2004) Effects of nanoscale exsolution in hematite-ilmenite on the acquisition of stable natural remanent magnetization. Earth Planet Sci Lett 224:461–475

Kittel C (1976) Introduction to Solid State Physics. John Wiley

Klausen SN, Lefmann K, Lindgård P-A, Clausen KN, Hansen MF, Bødker F, Mørup S, Telling M (2003) An inelastic neutron scattering study of hematite nanoparticles. J Magnetism Magnetic Mater 266:68–78

Klausen SN, Lefmann K, Lindgård P-A, Theil Kuhn L, Bahl CRH, Frandsen C, Mørup S, Roessli B, Cavadini N, Niedermayer C (2004) Magnetic anisotropy and quantized spin waves in hematite nanoparticles. Phys Rev B 70:214411

Kozlowski A, Kakol Z, Zalecki R, Knight K, Sabol J, Honig JM (1999) Powder neutron diffraction studies of Zn-doped magnetite. J Phys Condens Matter 11:2749–2758

Krueger S, Olson GJ, Rhyne JJ, Blakemore RP, Gorby YA, Blakemore N (1990) Small-angle neutron scattering from bacterial magnetite. J Appl Phys 67:4475–4477

Larson AC, Von Dreele RB (2000) General Structure Analysis System (GSAS). Los Alamos National Laboratory Report, LAUR 86-748

Lefmann K, Bødker F, Hansen MF, Vázquez H, Christensen NB, Lindgård P-A,Clausen KN, Mørup S (1999) Magnetic dynamics of small α-Fe_2O_3 and NiO particles studied by neutron scattering. Eur Phys J D 9: 491–494

Lefmann K, Bødker F, Klausen SN, Hansen MF, Clausen KN, Lindgård P-A, Mørup S (2001) A neutron scattering study of spin precession in ferrimagnetic maghemite nanoparticles. Europhys Lett 54: 526–532

Loong C-K (2006) Inelastic scattering and applications. Rev Mineral Geochem 63:233-254

Lovesey SW (1984) Theory of Neutron Scattering from Condensed Matter. Clarendon Press

Mangin PH, Dufour C, Rodmacq B (1993) Neutron investigations of magnetic multilayers. Physica B 192: 122-136

McEnroe SA, Brown LL, Robinson P (2004b) Earth analog for Martian magnetic anomalies: Remanence properties of hemo-ilmenite norites in the Bjerkreim-Sokndal Intrusion, Rogaland, Norway. J Appl Geophys 56:195−212

McEnroe SA, Harrison RJ, Robinson P, Golla U, Jercinovic MJ (2001) The effect of fine-scale microstructures in titanohematite on the acquisition and stability of NRM in granulite facies metamorphic rocks from Southwest Sweden: Implications for crustal magnetism. J Geophys Res 106:30523–30546

McEnroe SA, Harrison RJ, Robinson P, Langenhorst F (2002) Nanoscale hematite-ilmenite lamellae in massive ilmenite rock: an example of "Lamellar Magnetism" with implications for planetary magnetic anomalies. Geophys J Int 151:890–912

McEnroe SA, Skilbrei JR, Robinson P, Heidelbach F, Langenhorst F, Brown LL (2004c) Magnetic anomalies, layered intrusions and Mars. Geophys Res Lett 31:L19601

McEnroe, S. A. Langenhorst, F., Robinson P., Bromiley G., and C. Shaw (2004a) What's magnetic in the lower Crust? Earth Planet Sci Lett 226:175–192

McQueeney RJ, Yethiraj M, Montfrooij W, Gardner JS, Metcalf P, Honig JM (2005) Influence of the Verwey transition on the spin-wave dispersion of magnetite. J Appl Phys 97:10A902

Mezei F (1976) Novel polarized neutron devices - supermirror and spin component amplifyer. Comm Phys 1: 81–85

Moon RM, Riste T, Koehler WC (1969) Polarization analysis of thermal-neutron scattering. Phys Rev 181: 920–931

Morrall P, Schedin F, Langridge S, Bland J,Thomas MF, Thornton G (2003) Magnetic moment in an ultrathin magnetite film. J Appl Phys 93:7960–7962

Morrish AH (1994) Canted Antiferromagnetism: Hematite. World Scientific Publishing Co

Nathans R, Pickart SJ, Alperin HA, Brown PJ (1964) Polarized-neutron study of hematite. Phys Rev 136: 1641–1647

Nathans R, Shull CG, Shirane G, Andresen A (1959) The use of polarized neutrons in determining the magnetic scattering by iron and nickel. J Phys Chem Solids 10:138-146

Néel L (1948) Propriétés magnetiques des ferrites; ferrimagnétisme et antiferromagnétisme. Annal Phys 3: 137–198

Ott F, Cousin F, Menelle A (2004) Surfaces and interfaces characterization by neutron reflectometry. J Alloys Compnds 382:29–38

Parise JB (2006) Introduction to neutron properties and applications. Rev Mineral Geochem 63:1-25

Plant JS (1982) Neutron diffraction study of magnetisation processes in magnetite, Fe_3O_4, both multi-domain and magnetically hardened by milling. J Phys F 12:215–222

Redfern SAT (2006) Neutron powder diffraction studies of order-disorder phase transitions and kinetics. Rev Mineral Geochem 63:145-170

Robinson P, Harrison RJ, McEnroe SA, Hargraves RB (2002) Lamellar magnetism in the haematite-ilmenite series as an explanation for strong remanent magnetization. Nature 418:517–520

Robinson P, Harrison RJ, McEnroe SA, Hargraves RB (2004) Nature and origin of lamellar magnetism in the hematite-ilmenite series. Am Mineral 89:725–747

Robinson P, Heidelbach F, Hirt AM, McEnroe SA, Brown LL (2006) Crystallographic-magnetic correlations in single crystal hemo-ilmenite: New evidence for lamellar magnetism. Geophys J Int 165:17-31

Rodríguez-Carvajal J (1990) "FULLPROF: A Program for Rietveld Refinement and Pattern Matching Analysis". Abstracts of the Satellite Meeting on Powder Diffraction of the XV Congress of the IUCr, Toulouse, France, p. 127

Rodríguez-Carvajal J (1993) Recent advances in magnetic structure determination by neutron powder diffraction. Physica B 192:55–69

Rodríguez-Carvajal J (2001a) Magnetic structure determination from powder diffraction using the program FullProf. *In:* Proc XVIII Conf Appl Crystallography. Morawiec H, Stróz D (eds) World Scientific, p 30-36

Rodríguez-Carvajal J (2001b) Magnetic structure determination from powder diffraction. Symmetry analysis and simulated annealing. Mater Sci Forum 378-381:268–273

Samuelsen EJ (1969) Spin waves in antiferromagnets with corundum structure. Physica 43:353–374

Samuelsen EJ, Shirane G (1970) Inelastic neutron scattering investigation of spin waves and magnetic interactions in α-Fe_2O_3. Physica Status Solidi 42:241–256

Shirane G, Shapiro SM, Tranquada, JM (2002) Neutron Scattering with a Triple-axis Spectrometer. Cambridge University Press

Shull CG, Smart JS (1949) Detection of antiferromagnetism by neutron diffraction. Phys Rev 76:1256–1257

Shull CG, Strauser WA, Wollan EO (1951b) Neutron diffraction by paramagnetic and antiferromagnetic substances. Phys Rev 83:333–345

Shull CG, Wollan EO, Koehler WC (1951a) Neutron scattering and polarization by ferromagnetic materials. Phys Rev 84:912-921

Squires G L (1978) Introduction to the Theory of Thermal Neutron Scattering. Cambridge Press

Stahn J (2004) A switchable white-beam neutron polariser. Physica B 345:243–245

Watanabe H, Brockhouse BN (1962) Observation of optical and acoustical magnons in magnetite. Phys Lett 1:189–190

Wills AS (2000) A new protocol for the determination of magnetic structures using simulated annealing and representational analysis (SARAh). Physica B 276–278:680–681

Wills AS (2001a) Magnetic structures and their determination using group theory. J Phys IV 11:Pr9–133

Wills AS (2001b) Long-range ordering and representational analysis of the jarosites. Phys Rev B 63:064430

Wills AS (2001c) Conventional and unconventional orderings in the jarosites. Can J Phys 79:1501–1510

Wright JP, Attfield JP, Radaelli PG (2001) Long range charge ordering in magnetite below the Verwey transition. Phys Rev Lett 87:266401

Wright JP, Attfield JP, Radaelli PG (2002) Charge ordered structure of magnetite Fe_3O_4 below the Verwey transition. Phys Rev B 66:214422

Wright JP, Bell AMT, Attfield JP (2000) Variable temperature powder neutron diffraction study of the Verwey transition in magnetite Fe_3O_4. Solid State Sci 2:747–753

Yang JB, Zhou XD, Yelon WB, James WJ, Cai Q, Gopalakrishnan KV, Malik SK, Sun XC, Nikles DE (2004) Magnetic and structural studies of the Verwey transition in $Fe_{3.8}O_4$ nanoparticles. J Appl Phys 95: 7540–7542

Reviews in Mineralogy & Geochemistry
Vol. 63, pp. 145-170, 2006
Copyright © Mineralogical Society of America

Neutron Powder Diffraction Studies of Order-Disorder Phase Transitions and Kinetics

Simon A. T. Redfern

Department of Earth Sciences
University of Cambridge
Downing Street
Cambridge, CB2 3EQ, United Kingdom
e-mail: satr@cam.ac.uk

INTRODUCTION

One of the major applications of neutron diffraction in mineral sciences has been in the study of order-disorder processes and phase transitions. Neutron scattering methods provide unique insights into the origins and mechanisms of these processes, and have enabled mineralogists to develop new models of phase transformation behavior. The time-temperature dependence of processes such as phase transformations, exsolution, cation ordering and disordering in minerals has considerable potential geophysical, geochemical and petrological importance. Let's take atomic ordering as an example. Order-disorder transformations and similar structural phase transitions are typically some of the most efficient ways a mineral can adapt to changing temperature or chemical composition. Disorder of distinct species across different crystallographic sites at high temperature provides significant entropic stabilization of mineral phases relative to low-temperature ordered structures. Positional or orientational disordering can have similar drastic effects. For example, the calcite-aragonite phase boundary shows a significant curvature at high temperature due to disorder of CO_3 groups within the calcite structure, associated with an orientation order-disorder phase transition (Redfern et al. 1989). This leads to an increased stability of calcite with respect to aragonite over that predicted by a simple Clausius-Clapeyron extrapolation of the low pressure-temperature thermochemical data. Understanding of the structural characteristics of the phase transition in calcite developed in tandem with studies of the analogous transition in nitratine, $NaNO_3$, but it was not until Dove and Powell (1989) carried out high-temperature neutron diffraction experiments on powdered calcite that there was direct experimental evidence linking the thermodynamic and structural nature of the transition in $CaCO_3$. The key to the success of their study was the fact that neutrons penetrate the entire volume of samples held at extreme conditions (in this case very close to the melting temperature and under a confining CO_2 pressure).

Cation ordering in minerals may or may not involve a change in the symmetry of the crystal. This distinction was outlined by Thompson (1969), who defined the two cases as convergent and non-convergent ordering. In convergent ordering two or more crystallographic sites become symmetrically equivalent when their average occupancy becomes identical, and the order-disorder process is associated with a symmetry change at a discrete phase transition. This usually occurs (as a function of temperature) at a fixed temperature, the transition temperature (T_c), or on a phase diagram at a fixed composition in a solid solution, defined by the relative free energies of the two phases. In non-convergent ordering the atomic sites over which disordering occurs never become symmetrically equivalent, even when the occupancies are identical on each. It follows that no symmetry change occurs on disordering, and no phase transition exists. A convergent order-disorder phase transition occurs when the low-temperature

     DOI: 10.2138/rmg.2006.63.7

phase of a system shows a regular (alternating) pattern of atoms with long-range correlations, but the high-temperature phase has atoms arranged randomly with no long-range correlation. Experimentally, the distinction between the two can often be characterized straightforwardly using diffraction techniques, since diffraction measures the long-range correlation of structure. Usually, a minimum enthalpy is achieved by an ordered distribution (for example, in low-albite the preference of Al for one of the four symmetrically distinct tetrahedral sites), but configurational entropy (and often the coupled vibrational entropy) above 0 K results in a situation in which the free energy is a minimum for partially disordered distributions.

Order-disorder in minerals may occur over a variety of length scales and via a number of mechanisms. Substitutional disorder is typically observed, usually of cations over shared sites (such as Al/Si order-disorder in aluminosilicates), but certain molecular groups may display orientational order-disorder behavior, for example carbonate oxy-anions or ammonium ions in relevant phases. Most commonly only the long-range order is considered, because this is what is most obviously observed by structural diffraction methods, either through the direct measurement of scattering amplitudes at crystallographic sites or bond-lengths in the solid, or less directly through the measurement of coupled strains which may arise through the elastic interplay between the degree of order and the shape and size of the unit cell. Ordering over short length scales can also be detected through neutron methods, including inelastic neutron scattering (Loong 2006, this volume). Recently, computational methods have also been employed successfully to elucidate and illuminate experimental observations of ordering and to begin to separate and compare short- and long-range ordering effects (Meyers et al. 1998; Warren et al. 2000a,b; Harrison et al. 2000b).

The time-temperature dependence of cation ordering and disordering in minerals has considerable petrological importance. Not only does such order/disorder behavior have significant consequences for the thermodynamic stabilities of the phases in which it occurs, it can also play a significant role in controlling activity-composition relations for components, hence influencing inter-mineral major-element partitioning. Furthermore, since time-temperature pathways affect the final intra-mineral partitioning of (typically) cations within the structure of minerals, inverse modeling may be employed to infer the thermal histories of minerals from measured site occupancies. A quantitative knowledge of the temperatures and pressures of mineral assemblage formation in the crust and mantle is, therefore, fundamental to understanding the thermal evolution of the earth, and to the development of well-constrained petrological and geophysical models. In many cases the time scales of ordering mean that they must be measured *in situ*, rather like the requirement for high-speed photography to capture familiar fast kinetic processes (Fig. 1). For some time, geothermometric and geobarometric deductions have been based on the compositional variations of coexisting rock-forming minerals (e.g., cation partitioning between orthopyroxene/clinopyroxene; orthopyroxene/garnet; magnetite/ilmenite). Information on cooling rates (geospeedometry) is also potentially available from knowledge of intracrystalline cation partitioning. The convergent ordering of (for example) Al and Si on tetrahedral sites in feldspars has been used in this way as a thermometric indicator and marker of petrogenesis (Kroll and Knitter 1991), as has the non-convergent ordering of Mg and Fe on the M-sites of pyroxenes, which has been shown to be useful in the interpretation of the petrological history of the host rock (Carpenter and Salje 1994a).

Measurements of time-temperature-pressure dependent phenomena by neutron diffraction methods generally employ *in situ* techniques, using high temperature (T) furnaces, pressure (P) cells, combined P and T apparatus, and potentially sample environments that incorporate gas or vapor controls. Time dependent studies usually involve the measurement of the response of a system to a perturbation in the external conditions, and demand rapid data collection (Fig. 2). Problems of dead time and data capture can become important for particularly rapid processes, and there are possibilities for measurements at pulsed sources that adopt a stroboscopic approach, although such a methodology has not yet been exploited for mineralogical studies

Figure 1. *In situ* studies of order-disorder behavior in minerals have become possible using rapid data collection at high-flux neutron scattering instruments. In a number of cases the nature of the controls on ordering or disordering behavior is only revealed by such *in situ* study, made possible by technological developments. The situation is analogous to the discovery of how a horse moves in a gallop. Original perceptions (left) were proven inaccurate with the advent of "*in situ*" high speed photography (right) pioneered by Eadward Muybridge at Stanford University in the late 19th century.

•Perturbation (T,P, pH etc.):

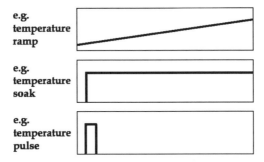

•Time-dependent measurement:

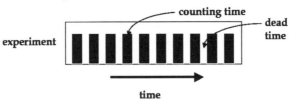

Figure 2. Features of *t-P-T-* studies.

(see Eckold et al. 1998, for an example of rapid transformation processes studied in a ferroelectric system).

Here, I highlight the advantages of using neutron powder diffraction for the study of typical order-disorder in minerals at high temperatures and pressures. I review recent progress in the application of neutron powder diffraction to the study of high-temperature order-disorder behavior in rock-forming minerals, focusing particularly on results obtained on non-convergent ordering in olivines, convergent order-disorder in Fe-Ti oxides and non-convergent ordering in spinels, and on potential for the same approach to study the inter-site partitioning of cations in the amphibole structure. In particular, we find that *in situ* studies made possible by the use of neutron methods have proven very valuable in determining the processes responsible for these sorts of phase transitions at the temperatures and pressure at which they occur. I finish by suggesting further possible future routes to study ordering in mineral systems at high-temperature and high-pressures using a novel apparatus designed to allow *in situ* high-T/P neutron diffraction.

NEUTRON DIFFRACTION CHARACTERISTICS RELEVANT TO ORDER-DISORDER STUDIES

It is worthwhile beginning our discussion of the use of neutrons for studying transformation processes in minerals by first rehearsing the roles and natures of diffraction, both of X-rays (the tool of mineralogists for approaching a century now) and neutrons (the theme of this volume) from periodic and aperiodic solids. These methods can be used to provide information about the atomic scale structure of materials. Each has seen substantial application in developing our understanding of the nature of materials' response to changes in variables such as temperature, pressure and composition, important in interpreting phase stabilities and phase transitions in minerals. It is clear that the two techniques have different, and somewhat complementary, advantages. The differences are apparent from first considerations. In contrast to X-rays, neutrons have significant mass (1.675×10^{-27} kg), spin (½) and magnetic moment (1.91 μ_N). But neutrons in thermal equilibrium adhere to a Maxwellian distribution of energies such that around room temperature there is a peak flux at around 25 meV, an energy which may be translated, by considering the kinetic energy of the neutron ($E = h^2/[2m\lambda^2]$), to a wavelength of $\lambda = 1.8$ Å. Bragg scattering of neutrons from crystals may be considered in much the same way as X-rays, therefore (see also Parise 2006, this volume).

In X-ray diffraction, the X-rays are scattered by electrons surrounding atoms. The atomic scattering factor, f, is determined by summing the contributions from all electrons, taking into account the path difference of the scattered waves (Fig. 3) and the electron density is obviously spread out over the entire volume of the atom. From Figure 4 we see that the phase difference between waves scattered from different parts of the atom increases with scattering angle or scattering vector $\mathbf{Q}$, where the modulus $Q = 4\pi \cdot \sin\theta/\lambda$ (when we apply this relation to Bragg scattering from lattice planes of spacing d [$\lambda = 2d\sin\theta$]), we clearly have $Q = 2\pi/d$, and this difference influences the $\mathbf{Q}$-dependence of f, $f(\mathbf{Q})$.

When we consider scattering of X-rays by atoms we can think of scattering from the continuous distribution of electrons around the atom. If we denote the electron density as $\rho_{el}(\mathbf{r})$, the X-ray atomic scattering factor is given as

$$f(\mathbf{Q}) = \int \rho_{el}(\mathbf{r}) \exp(i\mathbf{Q} \cdot \mathbf{r}) d\mathbf{r}$$

In the limit $\mathbf{Q} \rightarrow 0$, where the X-rays are scattered without deflection, we have

$$f(\mathbf{Q} = 0) = \int \rho_{el}(\mathbf{r}) d\mathbf{r} = Z$$

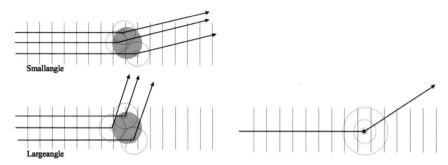

Figure 3. The phase difference of waves scattered from a large scattering object increases with scattering angle (left). For scattering of radiation from small scattering objects (e.g., the scattering of neutrons from the nucleus) the path differences will be minimal (right) and the scattering strength will not suffer destructive interference at large angles.

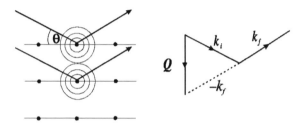

Figure 4. The relationship between Bragg angle, θ, and scattering vector, $\mathbf{Q}$, for diffraction from a lattice. $\mathbf{Q}$ may be defined as the difference between the wave vectors of the incoming and scattered rays, k_i and k_s.

where Z is the total number of electrons in the atom or ion. For all atoms and ions of interest $\rho_{el}(\mathbf{r})$ can be calculated using quantum mechanics in order to obtain the scattering factor. In practice there will not be an analytical function for $f(\mathbf{Q})$, and so for practical use the numeric values of $f(\mathbf{Q})$ are fitted to an appropriate functional form.

This fall off, or form factor, results in attenuation of the diffracted intensity at high scattering angles, θ, for constant wavelength (λ) diffraction from crystalline solids, giving rise to the characteristic weakening of signal at high scattering vectors $\mathbf{Q}$. At zero scattering angle all the electrons are in-phase and the atomic scattering factor is equal to the number of electrons. Therefore the heavier the element the higher the X-ray atomic scattering factor, which leads to X-ray scattering from a mineral being dominated by the heavier elements. It is, therefore, sometimes a challenge to obtain information about lighter atoms such as hydrogen and oxygen, using X-ray diffraction. As scattering angle increases the atomic scattering factor decreases (see Fig. 5) due to increasing destructive interference between X-rays scattered from electrons in different parts of the atom.

In application of these ideas to the X-ray scattering function, the width of the scattering function is given by the inverse of the atomic/ionic radius. Transferring these ideas to neutron scattering, where the neutrons are scattered by the nuclei which are typically $10^{-5}\times$ the size of the atom, the scattering function is so wide in Q that it can be treated as a constant for all values of Q of practical interest. This means that information in diffraction space out to high scattering vectors (small d-spacings) may be obtained without the problem of form factor attenuation.

The constancy of the neutron scattering length to high scattering vectors is particularly useful in powder diffraction studies where the increase in the number of observable reflections

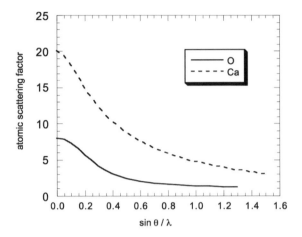

Figure 5. The variation of X-ray atomic scattering factor with angle for oxygen and calcium atoms.

that can be obtained at good signal-to-noise levels means that temperature factors and occupancies may be refined with greater confidence. In addition, for neutrons the scattering power (equivalent quantity to atomic scattering factor) varies irregularly across the periodic table (shown in Fig. 6 for natural abundance isotopes). It turns out that this is very important when it comes to considering how neutrons may be employed to study cation ordering processes. Neutrons also interact with magnetic fields from unpaired electrons (Von Dreele 2006, this volume). From Figure 6 it can be seen that it is therefore easier obtain information about lighter atoms, and even different isotopes of the same atom, using neutron diffraction.

The details of the interaction of neutrons with nuclei are discussed by Parise (2006, this volume), but it is worth reiterating here that the scattering length for neutrons may involve a spin-dependent component. While, for many nuclei, the nuclear spin is zero (and thus spin-

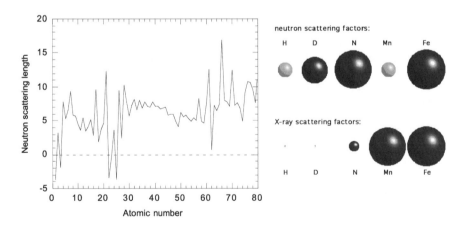

Figure 6. (left) The variation of mean coherent neutron scattering length (in fm, 10^{-15}m) with atomic number for natural abundance isotopes. (right) These data are represented graphically for a few examples and compared with scattering factors for X-rays. We see that Fe and Mn have sharply contrasting neutron scattering lengths (Mn's is negative, Fe's is positive) but very similar X-ray atomic scattering factors.

dependent components of the nucleus-neutron interactions are absent) even when there is a spin the effects are relatively weak. Isotope effects are usually ignored. For many important elements only one isotope occurs in significant quantities. For example, the natural abundances of the isotopes of oxygen, ^{15}O, ^{16}O and ^{17}O, are 1%, 98% and 1% respectively and thus we may assume (which is valid to a very good approximation) that the scattering lengths of all oxygen atoms in a structure are constant. There are, however, a few notable exceptions to this assumption. One such is nickel, for which there are two major isotopes ^{58}Ni (relative abundance 68.3%) and ^{60}Ni (relative abundance 26.1%) with respective scattering lengths of 14.4 and 2.8 fm, with neither of these nuclei having a non-zero spin. A useful resource for information on scattering lengths of elements and their isotopes is the NIST neutron scattering length and cross section website at *http://www.ncnr.nist.gov/resources/n-lengths/*. There one finds other pointers to potential problems associated with particular isotopes. For example, Li is naturally 92.5% ^7Li, which absorbs neutrons weakly, but the 7.5% abundant ^6Li has an absorption cross section that is more than twenty thousand times greater and which has a strong influence on the overall absorption of Li.

In certain cases the isotopic variability of neutron scattering cannot, therefore, be wholly ignored, and in some cases there is cause for using specific isotopes of elements of interest when synthetic samples are to be studied. Perhaps the most significant of these in the application of neutrons to mineralogy is the behavior of hydrogen (^1H) where the spin-dependence of the nucleus-neutron interaction (or proton-neutron interaction) is very important, more so since ^1H is more than 99.9% abundant in natural hydrogen. The proton and neutron both have spin ½, and can be aligned in four ways. The scattering function for atoms like hydrogen depends upon averages of the scattering lengths for the parallel and anti-parallel configurations of spins that can occur. Deviations from these averages lead to a high incoherent background in diffraction experiments involving hydrous phases (Fig. 7). This can be avoided by substituting hydrogen ^1H by deuterium, perhaps the most common example of the use of isotopic substitution employed in neutron diffraction studies. This is discussed in more detail by Kuhs and Hansen (2006, this volume).

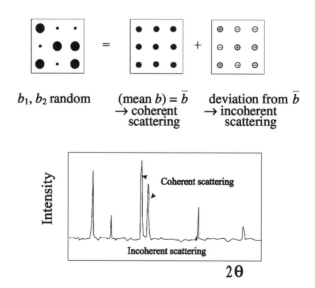

Figure 7. For minerals containing atoms with different isotopes or for which the scattering length can differ the Bragg scattering is determined by the average scattering length (mean *b*) whereas incoherent scattering gives rise to an increase in the background to a diffraction pattern.

In transferring these concepts to geosciences we can first observe that the application of neutron scattering to the study of earth materials is rather less mature than its use in other branches of the physical sciences, but this volume is testament to the increasing recognition now being given to the role it may play in solving problems of mineral behavior. Neutrons provide a unique probe for the study of minerals, since their wavelengths (sub-Ångstrom to tens of Ångstroms) are of the same order of magnitude as the inter-atomic spacings in minerals and correspond to energies similar to many electronic and atomic processes. The fact that the scattering power of neutrons does not depend upon the number of electrons surrounding the atom, but rather upon the nuclear cross section makes neutrons particularly suitable for studying order-disorder and mixing processes in minerals, where chemically similar atoms (which may also therefore tend to have similar X-ray scattering powers) substitute on crystallographic sites. For example, the nuclear scattering contrast for Mg and Al is almost five times greater than the contrast in X-ray atomic scattering factors. For Mn and Fe the difference is even more marked: the neutron scattering contrast is more than 36× greater than the X-ray scattering contrast, since Mn has a negative neutron scattering length while Fe's is large and positive, while their atomic scattering factors differ in magnitude by less than 4%.

Thus, elements that are difficult to distinguish by X-ray diffraction can show huge contrasts in neutron diffraction experiments. For example, one anticipates significant changes in the intensities of reflections within a powder diffraction pattern of a mineral in which Mn and Fe may interchange between sites (Fig. 8). Of particular additional interest to earth scientists is the possibility to detect hydrogen (or, more correctly, deuterium) in crystal structures using neutron diffraction, a task that is well nigh impossible by X-ray diffraction.

Aside from strong scattering contrasts between chemically similar pairs of substituting atoms, the characteristics of neutron scattering provide further advantages that may be exploited in studies of high-temperature order-disorder. As the fall-off in scattering power with scattering vector, **Q**, is negligible, data may be obtained out to high scattering vectors (corresponding to small *d*-spacings), and complex structures may confidently and routinely be refined from powder data using Rietveld methods with high precision and accuracy.

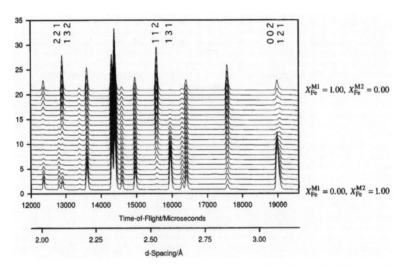

Figure 8. Simulated neutron powder diffraction patterns of $(Fe,Mn)_2SiO_4$ olivines (kneblites) as a function of *M*-site occupancy over the two sites of the olivine structure. Note the large changes in intensity of certain reflections, due to the large contrast in coherent scattering length for Mn and Fe.

Combining these facets of neutron powder diffraction (using fixed geometry time-of-flight methods) with the fact that stable sample environments may be constructed around the sample without the worry of overly attenuating the incident and diffracted beams, we quickly reach the conclusion that neutron powder diffraction is a powerful tool for observing structural changes at extremes of temperature. This much has been demonstrated by recent studies of the temperature dependence of inter-site partitioning of metal cations in olivines and spinels (Henderson et al. 1996; Redfern et al. 1996, 1997, 1998, 1999; Harrison et al. 1998), and the work on more complex hydrous silicates that is presented below. That said, the interaction of neutrons with matter is very weak. This has some positive consequences as well as the inherent disadvantage that scattering is generally weaker than for X-rays. In particular we find that neutrons probe the bulk of samples (penetrating 1 cm or more through the sample), they do not damage the sample, they will only be a small perturbation on the system, and systems respond linearly so that neutron scattering theory is very quantitative.

CASE STUDIES OF CATION ORDER-DISORDER PROCESSES PROBED BY POWDER NEUTRON DIFFRACTION

Non-convergent cation order-disorder in olivines and spinels

As early as 1983, Nord and co-workers had employed neutron diffraction for the study of Ni-Fe cation distributions in the olivine-related phosphate sarcopside (Nord 1983; Ericsson and Nord 1984). The temperature dependence of non-convergent cation exchange between the $M1$ and $M2$ octahedral sites of silicate olivines (Fig. 9) has also been the subject of a number of recent neutron diffraction studies, from the single crystal studies of members of the forsterite-fayalite solid solution (Untersteller et al. 1986; Artioli et al. 1995; Rinaldi and Wilson 1996) to powder diffraction studies of the same system (Redfern et al. 2000). The strong contrast between Mn (negative scattering length) and other cations has led to interest in the Fe-Mn, Mg-Mn, and Mg-Ni systems as model compounds (Ballet et al. 1987; Henderson et al. 1996; Redfern et al. 1996, 1997, 1998), but there have also been studies of Fe-Zn olivines (Krause et al. 1995) and most recently the Co-Mg system (Rinaldi et al. 2005). Millard et al. (2000) used neutron diffraction to determine cation distributions in a number of germanate olivines.

The high-temperature behavior of Fe-Mg order-disorder appears to be complicated by crystal field effects, which influence the site preference of Fe^{2+} for $M1$ and $M2$, but the cation exchange of the Fe-Mn, Mg-Mn, and Mg-Ni olivines is dominantly controlled by size effects: the larger $M2$ site accommodating the larger of the two cations in each pair (Mn or Ni, in these cases).

Figure 9. The structure of olivine can be thought of as edge-sharing slabs of $M1$ (dark) and $M2$ (light) octahedra, shown here aligned vertically with a completely ordered arrangement of cation occupation. Upon heating exchange between these sites may occur to give rise to disordered configurations. At no point do the two sites become symmetrically equivalent, and hence the disordering is non-convergent.

In all of the recent *in situ* experiments, the use of time-of-flight neutron powder diffraction allowed the measurement of states of order at temperatures in excess of 1000 °C under controlled oxygen fugacities (especially important given the variable oxidation states that many of the transition metal cations of interest can adopt). In the majority of the powder diffraction studies mentioned above, diffraction patterns were collected on the POLARIS time-of-flight powder diffractometer at the ISIS spallation source (Fig. 10; Hull et al. 1992). The diffraction patterns of the Fe-Mn and Ni-Mn olivines (Henderson et al. 1996; Redfern et al. 1996, 1997, 1998) were collected in four 30 minute time bins over two hours at each isothermal temperature step on heating, and over a single 30 minute period on cooling. Diffraction patterns of the Mg-Mn olivine sample were collected over one hour time intervals at each isothermal step on heating. Thus time-temperature pathways were investigated. Data were collected to rather large scattering vectors, corresponding to refineable information *d*-spacings at 0.5 Å or less (Fig. 11). Structural data were then obtained by Rietveld refinement of the whole patterns giving errors in the site occupancies of around 0.5% or less. The low errors in refined occupancies result principally from the fact that the contrast between Mn (with a negative scattering length) and the other cations is very strong for neutrons (cf. Fig. 8).

All experiments showed the same underlying behavior of the degree of order as a function of temperature. This can be modeled according to a Landau expansion for the excess Gibbs free energy of ordering, of the type:

$$\Delta G = -h\eta + \frac{a}{2}(T - T_c)\eta^2 + \frac{b}{4}\eta^4$$

where *h, a, b* and T_c are material-dependent parameters and an order parameter, η, describes the degree of cationic order/disorder over the two sites. This expression, chosen to describe the free energy change due to ordering, is formally equivalent to the reciprocal solution model at lowest order, although the manner in which free energy is partitioned between entropy and enthalpy differs between the two approaches (Carpenter et al. 1994; Kroll et al. 1994). The

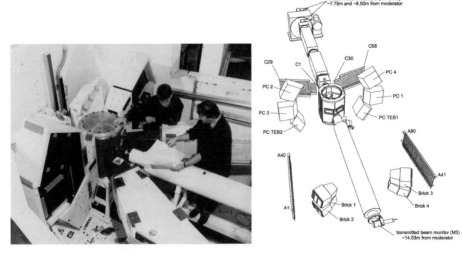

Figure 10. (left) Ron Smith and Steve Hull at the POLARIS time-of-flight diffractometer at ISIS (photo courtesy of CCLRC). (right) Line drawing of the fixed detectors banks surround the sample chamber allowing data to be collected for all relevant scattering vectors simultaneously, which is particularly beneficial for kinetic high-*T* studies. The large sample tank can accommodate a range of furnaces, cryostats, pressure cells and other environmental chambers.

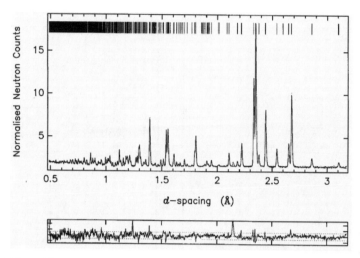

Figure 11. Time-of-flight neutron powder diffraction pattern of $(Fe_{0.3}Mn_{0.7})_2SiO_4$. Vertical bars represent the positions of reflections. The difference between the fit and the experiment is shown in the lower part of the figure (amplified: the peak at 2.1 Å is due to scattering from the vanadium sample can). Data down to 0.5 Å yield useable information, and aid the refinement of site occupancies across the structure (from Redfern at el. 1998).

Landau formulation essentially treats entropy as vibrational rather than configurational. Kroll et al. (1994) have shown that the addition of a configurational entropy term models the entropy at high η more accurately, in particular for the non-convergent ordering behavior of Mg and Fe on M-sites in pyroxene.

In each case studied by neutrons (Fig. 12) the order parameter remains constant at the start of the heating experiment, then increases to a maximum before following a steady decline with T to the highest temperatures. This general behavior reflects both the kinetics and thermodynamics of the systems under study: at low temperatures the samples are not in equilibrium and the results of the refinements reflect the kinetics of order-disorder, at high temperatures the states of order are equilibrium states, reflecting the thermodynamic drive towards high-temperature disorder. The initial increase in order results from the starting value being lower than equilibrium, and as soon as the temperature is high enough for thermally activated exchange to commence (on the time scale of the experiments), the occupancies of each site begin to converge towards the equilibrium order-disorder line. Using Ginzburg-Landau theory, which relates the driving force for ordering to the rate of change of order, one can obtain a kinetic description of the expected t-T-η pathway which relates to the thermodynamic description of the non-convergent disordering process:

$$\frac{d\eta}{dt} = \frac{\gamma \exp(-\Delta G^* / RT)}{2RT} \frac{\partial G}{\partial \eta}$$

Since the low temperature data, which lie below the thermodynamic disordering curve give information about the kinetics of order-disorder, and the high-temperature data on the equilibrium ordering curve provide the thermodynamic description of the process, the entire ordering process may be derived from a single neutron diffraction experiment (e.g., Redfern et al. 1997). In other words, the cation occupancies provided by these sorts of time-temperature *in situ* measurements provide data on the free energy surface for the order-disorder process, and the minima of this surface denote the equilibrium ordering behavior while its gradient is indicative of the kinetic behavior.

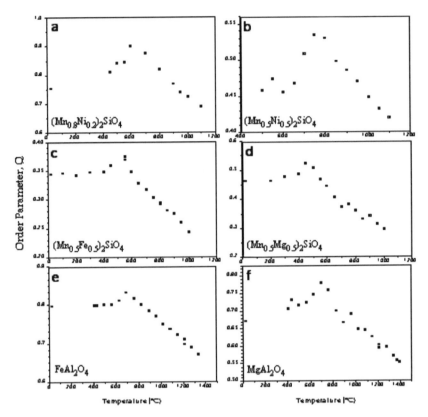

Figure 12. Temperature dependence of non-convergent metal cation-ordering in several olivines and spinels, all measured by Rietveld refinement of neutron powder diffraction data.

Similar behavior has been observed in spinels (Fig. 13), where a large number of studies have been carried out exploiting the scattering characteristics of neutrons. Neutrons have been valuable for determining Mg-Al distributions, since this atom pair have very similar X-ray atomic scattering factors but quite different neutron scattering lengths. Neutron scattering studies of this phase date back to the mid-70s, with the investigation by Rouse et al. (1976). Peterson et al. (1991) were the first to carry out a time-of-flight powder diffraction study of the nature of disordering in this phase, and their investigations have been extended and built upon in subsequent studies at higher T as well as high P (Pavese et al. 1999; Redfern et al. 1999; Méducin et al. 2004).

These studies of cation ordering in olivines and spinels have shown that, in most cases, the degree of cation order measured at room temperature is an indication of the cooling rate of a sample, rather than the temperature from which it has cooled. Calculated η-T cooling pathways for a Fe-Mn olivine are shown in Figure 14, where it is shown that variations in cooling rate over 13 decades might be ascertained from the degree of order locked in to room temperature.

The influence of processes of intra-mineral partitioning (order-disorder) on inter-mineral partitioning in olivine solid solutions has been pointed out by Bish (1981). Data obtained by neutron powder diffraction on the composition dependence of Ni-Mg ordering in olivines (Henderson et al. 2001) illustrate his argument. In Figure 15 the results were presented in

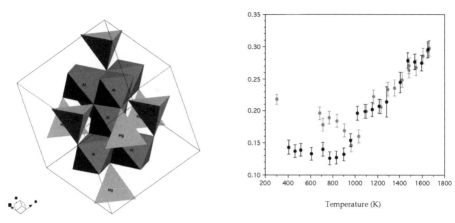

Figure 13. (left) Al-Mg order-disorder in spinel is non-convergent, involving exchange between the tetrahedral and octahedral sites which are occupied by Mg and Al respectively in fully ordered normal spinel. (right) Data collected by *in situ* neutron powder diffraction of $MgAl_2O_4$ spinel, demonstrating the increased disorder (occupancy of Al into tetrahedral sites) on heating (dark data points). The cooling path (lighter data points) in this experiment resulted in a more ordered spinel at the end of the experiment, since cooling in the diffractometer was slower than in the original synthesis quench (from Redfern et al. 1999).

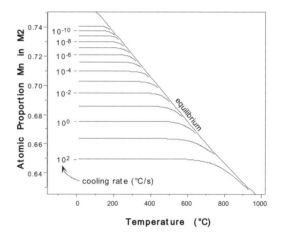

Figure 14. Calculated cooling paths over thirteen decades of cooling rate showing the dependence of the room-temperature site occupancy of $FeMnSiO_4$ on cooling rate. The room temperature site occupancy given by neutron diffraction is a direct measure of the cooling rate of the sample.

terms of a distribution coefficient, K_D, for disorder, which relates to the exchange reaction $Ni^{(M2)}Mg^{(M1)}SiO_4 \, Mg^{(M2)}Ni^{(M1)}SiO_4$, such that $K_D = [Mg^{(M2)}Ni^{(M1)}]/[Ni^{(M2)}Mg^{(M1)}]$. We see that the magnesium-rich sample shows a higher degree of order, with Ni ordering onto $M1$.

If one considers the partitioning of Ni between, say, a melt and olivine then we can write equations which depend upon the state of order of Ni in olivine. For a disordered sample:

$$2NiO_{\text{ liquid}} + Mg_2SiO_{4 \text{ olivine}} \leftrightarrow 2MgO_{\text{ liquid}} + Ni_2SiO_{4 \text{ olivine}}$$

but for an olivine in which all the Ni resides in $M1$ this must be re-written as

$$NiO_{\text{ liquid}} + Mg^{(M1)}Mg^{(M2)}SiO_{4 \text{ olivine}} \leftrightarrow MgO_{\text{ liquid}} + Ni^{(M1)}Mg^{(M2)}SiO_{4 \text{ olivine}}$$

Thus, for a disordered olivine the activity is simply equal to the molar proportion of the $NiSi_{0.5}O_2$ component, $X[NiSi_{0.5}O_2]$, but for an (intermediately) ordered olivine the system deviates from

Raoult's law. Neutron powder diffraction has been used to measure such changes in K_D with composition directly from determinations of the site scattering (and hence cation occupancies) at each site (Fig. 15). It is clear that the extents of ordering must be quantified in order to generate accurate models of mineral behavior. The *in situ* studies that have been performed in recent years have allowed the temperature dependence of this ordering to be determined accurately to high T. In these cases *in situ* study has been essential, since high-temperature disordered states are generally non-quenchable (cf. Fig. 1), due to the fast kinetics of cation exchange in olivines and spinels, and the unavoidable re-equilibration of samples on quenching from annealing conditions. Thus, neutron diffraction techniques are invaluable for directly determining the long range ordering characteristics of these important rock-forming minerals.

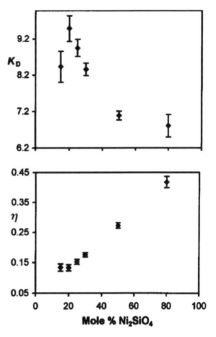

Figure 15. Trends for K_D and degree of order η vs. bulk composition at room temperature in $(Ni,Mg)_2SiO_4$ olivines, measured by neutron powder diffraction (from Henderson et al. 2001).

Cation ordering in crystal-chemically complex minerals: pushing the limits of powder data

Here, we consider two studies of cation ordering that show the strengths of neutron diffraction in tackling low-contrast element pairs in crystal-chemically complex minerals. In these studies the ordering involves Mg/Al on octahedral sites and Al/Si on tetrahedral sites. The neutron (scattering-length) contrasts for Mg/Al and Al/Si are not as great as for (say) Mn/Fe or Mn/Mg exploited in olivines (above), but are rather better than is the case for X-rays (scattering factors). The possibility arises, therefore, of being able determine site occupancies directly from site scattering; this is not possible by X-ray diffraction, where recourse must be made to mean bond length arguments. Both studies outlined below used the Rietveld refinement program GSAS (Larson and Von Dreele 2004).

Mg/Al ordering in dioctahedral micas. Micas (hydrous sheet silicates) with a high phengite component, $K[MgAl][Si_4]O_{10}(OH)_2$, are characteristic of high pressures of formation (>1 GPa) . These micas are "dioctahedral", having two out of three sites in the octahedral sheet occupied by divalent or trivalent cations, commonly Mg, Fe^{2+} and Al. Various polytypes exist that arise from different ways of stacking the 2:1 (tetrahedral:octahedral sheet ratio) slabs. A segment of the trigonal $P3_12$ phengite structure ($3T$ polytype) is shown in Figure 16. There are two non-equivalent octahedral sites ($M2$, $M3$) and two non-equivalent tetrahedral sites ($T1$, $T2$). The site topology of the octahedral layer consists of six-membered rings of alternating $M2$ and $M3$ sites surrounding a central vacant octahedral site. Similarly, the tetrahedral layers comprise six-membered rings of alternating $T1$ and $T2$ tetrahedra. The distribution of cations over the octahedral sites may be a function of pressure and/or temperature and may also be correlated with polytype.

The long-range ordering of octahedral cations in these micas is not well understood. This uncertainty is largely a consequence of the very similar X-ray scattering factors of Mg and Al. Single-crystal X-ray structure refinements of phengitic micas imply a contradiction between

Figure 16. The dioctahedral structure of phengite, viewed parallel to the triad axis. *M*2 and *M*3 octahedra are marked, and lie between two tetrahedral sheets of *T*1 (light) and *T*2 (dark) tetrahedra. The lower sheet is omitted for clarity.

octahedral-site occupancies (Mg, Al) refined from site-scattering values (electrons per site) and those indicated by <*M*-O> bond lengths. For example, in the single-crystal X-ray study by Amisano-Canesi et al. (1994), of a typical high-pressure phengitic mica of composition $K_{0.9}[Mg_{0.58}Al_{1.43}][Si_{3.57}Al_{0.43}]O_{10}(OH)_2$, site-scattering values require all Mg to be at *M*2 (with some Al) and *M*3 to be fully occupied by Al, whereas the <*M*2-O> average bond length is considerably shorter than <*M*3-O>. The site scattering values were not well constrained, due to the very similar X-ray atomic scattering factors of Mg and Al. The tetrahedral-site occupancies were also not well defined because of the low contrast between Al and Si for X-rays, and the variable correlations between <*T*-O> and Al/Si on sites. The scattering contrasts between Al and Mg and between Al and Si are better for neutrons than X-rays. Hence, in principle, neutron diffraction could resolve the site occupancy problem of phengites. With this

in mind, Pavese et al. (1997) used neutron powder diffraction to investigate cation ordering in the same phengite as studied by Amisano-Canesi et al. (1997). The powder route was used because crystals of phengite large enough for single-crystal neutron diffraction are unavailable. They used the high-resolution powder diffractometer (HRPD) at ISIS. This diffractometer has a very high resolution ($\Delta d/d \sim 0.0004$) in 90° 2θ and back-scattering (168° 2θ) configurations which is uniform over the entire diffraction pattern, and so it is well suited to investigating Mg/Al and Al/Si ordering in complex structures. For micas, neutron powder diffraction has the added advantage of reducing preferred orientation because a large volume of powder (3-4 cm³) can be loosely packed into the sample holder. A 3 cm³ sample of phengite powder (40-50 μm grain size) was used by Pavese et al. (1997). Four experiments were carried out at 293 K, 493 K, 696 K and 893 K, each lasting twelve hours. Using a starting model for Rietveld refinement in which (a) the tetrahedral bond lengths were constrained to be those found in the single-crystal X-ray study of Amisano-Canesi et al. (1994), and (b) refined site occupancies had to be consistent with the total octahedral composition as given in the chemical formula, Pavese et al. (1997) were able to refine the occupancies of *M*2 and *M*3 sites. It did not prove possible to refine occupancies for the *T*1 and *T*2 sites (unstable refinements). The neutron experiment clearly showed that, in contrast to all X-ray refinements of phengites, Mg orders at *M*3 (not *M*2) and *M*2 is fully occupied by Al. The refined occupancies they obtained at 293 K are *M*2(Al) = 1.06(5) and *M*3(Al) = 0.36(5). The <*M*2-O> and <*M*3-O> bond lengths of 1.964(5) Å and 1.979(5) Å are qualitatively consistent with these occupancies (unlike the X-ray results). At 893K the occupancies are *M*2(Al) = 0.95(6) and *M*3(Al) = 0.47(6), and so, within error, there is little or no Mg/Al disordering on the time scale of the experiment. This was the first experiment that indicated that neutron powder diffraction is capable of providing valuable information on octahedral site occupancies in such complex silicates, information that was unobtainable by X-ray diffraction.

Subsequent high temperature studies on phengite (Mookherjee et al. 2001; Fig. 17) have focused on the behavior of the hydroxyl group, and have been able to correlate infrared observations with structural changes at the hydroxyl site at extreme temperatures using constant wavelength angle dispersive neutron powder diffraction at ILL, D2B. Here a large

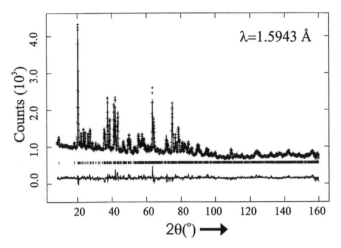

Figure 17. Constant wavelength neutron powder diffraction data for phengite (at 100 °C) collected at D2B, ILL (from Mookherjee et al. 2001). Data were collected at 100 °C intervals up to 1000 °C. Note the high background that results from the incoherent scattering off hydrogen in this natural (undeuterated) sample (cf. Fig. 7).

angle dispersive bank of detectors is arranged on an air bed and sweeps through a small range of angles to collect an entire diffraction pattern relatively rapidly (Fig. 18), in contrast to the fixed arrangement of detectors and wide distribution of neutron energies at the ISIS spallation source. Recent improvements ("super" D2B) have increased the number, resolution, and height of the detectors such that the detector bank operates as a quasi-2D system, resulting in an order of magnitude increase in counts at the detector. Radial collimators are used to good effect to improve the signal-to-noise ratio.

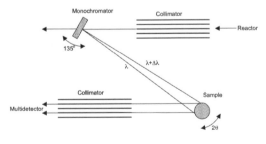

Figure 18. The constant wavelength powder diffractometer D2B at ILL showing the arrangement of radial detector collimators (lower left) and a schematic of the sample and diffractometer geometry.

Mg/Al and Al/Si ordering in amphiboles. Amphiboles are a major group of hydrous minerals, occurring in a wide range of geological environments and earth history. Their chemistry and cation ordering behavior is often diagnostic of the conditions of crystallization (pressure, temperature, cooling rates). As such, they have enormous potential as geological "indicators," provided that their cation ordering behavior can be quantified. However, these minerals are very complex structurally and chemically. A very successful approach to understanding the major crystal-chemical principles controlling cation ordering in amphiboles and their pressure-temperature stability has been to use synthetic analogues, where the complex chemistry of natural amphiboles is modeled using key condensed phases in systems of components such as Na_2O-CaO-MgO-Al_2O_3-SiO_2-H_2O. Unfortunately, synthetic amphibole crystals are almost always too small for single-crystal X-ray structure refinement. Experience has shown that Rietveld refinement of X-ray powder diffraction data does not give reliable site occupancies unless very contrasting element pairs are involved (e.g., Mg/Co, Mg/Si, Ga/Si). These compositions are often of little direct relevance to natural systems. Subtle problems such as Mg/Al and Al/Si ordering are certainly beyond the capabilities of X-ray powder diffraction because occupancies from site-scattering are likely to be unresolved and bond lengths are not sufficiently well constrained to derive meaningful occupancies from established bond length vs. site occupancy relationships.

Welch and Knight (1999) investigated Al/Si and Mg/Al ordering in the geologically important high-temperature amphibole *pargasite* $NaCa_2[Mg_4Al][Si_6Al_2]O_{22}(OH)_2$ by neutron powder diffraction. There are four $T1$ and four $T2$ tetrahedral sites and two $M1$, two $M2$ and one $M3$ octahedral sites per formula unit. Cation ordering involves the distribution of $4Mg + 1Al$ over the five octahedral sites and $2Al + 6Si$ over the eight tetrahedral sites. Welch and Knight used the high-flux, medium resolution time-of-flight diffractometer POLARIS at ISIS (Hull et al. 1992). Data were collected using the $90°$ 2θ and back-scattering ($145°$ 2θ) detector banks, giving a total *d*-spacing range from 0.5 Å to 4.3 Å. Pargasite has space group $C2/m$ and 15 atoms in its asymmetric unit. As such, it presents a considerable challenge to Rietveld refinement if we hope to determine the state of Mg/Al and Al/Si order. It is one of the most structurally complex minerals yet studied by neutron powder diffraction. A total of 64 parameters were refined: 45 structural (atomic coordinates, displacement parameters), 4 cell (a, b, c, β), 5 profile (scale factor, peak shape) and 10 background. The only "hard" constraints used in the refinements were those forcing refined site occupancies to be consistent with the known octahedral (Mg_4Al) and tetrahedral (Al_2Si_6) chemistries. Pargasite was synthesized hydrothermally at 0.1 GPa, 1200 K. The sample size for neutron diffraction was 2.4 g, which is large from a synthesis viewpoint, but modest for neutron diffraction. The neutron experiment was done at 295 K and lasted sixteen hours. The experimental, simulated and difference patterns are shown in Figure 19.

Notice that the powder diffraction pattern of pargasite (Fig. 19) shows many well resolved peaks below 1 Å - a favorable consequence of the high-Q Maxwellian "hump" in the energy spectrum of neutrons produced at ISIS. The structure of pargasite was refined to an R_{wp} value of 1.9%, which is much lower than R factors obtained for synthetic amphiboles from Rietveld refinements using X-ray powder data, which are typically around 10-15% (e.g., Raudsepp et al. 1987; Della Ventura et al. 1993, 1997; Hawthorne et al. 1997). However, it did not prove possible to refine site occupancies of tetrahedra and octahedra. Evidently, the scattering contrast for Al/Si and Mg/Al is not high enough to enable this to be done for this complex structure, compared with simple structures such as $MgAl_2O_4$ spinel for which site occupancies can be found from site scattering values (Redfern et al. 1999). However, the bond lengths are very well constrained (±0.002 Å to ±0.003 Å) and allow occupancies for tetrahedral and octahedral sites to be deduced using the well defined correlations between site occupancy and mean $T1$-O and M-O bond lengths (Hawthorne 1983; Oberti et al. 1995b). For example:

$$^{T1}Al = 33.2055 \, [<T1\text{-}O> - 1.6187] \qquad <M2\text{-}O> = 1.488 + 0.827(8)^{M2}<r>$$

where $<T1\text{-}O>$ and $<M2\text{-}O>$ are mean bond lengths and $^{M2}<r>$ is the mean radius of the $M2$

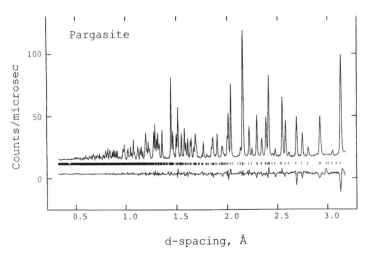

Figure 19. Time of flight powder diffraction pattern of the amphibole pargasite, collected at the POLARIS diffractometer, ISIS. Note the high quality data below 1Å *d*-spacing (from Welch and Knight 1999).

cation. The standard errors on bond lengths obtained from the neutron powder diffraction data approach those of single-crystal refinements and this is an important consequence of using neutrons: as neutron diffraction involves essentially point scattering (unlike X-ray diffraction), inter-nuclear distances can be determined very precisely, even from powder data. In the case of pargasite, the <*T*1-O> value implies that there is 30±13% Al/Si disorder over *T*1 and *T*2 sites. This amount of disorder is consistent with that found in a ^{29}Si MAS NMR study of the same sample (45±10% measured by Welch et al. 1998), and is also comparable to that in natural high-temperature pargasitic amphiboles by single-crystal X-ray refinement (42±10% determined by Oberti et al. 1995b). The <*M*2-O> and <*M*3-O> bond lengths, 2.051(3) Å and 2.048(3) Å, respectively, indicate fractional site occupancies on *M*2 of 0.25(4) Al and *M*3 of 0.50(8) Al. The <*M*1-O> bond length of 2.072(3) Å is consistent with *M*1 being filled by Mg. This ordering scheme is qualitatively similar to that observed in natural high-temperature pargasitic amphiboles by single-crystal structure refinement (Oberti et al. 1995a). However, while the Mg/Al distribution over *M*2 and *M*3 in the natural samples is statistical ($^2/_3$Mg + $^1/_3$Al on each *M*2 and *M*3 site), in synthetic pargasite it clearly is not. That Al orders at *M*2 and *M*3 (and not *M*1) in both natural and synthetic pargasites indicates that they share a common crystal chemistry. Nonetheless, the fact that the distribution of [6]Al is different in natural and synthetics may point to fundamentally different growth mechanisms. Hence, neutron powder diffraction has provided important new information about the relationship between cation distributions in natural amphiboles and their synthetic analogues.

Fe-Ti oxides. Our last case study of cation ordering is that of members of the $(FeTiO_3)_x(Fe_2O_3)_{1-x}$ solid solution, which have large saturation magnetizations and contribute significantly to the palaeomagnetic record. Often such material is observed to acquire self-reversed remnant magnetization. In all cases, the high-temperature $R\bar{3}c$ to $R\bar{3}$ cation ordering transition plays a crucial role in determining the thermodynamic and magnetic properties. This transition involves the partitioning of Ti and Fe cations between alternating (001) layers of the hexagonal-close-packed oxygen sublattice (Fig. 20).

Above the transition temperature (T_c) the cations are distributed randomly over all (001) layers. Below T_c the cations order to form Fe-rich A-layers and Ti-rich B-layers. Harrison et al. (2000a) carried out an in-situ time-of-flight neutron powder diffraction study of synthetic

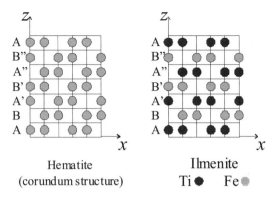

Figure 20. Schematic representation of the ordering scheme of low-temperature ilmenite compared to hematite.

Hematite
(corundum structure)

Ilmenite

Ti ●   Fe ●

samples of the $(FeTiO_3)_x(Fe_2O_3)_{1-x}$ solid solution with compositions $x = 0.7, 0.8, 0.9$ and 1.0 (termed ilm70, ilm80, ilm90 and ilm100), which provides an excellent illustrative case study to augment the discussions of models presented above. Harrison et al. (2000a) obtained cation distributions in members of the solid solution at high temperatures directly from measurements of the site occupancies using Rietveld refinement of neutron powder diffraction data. This proved especially powerful in this case because of the very large neutron scattering contrast between Ti (−3.482 fm) and Fe (+9.45 fm). The measurements offered the first insight into the equilibrium cation ordering behavior of this system over this compositional range and allow the simultaneous observation of the changes in degree of order, spontaneous strain and the cation-cation distances as a function of temperature. A qualitative interpretation of the observations was provided in terms of the various long- and short-range ordering processes which operate.

In discussing the changes in cation distribution which occur as a function of temperature, T, and composition, x, it is useful to define the long-range interlayer order parameter, η, as $(X_{Ti}^B - X_{Ti}^A)/(X_{Ti}^B + X_{Ti}^A)$. According to this definition, the order parameter takes a value of $\eta = 0$ in the fully disordered state (with Fe and Ti statistically distributed between the A- and B-layers) and a value of $\eta = 1$ in the fully ordered state (with the A-layer fully occupied by Fe and all available Ti on the B-layer). Values of η are shown in Figure 21. In all cases the estimated standard deviation in η measured by neutron diffraction is smaller than the size of the symbols. The value of η measured at room temperature represents the degree of order maintained after quenching the starting material from the synthesis temperature of 1300 °C. In ilm80, ilm90 and ilm100 the quenched starting material is almost fully ordered, with $\eta = 0.98$ in all three cases. This apparently low value of η in ilm70 may be due to the presence of chemical heterogeneities that develop on heating the sample below the solvus that exists at intermediate composition.

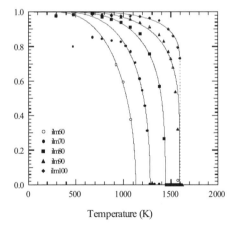

Figure 21. T dependence of the order parameter η for members of the ilmenite-hematite solid solution, determined from neutron powder diffraction (solid symbol: Harrison et al. 2000a) and quench magnetization (open symbols: Brown et al. 1993). Solid lines are fits using a modified Bragg-Williams model.

The ordering behavior in ilm80, ilm90 and ilm100 appears to be fully reversible, but the data close to T_c can only be fitted with a critical exponent for the order parameter, β, which is of the order of 0.1. This does not correspond to any classical mean field Landau-type model. Instead, a modified Bragg-Williams model is required, that describes the free energy phenomenologically in terms of a configurational entropy alongside an enthalpy that contains terms up to η^4:

$$\Delta G = RT \ln \Omega + \frac{1}{2} a \eta^2 + \frac{1}{4} b \eta^4$$

The hexagonal unit cell parameters provided by the diffraction measurements, a and c, and the cell volume, V, are plotted in Figure 22 for all temperatures and compositions measured by Harrison et al. (2000a). There are significant changes in both the a and c cell parameters correlated with the phase transition. Such changes are usually described by the spontaneous strain tensor, ε_{ij}. In the case of the $R\bar{3}c$ to $R\bar{3}$ transition, where there is no change in crystal system, the only non-zero components are changes in a and b, $\varepsilon_{11} = \varepsilon_{22}$ and in c, ε_{33}. The estimated variation in a_0 and c_0, the paraphase cell parameters, as a function of temperature, is shown by the dashed lines in Figure 22. From Figure 22a one sees that ε_{11} is negative, and that its magnitude increases with increasing Ti-content. The changes in a occur smoothly over a large temperature range and there is no sharp change in trend at $T = T_c$ in any of the samples. In contrast ε_{33} is positive. In ilm70, it is relatively small and c varies smoothly through the transition. In ilm80 and ilm90, ε_{33} is larger and the decrease in c occurs very abruptly at the phase transition. It should be noted that the magnitude of all spontaneous strains associated with ordering in these samples is relatively small, yet easily discernable using the ISIS time-of-flight instruments. This is consistent with the observations of Nord and Lawson (1989), who studied the twin-domain microstructure associated with the order-disorder transition. The twin boundaries have wavy surfaces, as is expected if there is no strain control over them. Furthermore, the fact that the spontaneous strain on ordering is very small provides the first

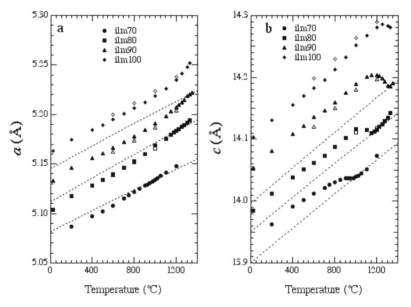

Figure 22. Variation in the cell parameters (a) a, and (b) c as a function of temperature, from Harrison et al. (2000a). Dashed lines are the estimated variation in a_0, and c_0 as a function of temperature.

hint that the length scale of the ordering interactions may not be very long-range. Generally, systems that display large strains on ordering tend to behave according to mean field models, as the strain mediates long-range correlations, whereas systems with weak strain interactions tend to show bigger deviations from mean-field behavior.

The spontaneous strain for long-range ordering in ilmenite is approximately a pure shear (with ε_{11} and ε_{33} having opposite sign and in the original data it is clear that the volume strain, $\varepsilon_V \approx 0$). It seems reasonable to assume that short-range ordering, which is often an important feature of such transitions, will play a significant role in determining the structural changes in the vicinity of the transition temperature in ilmenite. Here, short-range order may be defined by a parameter, σ, which is a measure of the degree of self-avoidance of the more dilute atom (e.g., Al-Al avoidance in aluminosilicates such as feldspars, or Ti-Ti avoidance in the case of ilmenite). The short-range order parameter, σ, is 1 for a structure with no alike nearest neighbors, and 0 for a totally random structure. For example, in ilmenite it may be defined as:

$$\sigma = 1 - \left(\frac{\text{proportion of Ti-O-Ti bonds}}{\text{proportion of Ti-O-Ti bonds in random sample}} \right)$$

Below T_c, σ includes a component due to long-range order. Meyers et al. (1998) therefore defined a modified short-range order parameter, σ', that excludes short-range order arising from long-range order: $\sigma' = (\sigma - \eta^2)/(1 - \eta^2)$. Short-range order will become important at temperatures close to and above T_c, where there is mixing of Fe and Ti on both the A- and B-layers. In addition, one expects that short-range ordering above T_c will be more important at compositions close to ilm100, where the Fe:Ti ratio approaches 1:1. Evidence of both these effects can be seen in the cell parameter variation as a function of temperature and composition, as illustrated schematically in Figure 23.

There is a rapid increase in the degree of short-range order at temperatures approaching T_c, which correlates with the rapid decrease in long-range order. Above T_c, σ decreases slowly, driven by the increase in configurational entropy at higher temperatures. The thin solid lines in Figure 23b show the effect of long-range ordering on the a and c cell parameters, the dashed lines show the effect of short-range ordering. The thick solid line shows the sum of the long- and short-range effects. In the case of the a cell parameter, the strains due to decreasing η and

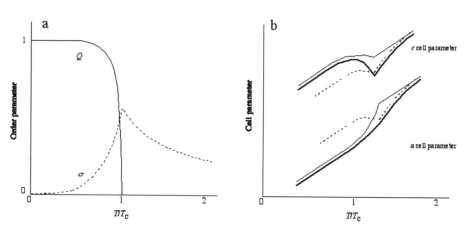

Figure 23. (a) Variation in long-range order, η, and short-range order, σ, as a function of T. (b) Effect of competing long- and short-range order on a and c parameters. Thin solid lines show long-range ordering effects, dashed lines show short-range effects. The thick lines give their sum (from Harrison et al. 2000a).

increasing σ compensate each other as the transition temperature is approached. This leads to a rather smooth variation in *a* as function of *T*, with no sharp change in *a* at $T = T_c$. In the case of the *c* cell parameter, the two strain components reinforce each other, leading to a large and abrupt change in *c* at $T = T_c$, as is observed in Figure 23. According to the arguments above, one expects this effect to be more obvious for bulk compositions close to ilm100, as indeed can be seen. Monte Carlo simulations of this system (Harrison et al. 2000b) confirm this interpretation of the strain effects, with the short-range order parameters due to nearest and next-nearest neighbor cation interactions behaving much as is shown schematically in Figure 23a.

CONCLUSIONS AND FUTURE PROSPECTS

Neutron scattering has proven invaluable in determining atomic occupancies over sites. Aside from the examples cited above, the power of neutrons to discern anion occupancies of relatively light elements, including (most importantly) oxygen places neutron diffraction in a unique position. An essential tool in the arsenal of the mineralogist, we are likely to see a further increase in the use of these methods for the determination of order-disorder phenomena and partial occupancies over sites. Our very first example was that of oxy-anion disorder in calcite. The latest neutron study of this phenomenon shows that the rotational disorder of the CO_3 groups is analogous to Lindemann melting (Dove et al. 2005; Fig. 24). Similarly, neutron diffraction has been used to investigate molecular orientational disorder in ammoniated silicates to great effect.

While we have noted the importance of neutron powder diffraction for the study of orientational order-disorder transitions, and have devoted much time to discussion of substitutional (alloy) order-disorder of cations in minerals, it should be noted that neutron

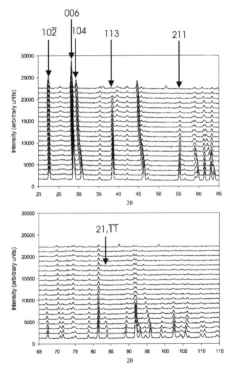

Figure 24. (left) stack plot of diffraction patterns of calcite as a function of temperature, collected at the DUALSPEC diffractometer, Chalk River. (above) Observed Fourier maps at height $z = 0$ of the hexagonal unit cell of calcite at 1189 K. The open circles show the carbon positions, with the location of the ordered bond to oxygen shown by the solid line. Oxygen distributions are banana-shaped around this vector, demonstrating the spread in orientations of the CO_3 group at this temperature (from Dove et al. 2005).

diffraction has also been employed in the study of positional disorder of both cations and anions in earth materials. Notable examples include the studies of Li-ion motion and disorder in β-eucryptite (Xu et al. 1999; Sartbaeva et al. 2004). While Li, being a light element, is difficult to detect using X-ray methods it shows strong scattering by neutrons, although isotopically enriched samples are used to avoid strong incoherent scattering. Other areas of study of disorder in minerals that neutron powder diffraction has addressed include numerous investigations of molecular and proton disorder in hydrous minerals, including a number of such studies of double-layer hydroxides as well as of inter-layer water in clay minerals. Such investigations have not, usually, involved the pursuit of phase transition phenomena, and therefore they have not been included in this chapter.

The extension of the high temperature studies, such as those outlined above, has begun with the development of *in situ* high-P/T methodologies adapted for use at neutron sources. Time-of-flight diffraction in 90° scattering geometry is particularly well suited for this purpose, and has been exploited for the study of cation order-disorder in spinel and in ilmenite as a function of both P and T simultaneously (Fig. 25; Méducin et al. 2004; Harrison et al. 2006). These studies have revealed that the thermodynamic controls on order-disorder transformations in minerals are not limited to the effects of the volumes of ordering/disordering on enthalpies, but that cation-cation (non-bonded) interactions, which drive ordering, change substantially with pressure and modify the order-disorder behavior significantly in the two examples investigate to date.

Such high-P/T experiments are particularly taxing, and push at the experimental limits of neutron powder diffraction. The samples studied in assemblies such as that shown in Figure 25 are typically very small (of the order of 4 mm in diameter and approximately the same dimension in length), and are surrounded by additional materials associated with the gasketing, anvils, and loading frame. Problems of small sample volumes are, therefore, compounded with problems of incident beam attenuation, incoherent scattering, and diffracted beam contamination in the case of poorly collimated set-ups. Such problems can be overcome by careful experiments, but the quality of data collected is necessarily compromised compared with the best ambient pressure datasets.

The extension of these studies to pressures and temperatures approaching the transition zone and lower mantle of the earth will require an order of magnitude improvement in neutron beam flux. Such an improvement is, fortunately, on the horizon with the next generation of

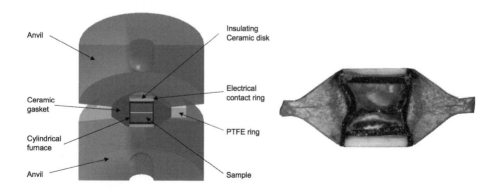

Figure 25. (left) The internal heating system employed in the Paris-Edinburgh loading frame high-P/T apparatus (LeGodec et al. 2001). Neutrons pass through the sample parallel to the vertical axis of the figure, scattering at 90° horizontally. (right) recovered sample assembly containing spinel heated to 1500 °C and pressurized to 3 GPa, used for *in situ* study by Méducin et al. (2004).

neutron sources currently under construction around the globe. Clearly the prospects for increased use of extreme sample environments in the application of neutron diffraction to systems at real earth interior conditions is likely to be substantial in the coming decade, with the advent of new spallation sources including the high pressure beam line SNAP at SNS and the extreme conditions beam line EXESS planned for target station 2 at ISIS.

ACKNOWLEDGMENTS

I gratefully and freely acknowledge the fruitful collaborations with many colleagues who have illuminated my studies of order-disorder in minerals by neutron diffraction. Much of the work described here was funded through CCLRC at ISIS.

REFERENCES

Amisano-Canesi A, Chiari G, Ferraris G, Ivaldi G, Soboleva SV (1994) Muscovite- and phengite-3T: crystal structure and conditions of formation. Eur J Minera. 6:489-496
Artioli G, Rinaldi R, Wilson CC, Zanazzi PF (1995) High temperature Fe-Mg cation partitioning in olivine: In-situ single-crystal neutron diffraction study. Am Mineral 80:197-200
Ballet O, Fuess H, Fritzsche T (1987) Magnetic structure and cation distribution in $(Fe,Mn)_2SiO_4$ olivine by neutron diffraction. Phys Chem Minerals 15:54-58
Bish DL (1981) Cation ordering in synthetic and natural Ni-Mg olivine. Am Mineral 66:770-776
Brown NE, Navrotsky A, Nord GL, Banerjee SK (1993) Hematite (Fe_2O_3) – ilmenite $(FeTiO_3)$ solid solutions: determinations of Fe/Ti order from magnetic properties. Am Mineral 78:941-951
Carpenter MA, Salje EKH (1994a) Thermodynamics of non-convergent cation ordering in minerals: II Spinels and orthopyroxene solid solution. Am Mineral 79:770-776
Carpenter MA, Salje, EKH (1994b) Thermodynamics of non-convergent cation ordering in minerals: III. Order parameter coupling in potassium feldspar. Am Mineral 79:1084-1098
Della Ventura G, Robert J-L, Raudsepp M, Hawthorne FC (1993) Site occupancies in monoclinic amphiboles: Rietveld structure refinement of synthetic nickel magnesium cobalt potassium richterite. Am Mineral 78: 633-640
Della Ventura G, Robert J-L, Raudsepp M, Hawthorne FC, Welch MD (1997) Site occupancies in synthetic monoclinic amphiboles: Rietveld structure refinement and infrared spectroscopy of (nickel, magnesium,cobalt)-richterite. Am Mineral 82:291-301
Dove MT, Powell BM (1989) Neutron diffraction study of the tricritical orientational order-disorder phase transition in calcite at 1260 K. Phys Chem Minerals 16:503-507
Dove MT, Sawinson IP, Powell BM, Tennant DC (2005) Neutron powder diffraction study of the orientational order-disorder phase transition in calcite, $CaCO_3$. Phys Chem Minerals 32:493-503
Eckold G, Hagen M, Steigenberger U (1998) Kinetics of phase transitions in modulated ferroelectrics: Time-resolved neutron diffraction from Rb_2ZnCl_4. Phase Transitions 67:219-244
Ericsson T, Nord AG (1984) Strong cation ordering in olivine-related (Ni,Fe)-sarcopsides – a combined Mössbauer, X-ray and neutron diffraction study. Am Mineral 69:889-895
Harrison RJ, Becker U, Redfern SAT (2000b) Thermodynamics of the R-3 to R-3c transition in the ilmenite-hematite solid solution. Am Mineral 85:1694-1705
Harrison RJ, Redfern SAT, O'Neill HStC (1998) The temperature dependence of the cation distribution in synthetic hercynite $(FeAl_2O_4)$ from in-situ neutron structure refinements. Am Mineral 83:1092-1099
Harrison RJ, Redfern SAT, Smith RI (2000a) *In situ* study of the R-3 to R-3c transition in the ilmenite-hematite solid solution using time-of-flight neutron powder diffraction. Am Mineral 85:194-205
Harrison RJ, Stone HJ, Redfern SAT (2006) Pressure dependence of Fe-Ti order in the ilmenite-hematite solid solution: implications for the origin of lower crustal magnetization. Phys Earth Planet Int 154:266-275
Hawthorne FC (1983) The crystal chemistry of the amphiboles. Can Mineral 21:173-480
Hawthorne FC, Della Ventura G, Robert J-L, Welch MD, Raudsepp M, Jenkins DM (1997) A Rietveld and infrared study of synthetic amphiboles along the potassium-richterite - tremolite join. Am Mineral 82: 708-716
Henderson CMB, Knight KS, Redfern SAT, Wood BJ (1996) High-temperature study of cation exchange in olivine by neutron powder diffraction. Science 271:1713-1715
Henderson CMB, Redfern SAT, Smith RI, Knight KS, Charnock JM (2001) Composition and temperature dependence of cation ordering in Ni-Mg olivine solid solutions: a time-of-flight neutron powder diffraction and EXAFS study. Am Mineral 86:1170-1187

Hull S, Smith RI, David WIF, Hannon AC, Mayers J, Cywinski R (1992) The POLARIS powder diffractometer at ISIS. Physica B 180:1000-1002

Krause MK, Sonntag R, Kleint CA, Ronsch E, Stusser N (1995) Magnetism and cation distribution in iron zinc silicates. Physica B 213:230-232

Kroll H, Knitter R (1991) Al, Si exchange kinetics in sanidine and anorthoclase and modeling of rock cooling paths. Am Mineral 76:928-941

Kroll H, Schlenz H, Phillips MW (1994) Thermodynamic modelling of non-convergent ordering in orthopyroxenes: a comparison of classical and Landau approaches. Phys Chem Minerals 21: 555-560

Landau LD (1937) On the theory of phase transitions, part I. Sov Phys JETP 7:19

Kuhs WF, Hansen TC (2006) Time-resolved neutron diffraction studies with emphasis on water ices and gas hydrates. Rev Mineral Geochem 63: 171-204

Larson AC, Von Dreele RB (2004) GSAS general structure analysis system. LAUR 86-748. Los Alamos National Laboratory, New Mexico, USA.

LeGodec Y, Dove MT, Francis DJ, Kohn SC, Marshall WG, Pawley AR, Price GD, Redfern SAT, Rhodes N, Ross NL, Schofield PF, Schooneveld E, Syfosse G, Tucker MG, Welch MD (2001) Neutron diffraction at simultaneous high temperatures and pressures, with measurement of temperature by neutron radiography. Mineral Mag 65:737-748

Loong C-K (2006) Inelastic scattering and applications. Rev Mineral Geochem 63:233-254

Méducin F, Redfern SAT, Le Godec Y, Stone HJ, Tucker MG, Dove MT, Marshall WG (2004) Study of cation order-disorder in $MgAl_2O_4$ spinel by *in situ* neutron diffraction up to 1600 K and 2.6 GPa. Am Mineral 89:981-986

Meyers ER, Heine V, Dove M (1998) Thermodynamics of Al/Al avoidance in the ordering of Al/Si tetrahedral framework structures. Phys Chem Minerals 25:457-464

Millard RL, Peterson RC, Swainson IP (2000) Synthetic $MgGa_2O_4$-Mg_2GeO_4 spinel solid solution and beta-$Mg_3Ga_2GeO_8$: chemistry, crystal structures, cation ordering, and comparison to Mg_2GeO_4 olivine. Phys Chem Minerals 27:179-193

Mookherjee M, Redfern SAT, Zhang M (2001) Thermal response of structure and hydroxylation of phengite 2M1: an *in situ* neutron diffraction and FTIR study. Eur J Mineral 13:545–555

Nord AG (1983) Neutron-diffraction studies of the olivine-related solid-solution $(Ni_{0.75}N_{0.25})_3(PO_4)_2$. Neues Jahrb Mineral Monatsh 9:422-432

Nord GL, Lawson CA (1989) Order-disorder transition-induced twin domains and magnetic properties in ilmenite-hematite. Am Mineral 74:160-176

Oberti R, Hawthorne FC, Ungaretti L, Canillo E (1995a) [6]Al disorder in mantle amphiboles. Can Mineral 33: 867-878

Oberti R, Ungaretti L, Canillo E, Hawthorne FC, Memmi I (1995b) Temperature-dependent Al order-disorder in the tetrahedral double chain of $C2/m$ amphiboles. Eur J Mineral 7:1049-1063

Parise JB (2006) Introduction to neutron properties and applications. Rev Mineral Geochem 63:1-25

Pavese A, Artioli G, Hull S (1999) *In situ* powder neutron diffraction of cation partitioning vs. pressure in $Mg_{0.94}Al_{2.04}O_4$ synthetic spinel. Am Mineral 84:905-912

Pavese A, Ferraris G, Prencipe M, Ibberson R (1997) Cation site ordering in phengite 3T from the Dora-Maira massif (western Alps): a variable-temperature neutron powder diffraction study. Eur J Mineral 9:1183-1190

Peterson RC, Lager GA, Hitterman RL (1991) A time-of-flight neutron powder diffraction study of $MgAl_2O_4$ at temperatures up to 1273 K. Am Mineral 76:1455-1458

Raudsepp M, Turnock AC, Hawthorne FC, Sherriff BL, Hartman JS (1987) Characterization of synthetic pargasitic amphiboles $(NaCa_2Mg_4M^{3+}Si_6Al_2O_{22}(OH,F)_2$: M^{3+} = Al, Cr, Ga, Sc, In) by infrared spectroscopy, Rietveld structure refinement and ^{27}Al, ^{29}Si, and ^{19}F MAS NMR spectroscopy. Am Mineral 72:580-593

Redfern SAT, Artioli G, Rinaldi R, Henderson CMB, Knight KS, Wood BJ (2000) Octahedral cation ordering in olivine at high temperature. II: An *in situ* neutron powder diffraction study on synthetic $MgFeSiO_4$ (Fa50). Phys Chem Minerals 27:630-637

Redfern SAT, Harrison RJ, O'Neill HStC, Wood DRR (1999) Thermodynamics and kinetics of cation ordering in $MgAl_2O_4$ spinel up to 1600 °C from *in situ* neutron diffraction. Am Mineral 84: 299-310

Redfern SAT, Henderson CMB, Knight KS, Wood BJ (1997) High-temperature order-disorder in $(Fe_{0.5}Mn_{0.5})_2SiO_4$ and $(Mg_{0.5}Mn_{0.5})_2SiO_4$ olivines: an *in situ* neutron diffraction study. Eur J Mineral 9: 287-300

Redfern SAT, Henderson CMB, Wood BJ, Harrison RJ, Knight KS (1996) Determination of olivine cooling rates from metal-cation ordering. Nature 381:407-409

Redfern SAT, Knight KS, Henderson CMB, Wood BJ (1998) Fe-Mn cation ordering in fayalite-tephroite $(Fe_xMn_{1-x})_2SiO_4$ olivines: a neutron diffraction study. Mineral Mag 62:607-615

Redfern SAT, Salje E, Navrotsky A (1989) High-temperature enthalpy at the orientational order-disorder transition in calcite: implications for the calcite/aragonite phase equilibrium. Contrib Mineral Petrol 101: 479-484

Rinaldi R, Gatta GD, Artioli G, Knight KS, Geiger CA (2005) Crystal chemistry, cation ordering and thermoelastic behaviour of $CoMgSiO_4$ olivine at high temperature as studied by *in situ* neutron powder diffraction. Phys Chem Minerals 32:655-664

Rinaldi R, Wilson CC (1996) Crystal dynamics by neutron time-of-flight Laue diffraction in olivine up to 1573K using single frame methods. Solid State Commun 97:395-400

Rouse KD, Thomas MW, Willis BTM (1976) Space group of spinel structure – neutron diffraction study of $MgAl_2O_4$. J Phys C: Solid State Phys 9:L231-L233

Sartbaeva A, Redfern SAT, Lee WT (2004) A neutron diffraction and Rietveld analysis of cooperative Li motion in β-eucryptite. J Phys Cond Matter 16:5267-5278

Thompson JB Jr (1969) Chemical reactions in crystals. Am Mineral 54 341-375

Untersteller E, Hellner E, Heger G, Sasaki S, Hosoya S (1986) Determination of cation distribution in synthetic (Mg,Fe)-olivine using neutron diffraction. Z Kristallogr 174:198-198

Von Dreele RB (2006) Neutron Rietveld refinement. Rev Mineral Geochem 63:81-98

Warren MC, Redfern SAT (2000a) Ab initio simulation of cation ordering in oxides: application to spinel. J Phys Cond Matter 12:L43-L48

Warren MC, Dove MT, Redfern SAT (2000b) Disordering of $MgAl_2O_4$ spinel from first principles. Mineral Mag 64:311-317

Welch MD, Knight KS (1999) A neutron powder diffraction study of cation ordering in high-temperature synthetic amphiboles. Eur J Mineral 11:321-331

Welch MD, Liu S, Klinowski J (1998) ^{29}Si MAS NMR systematics of calcic and sodic-calcic amphiboles. Am Mineral 83:85-96

Xu HW, Heaney PJ, Yates DM, Von Dreele RB, Bourke MA (1999) Structural mechanisms underlying near-zero thermal expansion in β-eucryptite: A combined synchrotron X-ray and neutron Rietveld analysis. J Mater Res 14:3138-3151

Reviews in Mineralogy & Geochemistry
Vol. 63, pp. 171-204, 2006
Copyright © Mineralogical Society of America

8

Time-resolved Neutron Diffraction Studies with Emphasis on Water Ices and Gas Hydrates

Werner F. Kuhs

Geowissenschaftliches Zentrum der Universität Göttingen
Abteilung Kristallographie
Goldschmidtstraße 1, 37077 Göttingen, Germany
e-mail: wf.kuhs@geo.uni-goettingen.de

Thomas C. Hansen

Institut Max von Laue-Paul Langevin, 6 rue Jules Horowitz, BP 156
38042 Grenoble Cedex 9, France
and
Geowissenschaftliches Zentrum der Universität Göttingen
Abteilung Kristallographie
Goldschmidtstraße 1, 37077 Göttingen, Germany
e-mail: hansen@ill.fr

INTRODUCTION

This chapter deals with applications of neutron diffraction to the understanding of transformation processes in geomaterials with a time-resolution from seconds to several days. The choice of neutrons rather than X-rays for such time-resolved studies is due to their specific advantages like e.g., the sensitivity to light elements, the detection of volume rather than surface phenomena or the low absorption in case of bulky sample environments. General features of diffractometers suitable for time-resolved studies are discussed and the high-flux diffractometer D20 at ILL is presented in some detail. Time-resolved processes usually are studied at non-ambient pressures and temperatures. The necessary sample environment needs to be provided and matched to the diffractometer set-up. Processes in water ices and gas hydrates are taken as examples and will be used to demonstrate the capabilities and limitations of the method.

Crystals are less static than what is usually suggested by immobile drawings of their atomic or molecular arrangements representing a time-space averaged picture of atomic positions as obtained from diffraction experiments. Rather, atoms or molecules exhibit thermal displacements and some even move inside the lattice as defects. These averaged displacements are described via atomic displacement parameters (e.g., Kuhs 2003) and are accessible also using diffraction techniques. Yet, there is even more movement in a crystal. Close to structural phase transitions the collective mobility of the constituents enables the formation of new structural arrangements to lower the free energy of the system (Redfern 2006, this volume). Such solid-solid transformations are not necessarily instantaneous, in particular when the new phase needs to be reconstructed by moving molecular and atomic constituents to a new position. Following such transitions is one of the main topics of this review.

Thermodiffractometry versus truly time-resolved studies

Phase transitions. If a material is capable of existing in more than one polymorphic form, the process of transformation from one polymorph to another is a phase transition (Redfern 2006, this volume). In the narrowest sense, these transitions are restricted to changes in structure only, in a wider one, the possibility of compositional changes is included. We shall

1529-6466/06/0063-0008$05.00   DOI: 10.2138/rmg.2006.63.8

call the latter a chemical reaction, to be distinguished from a purely structural, polymorphic transition. Time resolved work can be done in both cases and both cases are of interest here.

Thermodiffractometry. Sometimes, the formation of a new crystalline phase evolves in steps leading to short-lived intermediate phases, which may be accessed by diffractometry given that the data collection is sufficiently fast. A technique called thermodiffractometry (Pannetier 1985) gained popularity with the advent of one-dimensional position sensitive detectors, with which many powder diffraction patterns were taken while the samples were heated (or cooled) going through one or several phase transitions.

The term thermodiffractometry has been employed mainly for *in situ* diffraction studies on reversible phase transitions occurring with changing temperature. The temperature is continuously increased or decreased at a constant rate, and diffraction patterns are recorded during this temperature ramp at a constant pace. The dead time between two acquisitions has to be small, and the change of temperature during the acquisition of one pattern with sufficient counting statistics for later data treatment has to be sufficiently small.

In its narrowest sense, a thermodiffractometry experiment does not require a high time-resolution capacity, as the slope of the temperature ramp can be adapted to the necessary counting time. However, the term has been used as well, in a broader sense, for irreversible phase transitions or for experiments with a steep slope, not permitting to be in a thermodynamic but in a kinetic regime. One could, and sometimes has done, exploit the acquired data completely with respect to the kinetics of the phase transition.

Intermediate phases. In the frame of thermodiffractometry experiments in the latter sense, intermediate phases have unintentionally or purposely been observed (e.g., Kilcoyne and Manuel 2002) and studied, which have sometimes not been detected before (e.g., if the calorimetric fingerprint is not significant), and which may be only metastable and even not quenchable for *ex situ* studies. In this case the capability of doing time-resolved diffraction becomes important as the diffraction patterns must have sufficient counting rate for observation and acquisition of data sufficient for structural investigation in the short lapse of time the phase is present.

Truly time resolved studies

It is not the purpose to discuss thermodiffractometric experiments here, which have become a routine technique for exploring phase diagrams. We rather concentrate on truly time-resolved studies, in which a transformation process is followed at constant pressure and temperature to deduce kinetic information from diffraction data.

Studying the time dependency can be very instructive in giving quite detailed information on the mechanism of the transformation and delivering accurate reaction rate constants (see e.g., Levenspiel 1999). Time-resolved diffraction studies using powder samples give access to the relative amounts of the phases involved by a full pattern profile analysis (Knorr and Depmeier 2006, this volume). The formation of a new phase involves the formation of a critical nucleus and its subsequent growth. While the nucleus usually is so small that it yields only a very broad diffraction contribution, the growing crystallites produce increasingly sharp diffraction signals. The intensity of this evolving diffraction peak is proportional to the amount of new phase formed. A corresponding decrease of the Bragg peaks of the old phase should then be observed. The situation is more complicated in multiphase systems, in which constituents combine in heterogeneous chemical reactions to form new phases. Some of the constituents may be liquid or even gaseous. However, even in such cases the kinetics of the formation or decomposition of a crystalline phase may be followed by diffraction methods.

Meaningful information on the time-dependency of transformation processes in polycrystalline samples can only be obtained if the sample is sufficiently large to be representative. Transformations of different individual crystallites are not necessarily occurring simultaneously. Therefore it is often necessary to investigate large samples to limit boundary effects and to observe averages of large ensembles. For highly absorbing materials neutrons give frequently a more representative picture of the bulk behavior of the material, while standard X-rays laboratory experiments are biased towards near-surface effects. Neutrons are generally better suited for such studies due to their lower absorption as compared to X-rays. Even when hard synchrotron X-rays could be used, the analyzed sample volume usually is limited to a few mm^3 and not necessarily representative for the whole sample.

Geomaterials are formed over a large range of time-scales. Some processes are fast and could be followed at relevant time-scales in neutron diffraction experiments. However, as the access to neutron facilities is limited, it is hardly possible to study processes on a timescale longer than a few days. Work at the home laboratory allows extending the time-scales but often gives less quantitative evidence for the ongoing changes when compared to neutron diffraction. Processes may take as long as thousands or millions of years. Obviously, such processes cannot be studied directly by any means. Fortunately, under certain assumptions, and given an appropriate model, neutron experiments at much shorter time-scales may be used to extrapolate to a different regime with slower rates, usually assuming an Arrhenius behavior of an underlying rate-determining activated process (Lasaga 1998). Reactions close to the melting or decomposition point of materials are usually fast due to the increased mobility of the constituents. For most minerals the corresponding high temperature conditions are difficult to reach, however. Undoubtedly, neutron diffraction has advantages here as it allows the use of bulky furnaces without much loss of data quality.

Other important geomaterials like ices and gas hydrates become mobile and reactive at much lower temperatures and can be studied fairly easily using neutrons. Neutrons of course provide relevant details concerning hydrogen positions and ordering, but also have advantages, in particular for gas hydrates, where elevated gas pressures are needed which can more easily be handled using neutron diffraction as compared to X-rays. The kinetics of ice transformation during structural phase transitions or the kinetics of gas hydrate formation and decomposition were not well known despite their importance in geology and planetology. Time-resolved neutron diffraction proved to be a major tool to change this situation.

In the following we will concentrate on time-resolved studies of ices and gas hydrates. We first describe specific issues of these systems concerning neutron diffraction. Then we proceed by providing details of the experimental set-ups and conclude by discussing several examples in some detail. We concentrate here on technical aspects as a detailed discussion of the scientific implications is beyond the scope of the chapter.

HYDROGENOUS SAMPLES

Hydrogen and X-ray diffraction

Hydrogen as the lightest element contributes much less to the scattered intensity from X-ray diffraction than all elements it is associated with. It is therefore difficult to localize in combination with heavier elements. Although this contrast problem is less dramatic in ice than in compounds with metals, we have to deal with roughly an order of magnitude less scattering intensity contributed from hydrogen compared to oxygen.

Additionally, for X-rays, scattering occurs with electrons, which are not well localized around the hydrogen atom's nucleus. The center of the electron distribution in ice shifts considerably towards the oxygen atom bound to hydrogen. Even if the data quality allows the determination of the center of the hydrogen electron density position, corrections are necessary to obtain the hydrogen core position. Such corrections require a good knowledge of the nature of the chemical bonding.

Hydrogen and neutron diffraction

Nuclear neutron scattering is not sensitive to the electron distribution but to the nature of the nuclei. The magnitude of the atomic form factor determining the contribution of an atom to the scattered intensity is not a simple function of the number of protons and neutrons of a particular scattering nucleus. Apart from some exceptions, these form factors are in the same order of magnitude for all isotopes, therefore we do not have to struggle with the inherent imbalance problem of X-ray diffraction.

Incoherent scattering. In many cases we have to deal with incoherent scattering that does not carry positional information in contrast to coherent scattering. For structural diffraction work, we must consider incoherent scattering as an unwanted background contribution. Reasons for a high incoherent scattering cross section are mixtures of isotopes with strongly different scattering lengths b or, more likely, isotopes with strongly different scattering lengths b_- and b_+ for both possible spin orientations (Parise 2006a, this volume).

The dominant naturally occurring isotope of hydrogen is ^1H (99.985%), which has an incoherent scattering length of b_{inc} = 25.274 fm (Sears 1992). Consequently, hydrogen has an incoherent cross section of σ_{inc} = 80.27 barns (1 barn = 10^{-28} m), which is the highest one of all stable elements and isotopes, apart from three heavily absorbing samarium and gadolinium isotopes and the element gadolinium itself. The coherent scattering length of the element is b_{coh} = −3.739 fm and the cross section σ_{coh} = 1.7568 barns, which is rather low (similar to a weak scatterer like titanium, oxygen has a four times higher cross section, iron and nickel have a more than ten times higher one).

In addition to the extremely unfavorable signal to background ratio, the incoherent scattering contributes to an efficient absorption although the cross section for true absorption of hydrogen is relatively small. Any neutron scattered incoherently and afterwards scattered coherently will no longer be detected correctly in a diffraction experiment. On the other hand, coherently scattered neutrons are likely to experience incoherent scattering before leaving the sample causing deflections from the ideal direction from the point of view of a diffraction experiment. This efficient absorption depends on the wavelength λ and even on the chemical environment of the hydrogen atoms. Howard et al. (1987) measured the efficient absorption cross section, using a single crystal of a particular compound, to be $\sigma_H(\lambda)$/barns = 19.2(5)λ/Å + 20.6(9).

As a further drawback, especially in constant wavelength neutron powder diffraction experiments on larger samples, multiple scattering may become important which will derive the coherently diffracted neutron's path from the ideal one. This leads to a considerable broadening of all diffraction peaks in the powder diffraction pattern and thus lowers the angular resolution. Thus the determination of peak intensities becomes less precise.

Deuteration. One way to avoid this difficulty is the replacement of natural hydrogen with its isotope deuterium (^2H or D), called deuteration. Full deuteration is usually the best choice but cannot always be achieved due to experimental and/or financial reasons. If it is only for background and absorption reduction the replacement of a maximum of hydrogen against deuterium is aimed at. If detailed structural information on a hydrogen position is needed, one must aim for complete deuteration. In the case of partial deuteration, the exchange ratio should be known from the synthesis procedure or be determined by some means. This can be achieved e.g., by spectroscopy, by accurate density measurements of the constituents or by a crystallographic refinement of positional occupancies (see e.g., Chazallon and Kuhs 2002). Deuterium has a coherent scattering length of 6.671 fm, a coherent cross section of 5.592 barns. The different sign of the coherent scattering length means, that the corresponding scattering contributions of ^1H and ^2H will partly or fully annihilate when placed statistically on one crystallographic position. An occupation of such a site at a ratio of ^2H to ^1H of 3.739/6.671 will make it invisible to diffraction (35.9% ^2H and 64.1% ^1H).

A major drawback of deuteration, beside problems of feasibility, is the physico-chemical isotopic difference of hydrogen and deuterium. Normally, different isotopes of the same element behave chemically identically; there are only tiny mass-related differences in the thermodynamic and kinetic behavior of isotopic compounds. However, in the case of hydrogen, this mass difference is important, deuterium is a factor of about two heavier. Moreover, quantum effects lead to a better localization of the deuteron with a resulting slightly weaker H-bond for the deuterated compound translating into larger lattice constants (Röttger et al 1994). The molecular translational und librational motion is considerably reduced in the deuterated compound (corresponding to the relevant reduced masses) which in turn affects the localized and collective motions of water molecules. This is expressed, for example in the melting point of water ice, which is, slightly varying with pressure, several degrees higher for D_2O than for H_2O. Further to these thermodynamic effects, the kinetic constants differ considerably for natural and deuterated samples. The water molecule's local mobility in hydrogenated and deuterated ices Ih differs as evidenced in dielectric relaxation, the deuterated compound having a longer relaxation time (Johari 1976; Johari and Jones 1976, 1978). Although the role of water mobility in the clathration reaction or in the accompanying gas- and water- mass transport is not clear, the established differences suggest that the hydrate formation kinetics in the H_2O- and D_2O-systems could be different. This difference has been investigated by Staykova et al. (2003) and found to be rather small. Dramatic differences in the transition behavior between hydrogenated and deuterated systems could be expected when proton tunneling is involved (e.g., Matsuo 2003). Such a transition is likely to take place e.g., when approaching high pressure ice X (e.g., Pruzan et al. 2003), presently outside reach for neutrons due to the very high pressures needed. At more moderate pressures proton tunneling may also occur in other ices when ionic defects are present (Petrenko and Whitworth 1999), however the role of proton tunneling for kinetic phenomena remains largely unexplored.

Annular sample containers. In order to study the thermodynamics and the kinetics of natural samples directly without extrapolating data from deuterated samples, one has to take measures to reduce absorption and multiple scattering. In neutron powder diffraction experiments conducted in Debye-Scherrer geometry with a normally cylindrical sample in transmission, this can be achieved by using a double-walled sample can instead of a simple cylinder. The absorbing sample then has not a solid radius of, e.g., 2 to 10 mm, but is distributed cylindrically around an inner, hollow cylinder of a radius between, e.g., 2 and 5 mm with a thickness of 0.5 to 2 mm. With such a geometry, the incident and diffracted beam have roughly the same path length for all diffraction angles.

In the case of a solid cylinder, the path length is in the order of the sample diameter for low diffraction angles and distributed from close to zero to up to about twice the diameter for high diffraction angles. For strongly absorbing samples this means an increase of diffracted intensity from low to high angles, which can be—in principle—corrected for. However, this correction presents some major difficulties, especially when the efficient absorption coefficient cannot be calculated at high precision (which is the case for hydrogen as shown by Howard et al. 1987): There is an enormous correlation between the refinable absorption correction coefficients with structural parameters such as atomic displacement parameters, both, modulating the diffraction intensity from low to high angles. Even if the absorption correction is successful, the loss of information at low angles remains an insurmountable handicap for structural work.

The use of annular sample containers provides a more equal intensity distribution over the whole angular range with less need for absorption correction and a corresponding reduction of parameter correlations. Schmitt and Ouladdiaf (1998) provide an algorithm and a computer program for performing this correction.

The outer radius R and the ratio between the inner and the outer radius, ρ, need careful optimization, as a maximum radius cannot be exceeded for instrumental reasons (i.e., decrease of angular resolution) and as a minimum distance between the inner and the outer wall of the sample needs to be respected for reasons of mechanical feasibility. The equivalent radius R_{eq} of a full cylindrical sample with the same volume is given by $R_{eq}=R\cdot(1-\rho^2)^{\frac{1}{2}}$. The diffraction intensity gain at low diffraction angles can be very high (e.g., 15 for $\theta \approx 0°$ and $\mu R \approx 18$ with the linear absorption coefficient μ). For a full cylindrical sample, the optimized radius R_{opt} is given by $\mu R_{opt} \approx 1.5$ for $\theta \approx 0°$.

Finally, one should not forget that relatively long counting times or high incident neutron fluxes are necessary when working with non-deuterated hydrogenous samples: Even if one masters the problem of efficient absorption (and together with it the peak broadening due to multiple scattering), the problem of an extremely unfavorable low signal to background ratio remains. Only with deuteration incoherent scattering can be completely avoided, but the use of a high intensity diffractometer may produce useful data even under such unfavorable conditions in a reasonable measuring time. In the case of clathrate hydrate formation discussed below we found a decrease of precision of the refined phase fraction by roughly a factor of 40 when going from a partly deuterated CH_4-D_2O system to a fully hydrogenated CH_4-H_2O system, still far sufficient to refine meaningful phase fractions (Staykova et al. 2003).

INSTRUMENTS

Instrumentation at ILL

The ILL high flux reactor provides the world's highest thermal neutron flux as required for time-resolved experiments, particularly on hydrogen compounds. Two high-intensity powder diffractometers are of main interest D1B and D20 and were used in many of the experiments described below. They will be discussed in some detail.

The D1B powder diffractometer with a multi-wire gas chamber detector was used in the 1970's for the first truly time-resolved neutron diffraction experiments (Riekel and Schöllhorn 1976). The one-dimensional, curved position sensitive detector (PSD) covers a $\Delta 2\theta = 80°$ range of diffraction with a detector resolution of $0.1°$.

The successor of D1B (Convert et al. 1998, 2000; Hansen 2004) is the instrument D20 that provides a much higher flux at the sample position. To achieve a high, detected intensity, a stationary, curved linear PSD covers the whole 2θ range (Convert et al. 1997). The high detection stability of the micro-strip gas chamber detector enables D20 also to accomplish high precision in intensity measurement. Several monochromators are available.

A vertically focusing graphite monochromator provides $\lambda \approx 2.41$ Å at a take-off angle of $42°$ and is equipped with graphite filters to suppress harmonics. Alternatively a vertically and horizontally focusing copper monochromator Cu (200) in transmission geometry provides a monochromatic neutron beam with wavelengths of $\lambda \approx 0.82, 0.88, 0.94$ Å at take-off angles of $26°, 28°$ or $30°$. A second copper focusing monochromator Cu (200) in transmission provides $\lambda \approx 1.3$ Å at a take-off angle of $42°$; this configuration is shown in Figure 1. At $\lambda = 1.3$ Å the monochromatic beam has its highest flux of about 10^8 n·cm^{-2}·s^{-1}. Soller collimators allow to reduce the divergence of the incident polychromatic beam (27′) to $\alpha_1 = 10′$ or 20′. A germanium monochromator with variable vertical focusing allows obtaining higher resolution at higher take-off angles (65±2°, 90±2°, 118° and 120°, shown in Fig. 2).

The PSD of aluminum has a useful angular aperture of $153.6°$ (2θ). The detection zone is about 4 m long and 15 cm high. The gas filling of 3 bars ^{3}He and 1 bar CF$_4$ and the detection gap of 5 cm result in a neutron detection efficiency from 60% ($\lambda = 0.8$ Å) to 90% (2.4 Å). Each detector plate of electronically conducting glass (Oed 1988) carries 32 cells, 2.568 mm in size (0.1°), and composed of two basic chromium micro-strips (anodes and cathodes). To cover continuously $153.6°$, the PSD is composed of 48 juxtaposed, precisely cut plates in a polygonal

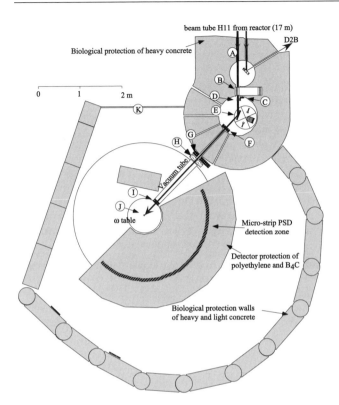

Figure 1. Schema of D20 in "high flux" configuration (take-off angle 42°). A: incident thermal neutron beam, B: optional 10' Soller collimator, C: primary fast thermal neutron beam shutter, D: highly oriented pyrolytic graphite (HOPG) filter, E: monochromator changer with HOPG (002) monochromator in beam, F: slits, G: secondary beam shutter, H: neutron beam monitor, I: slits, J: sample position, K: access gate.

Figure 2. D20 at 120° take-off angle (high resolution configuration). A part of the biological concrete protection wall is not in place. In the front the instrument control electronics. The detector is seen from behind, its counting chain being hidden behind mountable, neutron-shielding doors of polyethylene and boron carbide. Below the detector, on the movable platform, the data acquisition system is visible.

arrangement with a radius of 1471 mm, each plate covering 3.2°. The 1536 cells are connected to amplifiers and discriminators and anti-coincidence circuits (CLET). A cell, whose amplifier signal passed first the discriminator threshold, inhibits its neighbors for 1.5 μs, avoiding double detection.

This makes D20 an ideal tool for *in situ* diffraction studies with time constants even below a second. Its high intensity enables the use of complex sample environments, as the high count-rate may level out low signal-to-noise ratios. D20's high precision in intensity measurement is needed for differential measurements and studies on liquid and amorphous systems. The different options described above provide a large choice in Q-space, resolution, wavelength (0.8 to 2.4 Å), and flux. The continuous and simultaneous detection of series of complete diffraction patterns is necessary for systematic investigations of phase transitions during variation of a parameter such as pressure or temperature. The structural evolution of solids during a chemical reaction can be studied *in situ* with single diffraction patterns of down to 400 ms in highest intensity configuration, allowing quantifying short-living intermediate phases and elucidate subtle structural changes. High-resolution powder diffraction patterns can be obtained in a few minutes when the highest take-off angle is used.

Faster, but cyclic phenomena are observable in a stroboscopic data acquisition mode (Convert et al. 1990). A reversible process, which is reproducible in a cyclic, repeated way, can be observed using *stroboscopy*, allowing for a much better time-resolution, i.e., the shortest possible time slice of the phenomenon to be observed. This is possible by synchronizing a series of sequential acquisitions (*slices*) with the cycling of the observed process. The results of these acquisitions from corresponding slices in different cycles cumulate electronically in a register. The final data set corresponds then to one cycle of the process, but the counting rate for each data acquisition slice is as many times higher than there were cycles measured. One obtains sufficient counting statistics for a series of slices after each synchronization-cycle by repeating a sufficient number of individual cycles and accumulating the counting rates of corresponding slices. Unfortunately, for various reasons one cannot go for shorter and shorter slices and correspondingly higher cycling frequencies. There are electronic limitations to provide a good synchronization over a large number of cycles without any deviation, e.g., due to fatigue phenomena. The principal limitation, however, is due to the finite velocity of neutrons, resulting in uncertainty about the time of the scattering event in the sample of some dimension, and of the detection event in the commonly used gas chamber detectors. Compared to this effect, the $\Delta\lambda/\lambda$ distribution to the time-resolution is negligible. Time resolution is thus limited by the travel time of neutrons through sample and detection gap, at D20 about 10 μs. Brown et al. (1998, 2000) have used stroboscopic acquisitions for rheologic investigations of orientation changes of flat or rod like particles, suspended in water, during laminar flow.

A radial oscillating collimator with a focus aperture of 22 mm and an angular coverage of 156° in 2θ permits the use of most types of sample environment without any background contributions from, e.g., aluminum cryostat-calorimeters or niobium heating elements. D20 has dedicated sample environment devices like an electrical furnace with a cylindrical vanadium-foil heating element of 22, 30 or 40 mm diameter around the sample cylinder allows investigations on powders and liquids from room temperature up to 1300 °C without significant background contributions from the sample environment.

With this instrument many time-resolved diffraction experiments have been performed. Walton et al. (2001) studied the hydrothermal crystallization of barium titanate, $BaTiO_3$. The formation of the ferroelectric ceramic could be investigated in real time and under genuine reaction conditions. A nucleation-growth model as proposed by Hancock and Sharp (1972), based on the expressions of Avrami (1939, 1940, 1941) and Erofeyev (1946), was able to simulate the growth curves of barium titanate. For further details on a description of nucleation and growth processes, see Lasaga (1998) and Kashchiev (2000). Fehr et al. (2003) and Walk-

Lauffer (2003) designed an autoclave cell to perform time-resolved neutron diffraction experiments on dynamic processes during hydrothermal reactions in the presence of a (deuterated) hydrous fluid as they occur in the synthesis of aerated concrete. They investigated the influence of different additives like anhydrite or potassium sulfate on the formation of 1.13 nm tobermorite from portlandite and quartz at 190 °C and saturation pressure. The reaction time was set to about 6 hours and the reaction product consisted of tobermorite, semi-crystalline calcium-silicate-hydrate C-S-H(I) and quartz. Tobermorite forms at the expense of portlandite and quartz and by the reaction of semi-crystalline calcium-silicate-hydrate C-S-H(I) with quartz. The reaction progress α is inversely proportional to the remaining amount of portlandite or quartz, and its evolution expressed as a product of a function of α and a kinetic constant k (Bray and Redfern 1999).

In situ neutron diffraction traced the reaction mechanism during the self-propagating high-temperature synthesis (SHS) of Ti_3SiC_2 and related ceramics from furnace ignited stoichiometric Ti/SiC/C mixtures (Riley et al. 2002, 2006). The diffraction patterns indicate five stages, visible in Figure 3: (1) pre-heating of the reactants, (2) the $\alpha \rightarrow \beta$ phase transformation in Ti, (3) pre-ignition reactions, (4) the formation of a single solid intermediate phase in less than 0.9 s and (5) the rapid nucleation and growth of the product phase Ti_3SiC_2. The $\alpha \rightarrow \beta$ phase transformation in Ti is a necessary precursor to the reaction as is the subsequent reaction of Ti and C just before SHS ignition. This latter process is exothermic and provides the extra heat required for sustained SHS to occur. No amorphous contribution to the diffraction patterns from a liquid phase was detected and as such, it is unlikely that a liquid phase plays an important role in this SHS reaction. The intermediate phase is believed to be a solid solution of Si in TiC such that the overall stoichiometry is approximately 3Ti:1Si:2C. Lattice parameters and known thermal expansion data were used to estimate the ignition temperature as 901 ± 8 °C (confirmed by the $\alpha \rightarrow \beta$ phase transformation in Ti) and the combustion temperature as 2320 ± 50 °C. In further experiments, the data acquisition time could be reduced to 300 ms with a dead time of 40 ms between two acquisitions.

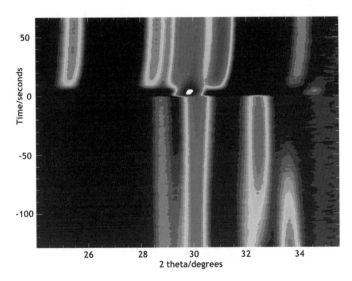

Figure 3. Contour plot of an angular range of 12° in 2θ during 200 s of the SHS of Ti_3SiC_2, representing about 220 diffraction patterns. 100 s before ignition ($t = 0$) the $\alpha \rightarrow \beta$ phase transition of Ti occurs (peaks at 32 and 34°). 30 s before ignition β-Ti is partially consumed to form TiC (peak at 30°). During the ignition (< 0.5 s), a solid solution of Si in TiC is formed. 2 s later, the intermediate product starts to decompose to Ti_3SiC_2 with a time constant of 5 s.

New electronic detectors developed at ILL permit much larger areas to be covered with two-dimensional PSDs. Hewat (2006) designed a new Diffractometer for Rapid ACquisition with Ultra-Large Area detector (DRACULA), to collect a 5 times larger solid angle than now possible with D20. In this it will approach the coverage of instruments at pulsed neutron sources. The advantage of using a reactor rather than a pulsed source is still that the time averaged flux on the sample can be an order of magnitude greater. DRACULA will become the fastest machine to study fast solid state reactions, with times slices of less than 100 ms.

Of course there are other facilities to perform time-resolved neutron diffraction studies. Time-resolved neutron diffraction is performed routinely on instruments such as GEM at ISIS (Williams et al. 1998; Radaelli et al. 2003; Hannon 2005) and HIPPO at LANSCE, though data acquisition times are considerably slower than with D20. PowGen3 under construction at SNS will set new standards for high-intensity TOF-diffraction with considerable promise for time-resolved investigations (Hodges 2004).

SAMPLE ENVIRONMENT

In almost all cases scientifically relevant time-resolved studies need to be performed at non-ambient temperatures and/or pressures. Often they need in addition a close control of the chemical activity of the involved species by adjusting the partial pressures of the gaseous constituents or surrounding atmospheres. In many cases these conditions need some modifications to standard equipment available at neutron scattering centers or even the development of purpose-built devices. It is beyond the scope of this review to give a full account of the sample environment options for neutron diffraction and it is impossible to give details on those purpose-built devices. Rather, we refer to the examples given further below to illustrate the preparations needed to do time-resolved studies. What should be said, however, is that complicated sample environments are usually more easily adapted to neutron scattering than to X-ray diffraction. General consideration for neutron scattering sample environments can be found in literature. Concerning high temperatures and high pressures very useful information is given in Part III of Reviews in Mineralogy and Geochemistry Vol. 41 (Peterson and Yang 2001). Fei and Wang (2001) focus on the state of art in synchrotron powder diffraction, but some of the large volumes presses, e.g., the so-called Paris-Edinburgh press (Besson et al. 1992), using developments of a toroidal high pressure device by Khvostantsev et al. in the 1970's, reviewed in Khvostantsev et al. (2004), found widespread application in time-of-flight and constant wavelength neutron diffraction (Parise 2006b, this volume). The current state of gas pressure cells used for some of the examples given below has been reviewed in a special issue of the proceedings of an International Workshop on Medium Pressure Advances for Neutron Scattering (Kuhs et al. 2005). A new design of a gas pressure cell combined with a closed-cycle helium cryostat is described by Lokshin and Zhao (2005).

Electrical furnaces with cylindrical heating elements of vanadium- or niobium-foil and concentric heat shields of the same material around the sample cylinder allow investigations on powders and liquids from room temperature up to about 1900 K (Bletry et al. 1984). Vanadium is preferred for its low background contribution in terms of parasitic Bragg peaks resulting from its low coherent scattering cross section, whilst niobium permits to obtain higher temperatures. Even higher temperatures, up to 2300 K, can be obtained with so-called mirror furnaces (Lorenz et al. 1993). Holland-Moritz et al. (2005) developed an electromagnetic levitation furnace, which allows container-less investigations, e.g., on super-cooled metallic melts.

To achieve low temperatures, the so-called *Orange Cryostat* has been developed at ILL in 1975. Cryogenic liquids such as nitrogen or helium do not interact with the neutron beam, and the only metal, the neutron beam is passing through, is aluminum or vanadium. The cryostat has a liquid nitrogen jacket, its liquid helium flow is controlled with a cold valve, it shows no interference between the calorimeter and the flow of the cryogenic liquid, and the sample is in thermal equilibrium with an exchange gas. The standard orange cryostat has a sample access of 49 mm diameter and is operational in a temperature range from 1.4 to 320 K. At D20, the calorimeter is of a smaller diameter of 25 mm and made of vanadium to allow for cleaner powder diffraction patterns.

The so-called cryofurnace based upon the Orange Cryostat design has been developed at ILL in 1983. It allows extending the accessible temperature range up to 600 K by avoiding indium seals at the calorimeter. Lower temperatures can be achieved with dilution inserts. Today, due to the high price of helium price and for ease of operation, closed-cycle refrigerators more and more replace Orange Cryostats, apart from very low temperature applications.

Various kinds of chemical reactors have been developed mostly for particular experiments, e.g., hydrothermal autoclaves either receiving the heating from an existing furnace (Fehr et al. 2003; Walk-Lauffer 2003) or equipped with its own heating system (Walton et al. 1999). Latroche et al. (2002) developed dedicated electrochemical cells.

For time-resolved investigations it is often crucial to reach the wanted thermodynamic conditions in the shortest possible time and to control them closely once the reaction is started. Environments of high inertia, often linked to high accuracy, can limit considerably the time-resolution and complicate the data treatment. In any case, it must be possible to change the thermodynamic conditions on beam, which is a particular challenge for time-resolved high pressure studies using mechanical clamps.

EXAMPLES

Water ices and gas hydrates are important materials on earth, the solar system and beyond. Hydrogen is the most common element in the Universe. Ice is widespread in many the outer smaller planets, the moons of Jupiter and Saturn or the nuclei of comets (Schmitt et al. 1998; Max 2003; Yershov 2004; Max et al. 2006). These compounds represent traditional fields of investigation for neutron diffraction. Neutrons allow to obtain details on the hydrogen positions and to establish the degree of proton ordering which may vary as a function of pressure and temperature. Neutrons also have advantages in high pressure studies; the medium pressure range in particular was object of numerous structural studies by neutron diffraction. The complex phase diagram of water is shown in Figure 4 exhibiting stable and metastable phases, some of which have no regime of thermodynamic stability at all (Petrenko and Whitworth 1999). While the melting curves are fairly well known, most of the phase boundaries between the solid phases are not well established, largely due to a considerable hysteresis, in particular at low temperatures. Even less is known in terms of the kinetics of the phase transitions between the different forms of ice and the reactions of various gases with ice leading to so-called "filled" ices (in which gas molecules enter and leave without breaking H-bonds) or to clathrate hydrates with encaged gas molecules (Kuhs 2004). Time-resolved neutron diffraction provides an excellent way to gain deeper insight into the transformation mechanisms on both topological or proton ordering transitions in ice as well as the uptake/release reactions with various gases and the underlying rate limiting processes.

The following three examples for time-resolved studies concentrate on the transformation kinetics of ices and gas hydrates. They include a number of experimental details, which should highlight some of the necessities and technical constraints typical for time-resolved work; some limitations will also be discussed. Each example closes with some more generally valid remarks concerning time-resolved work. For any more detailed information on the scientific implications of the results the reader is referred to the original literature cited.

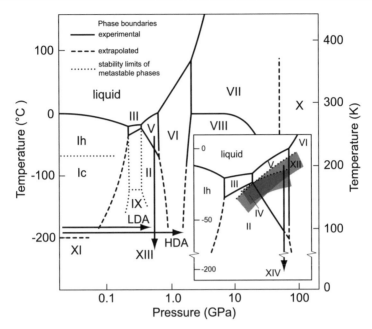

Figure 4. Phase diagram of water and water ices including some recently discovered stable and metastable forms (Lobban et al. 1998; Salzmann et al 2006). Arrows indicate the typical preparation paths (low-temperature pressurization or controlled cooling) for LDA; HAD, ice XIII and XIV. The insert shows an enlarged part of the medium pressure range where several forms of ice co-exist.

Kinetics of the high–density to low–density amorphous ice transition

Background. At least two different disordered states of water ice at low temperatures are known (Mishima et al. 1984), a *high-density amorphous* (HDA) ($\rho_{H_2O} \approx 1.17$ gcm^{-3}) and a *low-density amorphous* ice (LDA) state ($\rho_{H_2O} \approx 0.93$ gcm^{-3}). This phenomenon is known as *polyamorphism* (Mishima et al. 1985; Bosio et al. 1986; Floriano et al. 1986; Bellissent-Funel et al. 1987; Bizid et al. 1987; Mishima and Stanley 1998). Both states transform into each other when heating the sample above $T \approx 100$ K at ambient pressure (HDA → LDA) or when compressing the sample above $P \approx 1$ GPa at $T < 160$ K (LDA → HDA) (Mishima et al. 1991; Mishima 1994, 1996). Calorimetric experiments revealed the release of latent heat at the HDA → LDA transformation (Handa 1986b), indicating a first-order phase transition. On the other hand, LDA shows upon heating an onset of an endothermic transition before it crystallizes exothermally to cubic ice. This has been interpreted as a glass transition of LDA with $T_g \approx 140$ K (Johari et al. 1987; Handa and Klug 1988; Johari et al. 1991; Johari 1995), whilst Velikov et al. (2001) revised this value to $T_g \approx 165$ K.

Although the described experimental results on HDA and LDA are compatible with a first-order transition between two supercooled liquid states and experimental results are often referred to as due to a first-order phase transition, some experimental results contradict the idea of HDA and LDA being supercooled liquids. It lead to the discovery of a third disordered modification, a *very high density amorphous* ice (vHDA) with $\rho_{H_2O} \approx 1.25$ gcm^{-3} (Loerting et al. 2001).

Several distinct models are discussed to account for the existence and properties of the amorphous phases. Poole et al. (1994) introduced a model, which brought the proposed second critical point ($T_c \approx 220$ K, $P_c \approx 100$ MPa) with an emerging line of first-order transition towards lower temperatures (Poole et al. 1992) and the scenario of a retracing of the vapor-liquid spinodal in the supercooled region of water's phase diagram (Speedy 1982a,b; Sastry et al. 1993) together. In contrast to this model, in which intermediate states of the transition should reflect the superposed properties of the involved phases, a singularity-free model (Stanley and Teixeira 1980; Geiger and Stanley 1982) implies a continuous change of observables. The *in situ* neutron diffraction sampling of the static structure factor gave Koza et al. (2003) the opportunity of extracting information on the kinetics of the transformation and on the properties of the intermediate states, i.e., to distinguish between continuous variation of the amorphous structure and a superposition of the HDA and LDA structures in the course of the transformation.

Experimental. All HDA samples were prepared by slow compression of deuterated crystalline ice Ih ($V \approx 2.5$ ml) in a piston–cylinder apparatus at $T = 77$ K up to $P \approx 1.7$ GPa (Koza et al. 1999, 2000). After HDA had been formed the pressure was released and the sample retrieved from the pressure device under liquid nitrogen and transferred in cylindrical sample containers into an Orange Cryostat at $T = 77$ K. The evaporation of nitrogen from the sample containers was carried out at $T = 78–79$ K and controlled via the scattering signal of the samples. For each sample, the static structure factor of HDA and, after the *in situ* measurements, of LDA was determined at $T = 77$ and 127 K, respectively. At these conditions both amorphous states could be well reproduced.

Two different types of thermal treatment were applied to the samples. First, measurements were performed on samples that were heated directly from the HDA state to the nominal temperature T. A second set of experiments was performed on samples that had been pre-annealed at a temperature T_{an} and heated subsequently to the nominal T with 77 K < T_{an} < T < 127 K. The data collection started in all cases before the heat treatment was applied. This allowed us to monitor the transformation process continuously. Data were collected on the diffractometers D4 (incident neutron wavelength $\lambda = 0.7$Å) and D20 ($\lambda = 1.3$ and 2.4 Å) and the time-of-flight spectrometer IN6 ($\lambda = 4.1$ Å) at ILL. The raw data are corrected for different detector efficiencies and empty container scattering. To compare the energy-resolved time-of-flight data (IN6) with those from the diffraction instruments (D4 and D20) an energy integration is performed as $S(2\theta) = S(2\theta, t = 0) = S(2\theta, \omega)\, d\omega$. All data sets are finally converted via the mapping $S(2\theta) \rightarrow S(Q)$. Figure 5 shows the obtained $S(Q)$ of HDA and LDA as measured on D20.

Analysis. The data analysis is done on the *relative* changes of the structure factors observed while the sample passes from the HDA into the LDA state. The difference profile obtained from subsequent data sets with respect to a reference $S(Q)$, e.g., the $S(Q)$ of LDA, provides a convenient mathematical expression for monitoring these changes (Schober et al. 1998). $S(Q;\text{HDA})$ and $S(Q;\text{LDA})$ are taken as the static structure factors of HDA and LDA at 77 K and at 127 K, respectively. $I(t,T)$ is one for pure HDA and zero for pure LDA.

$$I(t,T) = \frac{\int_{Q_1}^{Q_2} S(Q;\text{LDA}) - S(Q;t,T)\,dQ}{\int_{Q_1}^{Q_2} S(Q;\text{LDA}) - S(Q;\text{HDA})\,dQ} \tag{1}$$

In a pure first-order transition the structure factors of the intermediate states can be expressed as a superposition of those of the initial and the final phases. In that case, $I(t,T)$ stands for the fraction of initial material HDA still present at time t and temperature T. For practical reasons, the integration limits have been chosen to $Q_1 = 1$ Å^{-1} and $Q_2 = 2.65$ Å^{-1}, a range covered by the whole series of experiments on different instruments (mainly D20, but also D4 and IN6 for some temperatures). Figure 6 shows $I(t,T)$ as determined from an experiment at D20 at two different temperatures.

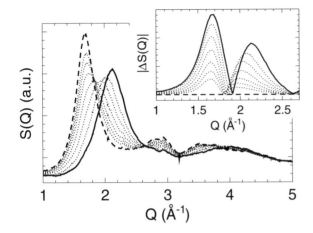

Figure 5. Evolution of the static structure factor $S(Q)$ during the transition from HDA (solid) to LDA (dashed) ice. Intermediate data (dotted) are selected from an *in situ* measurement of the transition performed at D20 at $T = 110$ K. The inset shows the difference of $S(Q;T)$ at corresponding conversion stages with respect to $S(Q;\text{LDA})$: $|\Delta S(Q)| = |S(Q;\text{LDA}) - S(Q;t)|$ [Used by permission of Institute of Physics Publishing, from Koza et al. (2003), *Journal of Physics: Condensed Matter*, Vol 15, Fig. 1, p. 323].

Figure 6. Temperature dependence of the transformation kinetics of HDA to LDA. The sigmoid-shape of $I(t,T)$ can be represented by superposition of the logarithmic HDA annealing and an *Avrami-Kolmogorov* behavior, reminiscent of a first-order transition. Such a fit to the data corresponds to the solid line in the graph for data at 105 K [after Koza et al. (2003)].

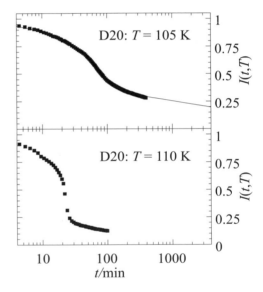

The time dependence of $I(t,T)$ could be fitted with a superposition of a logarithmic response and an *Avrami-Kolmogorov* equation (AKE) (Avrami 1939,1940,1941; Kolmogorov 1937), as described by Doremus (1985).

$$I(t,T) = (1-C) + C \exp\left[-\left(\frac{t}{\tau_0(T)}\right)^n\right] + B\ln(t) \qquad (2)$$

The logarithmic relaxation—the left part of Equation (2) (B is increasing with temperature)—is observed in other amorphous materials as well (Primak 1975; Grimsditch 1986; Tsiok et al. 1998) and is expected for the relaxation of materials with double-well potentials (Karpov and Grimsditch 1993) and from simulations of aging of liquids (Sciortino and Tartaglia 2001).

However, no additional transformation stages beyond the logarithmic relaxation have yet been revealed. The AKE behavior above 103 K is an additional transformation process consistent with a first-order transition, as it indicates nucleation and growth of a homogeneous low-density phase within a homogeneous high-density matrix. The exponent n in Equation (2) refines to $n \approx 1.5$, as expected for diffusion-controlled transitions with short nucleation times and isotropic growth (Doremus 1985; see also Pradell et al. 1998).

Apart from the fact that a well annealed HDA state does exist, after logarithmic relaxation, which will transform to LDA following a pure AKE behavior, the final state of this transition does not coincide with the LDA state one obtains after annealing at $T = 127$ K. The results show three states of the HDA-LDA transformation, an annealing stage of HDA, then the transition which seems to be a first order transition with coexistence of HDA and LDA and not a continuous change from HDA to LDA (to check this, the total $S(Q)$ must be a linear combination of the individual $S(Q)$ for each phase). Once the LDA is formed, it still anneals in a third stage, annealed LDA. To distinguish the later, the un-annealed LDA is hereafter called LDA'.

The activation energy ΔE results from exploiting the Arrhenius Equation (3); Figure 7 shows the corresponding Arrhenius plot.

$$\tau(T) = \tau_{\infty} \exp\left(\frac{\Delta E}{RT}\right) \tag{3}$$

ΔE has been determined to $\Delta E = (33\pm2)$ kJ·mol^{-1} and $\tau_{\infty} \approx 10^{-13\pm1}$ s which is a physically reasonable attempt time for crossing the energy barrier ΔE.

Further investigations. Koza et al. (2005) showed, that HDA ices obtained from compression of crystalline ice Ih or from vHDA ice as an intermediate stage (hereafter called HDA') of its transition into LDA' ice seem to be identical in terms of $S(Q)$ obtained from neutron diffraction at D20. The kinetics of the vHDA to LDA' transition at different temperatures is shown in Figure 8. During the annealing of vHDA, $S(Q)$ increases below $Q \approx 0.6$ Å. This small angle scattering signal reaches its maximum (the state of strongest heterogeneity, SSH) briefly after having passed the state of HDA' (HDA, obtained from ice Ih, shows the same signal level at low Q), before dropping drastically in the apparent first-order transition of HDA' to LDA'. The enhanced low-Q signal coincides with the behavior

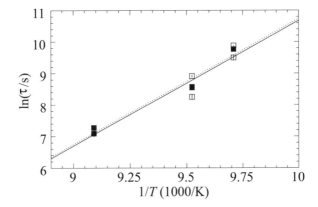

Figure 7. Arrhenius plot of time constants for the transformation of HDA to LDA calculated by different techniques – (open squares) determined with Equation (1) and (full squares) calculated from position of maximum in $S(Q)$. Lines correspond to linear fits, estimating the fit uncertainty by the deviation of all data points (solid line) and by taking the experimental error and temperature step width into account (dashed line) [Used by permission of Institute of Physics Publishing, from Koza et al. (2003), *Journal of Physics: Condensed Matter*, Vol 15, Fig. 7, p. 329].

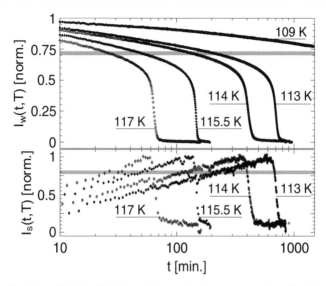

Figure 8. Kinetics of the vHDA into LDA′ transition, nominal temperatures given in the figure. The gray shaded area indicates the position of HDA′ and, respectively, HDA in the plot. The kinetics of the transient intensity at low-Q is shown below. Equation (1) defines I_w and I_s with $Q_1 = 0.6$ Å^{-1} and $Q_2 = 2.75$ Å^{-1} for wide-angle data (I_w), respectively $Q_1 = 0.35$ Å^{-1} and $Q_2 = 0.55$ Å^{-1} for small-angle data (I_s). [Reprinted figure with permission from Koza et al. (2005), *Physical Review Letters*, Vol. 94, Fig. 4, p. 125506-3. *http: //link.aps.org/abstract/PRL/v94/e125506*. Copyright 2005 by the American Physical Society.].

of the wide-angle $S(Q)$, its Fourier transform $D(r)$ shows in contrast to vHDA and LDA′ no oscillations beyond $r > 10$ Å, indicating a reduction of spatial correlations. Thus, HDA and HDA′ are heterogeneous structures. Although they look the same, the kinetics of their transition to LDA′ seems to be different. However, the fact that the transitions shown in Figure 8 are so fast and hard to fit with the Avrami-Kolmogorov expression, Equation (2), is also due to the fact that the speed of the transition at temperatures above $T \approx 110$ K is so high, that the evacuation of latent heat from the exothermic reaction cannot be assured, isothermal conditions are no longer fulfilled, and the exponent n in Equation (2) increases artificially—if a fit is possible at all—as the non-evacuated heat accelerates the transition due to a locally higher temperature.

In a recent series of experiments, Koza et al. (2006) combined neutron powder diffraction and small angle neutron scattering (SANS) to show that there are only two homogeneous disordered ice structures, vHDA and LDA. They come to the conclusion that the HDA state does not constitute any particular state of the amorphous water network, while Loertig et al. (2006) interpret their results otherwise. It is formed due to the preparation conditions, which determine the degree of heterogeneity. This can be quantified by SANS, using the DBM model proposed by Debye and Bueche (1949) for a mixture of two statistically distributed phases. If both phases present 50% of the total, an average domain size D results easily from twice the correlation length. This is exactly the situation for the SSH, where domain sizes from 11 to 13 Å have been determined for all investigated samples. The transformation speed increases systematically with a sample being of increasing heterogeneity. The precise reason for this is still subject to scientific discussion; possible models considered are those given by Adam and Gibbs (1965) or by Tanaka (2000).

In summary, kinetic studies of transitions between amorphous ices provided key information to understand the complex phenomenon of polyamorphism and they will be necessary in the future to quantify the nature of the heterogeneous states. This example also shows that a single experimental technique often is insufficient to get a complete understanding of the involved processes; here SANS proved to be a very powerful addition and, indeed, SANS is very apt to see the onset of time-dependent transformations better than wide angle diffraction, given that a scattering contrast develops (see Cole et al. 2006, this volume; Radlinski 2006, this volume; Wilding and Benmore 2006, this volume).

Synchrotron X-ray diffraction presents an alternative technique to neutron diffraction in this context, able to cover a wider Q-range at once and thus combining wide-angle- and small-angle-scattering. Also, the incoherent scattering from protons or deuterons is no longer a problem with X-ray diffraction, and the insensitivity to scattering from protons and deuterons is even an advantage in this particular case, as this polyamorphism is sufficiently expressed by the oxygen positions alone. Nevertheless, drawbacks of X-ray scattering are the low scattered intensity from light atoms and thus relatively long counting times and at the same time the enormous heat-load from the incident X-ray beam on the sample. The latter makes it very difficult, if not impossible, to keep the temperature stable and to have the sample in a thermodynamically well defined and stable state. In particular kinetic experiments suffer from this a lot and neutrons appear still as a good first choice.

Gas hydrate formation kinetics

Background. Gas hydrates are non-stoichiometric inclusion compounds encaging small, usually apolar guest molecules in a host-framework of hydrogen bonded water molecules. They exist as a stable solid phase at high gas pressures and/or low temperatures. Two main crystallographic structures of gas hydrates, the von Stackelberg cubic structure I and II (von Stackelberg and Müller 1954), are distinguished (Fig. 9) both consisting of two types of cavities, small and large cages, that can be occupied by the guest molecules to various degrees in a non-stoichiometric manner. It is generally assumed that the encaged gas molecules cannot exchange with the environment after formation. Rather, the guest molecules have to be built into the crystal structure during the growth process according to their chemical activity at the reaction site.

Since the 1950's, a large number of gas hydrate systems have been studied. The geologically and economically most important methane-water phase diagram is shown in Figure 10. Despite some 6000 publications, several physico-chemical properties of gas hydrates as well as their

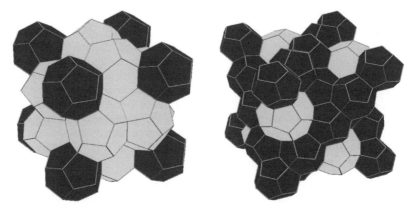

Figure 9. Schematic representation of clathrate hydrate structures of type I (left) and type II (right). Only the polyhedrons formed by oxygen are displayed. One unit cell is shown in each case. The smaller 5^{12} pentagon-dodecahedral cages (2 per unit cell, bcc packed, in type I; 16 per unit cell, ccp in type II) are drawn in dark grey, the larger $5^{12}6^2$ 14-faced polyhedrons (6 per unit cell in type I) and $5^{12}6^4$ 16-faced polyhedrons (8 per unit cell in type II) are drawn in light grey.

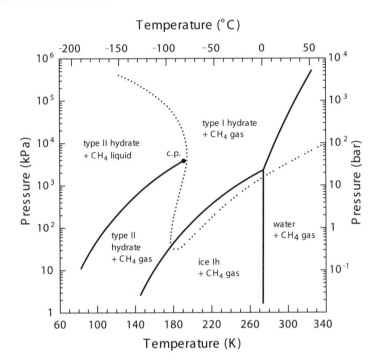

Figure 10. Phase diagram of the system water-methane. Bold lines give the established phase boundaries of water, ice, methane hydrate as well as liquid and gaseous methane with its critical point (c.p.). The dotted line represents a calculated phase boundary for structure I and structure II methane hydrate as calculated by Lundgaard and Mollerup (1992); the experimental separation line between structure I and II methane hydrate is not established.

formation and decomposition kinetics are neither well known nor properly understood, though they are of primary importance for a number of reasons. With traces of water in gas and oil pipelines operated at gas pressures well within the hydrate stability field, gas hydrates can form leading to a complete blockage. Likewise, the kinetics of CH_4-hydrate formation and decomposition is of major significance in geological settings, for our understanding of the role of methane gas in climate change, for a possible use of natural gas hydrate deposits as a future source of energy or simply for a more economic transport and storage of gas. CO_2 clathrate hydrates could also be a possible form of sequestering CO_2 into geological formations to reduce global warming. They may also play a major role for a number of terra-forming processes on Mars (Kargel 2004). Not much is known on the formation kinetics of CO_2 hydrates under Martian conditions and it is not certain that they form at the surface in any substantial amount due to the diurnal or annual temperature cycles. Our neutron diffraction work was intended to establish the formation and decomposition kinetics and in this way build a solid physico-chemical basis for the cases listed above.

In the Martian context, the most relevant formation process is the reaction of ice Ih with CO_2 gas to hydrate. A strong dependence of the transformation rates on the surface area of the gas-ice contact was demonstrated by Barrer and Edge (1967). Later, Hwang et al. (1990) studied the methane-hydrate growth on ice as a heterogeneous interfacial phenomenon and measured the clathrate formation rates during ice melting at different gas pressures. Sloan and Fleyfel (1991) discussed molecular mechanisms of the hydrate-crystal nucleation on ice surface, emphasizing the role of the quasi-liquid-layer (QLL) on ice. Takeya et al. (2000) made in situ observations of the CO_2-hydrate growth from ice powder for various thermodynamic conditions using laboratory X-ray diffraction. They distinguished the initial ice-surface coverage stage and a subsequent stage which was assumed to be controlled by gas and water diffusion through the hydrate shells surrounding the ice grains. The process was modeled following Hondoh and Uchida (1992) and Salamatin et al. (1998) in a single ice particle approximation. The respective activation energies of the ice-to-hydrate conversion were estimated as 0.2 and 0.4 eV (4.6 and 9.2 kcal/mol). The first in situ neutron diffraction experiments on kinetics of the clathrate formation from ice powders were presented by Henning et al. (2000). They studied the CO_2-hydrate growth on D_2O ice Ih, using the high intensity powder diffractometer HIPD at IPNS for temperatures from 230 to 263 K at a gas pressure of approximately 6.2 MPa. The starting material was crushed and sieved ice with unknown but most likely irregular shape of the grains. To interpret their results, the authors applied a simplified diffusion model of the flat hydrate-layer growth developed for the hydration of concrete grains. The activation energy of 6.5 kcal/mol was determined for the later stage of the hydrate formation process. This work has been continued by Wang et al. (2002) to study the kinetics of CH_4-hydrate formation on deuterated ice particles using neutrons. Their more sophisticated shrinking-ice-core model can actually be reduced to the diffusion model of Takeya et al. (2000, 2001). A higher activation energy of 14.7 kcal/mol was deduced for the methane hydrate growth on ice. Based on findings by Mizuno and Hanafusa (1987), the authors suggested that the quasi-liquid layer of water molecules at the ice-hydrate interface may play a key role in the (diffusive) gas and water redistribution although a definite proof could not be given. For further details on shrinking core models the reader is referred to detailed accounts by Froment and Bischoff (1990) and Levenspiel (1999).

In accordance with numerous experimental observations (Uchida et al. 1992, 1994; Stern et al. 1998; Henning et al. 2000; Kuhs et al. 2000; Takeya et al. 2000; Staykova et al. 2003), a thin gas-hydrate film rapidly spreads over the ice surface at the initial stage (stage I) of the ice-to-hydrate conversion. Subsequently, the only possibility to maintain the clathration reaction is the transport of gas molecules through the intervening hydrate layer to the ice-hydrate interface and/or of water molecules from the ice core to the outer hydrate-gas interface. As mentioned above, a diffusion-limited clathrate growth was assumed and simulated by Takeya et al. (2000), Henning et al. (2000), and Wang et al. (2002) on the basis of the shrinking-core models formulated for a single ice particle. However, in the case of porous gas hydrates, the gas and water mass transport through the hydrate layer becomes much easier, and the clathration reaction itself may be the rate-limiting step at a second stage of the hydrate formation that proceeds after the ice-grain coating. Certainly, we still can expect the onset of the further stage of the hydrate formation process completely or, at least, partly controlled by the gas and water diffusion through the hydrate phase, especially when a highly consolidated ice-hydrate structure develops with thick and dense hydrate shells surrounding ice cores. As a result, the hydrate-phase growth and expansion beyond the initial ice-grain boundaries into the sample voids and the corresponding reduction of the specific surface of the hydrate shells exposed to the ambient gas can be a principal factor which slows down the hydrate formation rates at the later stages of the clathration reaction, as predicted in Staykova et al. (2003) and confirmed in Kuhs et al. (2006). It should also be noted that in some cases a metastable formation of gas hydrates in a different structure (type I or type II) can be observed with a slow transformation into the stable phase (Halpern et al. 2001; Chazallon and Kuhs 2002; Staykova et al. 2003). The kinetics of a thermodynamically driven type I/ type II transition upon gas exchange has been studied using time-resolved neutron diffraction by Halpern et al. (2001).

Several studies were performed to quantitatively describe the formation process of CH_4 and CO_2 gas hydrates, both by in situ neutron diffraction on the high-flux diffractometer D20 of ILL as well as by in-house gas consumption experiments. While neutron experiments give unique access to the fast initial part of the clathration reaction, in-house gas consumption experiments are indispensable for the later, slower stages of the hydrate formation. Together with our kinetic diffraction studies, *ex situ* FE-SEM (field-emission scanning electron microscopy) observations of the formation of gas hydrates proved to be very helpful to better understand the evolution the ice microstructure during the clathration reaction and to construct a phenomenological multi-stage model of the gas-hydrate growth from ice powders (Staykova et al. 2003).

Experimental. Here, we summarize our above-cited recent work in which we attempted to quantitatively describe all stages of the formation process of CH_4 and CO_2 gas hydrates as followed by in-situ neutron diffraction and gas consumption experiments, starting from a well characterized ice-powder of known structure, grain size, and specific surface area. The high gas pressure needed to form and investigate gas hydrates in situ makes neutron diffraction a prominent tool for their investigation as neutrons have no problem to penetrate through thick-walled gas pressure cells located in a cryostat. While neutron experiments give unique access to the fast initial part of the clathration reaction, in-house gas consumption experiments are indispensable for the later much slower stages of the hydrate formation due to beam-time limitations. These in-house experiments were calibrated using neutron and X-ray (synchrotron as well as laboratory) diffraction techniques; without such a calibration the determination of the absolute degree of transformation by gas consumption methods is bound with large errors.

Spherical D_2O ice Ih grains with a typical diameter of several tens μm were prepared using a spraying technique. In order to quantify the morphology of the starting material, a representative part of the sample was investigated by cryo-FE-SEM. The pictures obtained were used to estimate the size distribution of the ice spheres. Measurements on different batches showed that the size distribution of ice spheres sprayed with the same nozzle is well reproducible and has a lognormal shape. The mean radius has been determined as 27 μm at relative standard deviation of 0.8. Samples with a packing density of about 65-70% were placed in high strength auto-frettaged aluminum gas pressure cells (Kuhs et al. 2005). The sample temperature reading was obtained from a calibrated temperature sensor fixed to the pressure cell wall. The Al sample cans were inserted into the pressure cell, already fixed to the sample stick, and the Bridgman seal was closed. This filling-operation was performed with a small stream of gas applied to ensure a complete gas filling of the system. Subsequently, the pressure cell was inserted into the orange cryostat and the temperature was equilibrated at the chosen value. Then the wanted gas pressure was applied within a few seconds while data collection was started concomitantly. As the settling of the pressure takes a few seconds, there is a corresponding uncertainty of the start of the reaction. In the case presented here this is within the chosen time resolution of the instrumental set-up and did not present an additional limitation.

To observe the changes of the diffraction patterns during gas hydrate formation we used D20 at its highest intensity setting at a wavelength of λ = 2.414 Å. The reaction of gas (at constant applied pressure and temperature) with ice grains was followed over a period of typically 10 to 20 h. Data were collected with a time resolution of 30 s or 1 min for the initial fast reaction and with a resolution of 5 min for the slower later part of the reaction. In this way data of good statistical precision were obtained, suggesting that even time-slices of several seconds would deliver useful information. To all data an efficiency correction and a background subtraction were applied. Subsequently, the measurements were analyzed in an automated way with the use of the full-pattern Rietveld refinement program GSAS (Larson and van Dreele 1990) delivering quantitative information on the amount of gas hydrate formed as a function of time with an accuracy of about 0.1 %. A two-phase (ice Ih + gas hydrate) Rietveld fit of the powder diffraction pattern obtained for each time slice was performed (Fig. 11). Refined parameters were the lattice constants of ice Ih and gas hydrate, the phase fractions and five to six background parameters; scale factor and absorption coefficient were fixed. The atomic positions and displacement parameters for D_2O ice Ih and CH_4- or CO_2-hydrate phases were taken from Klapproth (2002) and Klapproth et al. (2003) and were also kept fixed as no changes were expected. The weight fraction of the gas hydrate phase α (mole fraction of ice converted to

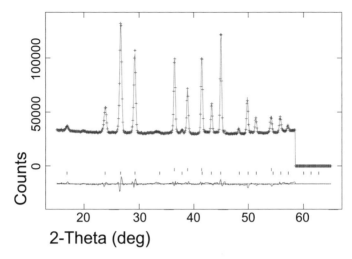

Figure 11. Low-angle part of a powder diffraction pattern obtained at a neutron wavelength of 2.414 Å of one time slice of a methane hydrate formation run at about 70% conversion at a temperature of 245 K with Rietveld-refinement result overlaid; crosses correspond to data points with a continuous line representing the calculated profile; top sequence of tick-marks represent ice Ih peak positions, bottom sequence to structure I hydrate peak positions; the bottom continuous line gives the difference between observed and calculated data points. The excluded region at angles > 59° correspond to a Bragg reflection of the Al pressure cell.

the gas hydrate) was extracted from the refinement for each time slice and plotted as a function of time (Fig. 12). An excellent statistical precision of the refined α's of 0.1% was obtained, which is hardly achieved by any other method. Moreover, also the accuracy of the obtained α's is distinctly superior to e.g., our gas consumption runs as the amount of formed gas hydrate is determined rather directly from a Rietveld refinement. Admittedly, what helped here was the fact that the hydrate formed is free of any texture and consists of an ideal powder with a typical crystallite size of a few μm to a few tens of μm. If the materials involved in the transformation are less ideal powders, this favorable situation for the determination of a reaction degree may deteriorate considerably.

In order to extend the time covered by the neutron runs with their inherent high data quality so-called "intermittent" experimental runs were performed which lasted 5-6 days (Kuhs et al. 2006). In both cases, the clathration reaction was started in the orange cryostat on the diffractometer and was tracked by neutron diffraction for about 20 h. Then the data acquisition was interrupted and the gas-pressure cell (stick) with the sample was moved to a separate low temperature bath where the reaction was continued out of the beam for a few days under identical conditions. During this time other users could perform their experiments on D20. Then each stick was placed back into the orange cryostat to measure the current reaction degree two more times—in the middle and at the end of the observation period.

Analysis. A detailed discussion of the models for analyzing the diffraction data is not given here; the interested reader is referred to the original publications (Staykova et al. 2003; Genov et al. 2004; Kuhs et al. 2006). It is sufficient to say that the starting material is reasonably described as a random dense packing of ice spheres as evidenced by cryo scanning electron microscopy. The neutron data were used to establish and refine a multi-stage model of the gas hydrate growth from ice Ih (Salamatin and Kuhs 2002) considering the influence of temperature and gas fugacity (which is related to the mechanical gas pressure) and using approximations for random dense packings introduced by Arzt (1982). A relatively short initial

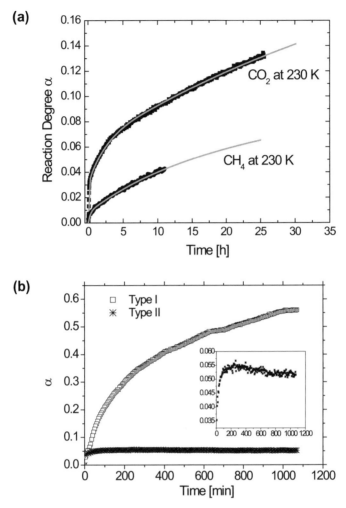

Figure 12. (a) Temperature dependency of the weight fraction of the transformed hydrate phase (reaction degree α): experimental points by symbols, fitted shrinking core model shown as gray curve [from Genov et al. (2004)]. (b) Simultaneous formation of a type I and type II structure of CO_2 hydrate. The insert shows the fraction of the metastable phase II, which after initial formation slowly back transforms into the stable structure I phase. Diffraction is the best tool to study the kinetics of this transformation. [Used by permission of Kluwer, from Kuhs (2004), in: High Pressure Crystallography, A. Katrusiak and P. McMillan (eds.), Fig. 7, p. 484]

stage I of hydrate film formation on the ice surface was distinguished followed by a stage II which generally includes two steps (sub-stages) presented by the clathration reaction at the ice-hydrate and gas-hydrate interfaces and by the diffusive gas and water transport through the hydrate shells surrounding shrinking ice cores. While the initial model was assuming monodispersity of the ice spheres, recently by generalizing Arzt's (1982) results an improved model for the hydrate growth in a polydisperse ensemble of randomly packed ice spheres was developed (Kuhs et al. 2006). This is of particular importance for the later stages of the gas hydrate formation. The difference in size of spherical ice particles in polydisperse samples results in different rates of their conversion to hydrates and part of the larger-size fraction becomes frequently isolated and switched out of the reaction. This additionally slows down the ice-to-hydrate conversion and stops the hydrate growth in the sample before the complete transformation is achieved. It could be shown that stage II, being dominated by the diffusive gas/water transport through the growing hydrate layer, still may be noticeably influenced by the interfacial clathration reaction. The principal kinetic parameters, reaction rate constant k_R and permeation (mass-transfer) coefficient D were inferred from the neutron diffraction and gas consumption data. The respective activation energies obtained from a the least-squares Arrhenius approximation are $Q_R \approx 92.8$ kJ/mol and $Q_D \approx 52.1$ kJ/mol. The rather high uncertainty level of the inferred reaction rate constant and corresponding activation energy Q_R should be emphasized. This most likely is a consequence of the observed domination of the diffusion mechanism additionally enhanced by a development of quasi-liquid layers on ice-hydrate and gas-hydrate interfaces at temperatures above 263 K. The diffusion activation energy for CH_4-hydrate formation is estimated with an uncertainty not higher than ±20%. Being comparable with previous results from the monodisperse approximation (Staykova et al. 2003), it is close to Wang's et al. (2002) estimate of 14.7 kcal/mol (61.3 kJ/mol) and practically coincides with $Q_D \approx 54.6$ kJ/mol deduced in (Genov et al. 2004) for CO_2-hydrate growth from ice powders. The similarity of activation energies for the methane and CO_2 case and the fact that the activation energy found is close to the energy of breaking hydrogen bonds in ices and hydrates suggests that the rate limiting process is the mobility of water molecules rather than the in-diffusion of the gases.

In summary, time-resolved neutron diffraction has contributed here in a crucial way to formulate a model of gas hydrate growth from ice powders which has predictive power and allows the calculation of the reaction degree as a function of temperature, pressure and grain size of a polydisperse starting material. Starting the reaction by changing the gas pressure can be done in a much faster and more precise way than changing the temperature, which permits a well-defined start of the reaction and allows to follow the first phase of the transformation with a time resolution of a few seconds. This example also demonstrates that often a detailed microstructural characterization (e.g., in terms of size distribution, particle shape and/or specific surface area) of the starting material is essential for a quantitative analysis of the diffraction results.

Anomalous preservation

Background. The last example for time-resolved neutron diffraction experiments is taken from our work on "anomalous preservation" of gas hydrates, which turned out to be related to the perfection of the ice formed during decomposition (Kuhs et al. 2004). "Anomalous preservation" (sometimes also called "self-preservation") of gas hydrates is a very intriguing phenomenon of considerable scientific and practical interest. Early observations of this effect were made independently by Davidson et al. (1986) and, more detailed, by Yakushev and Istomin (1992). These authors observed an unexpected persistence when gas hydrates were brought outside their field of stability at temperatures below the melting point of ice. More recently, Stern et al. (2001) and Takeya et al. (2001) have investigated the temperature dependency of the effect for the case of methane hydrate and found that the effect also had a lower limit. According to Stern et al. (2001) the "anomalous preservation window" extends from 240 K to the melting point of ice, while at temperatures below 240 K the decomposition is rapid and appears to be thermally activated. Within this window, the decomposition rates vary considerably by several orders of magnitude in a reproducible way (Fig. 13) with two minima at around 250 and 268 K. Takeya et al. (2002) confirmed this effect and suggested a diffusion limitation to explain the slow decomposition kinetics of gas hydrates within the anomalous preservation window. A similar, but not identical behavior was observed for CO_2 hydrate (Circone et al. 2003). Still, the deeper physical origin of "anomalous preservation" remains obscure and the controlling parameters elusive (Wilder and Smith 2002; Stern et al. 2002; Circone et al. 2004). The effect is of potential

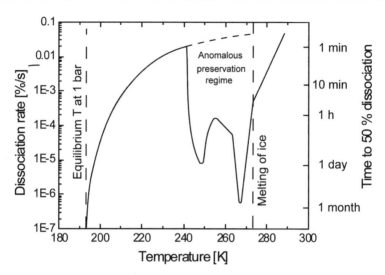

Figure 13. Temperature dependency of the decomposition rate of methane hydrate showing the region of anomalous preservation for a situation at 0.1 MPa (1 bar). [Used by permission of the Owner Societies, from Kuhs et al. (2004), Phys. Chem. Chem. Phys., Vol. 6, Fig. 1, p. 4917]

economic interest as it would allow for a low-cost compact and normal-pressure storage of gas in the form of hydrate by simple cooling to temperatures below 0°C (Gudmundsson et al. 2000) as well as it may have geological significance for some metastable surface-near gas hydrate occurrence Russian permafrost (Yakushev and Chuvilin 2000).

To be able to appreciate this connection we first turn to a discussion of the solid-solid transition of various metastable condensed forms of the water substance into ambient pressure ice. The decomposition of gas hydrates yields apparently normal ambient pressure hexagonal ice, so-called ice Ih as confirmed by Takeya et al. (2001, 2002) by laboratory X-ray diffraction. It is interesting to note that at lower temperatures various metastable solid water phases transform not into ice Ih but into so-called cubic ice, ice Ic. This form of ambient pressure ice is produced from amorphous forms of the water substance and from high-pressure ices when they are heated after a recovery at low temperature and ambient pressure (König 1943; Bertie et al. 1963) The transformation is ascribed to the on-set of mobility of Bjerrum defects promoting an ice-like crystal growth (Wooldridge et al. 1987). It was noticed early on (Arnold et al. 1968) that the diffraction patterns for ice Ic obtained from different starting materials were different. These differences were explained by Kuhs et al. (1987) in terms of various degrees of stacking faulting for ice Ic from different origins. The faults were identified as deformation stacking faults, which in diffraction experiments lead to the appearance of broad reflections at Bragg angles typical for ice Ih as well as to high- and/or low-angle shoulders on the Bragg peaks at genuine ice Ic positions. The width of the cubic reflections was used to estimate the particle size of ice Ic produced from ice II as 160 Å. Stacking faults in ice Ih and their creation by rapid temperature changes were also described by other authors (Higashi 1988) Some authors have investigated the transition of ice Ic into the normal hexagonal ice (ice Ih) by diffraction, which was found to take place over a seemingly large temperature range starting slowly at 150 K with a rapid progress of the transformation between 190 and 210 K (Kohl et al. 2000); at higher temperatures this work showed apparently pure ice Ih (Handa et al. 1986a). Differential thermal analysis of the ice Ic – ice Ih transition revealed the main transition region again at 190–205 K, with small but detectable events starting at ≈176 K (Handa et al. 1986a) and ending at ≈240 K

(Mayer and Hallbrucker 1987). A change of activation energy was observed at 185 K, and the transformation was observed to be complete at 210 K by Sugisaki et al. (1968). Already an earlier review of the situation showed a confusing picture (Hobbs 1974) in which the ice Ic – ice Ih transition was located at temperatures between 160 and 205 K, a situation which has not much improved since. Recently, additional evidence for the formation and existence of ice Ic at unusually high temperatures was established (Murray et al. 2005, Murray and Bertram 2006). The reason for such variability of the ice Ic – ice Ih transition temperatures are not clear, yet there are indications that in addition to the molecular arrangement of the parent phase mentioned above, the surface area of the ice Ic crystallites (Kumai 1968) has a significant influence. Thus it appeared worthwhile to look in some more detail at the transformation behavior of ice Ic into ice Ih in the temperature range in question with a well-defined starting material. Diffraction is the most promising tool as it permits insight into changes of the molecular arrangement not only at long ranges but also into a number of more local defect structures. In particular, time-resolved diffraction experiments allow for *in situ* studies of this transition and can give access to the transformation kinetics. Likewise, we have studied the crystallographic nature of the ice produced in decomposition of gas hydrates at temperatures below and within the anomalous preservation window by time-resolved neutron diffraction.

Experimental. The experiments were performed on the high-resolution scanning powder diffractometer D2B (wavelength 1.6 Å) as well as the high-intensity powder diffractometer D1B and D20 equipped with a linear position sensitive detector (wavelength 2.4 Å) at ILL. The first series of measurements consisted of studies of the ice Ic – ice Ih transition. The starting material used was high-pressure ice V (Lobban et al. 2000) recovered to ambient pressure at liquid nitrogen temperatures. Upon further heating at ambient pressure the recovered ice transformed into ice Ic at a temperature of 143 K within 15 h (Gotthardt 2001) Further temperature increase led to a gradual transition into ice Ih. The main structural rearrangements takes place at temperatures below 205 K in agreement with a number of previous observations (Dowell and Rinfret 1960; Kumai 1968; Handa et al. 1986a). Detailed observations were made in the temperature range between 180 and 265 K; for the experiment performed on D2B the temperature was increased at a rate of 10°/h in steps with a holding time of 30 min after reaching each target value. During both ramping and holding the temperature diffraction data were collected with a time resolution of 15 min; this is the time needed to collect data of sufficient quality to observe changes in the stacking fault arrangements by scanning with 0.05° steps in 2Θ. Particular attention was given to the peak intensity and peak shape of the cubic 111 reflection as well as the neighboring and partly overlapping hexagonal 100, 002 and 101 reflections. Intensity changes of these reflections upon heating were observed to take place first rapidly, then slowing down and coming essentially to hold. The time spent at each temperature was chosen to cover the period in which significant differences between adjacent data sets could be detected as established from exploratory runs on D20; with the good counting statistics of D20, intensity changes of 1% were detectable. The analysis of the intensity ratio of the hexagonal 100 and 002 reflection indicated the persistence of some cubic component at temperatures as high as 237 K. Following our earlier analysis (Kuhs et al. 1987) this was interpreted as cubic stacking sequences and represents the first unequivocal crystallographic evidence for the persistence of significant two-dimensional defects at temperatures above 205 K. Only at temperatures close to 240 K these imperfections finally disappear (Kuhs et al. 2004).

Independently, a number of time-resolved neutron diffraction runs were performed on the high-flux diffractometer D20 at ILL in order to study the decomposition behavior of gas hydrates. Custom-made gas pressure cells were used (Kuhs et al. 2005) which were filled with almost pure gas hydrates formed from hexagonal ice (Staykova et al. 2003). Samples were equilibrated at the desired pressure and temperature conditions. Concomitant with a pressure release to the designed end pressure, data collection was started. Again, as in the case of hydrate formation, the pressure release to the desired value (located between a few mbar and 1 bar) took a few seconds. Consequently, this time is then the achievable time-resolution at the beginning of the process. Complete diffraction patterns were recorded with a time resolution of 10 s up to 1 min for the initial part of the reaction and slower acquisition rates of typically 5 min for the later part of the decomposition process (Fig. 14). The complete sample of typically 1 cm³ was intercepted by the neutron beam. The analysis of the numerous diffraction data was performed in an automated fashion using the Rietveld program GSAS (Larson and von Dreele 1994) similar to the approach described for the gas hydrate formation reactions. Between 50 and 300 individual diffraction patterns were collected as a function of time for each of the decomposition runs. Beam-time restrictions limited the duration of each run to typically less than half a day.

Analysis. Quantitative information on the progress of the reaction was obtained from the phase fractions of ice and gas hydrate for each data set. The results are shown in Figure 15 for CH_4- and CO_2-hydrate. In agreement with earlier observations (Stern et al. 2001, Takeya et al. 2001) the initial decomposition was always fast, but slowed down for temperatures above approximately 240 K in the anomalous preservation regime. A phenomenological model combining an initial reaction-limited and a later diffusion-limited process can quantitatively explain the decomposition (Genov 2004). Similar successive processes also take place during the gas hydrate formation from ice as described above. More interesting in the present context is the detailed nature of the ice formed below and above the onset of anomalous preservation. A detailed inspection of the diffraction features of the ice obtained gives unequivocal evidence for the existence of stacking faults. They are clearly born out in shoulders of the main hexagonal diffraction peaks as well as in the non-ideal intensity ratio of the hexagonal 100 and 002 reflections. Insufficient crystallite statistics or textural effects can be safely excluded as an explanation of the non-ideal intensity ratios. It is noteworthy that the non-ideal character of the hexagonal ice formed is more pronounced at lower temperatures and disappears completely for all data sets in the self-preservation regime. Moreover, ice formed from CO_2- and CH_4-hydrate appears to be different in that the latter shows more pronounced features for deformation stacking faults born out in the high-angle shoulders of the hexagonal 100 reflection (Kuhs et al. 1987).

Clear evidence is found that the ice formed upon gas hydrate decomposition at temperatures below the anomalous preservation window is defective. It forms small crystallites

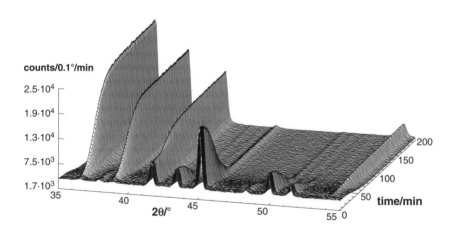

Figure 14. Stack of sequential powder diffraction patterns during decomposition of CO_2 gas hydrates. The decomposition took place at 210 K. The first 20 data acquisitions were done in 30 s each, afterwards, 199 patterns were taken in 1 min each. Thus, the whole series of sequential data acquisitions took 210 min. The counting rate per detector cell of 0.1° width in 2θ was normalized to acquisitions of 1 min.

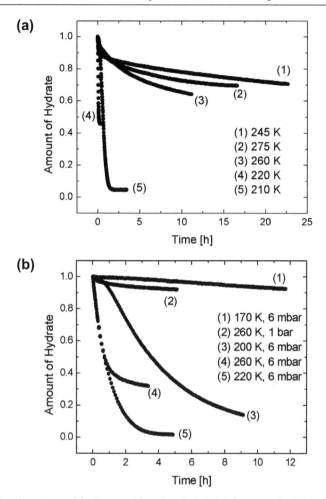

Figure 15. Time dependency of the decomposition of gas hydrate into ice shown for different temperatures. At higher temperatures the decomposition is slower due to the effect of anomalous preservation (a) CH_4 hydrate (b) CO_2 hydrate. [Used by permission of the Owner Societies, from Kuhs et al. (2004), Phys. Chem. Chem. Phys., Vol. 6, Fig. 3, p. 4919.]

of a few μm, which do not combine to larger, more homogeneous assemblies below 240 K. A complete annealing of stacking faults and appreciable grain growth of the ice crystallites sets in at temperatures of approximately 240 K. The initial amount of faults and the details of the step-wise disappearance of these stacking-faults upon temperature increase apparently depend on the parent phase as well as the speed of transformation into ice. Remarkable differences in lattice defects were established earlier on for the various high-pressure ices as parent phases (Arnold et al. 1968) In a similar way, differences in the degree of perfection were found for ice produced from decomposing CO_2 and CH_4 hydrate, with the latter showing more imperfections. As the water topology of both hydrates is identical (both form a cubic type I hydrate structure) the difference must arise from the different transformation kinetics, with CO_2 hydrate decomposing distinctly slower with a resulting less defective ice.

This example shows that a wealth of information going far beyond the perfect crystal picture is contained in diffraction data. Time-resolved studies on phase-transitions and chemical reactions usually imply changes of the microstructure of the crystallites translating into changes in the diffraction profiles of the newly formed phases. A closer look at these details is quite often a worthwhile undertaking, but certainly also is more involved and not necessarily straightforward using existing structure analysis and refinement programs.

CONCLUSIONS AND OUTLOOK

The time-resolved work on ices and gas hydrates is by far not completed and many problems remain unsolved. So far, these studies have shown that much can be learned from truly time-resolved studies about the processes controlling structural phase transitions and chemical reactions in ices and gas hydrates. Here, we can only indicate a few further research lines of interest. Concerning the polyamorphism of ice it will be important to prove or disprove the ergodicity (time-average over one molecule in phase space identical to the space average over the ensemble considered) of the different states and time-resolved work on samples with well documented history will certainly shed further light on this issue. Most likely, diffraction, small-angle scattering and inelastic work need to be combined to really advance our understanding. Of the numerous phase transitions between the various crystalline phases of ice very few have been studied in detail. Transitions involving the restructuring of the H-bond network (usually pressure-driven) are strongly first order with considerable hysteresis. Yet, there is some evidence that topotactic relationships exist (Bennett et al. 1997), an observation which is far from being understood in detail. Time-resolved diffraction studies of textured samples may provide further insight. The usually temperature-driven proton orderings are gradual without symmetry change in some cases (ice III, ice V; Lobban et al. 2000) or abrupt and first order with symmetry changes in others (ice VII/VIII; Kuhs et al 1984). Sometimes doping is necessary to promote the proton ordering leading also to changes in symmetry as in the case of ice Ih/ice XI doped with KOH (Matsuo et al. 1986) or ices V/XIII and XII/XIV doped with HCl (Salzmann et al. 2006). It is generally accepted that Bjerrum and ionic defects in the ice lattice play an important role in the proton ordering processes (e.g., Wooldridge et al. 1987; Petrenko and Whitworth 1999) yet very little work is done to quantify the ordering processes via activation energies or molecular transport processes. Here again, time-resolved neutron diffraction studies carry considerable promise. Many open questions also prevail in the field of gas hydrates and filled ices. Hydrate formation from liquid water is certainly an issue worth looking at with diffraction methods in combination with light scattering to study the evolution of grain sizes. Two-dimensional detectors may allow in the future to follow grain sizes at larger scales also by neutron diffraction to study Ostwald ripening and related phenomena. Furthermore, it seems worthwhile to study not only the changes in Bragg intensities during a transition but also to look at phenomena like particle size broadening, induced strain or the signatures of defects like stacking faults in the peak profiles. These features call for a very good angular resolution at neutron fluxes acceptable for kinetic work. Such studies will generally need improved sources, instruments and detectors as they are under development both at reactor neutron sources such as ILL and time-of-flight facilities at LANSCE, ISIS and SNS.

Powder diffraction is a classical entrance door to neutron scattering. Experiments are easy to set up and powder samples normally readily available. Constant wavelength diffraction does not stand in direct competition to time-of-flight techniques, both techniques are complementary, and one particular advantage for time-resolved work of reactor-based diffraction is the constancy and reliability of source. The data treatment such as Rietveld refinement (Knorr and Depmeier 2006, this volume) or pair distribution function analysis (Proffen 2006, this volume) have become standardized and work efficiently with neutron powder diffraction data. Due to the penetration of neutrons, complex and unique sample environment conditions can be fairly easily adapted. Applications of time-resolved neutron diffraction techniques to unresolved problems in geology or mineralogy are numerous (see also Redfern 2006, this volume). One has only started to discover their potential, building on earlier experiences from material science. Yet, to fully take advantages of neutron scattering one has not only to improve the performance of the instruments but also advance sample environments. Some devices are available at neutron scattering centers, others need to be built or adapted by the user to match the specific needs of kinetic work.

Finally, it should be emphasized that neutron diffraction undoubtedly gives very valuable insight into transformation kinetics, however, it provides only a time-space averaged picture of the processes. Often complementary studies using locally resolving methods like electron microscopy, tomography or other imaging techniques are needed for a complete understanding. Moreover, in a number of cases neutron wide angle diffraction may find important complements in other neutron scattering techniques, like SANS (Radlinski 2006, this volume) or quasi-elastic and inelastic scattering (Loong 2006, this volume).

ACKNOWLEDGMENTS

One author (WFK) thanks the Deutsche Forschungsgemeinschaft (DFG) and the Bundesministerium für Bildung und Forschung (BMBF) in its program GEOTECHNOLOGIEN for multiple financial support (grants 03G0553A and 03G0605B); he also thanks the ILL for hospitality during a sabbatical leave. To the work presented here a number of people have made substantial contributions which are gratefully acknowledged: Andrzej Falenty, Georgi Genov, Alice Klapproth, Doroteya Staykova from the University of Göttingen, Andrey Salamatin from Kazan State University as well as Pierre Convert and Michael M. Koza from the Institut Laue-Langevin in Grenoble. This is publication no. GEOTECH-229 in the BMBF research initiative GEOTECHNOLOGIEN.

REFERENCES

Adam G, Gibbs JH (1965) On temperature dependence of cooperative relaxation properties in glass-forming liquids. J Chem Phys 43:139-146
Arnold GP, Finch ED, Rabideau SW, Wenzel RG (1968) Neutron diffraction study of ice polymorphs III. Ice Ic. J Chem Phys 49:4365-4369
Arzt E (1982) The influence of an increasing particle coordination on the densification of spherical powders. Acta Metall 30:1883-1890
Avrami M (1939) Kinetics of phase change. I. General theory. J Chem Phys 7:1103-1112
Avrami M (1940) Kinetics of phase change. II. Transformation-time relations for random distribution of nuclei. J Chem Phys 8:212-224
Avrami M (1941) Kinetics of phase change III. Granulation, phase change, and microstructure. J Chem Phys 9:177-184
Barrer RM, Edge AVJ (1967) Gas hydrates containing argon, krypton and xenon: kinetics and energetics of formation and equilibria. Proc Royal Soc London A 300:1-24
Bellissent-Funel M-C, Teixeira J, Bosio L (1987) Structure of high-density amorphous water. II. Neutron scattering study. J Chem Phys 87:2231-2235
Bennett K, Wenk HR, Durham WB, Stern LA, Kirby SH (1997). Preferred crystallographic orientation in the ice I to II transformation and the flow of ice II. Philos Mag A 76:413-435
Berliner R, Popovici M, Herwig KW, Berliner M, Jennings HM, Thomas JJ (1998) Quasielastic neutron scattering study of the effect of water-to-cement ration on the hydration kinetics of tricalcium silicate. Cement Concrete Res 28:231-243
Bertie JE, Calvert LD, Whalley E (1963) Transformations of ice II, ice III, and ice V at atmospheric pressures. J Chem Phys 38:840-846
Besson JM, Nelmes RJ, Hamel G, Loveday JS, Weill G, Hall S (1992) Neutron powder diffraction above 10 GPa. Physica B 180-181: 907-910
Bizid A, Bosio L, Defrain A, Oumezzine M (1987) Structure of high-density amorphous water. 1. X-Ray-diffraction study. J Chem Phys 87:2225-2230
Bletry J, Taverniere P, Senillou C, Desre P, Maret M, Chieux P (1984) High temperature furnaces for small and large angle neutron scattering of disordered materials. Rev Phys Appl 19:725-730

Bosio L, Johari GP, Teixeira J (1986) X-ray study of high-density amorphous water. Phys Rev Lett 56:460-463

Bray HJ, Redfern SAT (1999) Kinetics of dehydration of Ca-montmorillonite. Phys Chem Minerals 26:591-600

Brown ABD, Clarke SM (2000) Orientational order in concentrated dispersions of plate-like kaolinite particles under shear. J Rheology 44(2):221-233

Brown ABD, Clarke SM, Rennie AR (1998) Shear induced alignment of kaolinite: studies using a diffraction technique. Prog Colloid Polymer Sci 110:80-82

Chazallon B, Kuhs WF (2002) *In situ* structural properties of N_2-, O_2-, and air-clathrates by neutron diffraction. J Chem Phys 117:308-320

Circone S, Stern LA, Kirby SH (2004) The role of water in gas hydrate dissociation. J Phys Chem B 108:5747-5755

Circone S, Stern LA, Kirby SH, Durham WB, Chakoumakos BC, Rawn CJ, Rondinone AJ, Ishii Y (2003) CO_2 hydrate: synthesis, composition, structure, dissociation behavior and comparison to structure I CH_4 hydrate. J Phys Chem B 107:5529-5539

Cole DR, Herwig KW, Mamontov E, Larese J (2006) Neutron scattering and diffraction studies of fluids and fluid-solid interactions. Rev Mineral Geochem 63:313-362

Convert P, Berneron M, Gandelli R, Hansen TC, Oed A, Rambaud A, Ratel J, Torregrossa J (1997) A large high counting rate one-dimensional position sensitive detector: the D20 banana. Physica B 234:1082-1083

Convert P, Hansen T, Oed A, Torregrossa J (1998) D20 high-Flux two-axis neutron diffractometer. Physica B 241-243:195-197

Convert P, Hansen T, Torregrossa J (2000) The high intensity two axis neutron diffractometer D20 first results. Mater Sci Forum 321-324:314-319

Convert P, Hock R, Vogt T (1990) High-speed time-resolved crystallography with neutrons - a feasibility study. Nucl Instrum Methods Phys Res A 292:731-733

Crank J (1975) The Mathematics of Diffusion. Clarendon Press

Davidson DW, Garg SK, Gough SR, Handa YP, Ratcliffe CI, Ripmeester JA, Tse JS, Lawson WF (1986) Laboratory analysis of a naturally occurring gas hydrate from sediments of the Gulf of Mexico. Geochim Cosmochim Acta 50:619-623

Debye P, Bueche AM (1949) Scattering by an Inhomogeneous Solid. J Appl Phys 20:518-525

Doremus RH (1985) Rates of Phase Transitions. Academic

Dowell LG, Rinfret AP (1960) Low-temperature forms of ice as studied by X-ray diffraction. Nature 188:1144-1148

Erofeyev BV (1946) A generalized equation of chemical kinetics and its application in reactions involving solids. Dokl Akad Nauk SSSR 52:511-514

Fehr KT, Huber M, Zuern SG, Peters E. (2003) Determination of the reaction kinetics and reaction mechanisms of Al-tobermorite under hydrothermal conditions by in-situ neutron diffraction. *In:* Hydrothermal Reactions and Techniques. Feng SH, Chen JS, Shi, Z (eds) World Scientific, p 19-26

Fei Y, Wang Y (2001) High-pressure and high-temperature powder diffraction. Rev Mineral Geochem 41:521-558

Floriano MA, Whalley E, Svensson EC, Sears VF (1986) Structure of high-density amorphous ice by neutron diffraction. Phys Rev Lett 57:3062-3064

Froment GF, Bischoff KB (1990) Chemical Reactor Analysis and Design. Wiley & Sons

Fujii K, Kondo W (1974) Kinetics of hydration of tricalcium silicate. J Am Ceram Soc 57:492-497

Geiger A, Stanley HE (1982) Low-density patches in the hydrogen-bonded network of liquid water: Evidence from molecular dynamics computer simulations. Phys Rev Lett 49:1749-1752

Genov G (2004) Physical processes of the CO_2 hydrate formation and decomposition at conditions relevant to Mars. Doctoral Thesis, Georg-August-Universität, Göttingen

Genov G, Kuhs WF, Staykova DK, Goreshnik E, Salamatin AN (2004) Experimental studies on the formation of porous gas hydrates. Am Mineral 89:1229-1239

Gotthardt F (2001) Phasenverhalten und Phasenumwandlungen im Eis bei Gegenwart der Gase Helium, Neon und Argon. Doctoral Thesis, Georg-August-Universität, Göttingen

Grimsditch M (1986) Annealing and relaxation in the high-pressure phase of amorphous SiO_2. Phys Rev B 34:4372-4373

Gudmundsson A, Andersson V, Levik OI, Mork M (2000) Hydrate technology for capturing stranded gas. Ann N Y Acad Sci 912:403-410

Halpern Y, Thieu V, Henning RW, Wang X, Schultz AJ (2001) Time-resolved *in situ* neutron diffraction studies of gas hydrate: Transformation of structure II (sII) to structure I (sI) J Am Chem Soc 123:12826-12831

Hancock JD, Sharp JH (1972) Method of comparing solid- state kinetic data and its application to the decomposition of kaolinite, brucite, and $BaCO_3$. J Am Ceram Soc 55:74-77

Handa YP, Klug DD (1988) Heat capacity and glass transition behavior of amorphous ice. J Phys Chem 92:3323-3325

Handa YP, Klug DD, Whalley E (1986a) Difference in energy between cubic and hexagonal ice. J Chem Phys 84:7009-7010

Handa YP, Mishima O, Whalley E (1986b) High-density amorphous ice. III. Thermal properties. J Chem Phys 84:2766-2770

Hannon AC (2005) Results on disordered materials from the General Materials diffractometer, GEM, at ISIS. Nucl Instrum Methods Phys Res A 551:88-107

Hansen TC (2004) Future trends in high intensity neutron powder diffraction. Mater Sci Forum 443-444: 181-186

Henning RW, Schultz AJ, Thien V, Halpern Y (2000) Neutron diffraction studies of CO_2 clathrate hydrate: formation from deuterated ice. J Phys Chem 104:5066-5071

Hewat AW (2006) High flux diffractometers on reactor neutron sources. Physica B 385-396, doi:10.1016/j.physb.2006.05.316

Higashi A (1988) Lattice Defects in Ice Crystals. X-ray Topographic Observations. Hokkaido Univ. Press

Hobbs PV (1974) Ice Physics, Clarendon Press

Hodges JP (2004) POWGEN3: A high resolution third generation TOF powder diffractometer under construction at the SNS. J Minerals Metals Materials Society 56:206

Holland-Moritz D, Schenk T, Herlach DM, Convert P, Hansen T (2005) Electromagnetic levitation apparatus for diffraction investigations on the short-range order of undercooled metallic melts. Meas Sci Technol 16:372-380

Hondoh T, Uchida T (1992) Formation process of clathrate air-hydrate crystals in polar ice sheets. Teion Kagaku [Low Temperature Science] A51:197-212

Howard JAK, Johnson O, Schultz AJ, Stringer AM (1987) Determination of the Neutron Absorption Cross Section for Hydrogen as a Function of Wavelength with a Pulsed Neutron Source. J Appl Cryst 20:120-122

Hwang MJ, Wright DA, Kapur A, Holder GD (1990) An experimental study of crystallization and crystal growth of methane hydrates from melting ice. J Inclus Phenom 8:103-116

Johari GP (1976) The dielectric properties of H_2O and D_2O ice Ih at MHz frequencies. J Chem Phys 64:3998-4005

Johari GP (1995) Phase transition and entropy of amorphous ices. J Chem Phys 102:6224

Johari GP, Hallbrucker A, Mayer E (1987) The glass-liquid transition of hyperquenched water. Nature 330: 552-553

Johari GP, Hallbrucker A, Mayer E (1991) Isotope and impurity effects on the glass-transition and crystallization of pressure-amorphized hexagonal and cubic ice. J Chem Phys 95:6849-6855

Johari GP, Jones SJ (1976) Dielectric properties of polycrystalline D_2O ice Ih (hexagonal). Proc Roy Soc London A349:467-495

Johari GP, Jones SJ (1978) The orientation polarization in hexagonal ice parallel and perpendicular to the c-axis. J Glaciology 21:259-276

Kargel JS (2004) Mars; a Warmer, Wetter Planet. Springer

Karpov VG, Grimsditch M (1993) Pressure-induced transformations in glasses. Phys Rev B48:6941-6948

Kashchiev D (2000) Nucleation. Basic Theory with Applications. Butterworth-Heineman

Khvostantsev LG, Slesarev VN, Brazhkin VV (2004) Toroid type high-pressure device: history and prospects. High Press Res 24: 371-385

Kilcoyne SH, Manuel P (2002) A kinetic neutron diffraction study of the crystallisation of α-Er_7Fe_3. Appl Phys A-Mater 74:S1166-S1168

Klapproth A (2002) Strukturuntersuchungen an Methan- und Kohlenstoffdioxid-Clathrat-Hydraten. Doctoral Thesis, Georg-August-Universität, Göttingen

Klapproth A, Goreshnik E, Staykova DK, Klein H, Kuhs WF (2003) Structural studies of gas hydrates. Can J Phys 81:503-518

Knorr K, Depmeier W (2006) Application of neutron powder-diffraction to mineral structures. Rev Mineral Geochem 63:99-111

Kohl I, Mayer E, Hallbrucker A (2000) The glassy water - cubic ice system: a comparative study by X-ray diffraction and differential scanning calorimetry. Phys Chem Chem Phys 2:1579-1586

Kolmogorov AN (1937) On the statistical theory of metal crystallization. Izv Akad Nauk SSR Ser Mat 3: 355-359

König H (1943) Eine kubische Eismodifikation. Z Kristallogr 105:279-286

Koza M, Schober H, Toelle A, Fujara F, Hansen T (1999) Formation of ice XII at different conditions. Nature 397:660-661

Koza MM, Geil B, Winkel K, Köhler C, Czeschka F, Scheuermann M, Schober H, Hansen T (2005) Nature of amorphous polymorphism of water. Phys Rev Lett 94:125506(4)

Koza MM, Hansen T, May R, Schober H (2006) Link between the diversity, heterogeneity and kinetic properties of amorphous ice structures. J Non-Cryst Solids, doi:10.1016/j.jnoncrysol.2006.02.162

Koza MM, Schober H, Fischer HE, Hansen T, Fujara F (2003) Kinetics of the high- to low-density amorphous water transition. J Phys Condens Matt 15:321-332

Koza MM, Schober H, Hansen T, Toelle A, Fujara F (2000) Ice XII in its second regime of metastability. Phys Rev Lett 84:4112-4115

Kuhs WF (2003) Atomic displacement parameters. *In:* International Tables for Crystallography, D: Physical Properties of Crystals. Authier A (ed), IUCr, Kluwer, p 228-242

Kuhs WF (2004) The high pressure crystallography of gas hydrates. *In:* High-Pressure Crystallography. Katrusiak A, McMillan P (eds) Kluwer, p 475-494

Kuhs WF, Bliss DV, Finney JL (1987) High-resolution neutron powder diffraction study of ice Ic. J Phys Coll C1 Paris 48:631-636

Kuhs WF, Finney JL, Vettier C, Bliss DV (1984) Structure and hydrogen ordering in ices VI, VII, and VIII by neutron powder diffraction. J Chem Phys 81:3612-3623

Kuhs WF, Genov G, Staykova DK, Hansen TC (2004) Ice perfection and onset of anomalous preservation of gas hydrates. Phys Chem Chem Phys 6:4917-4920

Kuhs WF, Hensel E, Bartel H (2005) Gas pressure cells for elastic and inelastic neutron scattering. J Phys Condens Matter 17: 3009-3015

Kuhs WF, Klapproth A, Gotthardt F, Techmer K, Heinrichs T (2000) The formation of meso- and macroporous gas hydrates. Geophys Res Lett 27:2929-2932

Kuhs WF, Staykova DK, Salamatin AN (2006) Formation of methane hydrate from polydisperse ice powders: Long-term experiments. J Phys Chem B 110:13283-13295

Kumai M (1968) Hexagonal and cubic ice at low temperature. J Glaciology 7:95-108

Larson AC, von Dreele RB (1990) Los Alamos National Laboratory. Report No. LAUR 86-748

Lasaga AC (1998) Kinetic Theory in the Earth Sciences. Princeton University Press

Latroche M, Chabre Y, Decamps B, Percheron-Guegan A, Noreus D (2002) *In situ* neutron diffraction study of the kinetics of metallic hydride electrodes. J Alloy Compd 334:267-276

Levenspiel O (1999) Chemical Reaction Engineering. Wiley & Sons

Lobban C, Finney JL, Kuhs WF (1998) The structure of a new phase of ice. Nature 391:268-270

Lobban C, Finney JL, Kuhs WF (2000) The structure and ordering of ices III and V. J Phys Chem 112:7169-7180

Loerting T, Salzmann C, Kohl I, Mayer E, Hallbrucker A (2001) A second distinct structural "state" of high-density amorphous ice at 77 K and 1 bar. Phys Chem Chem Phys 3:5355-5357

Loerting T, Schustereder W, Winkel K, Salzmann CG, Kohl I, Mayer E (2006) Amorphous ice: Stepwise formation of very-high-density amorphous ice from low-density amorphous ice at 125 K. Phys Rev Lett 96:025702-025704

Lokshin KA, Zhao Y (2005) Advanced setup for high-pressure and low-temperature neutron diffraction at hydrostatic conditions. Rev Sci Instrum 76:063909

Loong C-K (2006) Inelastic scattering and applications. Rev Mineral Geochem 63:233-254

Lorenz G, Neder RB, Marxreiter J, Frey F, Schneider J (1993) Mirror furnace for neutron diffraction up to 2300 K. J Appl Crystallogr 26:632-635

Lundgaard L, Mollerup J (1992) Calculation of phase diagrams of gas-hydrates, Fluid Phase Equil 76:141-149.

Matsuo T (2003) Quantum aspects of low-temperature properties of crystals: A calorimetric study in interaction with spectroscopy and diffraction. Pure Appl Chem 75:913-926

Matsuo T, Tajima Y, Suga H (1986) Calorimetric study of a phase transition in D_2O ice Ih doped with KOD: ice XI. J Phys Chem Solids 47:165-173

Max MD (ed) (2003) Natural Gas Hydrate in Oceanic and Permafrost Environments. Kluwer Academic Publishers

Max MD, Johnson AH, Dillon WP (eds) (2006) Economic Geology of Natural Gas Hydrate. Springer, Dordrecht

Mayer E, Hallbrucker A (1987) Cubic ice from liquid water. Nature 325:601-602

Mishima O (1994) Reversible first-order transition between two H_2O amorphs at 0.2 GPa and 135 K. J Chem Phys 100:5910-5912

Mishima O (1996) Relationship between melting and amorphization of ice. Nature 384:546-549

Mishima O, Calvert LD, Whalley E (1984) An apparently first-order transition between two amorphous phases of ice induced by pressure. Nature 310:393-395

Mishima O, Calvert LD, Whalley E (1985) Phase behaviour of metastable water. Nature 314:76-78

Mishima O, Stanley HE (1998) The relationship between liquid, supercooled and glassy water. Nature 396: 329-335

Mishima O, Takemura K, Aoki K (1991) Visual observations of the amorphous-amorphous transition in H_2O under pressure. Science 254:406-408

Mizuno Y, Hanafusa N (1987) Studies of surface properties of ice using nuclear magnetic resonance. J Phys Coll C1 Paris 48: 511-517

Murray BJ, Bertram AK (2006) Formation and stability of cubic ice in water droplets. Phys Chem Chem Phys 8:186-192

Murray BJ, Knopf DA, Bertram AK (2005) The formation of cubic ice under conditions relevant to Earth's atmosphere. Nature 434:202-205

Oed A (1988) Position-sensitive detector with microstrip anode for electron multiplication with gases. Nucl Instrum Methods Phys Res A 263:351-359

Pannetier J (1985) Time-Resolved Neutron Powder Diffraction. Chem Sci 26A:131-139

Parise JB (2006a) Introduction to neutron properties and applications. Rev Mineral Geochem 63:1-25

Parise JB (2006b) High pressure studies. Rev Mineral Geochem 63:205-231

Peterson RC, Yang H (2001) High-temperature devices and experimental cells for X-ray and neutron diffraction experiments. Rev Mineral Geochem 41:425-444

Petrenko VF and Whitworth RW (1999) Physics of Ice. Oxford University Press, Oxford.

Poole PH, Grande T, Austen Angell C, Sciortino F, Eugene Stanley H (1994) Effect of hydrogen bonds on the thermodynamic behavior of liquid water. Phys Rev Lett 73:1632-1635

Poole PH, Sciortino F, Essmann U, Stanley HE (1992) Phase behaviour of metastable water. Nature 360:324-328

Pradell T, Crespo D, Clavaguera N, Clavaguera-Mora, MT (1998) Diffusion controlled grain growth in primary crystallization. Avrami exponents revisited. J Phys Cond Matter 10:3833-3844

Primak W (1975) The Compacted States of Vitreous Silica. Gordon and Breach

Proffen T (2006) Analysis of disordered materials using total scattering and the atomic pair distribution function. Rev Mineral Geochem 63:255-274

Pruzan P, Chervin JC, Wolanin E, Canny B, Gauthier M, Hanfland M (2003) Phase diagram of ice in the VII-VIII-X domain. Vibrational and structural data for strongly compressed ice VIII. J Raman Spectroscopy 34:591-610

Radaelli PG, Hannon AC, Chapon LC (2003) GEM: A Shining Light in the ISIS Crown. Notiz Neut Luce Sinc 8:19-26

Radlinski AP (2006) Small-angle neutron scattering and the microstructure of rocks. Rev Mineral Geochem 63:363-397

Redfern SAT (2006) Neutron powder diffraction studies of order-disorder phase transitions and kinetics. Rev Mineral Geochem 63:145-170

Riekel C, Schöllhorn R (1976) A neutron diffraction study on the intercalation of ammonia into tantalum disulfide. Mater Res Bull 11:369-376

Riley DP, Kisi EH, Hansen TC, Hewat AW (2002) Self-propagating high-temperature synthesis of Ti_3SiC_2: I, ultra-high-speed neutron diffraction study of the reaction mechanism. J Am Ceram Soc 85:2417-2424

Riley DP, Oliver CP, Kisi EH (2006) In-situ neutron diffraction of titanium silicide, Ti_5Si_3, during self-propagating high-temperature synthesis (SHS). Intermetallics 14:33-38

Röttger K, Endriss A, Ihringer J, Doyle S, Kuhs WF (1994) Lattice constants and thermal expansion of H_2O and D_2O ice Ih between 10 and 265 K. Acta Crystallogr B 50:644-648

Salamatin AN, Hondoh T, Uchida T, Lipenkov VY (1998) Post-nucleation conversion of an air bubble to clathrate air-hydrate crystal in ice. J Cryst Growth 193:197-218

Salamatin AN, Kuhs WF (2002) Formation of porous gas hydrates. Proc Fourth Int Conf Gas Hydrates, Yokohama 766-770

Salzmann CG, Radaelli PG, Hallbrucker A, Mayer E, Finney, JL (2006) The preparation and structures of hydrogen ordered phases of ice. Science 311:1758-1761

Sastry S, Sciortino F, Stanley HE (1993) Limits of stability of the liquid phase in a lattice model with water-like properties. J Chem Phys 98:9863-9872

Schmitt B, de Bergh C, Festou M (eds) (1998) Solar System Ices. Astrophysics and Space Science Library 227. Kluwer Academic Publishers

Schmitt D, Ouladdiaf ,B (1998) Absorption correction for annular cylindrical samples in powder neutron diffraction. J Appl Crystallogr 31:620-624

Schober H, Koza M, Toelle A, Fujara F, Angell CA, Boehmer R (1998) Amorphous polymorphism in ice investigated by inelastic neutron scattering. Physica B241-243:897-902

Sciortino F, Tartaglia P (2001) Aging in simple liquids: A numerical study. J Phys Cond Matter 13:9127-9140

Sears VF (1992) Neutron scattering lengths and cross sections. Neutron News 3:29-37

Sloan ED Jr (1998) Clathrate Hydrates of Natural Gases. Marcel Dekker, Inc.

Sloan ED Jr, Fleyfel F (1991) A molecular mechanism for gas hydrate nucleation from ice. Inst Chem Eng J 37:1281-1292

Speedy RJ (1982) Limiting forms of the thermodynamic divergencies at the conjectured stability limit in superheated and supercooled water. J Phys Chem 86:3002-3005

Speedy RJ (1982) Stability-limit conjecture. An interpretation of the properties of water. J Phys Chem 86:982-991

Stanley HE, Teixeira J (1980) Interpretation of the unusual behavior of H_2O and D_2O at low temperatures: Tests of a percolation model. J Chem Phys 73:3404-3422

Staykova DK, Kuhs WF, Salamatin AN, Hansen T (2003) Formation of porous gas hydrates from ice powders: Diffraction experiments and multi-stage model. J Phys Chem B107:10299-10311

Stern LA, Circone S, Kirby SH, Durham WB (2001) Anomalous preservation of pure methane hydrate at 1 atm. J Phys Chem B105:1756-1762

Stern LA, Circone S, Kirby SH, Durham WB (2002) Reply to "Comments on 'Anomalous preservation of pur methane hydrate at 1 atm'." J Phys Chem B106:228-330

Stern LA, Hogenboom DL, Durham WB, Kirby SH, Chou IM (1998) Optical-cell evidence for superheated ice under gas-hydrate-forming conditions. J Phys Chem B102:2627-2632

Sugisaki M, Suga H, Seki S (1968) Calorimetric study of the glassy state. IV: Heat capacities of glassy water and cubic ice. Bull Chem Soc Japan 41:2591-2599

Takeya S, Ebinuma T, Uchida T, Nagao J, Narita H (2002) Self-preservation effect and dissociation rates of CH_4 hydrate. J Cryst Growth 237-239:379-382

Takeya S, Hondoh T, Uchida T (2000) In-situ observations of CO_2 hydrate by X-ray diffraction. Ann N Y Acad Sci 912:973-982

Takeya S, Shimada W, Kamata Y, Ebinuma T, Uchida T, Nagao J, Narita H (2001) In-situ X-ray diffraction measurements of self-preservation effect of CH_4 hydrate. J Phys Chem A105:9756-9759

Tanaka H (2000) Thermodynamic anomaly and polyamorphism of water. Europhys Lett 50:340-346

Tsiok OB, Brazhkin VV, Lyapin AG, Khvostantsev LG (1998) Logarithmic kinetics of the amorphous-amorphous transformations in SiO_2 and GeO_2 glasses under high pressure. Phys Rev Lett 80:999-1002

Uchida T, Hondoh T, Mae S, Duval P, Lipenkov VY (1992) In-situ observations of growth process of clathrate air-hydrate under hydrostatic pressure. *In:* Physics and Chemistry of Ice. Maeno N, Hondoh T (eds) Hokkaido University Press, p 121-125

Uchida T, Hondoh T, Mae S, Duval P, Lipenkov VY (1994) Effects of temperature and pressure on transformation rate from air-bubbles to air-hydrate crystals in ice sheets. Ann Glaciology 20:143-147

Velikov V, Borick S, Angell CA (2001) The glass transition of water, based on hyperquenching experiments. Science 294:2335-2338

von Stackelberg M, Müller HR (1954) Feste Gashydrate II: Struktur und Raumchemie. Z Elektrochemie 58: 25-39

Walk-Lauffer B (2003) Untersuchung des Einflusses von Sulfaten auf das System $CaO-SiO_2-Al_2O_3-K_2O-H_2O$ mittels Wärmeflusskalorimetrie und in-situ Neutronenbeugung unter hydrothermalen Bedingungen. Dissertation, Universität Siegen, Logos Verlag

Walton RI, Francis RJ, Halasyamani PS, O'Hare D, Smith RI, Done R, Humphreys RJ (1999) Novel apparatus for the *in situ* study of hydrothermal crystallizations using time-resolved neutron diffraction. Rev Sci Instrum 70:3391-3396

Walton RI, Millange F, Smith RI, Hansen TC, O'Hare D (2001) Real time observation of the hydrothermal crystallization of barium titanate using *in situ* neutron powder diffraction. J Am Chem Soc 123(50): 12547-12555

Wang X, Schultz AJ, Halpern Y (2002) Kinetics of methane hydrate formation from polycrystalline deuterated ice. J Phys Chem A 106:7304-7309

Warzinski RP, Lynn, RJ Holder GD (2000) The impact of CO_2 clathrate hydrate on deep ocean sequestration of CO_2. Ann N Y Acad Sci 912:226-234

Wilder JW, Smith DH (2002) Comments on "Anomalous preservation of pure methane hydrate at 1 atm." J Phys Chem B106:226-227

Wilding MC, Benmore CJ (2006) Structure of glasses and melts. Rev Mineral Geochem 63:275-311

Williams WG, Ibberson RM, Day P, Enderby JE (1998) GEM - General Materials Diffractometer at ISIS. Physica B241-243:234-236

Wooldridge PJ, Richardson HH, Devlin JP (1987) Mobile Bjerrum defects: A criterion for ice-like crystal growth. J Chem Phys 87:4126-4131

Yakushev VS, Chuvilin EM (2000) Natural gas and hydrate accumulation within permafrost in Russia. Cold Regions Sci Technol 31:189-197

Yakushev VS, Istomin VA (1992) Gas-hydrate self-preservation effect. *In:* Physics and Chemistry of Ice. Maeno N, Hondoh T (eds) Hokkaido University Press, p 136-140

Yershov ED (2004) General Geocryology. Cambride Univ Press

Reviews in Mineralogy & Geochemistry
Vol. 63, pp. 205-231, 2006
Copyright © Mineralogical Society of America

9

High Pressure Studies

John B. Parise

Department of Geosciences, Chemistry Department and
Center for Environmental Molecular Sciences
Stony Brook University
Stony Brook, New York, 11794-2100, U.S.A.
e-mail: john.parise@sunysb.edu

INTRODUCTION

In principle, the methods described elsewhere in this volume are available at high pressures (HP). Of course, "in principle" implies the availability of techniques for maintaining samples of a sufficient volume at the HP to allow observation of a signal. Compared to studies on the same material at ambient conditions, neutron scattering at HP is a flux hungry technique. The HP device, irrespective of its design, will necessarily force compromise between signal and parasitic scattering and absorption from the cell. Breakthroughs in HP neutron science therefore depend on novel HP cell and beamline design that maximize pressure, sample volume, and signal to noise discrimination.

HIGH PRESSURE CELLS FOR NEUTRON SCATTERING

Overview and historical context

Successful high-pressure structural studies using neutron scattering techniques date from at least the 1960s (Bloch et al. 1966; Lechner and Quittner 1966; Litvin et al. 1966; Brugger et al. 1967). Although early HP cell designs were installed at reactor sources (Paureau and Vettier 1975) it became clear the geometric restrictions associated with collecting data in the angular dispersive geometry lead to contributions from parasitic scattering from the high pressure cell. Typical of these early designs was the so-called "McWhan cell" (McWhan et al. 1974), a piston cylinder apparatus (Fig. 1) with an alumina cylinder radially supported by compressing two steel plates with a hydraulic ram. This left a neutron window between the plates ~6 mm high and allowed a number of very important studies, with pressures up to 3 GPa achieved routinely. Despite difficulties at the higher pressures with poor counting statistics, and severe peak overlap problems because of diffraction from alumina, several important studies were carried out including careful work on the high pressure phases of ice, D_2O-VI, -VII and VIII (Kuhs et al. 1984). Apart from parasitic scattering from the HP cell, another disadvantage was the requirement that loading of pressure be done off line; the cell was clamped to maintain pressure (Paureau and Vettier 1975) and then transported to the neutron beam for data collection. This was required at each pressure point. It makes possible the deployment of several cells at beamlines, while only having to invest in a single high pressure station, and echoes of this strategy are seen in several recent designs for off-line loaded HP cells (Zhao et al. 2005). Modern HP cell designs tend to incorporate on-line loading, where P can be varied on-line with the aid of computer controlled pumping stations, rather than having to break the experiment, unload the cell, pressurize it and replace it in the beam. On-line loading is more compatible with greater source brightness, added user pressure for beamtime and the convenience of doing measurements at variable PT, in cryostats for example. Several

    DOI: 10.2138/rmg.2006.63.9

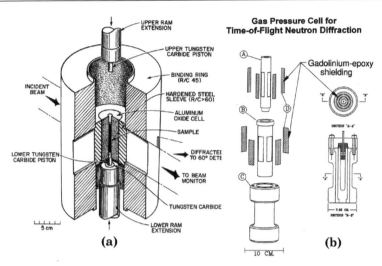

Figure 1. Pressure cells for neutron diffraction studies used at Argonne laboratories (JD Jorgensen, pers. Comm.) (a) piston cylinder type cell (McWhan et al. 1974). (b) Gas cell (Jorgensen 1978) in use at the IPNS, Argonne, USA. Carefully designed shielding and windows in the gas cell allow access to data banks at the spallation source and provide data free of parasitic scattering.

cell designs suitable for single crystal studies are summarized in earlier short course notes dedicated to high pressure research (Miletich et al. 2000).

The more restricted or fixed scattering geometry associated with time-of-flight techniques greatly simplifies the construction of pressure cells. Some of the earliest applications of high-pressure time-of-flight neutron powder diffraction were by the Brugger group who employed opposed anvil devices and alumina insert cells along with a chopper spectrometer at the Idaho Falls reactor and later at the Argonne National Laboratory at both reactor and spallation neutron sources (Vogel and Priesmeyer 2006, this volume). Some of the more important early studies include the crystal structure of Bi-II (Brugger et al. 1967) the first Rietveld-style refinement using high pressure data (Decker et al. 1974).

It was quickly realized that for studies at moderate pressures (<3 GPa) and low temperatures, a carefully designed gas pressure apparatus provided very clean data at a spallation source (Fig. 1). Although the pressure range is greatly restricted, careful analysis of data with superior signal-to-noise discrimination led to some fundamental discoveries. Excellent examples include the first experimental confirmation of what is now taken for granted in high-pressure silicate mineralogy. Measurements of interatomic distances and angles to about 2.5 GPa for quartz showed that the compression results solely from a cooperative rotation or tilting of the SiO_4 tetrahedra around their shared oxygen corners, with individual tetrahedra remaining relatively rigid (Jorgensen 1978). Another truly beautiful piece of work involves the pressure-induced soft-mode transition in ReO_3; one of the few cubic perovskite-related materials that do what perovskites are predicted to do at HP (Jorgensen et al. 1986).

More recently, development of high-pressure cells has concentrated on reliably achieving pressures in excess of 3 GPa using two approaches. One approach, initially advanced by Russian workers at JINR, Dubna, concentrated on the use of "gem cells" (Fig. 2) based on the use of sapphire or diamond anvils (Glazkov et al. 1989; Ivanov et al. 1995). These diamond- and sapphire-anvil cells are similar to those used in laboratory- and synchrotron-based X-ray experiments. Several advantages accrue from this approach, including the high pressures attainable with small samples and small culet sizes for the gems, and the transferability of infrastructure developed at synchrotron sources for flexible optics, and laser heating. The first HP neutron beamline constructed with flexibility built in is described in the final section of this chapter. The disadvantages of gem cells include the small sample volumes, of perhaps 1-5 mm³, which are small by neutron standards; typically volumes of 100 mm³ provide routine powder diffraction data. Despite these disadvantages several important studies have been carried out using gem cells, mainly in the area of magnetism (Aksenov et al. 1995; Goncharenko et al. 1995; Ivanov et al. 1995; Braithwaite et al. 1996; Link et al. 1998) where the observation of the appearance and disappearance of magnetic reflections at pressure provides enough information to deduce the magnetic structure (Von Dreele 2006, this volume). Some of the disadvantages of gem cells could be offset with development of higher flux sources, detectors of greater sensitivity, and tougher and larger (diamond) anvils. New developments in all these areas are discussed below and recent experimental results suggest these gem cells hold great promise for diffraction studies of single crystals, especially when coupled to focusing optics and sensitive area detectors (McIntyre et al. 2005).

The Paris-Edinburgh (PE) press

The PE press (Besson et al. 1992; Nelmes et al. 1993a) shown in Figure 3, is the most widely used, accepted and productive HP cell of the past decade. While the PE press is similar in concept to the opposed-anvil devices used by the Brugger group (Brugger et al. 1967; Worlton et al. 1967, 1968) it achieves pressures greater than 10 GPa with sample volumes of about 100 mm³ while maintaining a compact design. The "PE press" is continually updated and developed for both spallation (Loveday et al. 1995) and reactor (Klotz et al. 2005a) neutron sources. First installed at the ISIS spallation neutron source at the Rutherford Appleton Laboratory, Chilton, United Kingdom, the PE press is now used at several X-ray and neutron facilities world-wide for both elastic and inelastic scattering (Klotz 2001; Klotz

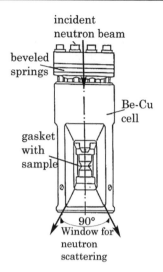

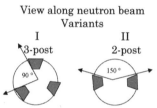

Figure 2. A variation of the "gem cell" (Ivanov et al. 1995) used by the Goncharenko group in Russia and used extensively for magnetic studies at high pressure at the LLB, Saclay, France (see text). The alloy BeCu is chosen for construction of the body of the cell since it is non-magnetic. There are two variants of the cell with 2 windows 150° apart in the plane perpendicular to the cell axis or 3 windows 90° apart (see plan view right). The cell can be fixed from the top by a screw or from the bottom by 4 screws. The cell is 88 mm high and 36.5 mm in diameter and can accommodate sapphire anvils 10 mm in height, which can be replaced with anvils such as moissanite (SiC) and diamond.

et al. 1995a, 2000, 2005b). Rather than compromise sample volume to achieve pressures above 3 GPa, the PE press was designed to maintain a sample volume comparable with that of the alumina insert cell (Fig. 1) while achieving high pressures by the use of a novel compact design. This was achieved by leveraging developments in the Russian HP community of a novel toroid anvil design (Khovstantsev et al. 1977, 2004), which allowed larger sample volume, with a press with a ram with small displacement, making it more compact. The hydraulic ram (Fig. 3) was a significant development allowing a press of 200 tonnes capacity within a device only 20 cm across (Fig. 3) and having an overall mass of only 50 kg. Powder samples of 80-100 mm³ are routinely taken to 12 GPa using tungsten carbide anvils and 25 GPa (and above) has been

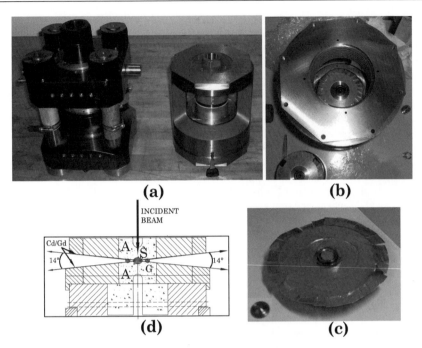

Figure 3. (a) The original Paris-Edinburgh design with 4-posts (left) with the new VX design, with a car key for scale. The VX design allows more open access to the incident and scattered beam, as would be required for greater detector coverage (see Fig. 16 and 17). The VX is shown open in (b) with the toroid exposed and the threaded breach holding the opposing anvil removed (lower left). (c) The anvil is coated with neutron absorbing material (cadmium in this case) and the TiZr gasket is shown surrounding the sample. This sample will be flooded with deuterated methanol-ethanol and the cap shown lower left placed on top with the opposing anvil from the breach bolt (b). The anvil package produced is shown schematically and in cross section in (d) where the sample (S) is contained between the two anvils (A) made of tough materials such as tungsten carbide, sintered diamond or cubic boron nitride. The toroid design anvil (Khvostantsev et al. 2004) is shown here but many other anvil profiles are possible and are adopted for various sample environments (Le Godec et al. 2003). The sample is girdled by a metal gasket (G), fully enclosed by a soft metal encapsulated (SME) gasket (Marshall and Francis 2002). The incident and scattered beam geometry is through the anvil and gasket, respectively, for most experiments at spallation sources, and through the gasket for incident and scattered beams at reactor sources.

reached with smaller samples of 40 mm^3 and anvils made of sintered diamond. More recently the design is adapted for studies of single crystals (Bull et al. 2005).

While the PE press accepts various anvil profiles, including multi-anvil packages, the anvil design most closely associated with the PE program is shaped to accommodate a toroidal gasket (Fig. 3) and is based on the work of Khvostantsev (Khvostantsev et al. 2004). In this design, the centre of the anvil is hollowed out to give a spheroidal sample volume. The original experiments at the ISIS spallation source with the 4-post PE cell (Fig. 3a) used null-scattering TiZr alloy as gasket material to surround the sample and a beam incident along the cell axis (Fig. 3d). The sample is loaded into the hemispherical toroid (Fig. 3b) with the TiZr gasket, either without pressure transmitting media such as Fluorinert® or without, and the threaded breach bolt containing the opposing anvil is screwed into place forming a quasi-spherical sample illuminated through the breach and anvils, modified to minimize absorption by the WC or sintered diamond (Fig. 3d). Scattering data are collected in banks of detectors positioned at $2\theta = \pm90°$ with access from the sloping profiles of the anvils restricted to $14°$ centered on $2\theta = 90°$ (Fig. 3d).

To minimize background and to prevent the detectors from viewing any illuminated anvil material the faces of the anvils are coated with cadmium sheet or gadolinium paint (Figs. 3b-d). This shielding allows the collection of data essentially free from contaminant peaks at lower pressures. It is difficult to exclude all parasitic scattering however, especially at higher P, where the anvil gap closes and illumination of the anvils is unavoidable. Most diffraction patterns are contaminated with some scattering from the cell. The through-anvil cell geometry also requires careful attenuation corrections, especially in the case where sintered diamond anvils are used (Wilson et al. 1995). In that case Bragg scattering from the diamond produces edges (Bragg edges) in the diffraction patterns and the corrections for these are non-trivial (Wilson et al. 1995).

A crucial innovation in gasketing the sample in the PE cell was introduced at ISIS (Marshall and Francis 2002). A deuterated methanol-ethanol-water mixture is the standard pressure-transmitting medium in high-pressure studies. However this mixture was known for many years to be detrimental to sintered anvil materials such as WC and diamond at pressure above about 2 GPa. By using machined TiZr alloys and an encapsulated design (Fig. 3c) that completely surrounded the sample, and so protected the anvils, quasi-hydrostatic conditions can be maintained to at least 10 GPa. The TiZr SME gasket (Fig. 3c) is composed of several pieces that fit neatly into the toroid holding the sample and are tight against leakage by quickly advancing the bottom anvil against the breach shortly after the sample is bathed in pressure transmitting fluid.

The PE cell at reactor sources. At reactor sources elastic scattering in the PE cell is carried out in the radial geometry with the beam entering and exiting through the gasket material (Fig. 3). The advantage of this geometry has been illustrated recently (Kernavanois et al. 2002; Klotz et al. 2005a; Parise et al. 2006) with studies employing BN anvils and null scattering TiZr gaskets. A preliminary investigation of ice-VII and of magnetic ordering in CoO (Kernavanois et al. 2002) with the original 4-post press (Fig. 3) and the BN anvil package produced very encouraging results. The data were collected with a 154° position sensitive detector at the D20 beamline of the ILL. Since the center of the press posts are 90° apart, the pattern was unaffected over only about a 70° range and care had to be taken to introduce the beam close to one of the posts.

In follow-up experiments a new generation PE cell, the VX series, where the cell no longer has tie rods but is instead machined from a monolith of metal that produces two "posts" centered at 180° degrees apart (Fig. 3a) allowed greater access to the angle dispersive scattering (Klotz et al. 2005a). The key innovation however, was the use of BN anvils that provide self-collimation at the reactor source and provide data essentially free of parasitic scattering without the need to coat the anvils with neutron absorbing paint. (One disadvantage of the coatings at high pressures is their tendency to shadow the detector as the anvil gap closes.) Of course, since boron is an excellent absorber of neutrons, the use of these anvils also means that the scattering observed from the sample is restricted to the portion not shadowed by the anvils; only scattering from the sample within the anvil gap is detected (Fig. 3d). The advantages however outweigh this limitation; this geometry is especially useful for studies of magnetic ordering at high pressure (Kernavanois et al. 2002; Parise et al. 2006) since the reactor source has considerably more time averaged flux at low Q, where magnetic peaks are strongest. To date useful data were collected in this mode at the D20 diffractometer at ILL (Kernavanois et al. 2002) at Chalk River (Parise et al. 2006) and at The HIFR reactor at Oak Ridge.

Finally a note on the much touted synchrotron/neutron synergy: the BN anvils were in common use at the ESRF, where the PE cell was used for high temperature work (Crichton and Mezouar 2002; Falconi et al. 2004; Crapanzano et al. 2005) and their use at the ILL reactor (Kernavanois et al. 2002) is a direct outgrowth of the foresight shown in situating the ESRF and ILL at the same campus in Grenoble.

There is now great variety in the types of cells available to pressurize samples (Figs. 1-3) for neutron scattering studies and the choice depends on the pressure range desired. To achieve the highest pressures will always require a smaller sample volume for a particular cell design. Apart from that inescapable condition, there are several other choices the researcher needs to consider. For example there is growing interest in studying materials of increasing structural complexity under pressure using neutron scattering. For materials of sufficient complexity (Colligan et al. 2005) the powder technique may need to be supplemented by careful single crystal studies (McIntyre et al. 2005). The following non-exhaustive and highly personal perspective on high pressure neutron diffraction studies includes examples of work carried out using a variety of cells and neutron sources.

EXAMPLES OF POWDER DIFFRACTION STUDIES AT HIGH PRESSURES

The reasons for using neutron scattering at high pressure are essentially those for neutron studies at ambient condition: location of low-Z atoms, particularly protons, determination of magnetic structure and site-by-site magnetic moments, and determination of phonon dispersion relationships. The neutron's sensitivity to hydrogen and magnetic structure are particularly important in the Earth sciences; hydrous minerals are implicated in water cycling in the mantle (Jacobsen et al. 2004) and magnetic studies have implications for paleomagnetism, for example (Von Dreele 2006, this volume).

Studies of magnetism

While a material's bulk susceptibility can be determined by other means only neutron scattering provides ready access to site-by-site determination of spin orientation and magnetic moment (Von Dreele 2006, this volume). Magnetic studies were amongst the first carried out with neutrons in the late 1940s and early 1950s (Shull et al. 1951). However, few studies of the effects of pressure on magnetic structure were reported (Werner et al. 1968) until cells (Figs. 1 and 2) became available at neutron sources (Litvin et al. 1966; Brugger et al. 1967; Worlton et al. 1967). Important studies in this period included the observation that spin-flip transitions, involving reorientation of spins in antiferromagnetic materials and well known from low temperature studies (Morin 1950) could also be induced by high pressure. Neutron scattering is particularly valuable in these studies since, for antiferromagnetic materials, extra peaks appear in the powder diffraction or single crystal patterns due to magnetic ordering (Von Dreele 2006, this volume). These peaks do not appear in the X-ray pattern and so they are relatively easily indexed. Two examples of high pressure studies serve to illustrate work at both reactor and spallation neutron sources.

FeS-II. Systematic structural studies of stoichiometric FeS are hampered by the pseudo-symmetry inherent in phases encountered in the high *P-T* phase diagram. All three of the high pressure structures discovered to date (Nelmes et al. 1999) are related to the NiAs structure-type and small displacements of atomic positions from those in this ideal structure produce subtle peak splitting and weak superlattice reflections in X-ray and neutron diffraction patterns. While changes in X-ray patterns are subtle, changes in magnetic structure result in dramatically different neutron powder diffraction patterns for the three high pressure phases (Nelmes et al. 1999).

At ambient conditions in the mineral troilite (FeS-I) the Fe^{2+} magnetic moments are aligned perpendicular to the basal plane (the *a*-direction for the MnP-related FeS-II in Fig. 4) and coupled antiferromagnetically between layers. In the neutron diffraction pattern the clearest indicator of the spin re-orientation transition accompanying transition to FeS-II at about 3.4 GPa (Fig. 4) is the appearance of a strong magnetic reflection—absent in the troilite pattern—at what corresponds to the (001) reflection of the NiAs substructure. At 6.7 GPa, FeS-II transforms to phase III (Marshall et al. 2000). The appearance of the $(001)_{NiAs}$ peak

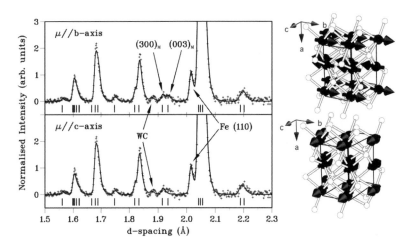

Figure 4. Simultaneous Rietveld refinement of the FeS phase II (MnP-related phase) using data collected data at 4.0 GPa for 2θ = 90° at ISIS. The curves have the usual meanings and the tick marks indicate the positions of allowed peaks for FeS nuclear peaks. The two refinement showing the refinements (smooth curves) for μ ‖ b (upper) and μ ‖ c (lower). Adapted from Marshall and Francis (2002).

above 3.4 GPa unambiguously identifies the magnetic transition as involving a spin flip of the moment from lying perpendicular to the basal plane lying somewhere in it (Fig. 4).

The neutron diffraction data shown in Figure 4 were obtained at the HiPr facility of ISIS using a V4 PE cell (Fig. 3). One disadvantage of the spallation source (Vogel and Priesmeyer 2006, this volume) for magnetic studies when the standard through anvil beam geometry is used is the inability to access high d-spacings in the 90° banks. A novel design feature of HiPr is the ability to collect diffraction data using the standard through-anvil geometry with scattering angle coverage 2θ = 90° ± 7° (Fig. 3d) and a through-gasket geometry, which permits data collection out to long d-spacings (10 Å) using the low angle detectors, 2θ = 30° ± 10°.

The higher resolution and peak-to-background discrimination afforded at HiPr was advantageous in determining the absolute spin direction within the *bc*-plane for the FeS-II structure (Fig. 4). Due to the well-known limitation of the powder method (Shirane 1959) it was thought initially that it would not be possible to determine the moment orientation within the *bc* plane. The lattice constant ratio *c/b* differs from √3—the ideal value for the NiAs subcell—by just 0.7%. The resolution of the low-angle data is not sufficient to determine the intensity distribution in the strong (100)/(001) magnetic doublet, by far the strongest of the few purely magnetic reflections. Nevertheless, there are sufficient statistics to discriminate between possible models for the orientation of the spin in the basal-plane by noting that the goodness-of-fit of the refinement with μ ‖ b was slightly better than with μ ‖ c.

Figure 4 shows data collected at 4.0 GPa, over the d-spacing range 1.5-2.3 Å, with the refinements for μ ‖ b and μ ‖ c in the upper and lower patterns, respectively. Of particular interest here is the weak feature between 1.90 and 1.95 Å d-spacing range, in which the proposed magnetic unit cell predicts two purely magnetic reflections (reflections which would not appear in the X-ray diffraction pattern)—the (300)/(003) doublet. Close inspection of Figure 4 shows that the μ ‖ c model is unable to account for all the observed intensity in this feature, for in this case the (003) reflection must have zero intensity - the magnetic moments and the scattering vector are parallel. By contrast, the μ ‖ b model seems to account for all of the observed intensity in this feature and a free refinement of the moment orientation in the *bc* plane does indeed converge to this configuration. The same result was obtained for the three other patterns obtained for FeS phase II. In particular, the weak peak attributed to the (300) magnetic reflection moved in step with the nuclear reflections, so eliminating the possibility that this feature was due to contaminant scattering from the pressure cell, sample tank, etc.

Studies of HP magnetic structure at reactor sources

The magnetic structure of hematite (α-Fe₂O₃). Hematite is important in paleomagnetism and the origin of its natural remnant magnetization, widely used in red sediments (Dunlop 1970) may not be due to intrinsic anisotropic superexchange interactions giving rise to spin-canting and a weak ferromagnetism (Moriya 1960). The structural and magnetic properties as a function of T are well known (Morin 1950; Shull et al. 1951). At ambient PT the nuclear structure is rhombohedral (shown in Fig. 5 in the hexagonal setting) and related to that of corundum (Al_2O_3). The magnetic structure, amongst the first determined by Shull (1951), has spins on the Fe atoms aligned ferromagnetically within each of the basal layers and coupled antiferromagnetically between them. Morin (1950) found that below 250 K, at the Morin transition, the NPD pattern can be accounted for on the basis of an antiferromagnetic lattice with iron spins oriented along the $[111]_{rhombohedral}$ (= $c_{hexagonal}$) direction (Fig. 5). The spin flop transition, with the change in moment direction from within the basal plane to perpendicular to it, is seen readily in the relative intensities of the first two magnetic peaks (Fig. 6). The $(111)_M$ $(100)_M$ are both pure magnetic (M) reflections and the $(110)_N$ is of pure nuclear (N) origin. As T is lowered at ambient P toward the Morin transition, the $(111)_M$ disappears at the transition. The

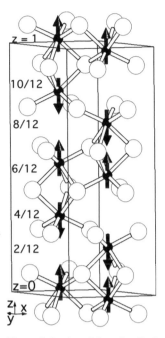

Figure 5. Portion of the unit cell of hematite showing the nuclear and magnetic structure below the Morin transition where spins are oriented along the $c_{hexagonal}$ axis = $[111]_{Rhombohedral}$. The hexagonal unit cell is used here to illustrate the planes of Fe atoms with parallel spins and their shifts, up and down from the plane (~ ± 0.02 from the values for the z-coordinates are given to the left).

moment strength and direction can be readily detected by measuring the relative intensities of these first 3 reflections in the NPD pattern (Fig. 6). Rietveld structure refinement is a convenient way to obtain the variation in moment along with the nuclear structure (see Von Dreele 2006, this volume; Knorr and Depmeier 2006, this volume).

These experiments are quite straight forward at a reactor source, where incident and scattered beams are in the equatorial plane for angular dispersive (AD) measurements (Kernavanois et al. 2002; Klotz et al. 2005a; Parise et al. 2006). For our AD measurements we used standard WC anvils, which need to be "dressed" with Cd-foil to minimize parasitic scattering from steel and WC in the anvils. This can be avoided by using BN anvils but at the expense of limiting scattering from the roughly spherical sample to the slit between the BN anvils (see text above and Fig. 3). We chose to maximize sample scattering since contamination from the steel and WC occur far from the low angle magnetic peaks of interest (Fig. 6).

Previous studies of the effect of P on the Morin transition reveal a rapid increase in T_M of about 3 °/kbar (1 kbar = 0.1 GPa) below 0.6 GPa in a hydrostatic environment (Searle 1967) suggesting the transition should reach room temperature at about 1.5 GPa. Later experiments carried out to 10 GPa in a non-hydrostatic environment (Goncharenko et al. 1995) show reorientation of the spin towards the axial direction (Fig. 5) but complete reorientation was not

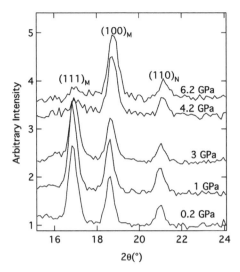

Figure 6. Low angle data ($\lambda = 1.3290(2)$ Å) collected at the C2 diffractometer at the Chalk River reactor, Canada, showing the magnetic (M) and nuclear (N) peaks for hematite. The indexing is with respect to the rhombohedral setting of the hexagonal unit cell shown in Figure 5. The equivalent peaks in the hexagonal setting, starting with the rhombohedral $(111)_M$ and moving to higher angle are (003), (101) and (102) in the hexagonal setting.

achieved even when the sample was cooled below 50 K, suggestive of a new magnetic phase, with a moment direction intermediate between those above and below T_M (Goncharenko et al. 1995). The authors (Goncharenko et al. 1995) note however, that some of the observed effects could be due to deviatoric stresses due to the absence of pressure media. Recent experiments carried out at the Chalk River reactor (Parise et al. 2006) using a variety of pressure media including Fluorinert® and deuterated methanol-ethanol mixtures (Marshall and Francis 2002) in the PE cell, support the hypothesis that the Morin transition is suppressed by non-hydrostatic conditions (Fig. 7). Although Fluorinert® is solid above 4 GPa we observe almost complete disappearance of the $(111)_M$ at about 6 GPa (Fig. 6).

The data collected at HP can be treated straightforwardly using Rietveld fitting with four-phases—hematite's magnetic and nuclear structure, and the parasitic scattering from WC and Steel. The magnetic moment orientation is described in a model using spherical polar coordinates. The results of this fitting are summarized in Figure 7 where the moment angles determined as a function of P from different studies are compared. The behavior of hematite at high P is quite different in different pressure media. In all cases the moment magnitude remains unchanged but in Fluorinert® the moment orientation with respect to

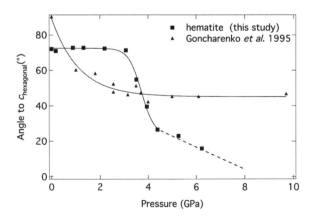

Figure 7. Variation with P of the angle between the moment vector and the c-axis for the hexagonal cell (rhombohedral [111]) of hematite shown in Figure 5. The data (squares) were obtained from Rietveld refinement using data obtained from a hematite sample in Fluorinert® in the PE cell. The lines are guides to the eye.

the $c_{hexagonal}$-axis remains constant until about 3.5 GPa, where upon it decreases markedly to the maximum pressure studied (6.3 GPa). The pressure where the steep decrease commences corresponds approximately to the freezing of Fluorinert®. If the trend predicted by the dotted line in Figure 3 were to continue T_M should reach room T at about 9 GPa. The moment does not change in hematite above 4 GPa when it is pressurized without pressure medium (Fig. 7). The results obtained thus far demonstrate the facility with which magnetic structure at HP can be surveyed using the PE cell on reactor sources and how rich are the magnetic phenomena encountered (Von Dreele 2006, this volume).

Studies of hydrogen bonding in minerals with extended structures

Hydrogen-bonded networks are responsible for the formation of ice and for building polyhedral cages around guest molecules to form solid clathrate hydrates at high pressures and low temperatures (Kuhs and Hansen 2006, this volume). There is no better example of the power of neutron studies than on-going research into the structures and the *PT* conditions of formation for the numerous stable and metastable polymorphs of ice. The *PT* phase diagram of ice is bewilderingly rich (Fig. 8) and was confused prior to work that combined neutron powder diffraction and *in situ* studies in HP cells. Neutron powder diffraction work over the past six decades has identified, and still is identifying many of the phases under their conditions of stability, or metastability (Wollan et al. 1949; Kuhs and Lehmann 1983; Jorgensen et al. 1984; Kuhs et al. 1984, 1987; Jorgensen and Worlton 1985; Floriano et al. 1986; Londono et al. 1988, 1993; Li et al. 1991; Nelmes et al. 1993b; Nield and Whitworth 1995; Jackson et al. 1997; Lobban et al. 1998; Demirdjian et al. 2002; Finney et al. 2002a,b; Salzmann et al. 2006). The neutron scattering work cited demonstrates that *in situ* high pressure neutron diffraction, on either single crystal or powders, allows easy detection of phase changes, the planning of various PT trajectories through the phase diagram, which sometimes leads to new phases, and definition of the hydrogen (deuterium) positions resulting from changes in H-ordering (Salzmann et al. 2006).

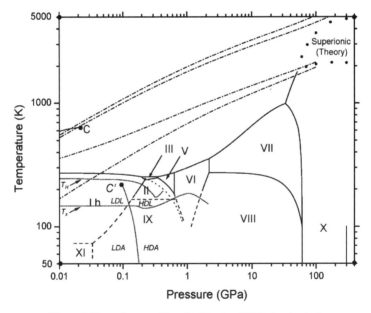

Figure 8. Phase diagram of ice after Hemley (2000) showing both stable and metastable phases to very high pressures.

Other ice-related research includes work on clathrates of hydrogen and methane, which may be important for energy storage and recovery (Lokshin et al. 2005). Neutron scattering carried out *in situ* using a gas pressure apparatus (Fig. 2) is the only structural probe sufficiently sensitive to hydrogen (and deuterium) to address questions of occupancy in the cages of the clathrates. For example (Lokshin et al. 2004) using deuterium rather than hydrogen to avoid the undesirable contributions from ^{1}H incoherent scattering, found ^{2}H-guest occupancy can be reversibly changed by variation of pressure and temperature and that the maximum density of deuterium in the clathrates is higher than the value in most metal hydrides being considered as hydrogen-storage materials. An excellent introduction to the subject of ices is provided in elsewhere in this volume (Kuhs and Hansen 2006, this volume) and will not be pursued further here other than to note the variety of elastic and inelastic neutron scattering studies being carried out in the area of high pressure ices (Nelmes et al. 1993b; Klotz et al. 1995b; Loveday et al. 1996, 2000; Tse et al. 2000; Le Godec et al. 2004; Lokshin et al. 2004, 2005; Strassle et al. 2006).

Hydrogen bond formation vs. hydrogen repulsion in extended solids. Studies at high pressure using spectroscopic and scattering techniques suggest that hydrogen bonds involving OH$^-$ ions in extended mineral structures behave differently from those in molecular systems. They are generally weaker, more prone to bifurcation and the formation of chelated structures (Jeffrey 1997) and are affected by H...H repulsion. An understanding of these differences in behavior at high pressure has been possible through the use of modern crystallographic (Parise et al. 1994, 1998a,b, 1999; Kunz et al. 1996; Peter et al. 1999; Kagi et al. 2000, 2003; Prewitt and Parise 2000; Friedrich et al. 2002; Lager et al. 2002, 2005; Liu et al. 2003; Colligan et al. 2005; Komatsu et al. 2005; McIntyre et al. 2005; Lee et al. 2006) and spectroscopic (Kruger et al. 1989; Duffy and Ahrens 1991; Cynn and Hofmeister 1994; Nguyen et al. 1994, 1997; Duffy et al. 1995a,b,c; Cynn et al. 1996; Hofmeister et al. 1999) techniques. These studies include work on both simple and more complex hydrogen-containing minerals and analogs.

High-pressure elastic neutron powder diffraction has a distinct advantage in uncovering the interplay between H-bond formation and H...H repulsion in the O-H...O moiety. While subtle stereochemical effects tend to mask changes attributed to hydrogen bonding as side groups vary in a sequence of molecular structures (Jeffrey 1997) pressure provides a means, in theory, of varying interatomic distances in an isochemical system. As Jeffrey (1997) points out, the scatter in correlation curves of the A-H IR stretch vs. A...B separation is far greater than errors associated with the experimental measurements. This is probably due to subtleties to do with chemistry, which changes from compound to compound. Clearly the removal of the chemical variable, and a means to systematically change the degree of hydrogen bonding, is attained through the use of pressure.

Because of its sensitivity to hydrogen, and because rigorous theory (see Parise 2006, this volume) allows the construction and testing of models ab initio, neutron diffraction is the tool of choice for the study of hydrogen bonding in extended solids. It is especially powerful when coupled with high pressure spectroscopic investigations (Peter et al. 1999; Kagi et al. 2000). The large incoherent cross section of hydrogen tends to overwhelm the signal with background in powder studies of hydrous materials. In this case substitution of hydrogen with deuterium, which has a much smaller, though still significant, incoherent cross-section, is effective. The small possible effects of this substitution are discussed in detail elsewhere (Jeffrey 1997). While the energetics and the stability of the O-H...O attractive interaction has been especially important in research involving ice, little attention has been paid to the H...H repulsive interaction. Neutron scattering at HP, on deuterated samples to provide D...D distances as proxies for the H...H distance, is an excellent means to study these interactions. Rather than using side groups or comparisons between unrelated structures, continuous variation of H...O (D...O) and H...H (D...D) distances can be achieved in an isochemical system. This removes ambiguities due to steric effects, introduced when systems with very different structural chemistries are compared (Jeffrey 1997). Further, in simple compounds such as the layered hydroxides, the O-H...O moiety is relatively isolated from the remainder of the structure; the interactions between the attractive and repulsive forces can be studied continuously as a function of the pressure variable.

The simple layered hydroxides related to the CdI_2-structure (Parise et al. 1994) possess a unique geometry with hydrogen atoms isolated between a framework of MO_6 octahedra (Fig. 9), where M = Mn, Fe, Co, Ni, Cd, Mg, Ca. The H...O and H...H interactions will vary as the interlayer separation is changed by the application of pressure, and the realization of this has prompted a number of spectroscopic (Kruger et al. 1989; Nguyen et al. 1994, 1997; Duffy et al. 1995a,b) and crystallographic investigations (Parise et al. 1994, 1998a,b, 1999; Catti et al. 1995). Some spectroscopic studies of $M(OH)_2$ hydroxides indicate "phase transitions" at high pressure. While most involve subtle rearrangements of the H (or D) atoms, there are reports of whole scale sub-lattice amorphization. Claims that correlations between the oxide and hydrogen sublattice, such as those shown schematically in Figure 8, are lost on pressurization involve more than semantics (Nguyen et al. 1997). Neutron diffraction definitively shows that, while there are changes in the configuration of the hydrogen (deuterium) sites, the O-D bond length remains essentially in tact and that all deuterons remain correlated with those oxygen sites (Parise et al. 1999).

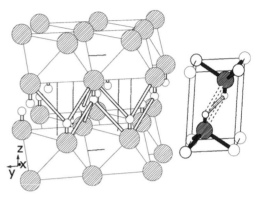

Figure 9. (left) Polyhedral representation of the structure of the layered hydroxides (and deuteroxides) related to brucite, $Mg(OH)_2$, which include $A(OH)_2$ A = Ca, Cd, Mn, Fe, Co, Ni. The unit cell contents are shown right with each hydrogen involved in three weak hydrogen bonds (>2.5 Å), shown as dashed lines and three H...H repulsive interactions (~2 Å) shown as hashed lines. Stripped balls represent oxygen while small and large circles represent hydrogen and magnesium, respectively.

The role of D...D repulsion, and D...M, where M is some metal (Williams and Guenther 1996; Peter et al. 1999), was not generally appreciated until reliable neutron scattering studies became possible at above 10 GPa. In a series of HP neutron powder diffraction studies of the $M(OD)_2$ compounds (M = Mg, Ca, Mn, Ni, Fe, Co) it became increasingly clear that, especially above 10 GPa, the structural response in these layered materials (Fig. 8) is dominated not by the formation of O-D...O bonds but by the D...D repulsion as deuterium is forced to occupy smaller volumes (Parise et al. 1999). The response of the $M(OD)_2$ compounds is to disorder the deuteron from the 3-fold site in space group $P\bar{3}m1$ to three equivalent sites which are $^1/_3$-occupied (Parise et al. 1998b, 1999).

Spectroscopic studies (Nguyen et al. 1997) suggest complete disorder and "amorphization" of the H-sublattice for one layered hydroxide, $Co(OH)_2$, but the results provided by neutron diffraction are different to what is expected for "amorphization". In silicate glasses for example, short-range order is maintained, recall the rigid tetrahedron first described by Jorgensen in his high pressure studies of quartz (Jorgensen 1978), while intermediate and long range order is lost. High pressure neutron powder diffraction (NPD) studies of $Co(OD)_2$ suggest exactly the opposite of this picture (Figs. 9 and 10) with evidence that *long* range order is maintained, while the deuteron is disordered over several partially occupied sites (Parise et al. 1999). The scattering has been modeled with three fold disorder (Fig. 10) and this is common to all of the $M(OD)_2$

compounds studied (Parise et al. 1998b). The role of H...H repulsion and the power of the joint use of NPD and spectroscopic techniques are also illustrated in recent studies of more complex hydrous minerals at high pressure (Kagi et al. 2000, 2003; Komatsu et al. 2005).

The importance of future total scattering studies at high pressures. While the NPD studies show clearly that long range order is maintained in the $M(OD)_2$ materials, and that the short range disorder is consistent with the broadening and shifting of peaks observed in spectroscopic studies, neither analysis of Bragg scattering nor analysis of the data from spectroscopic probes addresses intermediate range structures. Although models for

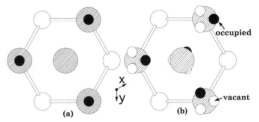

Figure 10. Portion of the structure of deuterated brucite $(Mg(OD)_2$, $P\bar{3}m1$; $a = 3.14$, $b = 4.71$ Å) projected down [001]. Large stripped circles are oxygen and the remaining large and small circles are magnesium and deuterium-sites, respectively. Because of D...D (< 2 Å) repulsive interaction at high pressure the single site at $^1/_3$, $^2/_3$, z; $z \sim 0.41$, shown as small filled circles in (a), splits into 3 equivalent sites at $(x, 2x, z)$ with z decreasing as x increases to preserve an O-D distance of about 1 Å (b). Small filled and open circles in (b) represent occupied and vacant sites in one possible model for the occupancy of the deuterium sites at high pressures.

intermediate range order are proposed (Parise et al. 1999) these are speculative (Fig. 10). In order to address the question of intermediate range order in these materials the *total* scattering (Proffen 2006, this volume) must be analyzed. This would include the diffuse scattering, which must arise if, even locally, structures such as those shown in Figure 10 exist. The corrections for parasitic scattering, absorption and inelastic scattering become especially important in these cases but protocols for such experiments with neutrons (Le Godec et al. 2004) and X-rays (Martin et al. 2005; Parise et al. 2005) are now available. An excellent discussion of the total scattering technique is given elsewhere in this volume (Proffen 2006, this volume).

The role of metal-hydrogen (deuterium) repulsion: Hydrogen bonding of goethite (α-FeOOD) at high pressure. The role of H...H repulsion in disorder is illustrated in systematic NPD studies of the layered hydroxides (Figs. 9 and 10) where H(D) is restricted to the layers, thereby maximizing H(D)...O and H(D)...H(D) interactions. At pressures above 20 GPa it is possible the M...H(D) interaction will become important but this interaction is more easily studied at modest pressures in compounds where the interaction is imposed by the crystal structure. This is the case in the compound Pb(OD)X; X = Cl, Br, I (Peter et al. 1999) and in goethite (Fig. 11). The oxygen atoms in goethite occupy two crystallographically

Figure 11. Polyhedral drawing of the structure of goethite (α-FeOOD, *Pbnm* a =9.91, b = 3.00, c = 4.57 Å at 7 GPa determined from a deuterated sample in the PE cell in d-methanol/d-ethanol pressure transmitting medium (Marshall and Francis 2002). Oxygen and deuterium are shown as stripped and white circles, respectively. Dotted lines represent H-bonds.

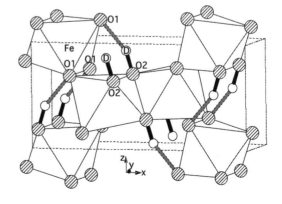

independent sites, O(1) and O(2), in a slightly distorted HCP arrangement (Fig. 11). The O(1) is surrounded by three iron atoms to form an almost planar triangle. The O(2) is hydrogenated, and surrounded by three iron atoms and a hydrogen atom, which form a distorted tetrahedral coordination (Fig. 12). A non-linear hydrogen bonding connects O(2)-H...O(1) in an adjacent octahedron. Prior spectroscopic (Williams and Guenther 1996) and high pressure synchrotron X-ray powder diffraction experiments (Nagai et al. 2003) showed the compression behavior of goethite is anisotropic with the *a*-axis almost twice as compressible as either *b*- and *c*-axes. These studies also showed the compression of goethite seems to be mainly controlled by shortening of the hydrogen bonded O...O distance (Figs. 11 and 12). Spectroscopic evidence (Williams and Guenther 1996) also indicated a role for the O(2)-H...O(1) bond angle and Fe...D repulsion. Precise determination of hydrogen (deuterium) positions in goethite with decreasing volume provided confirmation of these observations.

About 100 mg of deuterated goethite was loaded into the PE cell (Fig. 3) with null scattering TiZr gaskets and deuterated methanol/ethanol 4:1 mixture as pressure transmitting media in order to generate hydrostatic pressure up to 10 GPa (Marshall and Francis 2002). Time of Flight (TOF) neutron diffraction patterns (Vogel and Priesmeyer 2006, this volume) were obtained on the POLARIS beamline at ISIS. No pressure standard was included and the *P* (Fig. 12) was estimated from the compression data obtained from X-ray scattering (Nagai et al. 2003). Determination of the equation of state with X-rays rather than with neutrons, if possible, saves considerable neutron beamtime, which can then be dedicated to improving the statistics of HP neutron data, a much better use of this scarce resource.

TOF neutron diffraction data were collected at 0.1 MPa, 2.3, 5.2 and 7.2 GPa, typically for 5 hours at each pressure. No significant peak broadening occurred at all pressure conditions. Rietveld refinement provided structural parameters at each pressure, which were used to determine the interatomic distance shown in Figure 12 (Nagai et al. 2006). As expected all interatomic distances decrease smoothly with pressure with the exception of the O-D distance, which is constant within experimental error. It is interesting the O(2)-D...O(1) angle apart

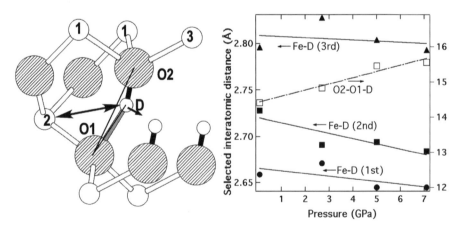

Figure 12. The environment surrounding the D-site in FeOOD (goethite) showing the important interactions between D (small open circles) O (stripped) and Fe (larger open circles). The numbered Fe-sites (1, 2, 3) refer to the 1st, 2nd and 3rd nearest neighbors to the D-site and correspond to the filled circles, squares and triangles, respectively shown right, where change in the D-Fe distances and O2-O1-D are plotted as a function of pressure. On the graph to the right, the left and axis referring to the Fe-D distances and the right hand axis to the O2-O1-D angle, which increases with pressure in response to the approach of the 2nd neighbor Fe and increased Fe-D repulsion (arrow). All symbols are larger than the estimated standard deviations of the distances and angle cited.

further from a linear geometry with shortening of the hydrogen bonded O...O distance at pressure. That seems to be due to repulsion between D atom and the 2nd neighbor Fe atom. Indeed, the distance becomes almost the same as that between the D atom and the 1st neighbor Fe atom at pressure.

Studies at high pressure and temperature

In-situ neutron diffraction while simultaneously maintaining high pressure and high temperature (HP-HT) conditions seek to address problems in Earth science such as the stability of minerals involved in the water and carbon cycles. *In situ* high *P-T* neutron diffraction measurements of shear and texture, accompanied perhaps by phase transitions, are dealt with in other chapters in this volume (Wenk 2006, this volume; Daymond 2006, this volume). This brief review will concentrate on some of the technical aspects of the carrying out studies at high pressures and temperatures.

A number of experimental difficulties accompany neutron diffraction studies at simultaneous high pressure and high temperature conditions, including: 1) A restricted diffraction geometry, further complicated by the additional components in the cell assembly required for the heating system, potentially adding parasitic scattering to the diffraction pattern; 2) compression of a large sample volume (30-100 mm^3) while maintaining hydrostatic pressure; 3) heating for long periods of time (6 ~ 12 hours) while maintaining a stable temperature with low gradient.

Neutron diffraction experiments under simultaneous HP-HT up to 10 GPa and 1500 K have been carried out successfully (Le Godec et al. 2001a,b, 2003, 2004, 2005; Martinez-Garcia et al. 2000, 2002; Meducin et al. 2004; Stone et al. 2005a,b; Zhao et al. 2005; Strassle et al. 2006). Many of these experiments were conducted over a period of days with typical data collection times of 6-12 hours; the diffraction data are of sufficient quality to allow full Rietveld refinements with simultaneous determination of site occupancy and displacement parameters for the best quality data (Meducin et al. 2004) albeit at modest pressures of around 3 GPa. Typical cell assemblies for toroidal anvils include designs from the Los Alamos group (Zhao et al. 2005). While these experiments have been on-going for quite a while at both synchrotron and neutron sources (Le Godec et al. 2001a; Martinez-Garcia et al. 2000), significant improvements are needed to push high *PT* to the 30 GPa and 2000 K to allow studies to extend to the conditions expected to exist in the Earth's transitions zone and beyond. All aspects, including press, anvil and furnace design, thermal and electric insulation, selection of gasket materials and re-enforcement rings, and protection of electric and thermocouple leads, need to be taken into account.

One step towards an integrated approach is the new toroidal anvil press, TAP-98, installed at the HIPPO beamline at Los Alamos (Zhao et al. 2005). This toroidal anvil press (designated TAP-98) was installed in 1998 and designed in a joint project between Los Alamos, University of Colorado and the Russian Academy of Sciences to design a compact press suitable for neutron experiments at a spallation source (Zhao et al. 2005). TAP-98 can be laid down into the HIPPO sample chamber with different diffraction settings It can also have "side" and "front" setting and use other diffraction banks besides the 90° bank, as is possible for the PE cell at the PEARL/HiPr facility at ISIS (See section on FeS magnetism above).

RECENT DEVELOPMENTS

The provision of new sources of neutrons, and X-rays is allowing innovation in the several areas of Earth science, including high pressure science (Parise and Brown 2006). High-pressure neutron experiments are "flux hungry." This is because for a given cell design and force, higher pressures can only be obtained by decreasing the sample volume. The need to obtain data on smaller samples is driving the development of ever more powerful neutron sources and continual instrument improvement to make the most of every neutron. Developments include improved area detectors with wider coverage, neutron focusing using monochromators, neutron guides, and adaptation of Kirkpatrick-Baez focusing optics by the synchrotron radiation community to neutron studies (Fig. 13).

New optics

As with X-rays, focusing is a technique used to concentrate and direct a neutron beam onto a small area. High-pressure experiments, which necessarily have small samples at the highest pressures, would benefit from concentrating the flux into smaller spot sizes while minimizing divergence. Although the use of large curvature mirrors is widespread in the neutron community, especially at cold neutron sources, adaptation of bent focusing optics and its potential for focusing thermal neutrons was not widely appreciated. While micro-capillary optics, first developed for X-rays and adapted to focus neutrons, often lead to considerable divergence, an alternative method (Ice et al. 2005) using Kirkpatrick-Baez neutron super mirrors, again adapted from work with X-rays (Ice et al. 2005), does not suffer the same divergence problems (Figs. 13 and 14).

A new prototype microfocusing system was recently tested at the Chalk River neutron reactor in Canada (Fig. 13) using a pink beam and a neutron-sensitive imaging plate (IP) detector set at the sample position. The mirrors, shown in Figure 13, were used to obtain a focal spot about 90 μm × 90 μm with a divergence of less than 30 μrad at the sample position (Fig. 14). Diffraction data were then obtained from a free standing crystal of forsterite (Mg_2SiO_4) about 0.3×0.3 mm^2 in cross-section with the optic inserted and removed. The diffraction patterns obtained with the 90 μm spot not only showed about an order of magnitude increase in signal (diffracted intensity) but a 2-fold decrease in background compared to pattern obtained with only slits defining the beam (Fig. 13). Further, a crystal of FeO measuring $0.2 \times 0.2 \times 0.1$ mm^3 inserted in a high pressure moissanite gem cell (Xu et al. 2002) and at a nominal pressure of 20 GPa provided single crystal diffraction patterns, which could be indexed and

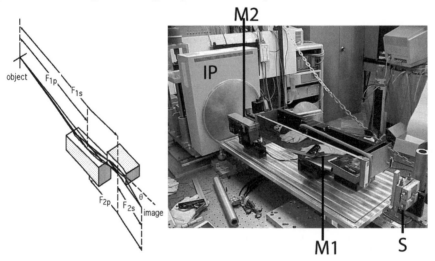

Figure 13. (left) Schematic of the neutron focusing mirrors developed at Oak Ridge National Laboratory (Ice et al. 2005). These mirrors were tested with a pink beam at Chalk River in the Spring of 2005 using the set-up shown at the right. The beam, entering from the lower right, is collimated with a set of slits (S) and then focused by a vertical mirror (M1) approximately 50 cm in length and a horizontal mirror (M2). The focused beam is imaged using an Imaging plate (IP) detector fitted with a plate doped with Gd sensitive to neutrons, and set at the sample position.

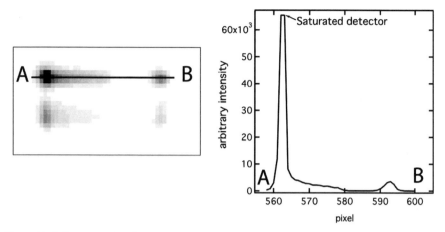

Figure 14. (left) Image taken from the IP detector (Fig. 12) showing the "gang-of-four" spots from the unfocused beam (lower right) the beam focused by the vertical mirror alone (lower left) horizontal mirror alone (upper right) and the intense spot focused by both mirror (see Fig. 12). A traverse across the detector (AB) shows the doubly focused spot saturates the detector and has a half width of about 90 μm.

integrated. This demonstration marks a major breakthrough and expansion of the science vision for what is possible for high-pressure neutron diffraction. Neutron focusing is used to improve spatial resolution and lower detection limits in neutron-based analytical methods. Beams smaller than those obtained in these early experiments are possible as is the production of more compact mirrors. With further optimization of the optics and the promise of at least an order of magnitude increase in flux provided by the Spallation Neutron Source (see Vogel and Priesmeyer 2006, this volume) we may be close to "routine" neutron crystallography on samples of a size comparable to those now used with sealed-tube X-ray sources. This should open up new science on small mineral samples, including those held at high pressure.

New detectors and single crystal techniques

It is clear from earlier HP studies of deuterium (Glazkov et al. 1989) that some of the disadvantages of the small sample sizes necessitated by the use of diamond anvil cells to obtain higher pressures was the use of large area detectors. The high signal-to-noise discrimination afforded by the use of the single crystal technique makes the effort to obtain and preserve them at high pressure well worthwhile. Recent developments in the construction of large volume capacity gem cells (Xu et al. 2002, 2004) and the synthesis of large CVD diamonds (Yan et al. 2002) suggest we may soon be able to routinely perform high pressure single crystal neutron scattering measurements on samples of a size (150 μm × 150 μm × 50 μm) close to that associated with DAC high-pressure studies in individual investigator laboratories.

One picture says it all. Diffractometers such as VIVALDI (Chung et al. 2004) at the ILL, which routinely collects data at ambient and low temperatures on crystals with edge dimensions of about 150 μm (Cole et al. 2001), are well suited to small crystals, rapid chemical crystallography, reciprocal-space surveys, and studies of structural and magnetic phase transitions. This diffractometer utilizes an *unfocused* pink neutron beam from a reactor source (Vogel and Priesmeyer 2006, this volume) and a curved cylindrical imaging plate detector to maximize coverage. The advantages of the large-solid-angle Laue technique can also be implemented in high-pressure single crystal crystallography.

Figure 15 shows a neutron diffraction pattern taken from a natrolite crystal in a "panoramic" moissanite anvil cell (Xu et al. 2002). On close inspection we identify three

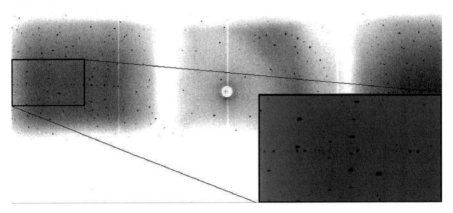

Figure 15. Laue diffraction pattern from a 0.5 mm³ natrolite, $Na_2[Al_2Si_3O_{10}]\cdot2(H_2O)$, single crystal sample in a moissanite-anvil cell: The HP cell (similar to that shown in Fig. 2) is set at $\varphi = 65°$, and an exposure time was 1 hr. The two vertical strips in the imaging plate result from absorption by the two vertical posts in the cell (Fig. 2). Taken from McIntyre et al. (2005).

diffraction patterns—one from the natrolite sample and two from the two tips of the conical moissanite anvils. Each pattern was indexed, and the data were integrated for normal single-crystal data analysis. The structure of natrolite refined well, with all coordinates agreeing within three estimated standard deviations with previously published neutron diffraction data (McIntyre et al. 2005); this result demonstrates that full crystallographic investigation in a pressure cell is quite feasible using the white-beam neutron Laue technique. This situation will improve dramatically when coupled with neutron focusing. VIVALDI is part of a wider program of detector upgrades at reactor and spallation neutron sources involving the use of electronic area detectors to increase both coverage and sensitivity.

Bringing it all together: the SNAP beamline

The SNS, being built in Oak Ridge, United States, by the US Department of Energy is designed to operate initially at 1.4 MW and to provide the most intense pulsed neutron beams in the world (Vogel and Priesmeyer 2006, this volume). For the high-pressure community, several decades of steady improvement in pressure capability, detector sensitivity, and accumulated community experience is being poured into construction of the world's first dedicated high-pressure instrument, the Spallation Neutron and Pressure (SNAP) beamline. The conceptual design for this instrument takes advantage of many recent breakthroughs in cell design and beamline optics discussed above.

The SNAP instrument will allow the study of the structural and dynamic properties of materials under extreme conditions, such as those found in the deep Earth and planetary interiors. A suite of high-pressure devices (Fig. 3) will cover the range of pressures (0.1-100 GPa) of interest to the Earth and materials science communities who are engaged in solving problems ranging from the formation and stability of gas clathrates at the low end of the pressure scale, to core-mantle interactions and planetary interiors at the more extreme conditions. The SNAP beamline is inspired, in part, by the success of high-pressure programs at X-ray sources, where flexible optics have led to versatile instruments adaptable to user needs as scientific opportunities arise. The SNAP instrument is designed to integrate recent progress in the design of high-pressure cells with flexible optics, including recently developed focusing methodologies (Ice et al. 2005). It is also optimized for the collection of single-crystal data while maintaining capabilities for powder diffraction (Fig. 16). Some of the unusual features of the beamline are outlined below.

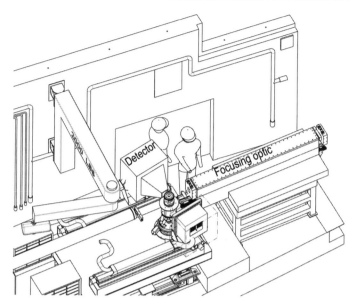

Figure 16. Schematic of the interior of the SNAP hutch shows the layout of the removable focusing optic and sample position. Two movable detectors, set in this case at close to 90° to the beam, are shown; one has the shielding removed. A closer view of he PE VX-5 cell (Klotz et al. 2005a) mounted on the positioning stage is shown in Figure 17.

Instrument overview. The SNAP instrument will be a high flux medium resolution diffractometer designed primarily for the study of very small samples under extreme conditions of pressure and temperature. The incident flight path is planned to be 15 m, the shortest possible given the target shielding constraints, with a secondary flight path of ~0.4 m. Operating at 60 Hz will result in frame overlap at approximately $\lambda \sim 4.0$ Å. The incident beam neutron optics and highly pixilated compact detector array will be integrated with the high-pressure cells and the hydraulic press assembly (Figs. 16 and 17).

High-pressure cell components. The design of the high-pressure cells may be partitioned into the three broad categories described above. The first consists of "large volume" cells with toroidal type tungsten carbide or sintered diamond opposed anvils with typical sample volumes upwards of 50 mm³ and operational in the 0-25 GPa pressure range (Fig. 3). These cells can be used for studies of weakly scattering systems that require very large sample volumes to maximize the sample signal relative to the background. Neutron scattering with isotope substitution from glasses is an excellent example of such studies (Wilding and Benmore 2006, this volume). The second category consists of large-sample opposed single crystal "gem" anvil cells (Fig. 2); these will be the workhorse cells for experiments at the highest pressures. Such cells are under development currently, and functioning prototypes that incorporate large single crystal moissanite (SiC) anvils weighing up to 20 ct are now in routine use (McIntyre et al. 2005; Xu et al. 2002, 2004). Useful sample sizes are currently >2 mm³; pressures as high as 65 GPa have been reached with moissanite anvils (Xu et al. 2002). The final category of pressure cells will consist of large volume gas pressure cells to be used when very large samples and very hydrostatic pressure conditions are required (Fig. 1). Practical pressures upwards of 3 GPa at about 20 K are currently possible using cells with supported maraging steel inner assemblies. While maraging steel contains too much Ni to be useful in neutron scattering because of activation problems, inner assemblies made from materials such as TiZr offer great

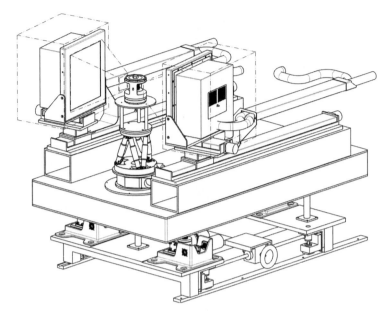

Figure 17. Detail of the proposed sample geometry and positioning stages for the SNAP instrument. Since experiments on both single crystals and powders are envisioned, sample positioning and detector placement are flexible. In this case a PE cell of the VX generation (Klotz et al. 2005a) is mounted on a hexapod and rotation stage, all of which are mounted on *xy* positioners to enable straightforward alignment to the focused beam that may be <100 μm in size. The detectors are also movable in order to either optimize for resolution or flux.

promise for background reduction. An as yet untested possibility, but one that is becoming increasing appealing, is the adaptation of standard diamond anvil cell technologies. This is possible because of the continuing advances in focusing optics (Figs. 13, 14).

Moderator choice. Given the extremely small samples used with modern high-pressure anvil type devices it is a primary design goal to maximize neutron flux on sample while maintaining sufficient timing resolution for diffraction studies. For this purpose the cryogenic poisoned decoupled hydrogen (PDH) moderator is currently considered superior to the ambient temperature water (AW) moderator. With a 15.4 m total flight path frame overlap occurs at $\lambda \sim 4$ Å. It is noted that the PDH moderator has superior flux beyond ~ 2 Å ($d = 1.5$ Å at 90°) and has approximately an order of magnitude more flux at 4 Å ($d = 2.83$ Å at 90°). Monte Carlo computer simulations comparing diffraction peaks produced by the two moderators have indicated dramatic increases in peak intensity for the PDH moderator beyond 2 Å. The PDH moderator also gives superior timing resolution over the entire first frame.

Detectors and detector orientation. The detectors used in the SNAP instrument must be matched as closely as possible to the solid angle defined by the scattered neutrons from high-pressure cells. At this time, the panoramic-style cells (Figs. 16, 17) provides the largest solid angle, defined by scattered neutrons that are not occluded by elements of the cell itself, roughly 140° in 2θ; it will extend from roughly 20° to 130° when placed vertically in the neutron beam and ± 37° out of the scattering plane (Figs. 16, 17). As such, two identical detector arrays can be placed on both sides of the instrument, such that each of the detector arrays is centered at a 2θ = ± 90° scattering angle. These detectors are to be constructed such that they provide continuous, or as nearly continuous, coverage as possible. Note however that these detectors will be designed to allow sufficient access to the high-pressure device for electronic and hydraulic cables, optical components, and for cell position/orientation manipulations while in the beam. As such, the detectors will be centered with 500 mm distance from the detector center to the sample position; this will allow enough room to place large-volume pressure cell goniometers and maintain enough space in the instrument enclosure for the detector itself, associated hardware, and detector shielding/collimation.

Two approximately 450×450 mm^2 arrays of Anger type detector are used. Anger camera type detectors, each with an approximately 150×150-mm active area, are chosen as the most suitable option. These may be tiled to the specified angular requirements in a 3×3 array centered at $2\theta = 90°$ will cover approximately the 450×450 mm^2 area. Apart from area, the detectors are designed to specific criteria. For greatest flexibility of positioning these detectors should be tiled in a plane, not on a sphere, which would fix the distance from the sample. The position resolution is 2.5 mm with an efficiency at $\lambda = 0.5$ Å of 68% for a 2-mm scintillator and 44% for a 1-mm scintillator. Time binning resolution will be 5 μs, with a maximum count rate capability of 3×10^5 neutrons per second per detector. The Detector thickness, including electronics packaging is < 400 mm with minimized gamma sensitivity. The detector and frame will have provisions for being lifted by a crane (Fig. 16) but in day-to-day operation each detector will be remotely positioned with accuracy at any position ± 0.002 mm with precise adjustment features in the X, Y and Z planes.

Frame overlap and T_0 choppers. To reduce the background a set of at least two disk choppers will be used, one located at 6 m and the other at 11 m from the moderator. When operating in the first frame background generated by neutrons with $\lambda \sim 15$ Å from previous pulses will pass through the chopper system, and likewise operating in the second frame neutrons with $\lambda \sim 18.5$ Å will pass. Note that the flux of the PDH moderator is reduced by three orders of magnitude at $\lambda \sim 15$ Å. This is however still expected to be the largest source of background. At a scattering angle of 90° the *d*-space coverage will range from 0.28 Å $\le d \le 3.00$ Å, in the second frame this is extended to $d = 5.94$ Å. At the lowest angles this extends from 0.58 Å $\le d \le 14.23$ Å and in the highest angle banks from 0.15 Å $\le d \le 4.35$ Å.

Very high-energy (prompt) neutrons often contribute to background by moderating in shielding material and/or scattering from neutron optical components in the beamline. To significantly reduce the number of prompt neutrons entering the SNAP beamline will include a T_0 chopper with a low wavelength cut off of approximately $\lambda = 0.4$ to 0.5 Å.

Neutron focusing mirrors. SNAP, like all SNS beamlines, has challenges built into the design for focusing, as the importance of imaging on the sample was not realized in the early design. Specifically there is a limited path length (Fig. 16) available for adjustable slits and other essential optical elements (2.5 m). There is also the need for about 250 mm of clearance between the end of the optical elements and the sample (Figs. 16, 17). Finally the imaging system must be removable to allow for other optical approaches. The use of a neutron focusing will be crucial for providing the maximum neutron flux on sample. The 2nd generation SNAP mirror design (as opposed to the mirrors shown in Fig. 13) is based on the use of nested mirrors. This design is much more compact than a traditional KB mirror system, which makes it ideal for the SNAP beamline. It also allows for a larger sample to optics distance and because both mirrors are relatively large, can work well up to a larger beam size when needed. As can be seen, in a normal KB mirror system (Fig. 13), if a second mirror is placed before its paired partner, it must be very long and very perfect (expensive and difficult to make). If it is placed after its partner mirror, then it restricts the clearance to the sample. The nested design will be more compact, provide a larger distance between optic and sample and a larger acceptance with uncompromised beams up to ~ 2 mm. It is believed that, with improved optics, the 90 μm beams obtained at Chalk River in 2004 (Fig. 14) can be reduced to 25 μm. The possibility of obtaining usable data on laboratory X-ray sized crystals is now a very real possibility and must be tested further.

Not only will the SNAP mirrors be good for SNAP; the hope is that they will be widely adopted and will be the first wave of a revolution in neutron science. The vision is that this form of non-dispersive focusing optics will become essential for all samples less than 2 mm in size. Progress in this area is rapid with designs evolving and the prospect of sub-20 μm spots a real possibility prior to the commissioning of the SNAP beamline in 2008. In all likelihood the devices available at the time of writing will be obsolete by the time SNAP is commissioned.

CONCLUSIONS

The past two decades have seen tremendous progress in the capabilities for high-pressure research. While the orders of magnitude flux gains available at synchrotron X-ray storage rings and LINACS are unlikely to ever be matched at neutron sources, the unique properties of the neutron make it an indispensable part of high pressure condensed matter research. Studies of H-bonding, magnetism under high-pressure conditions are clearly important studies to carry out with neutron scattering. Progress in cell design, neutron focusing and detectors subtending larger solid angles is happening just as higher flux sources are coming on line. This suggests the possibility of designing beamlines with more flexible optics capable of taking full advantage of these exciting developments. The Spallation Neutrons and Pressure (SNAP) beamline is the first attempt at an integrated approach to take full advantage of the frenzied development in optics, detectors and cells. In the future it is likely that high-pressure programs will spread throughout the suite of instruments being built at both spallation and reactor sources.

ACKNOWLEDGMENTS

Many collaborators made the work reported here possible and most are listed as coauthors in publications. I am especially grateful to the following collaborators who provided data and advice on as yet unpublished results: SNAP collaborators Chris Tulk, Steve Chae and Gene Ice (Oak Ridge) Darren Locke (Stony Brook) Dave Mao and Russell Hemley (Geophysical Laboratory of the Carnegie Institution) and Ian Swainson (Chalk River). The goethite data were collected in collaboration with Takaya Nagai (Hokkaido), and Hiroyuki Kagi (Tokyo) whom I gratefully acknowledge for allowing me to present data prior to publication. Partial financial support was provided by the U.S. National Science Foundation through its CHE (0221934) EAR (0510501) and DMR (0452444) programs and by the U. S. Department of Energy (DE-FG02-03ER46085) which supports the construction of the SNAP beamline and the Spallation Neutron Source at Oak Ridge National Laboratory in Tennessee, USA. The SNAP executive committee and IDT members have been invaluable sources of information and inspiration at annual meetings capably organized by Emily Vance (Stony Brook).

REFERENCES

Aksenov VL, Balagurov AM, Savenko BN, Glazkov VP, Goncharenko IN, Somenkov VA, Antipov EV, Putilin SN, Capponi JJ (1995) Neutron diffraction study of the high-temperature superconductor $HgBa_2CaCu_2O_{6.3}$ under high pressure. High Press Res 14:127-137

Besson JM, Nelmes RJ, Hamel G, Loveday JS, Weill G, Hull S (1992) Neutron powder diffraction above 10 GPa. Physica B 180:907-910

Bloch D, Chaisse F, Pauthene R (1966) Effects of hydrostatic pressure on magnetic ordering temperatures and magnetization of some ionic compounds. J Appl Phys 37:1401-1402

Braithwaite D, Goncharenko IN, Mignot JM, Ochiai A, Vogt O (1996) Magnetic phase diagram of USb at high pressure. Europhysics Letters 35:121-126

Brugger RM, Bennion RB, Worlton TG (1967) Crystal structure of bismuth-II at 26 kbar. Phys Lett A 24: 714-717

Bull CL, Guthrie M, Klotz S, Philippe J, Strassle T, Nelmes RJ, Loveday JS, Hamel G (2005) Toroidal anvils for single-crystal neutron studies. High Press Res 25:229-233

Catti M, Ferraris G, Pavese A (1995) Static compression and H disorder in brucite, Mg(OH)$_2$, to 11 GPa: A powder neutron diffraction study. Phys Chem Minerals 22:200-206

Chung EML, Lees MR, McIntyre GJ, Wilkinson C, Balakrishnan G, Hague JP, Visser D, Paul DM (2004) Magnetic properties of tapiolite (FeTa$_2$O$_6$); a quasi two-dimensional (2D) antiferromagnet. J Phys Condens Matter 16:7837-7852

Cole JM, McIntyre GJ, Lehmann MS, Myles DAA, Wilkinson C, Howard JAK (2001) Rapid neutron-diffraction data collection for hydrogen-bonding studies: application of the Laue diffractometer (LADI) to the case study zinc (tris)thiourea sulfate. Acta Crystallogr A57:429-434

Colligan M, Lee Y, Vogt T, Celestian AJ, Parise JB, Marshall WG, Hriljac JA (2005) High-pressure neutron diffraction study of superhydrated natrolite. J Phys Chem B 109:18223-18225

Crapanzano L, Crichton WA, Monaco G, Bellissent R, Mezouar M (2005) Alternating sequence of ring and chain structures in sulphur at high pressure and temperature. Nature Mater 4:550-552

Crichton WA, Mezouar M (2002) Noninvasive pressure and temperature estimation in large-volume apparatus by equation-of-state cross-calibration. High Temp-High Press 34:235-242

Cynn H, Hofmeister AM (1994) High-pressure IR spectra of lattice modes and OH vibrations in Fe-bearing wadsleyite. J Geophys Res 99:17717-17722

Cynn H, Hofmeister AM, Burnley PC, Navrotsky A (1996) Thermodynamic properties and hydrogen speciation from vibrational spectra of dense hydrous magnesium silicates. Phys Chem Minerals 23:361-376

Daymond MR (2006) Internal stresses in deformed crystalline aggregates. Rev Mineral Geochem 63:427-458

Decker DL, Beyerlei R, Roult G, Worlton TG (1974) Neutron-diffraction study of KCN III and KCN IV at high pressure. Phys Rev B 10:3584-3593

Demirdjian B, Ferry D, Suzanne J, Toubin C, Picaud S, Hoang PNM, Girardet C (2002) Structure and dynamics of ice Ih films upon HCl adsorption between 190 and 270 K. I. Neutron diffraction and quasielastic neutron scattering experiments. J Chem Phys 116:5143-5149

Duffy TS, Ahrens TJ (1991) The shock wave equation of state of brucite Mg(OH)$_2$. J Geophys Res 14:14,319-14,330

Duffy TS, Hemley RJ, Mao HK (1995a) Structure and bonding in hydrous minerals at high pressure: Raman spectroscopy of alkaline earth hydroxides. *In:* Volatiles in the Earth and Solar System. KA Farley (ed) American Institute of Physics, p 211-220

Duffy TS, Meade C, Fei Y, Mao H-k, Hemley RJ (1995b) High-pressure phase transition in brucite, Mg(OH)$_2$. Am Mineral 80:222-230

Duffy TS, Shu J, Mao H-k, Hemley RJ (1995c) Single-crystal x-ray diffraction of brucite to 14 GPa. Phys Chem Minerals 22:277-281

Dunlop DJ (1970) Hematite - intrinsic and defect ferromagnetism. Science 169:858-860

Falconi S, Crichton WA, Mezouar M, Monaco G, Nardone M (2004) Structure of fluid phosphorus at high temperature and pressure: An X-ray diffraction study. Phys Rev B 70:144109-1–144109-8

Finney JL, Bowron DT, Soper AK, Loerting T, Mayer E, Hallbrucker A (2002a) Structure of a new dense amorphous ice. Phys Rev Lett 89:205503-1–205503-4

Finney JL, Hallbrucker A, Kohl I, Soper AK, Bowron DT (2002b) Structures of high and low density amorphous ice by neutron diffraction. Phys Rev Lett 88:225503-1–225503-4

Floriano MA, Whalley E, Svensson EC, Sears VF (1986) Structure of high-density amorphous ice by neutron-diffraction. Phys Rev Lett 57:3062-3064

Friedrich, A, Lager, GA, Ulmer, P, Kunz, M, and Marshall, WG (2002) High-pressure single-crystal X-ray and powder neutron study of F,OH/OD-chondrodite: Compressibility, structure, and hydrogen bonding. Am Min, 87: 931-939

Glazkov VP, Goncharenko IN, Irodova VA, Somenkov VA, Shilstein SS, Besedin SP, Makarenko IN, Stishov SM (1989) Neutron-diffraction study of molecular deuterium equation of state at high-pressures. Z Phys Chem N Fol 163:509-514

Goncharenko IN, Mignot J-M, Andre G, Lavrova OA, Mirebeau I, Somenkov VA (1995) Neutron diffraction studies of magnetic structure and phase transitions at very high pressures. High Press Res 14:41-53

Hemley RJ (2000) Effects of high pressure on molecules. Ann Rev Phys Chem 51:763-800

Hofmeister AM, Cynn H, Burnley PC, Meade C (1999) Vibrational spectra of dense, hydrous magnesium silicates at high pressure: importance of the hydrogen bond angle. Am Mineral 84:454-464

Ice GE, Hubbard CR, Larson BC, Pang JWL, Budal JD, Spooner S, Vogel SC (2005) Kirkpatrick-Baez microfocusing optics for thermal neutrons. Nucl Instrum Methods Phys Res A539:312-320

Ivanov AS, Goncharenko IN, Somenkov VA, Braden M (1995) Phonon-dispersion in graphite under hydrostatic-pressure up to 60 Kbar using a sapphire-anvil technique. Physica B 213:1031-1033

Jackson SM, Nield VM, Whitworth RW, Oguro M, Wilson CC (1997) Single-crystal neutron diffraction studies of the structure of ice XI. J Phys Chem B 101:6142-6145

Jacobsen SD, Smyth JR, Spetzler H, Holl CM, Frost DJ (2004) Sound velocities and elastic constants of iron-bearing hydrous ringwoodite. Phys Earth Planet Inter 143-44:47-56

Jeffrey GA (1997) An introduction to hydrogen bonding. Oxford University Press

Jorgensen JD (1978) Compression mechanisms in alpha-quartz structures-SiO_2 and GeO_2. J Appl Phys 49: 5473-5478

Jorgensen JD, Beyerlein RA, Watanabe N, Worlton TG (1984) Structure of D_2O ice-VIII from *in situ* powder neutron-diffraction. J Chem Phys 81:3211-3214

Jorgensen JD, Worlton TG (1985) Disordered structure of D_2O ice VII from *in situ* neutron powder diffraction. J Chem Phys 83:329-333

Jorgensen JE, Jorgensen JD, Batlogg B, Remeika JP, Axe JD (1986) Order parameter and critical exponent for the pressure-induced phase-transitions in RO_3. Phys Rev B 33:4793-4798

Kagi H, Nagai T, Loveday JS, Wada C, Parise JB (2003) Pressure-induced phase transformation of kalicinite ($KHCO_3$) at 2.8 GPa and local structural changes around hydrogen atoms. Am Mineral 88:1446-1451

Kagi H, Parise JB, Cho H, Rossman GR, Loveday JS (2000) Hydrogen bonding interactions in phase A $[Mg_7Si_2O_8(OH)_6]$ at ambient and high pressure. Phys Chem Minerals 27:225-233

Kernavanois N, Hansen T, Ressouche E, Henry YJ, Strässle T, Klotz S, Hamel G, Parise JB (2002) Magnetism of CoO under pressures up to 7 GPa. *http://www.ill.fr/AR-03/site/02_scientific/021_magnetic/02_magn_10.htm*

Khovstantsev LG, Vereshchagin LF, Novikov AP (1977) Device of toroid type for high pressure generation. High Temp High Press 9:637-639

Khvostantsev LG, Slesarev VN, Brazhkin VV (2004) Toroid type high-pressure device: history and prospects. High Press Res 24:371-383

Klotz S (2001) Phonon dispersion curves by inelastic neutron scattering to 12 GPa. Z Kristallogr 216:420-429

Klotz S, Besson JM, Schwoererbohning M, Nelmes RJ, Braden M, Pintschovius L (1995a) Phonon-dispersion measurements at high-pressures to 7 GPa by inelastic neutron-scattering. Appl Phys Lett 66:1557-1559

Klotz S, Braden M, Besson JM (2000) Inelastic neutron scattering to very high pressures. Hyperfine Interact 128:245-254

Klotz S, Gauthier M, Besson JM, Hamel G, Nelmes RJ, Loveday JS, Wilson RM, Marshall WG (1995b) Techniques for neutron-diffraction on solidified-gases to 10 GPa and above - applications to ND_3 Phase-IV. Appl Phys Lett 67:1188-1190

Klotz S, Strässle T, Rousse G, Hamel G, Pomjakushin, V (2005a) Angle-dispersive neutron diffraction under high pressure to 10 GPa. Appl Phys Lett 86: article 031917, pp 3

Klotz S, Strassle T, Salzmann CG, Philippe J, Parker SF (2005b) Incoherent inelastic neutron scattering measurements on ice VII: Are there two kinds of hydrogen bonds in ice? Europhys Lett 72:576-582

Knorr K, Depmeier W (2006) Application of neutron powder-diffraction to mineral structures. Rev Mineral Geochem 63:99-111

Komatsu K, Kagi H, Okada T, Kuribayashi T, Parise JB, Kudoh Y (2005) Pressure dependence of the OH-stretching mode in F-rich natural topaz and topaz-OH. Am Mineral 90:266-270

Kruger MB, Williams Q, Jeanloz R (1989) Vibrational spectra of $Mg(OH)_2$ and $Ca(OH)_2$ under pressure. J Chem Phys 91:5910-5915

Kuhs WF, Hansen TC (2006) Time-resolved neutron diffraction studies with emphasis on water ices and gas hydrates. Rev Mineral Geochem 63: 171-204

Kuhs WF, Bliss DV, Finney JL (1987) High-resolution neutron powder diffraction study of Ice-Ic. J Phys 48: 631-636

Kuhs WF, Finney JL, Vettier C, Bliss DV (1984) Structure and hydrogen ordering in Ice-VI, Ice-VII, And Ice-VIII by neutron powder diffraction. J Chem Phys 81:3612-3623

Kuhs WF, Lehmann MS (1983) The structure of Ice Ih by neutron-diffraction. J Phys Chem 87:4312-4313

Kunz M, Leinenweber K, Parise JB, Wu T-C, Bassett WA, Brister K, Weidner DJ, Vaughan MT, Wang Y (1996) The baddeleyite-type high pressure phase of $Ca(OH)_2$. J High Press Res 14:311-319

Lager GA, Downs RT, Origlieri M, Garoutte R (2002) High-pressure single-crystal X-ray diffraction study of katoite hydrogarnet: Evidence for a phase transition from $Ia3d \rightarrow I43d$ symmetry at 5 GPa. Am Mineral 87:642-647

Lager GA, Marshall WG, Liu ZX, Downs RT (2005) Re-examination of the hydrogarnet structure at high pressure using neutron powder diffraction and infrared spectroscopy. Am Mineral 90:639-644

Le Godec Y, Dove MT, Francis DJ, Kohn SC, Marshall WG, Pawley AR, Price GD, Redfern SAT, Rhodes N, Ross NL, Schofield PF, Schooneveld E, Syfosse G, Tucker MG, Welch MD (2001a) Neutron diffraction at simultaneous high temperatures and pressures, with measurement of temperature by neutron radiography. Mineral Mag 65:737-748

Le Godec Y, Dove MT, Redfern SAT, Tucker MG, Marshall WG, Syfosse G, Besson JM (2001b) A new high P-T cell for neutron diffraction up to 7 GPa and 2000 K with measurement of temperature by neutron radiography. High Press Res 21:263-280

Le Godec Y, Dove MT, Redfern SAT, Tucker MG, Marshall WG, Syfosse G, Klotz S (2003) Recent developments using the Paris-Edinburgh cell for neutron diffraction at high pressure and high temperature and some applications. High Press Res 23:281-287

Le Godec Y, Hamel G, Martinez-Garcia D, Hammouda T, Solozhenko VL, Klotz S (2005) Compact multianvil device for in situ studies at high pressures and temperatures. High Press Res 25:243-253

Le Godec Y, Strassle T, Hamel G, Nelmes RJ, Loveday JS, Marshall WG, Klotz S (2004) Techniques for structural studies of liquids and amorphous materials by neutron diffraction at high pressures and temperatures. High Press Res 24:205-217

Lechner R, Quittner G (1966) Pressure-induced phonon frequency shifts in lead measured by inelastic neutron scattering. Phys Rev Lett 17:1259-1261

Lee Y, Hriljac JA, Parise JB, Vogt T (2006) Pressure-induced hydration in zeolite tetranatrolite. Am Mineral 91:247-251

Li JC, Londono JD, Ross DK, Finney JL, Tomkinson J, Sherman, WF (1991) An inelastic incoherent neutron-scattering study of ice II, Ice-IX, Ice-V, And Ice-VI - in the range from 2 To 140 Mev. J Chem Phys 94: 6770-6775

Link P, Goncharenko IN, Mignot JM, Matsumura T, Suzuki T (1998) Ferromagnetic mixed-valence and kondo-lattice state in TmTe at high pressure. Phys Rev Lett 80:173-176

Litvin DF, Losmanov AA, Ponjatov EG, Trapezni VA (1966) Apparatus for neutron-diffraction investigation of substances under high pressure. Acta Crystallogr S 21:A220

Liu ZX, Lager GA, Hemley RJ, Ross NL (2003) Synchrotron infrared spectroscopy of OH-chondrodite and OH-clinohumite at high pressure. Am Mineral 88:1412-1415

Lobban C, Finney JL, Kuhs WF (1998) The structure of a new phase of ice. Nature 391:268-270

Lokshin KA, Zhao YS, He DW, Mao W, Mao HK, Hemley RJ, Lobanov MV, Greenblatt, M (2005) Hydrogen clathrate hydrate - Novel hydrogen storage material: Crystal structure, kinetics, and phase diagram. Abstracts Am Chem Soc 229:U589-U589

Lokshin KA, Zhao YS, He DW, Mao WL, Mao HK, Hemley RJ, Lobanov MV, Greenblatt M (2004) Structure and dynamics of hydrogen molecules in the novel clathrate hydrate by high pressure neutron diffraction. Phys Rev Lett 93:125503-1–125503-4

Londono D, Kuhs WF, Finney JL (1988) Enclathration of helium in Ice-II - the 1st helium hydrate. Nature 332: 141-142

Londono JD, Kuhs WF, Finney JL (1993) Neutron-diffraction studies of ice-III and ice-IX on under-pressure and recovered samples. J Chem Phys 98:4878-4888

Loveday JS, Hamel G, Nelmes RJ, Klotz S, Guthrie M, Besson JM (2000) Neutron diffraction studies of hydrogen-bonded ices at high pressure. High Press Res 17:149-155

Loveday JS, Marshall WG, Nelmes RJ, Klotz S, Hamel G, Besson JM (1996) The structure and structural pressure dependence of sodium deuteroxide-V by neutron powder diffraction. J Phys Condens Matt 8: L597-L604

Loveday JS, Nelmes RJ, Marshall WG, Wilson RM, Besson JM, Klotz S, Hamel G, Hull S (1995) High pressure neutron powder diffraction: The neutron scattering aspects of work with the Paris-Edinburgh cell. High Press Res 14:7-12

Marshall WG, Francis DJ (2002) Attainment of near-hydrostatic compression conditions using the Paris-Edinburgh cell. J Appl Crystallog 35:122-125

Marshall WG, Nelmes RJ, Loveday JS, Klotz S, Besson JM, Hamel G, Parise JB (2000) High-pressure neutron-diffraction study of FeS. Phys Rev B 61:11201-11204

Martin CD, Antao SM, Chupas PJ, Lee PL, Shastri SD, Parise JB (2005) Quantitative high-pressure pair distribution function analysis of nanocrystalline gold. Appl Phys Lett 86:061910-1–061910-3

Martinez-Garcia D, Ferrer-Roca C, Le Godec Y, Munsch P, Itie JP, Munoz-Sanjose V (2002) X-ray diffraction study of $Cd_{0.8}Zn_{0.2}Te$ under high pressure and temperature. High Press Res 22:403-406

Martinez-Garcia D, Le Godec Y, Mezouar M, Syfosse G, Itie JP, Besson JM (2000) Equations of state of MgO at high pressure and temperature. High Press Res 18:339-344

McIntyre GJ, Melesi L, Guthrie M, Tulk CA, Xu J, Parise JB (2005) One picture says it all - high-pressure cells for neutron Laue diffraction on VIVALDI. J Phys Condens Matter 17:S3017-S3024

McWhan DB, Bloch D, Parisot G (1974) Apparatus for neutron-diffraction at high-pressure. Rev Sci Inst 45: 643-646

Meducin F, Redfern SAT, Le Godec Y, Stone HJ, Tucker MG, Dove MT, Marshall, WG (2004) Study of cation order-disorder in $MgAl_2O_4$ spinel by in situ neutron diffraction up 1600 K and 3.2 GPa. Am Mineral 89: 981-986

Miletich R, Allan DR, Kuhs WF (2000) High-pressure single-crystal techniques. Rev Mineral Geochem 41: 445-519

Morin FJ (1950) Magnetic susceptibility of α-Fe_2O_3 and α-Fe_2O_3 with added titanium. Phys Rev 78:819-820

Moriya T (1960) Anisotropic superexchange interaction and weak ferromagnetism. Phys Rev 120:91-98

Nagai T, Kagi H, Yamanaka T (2003) Variation of hydrogen bonded O...O distances in goethite at high pressure. Am Mineral 88:1423-1427

Nagai T, Parise JB, Kagi H, Yamanaka T (2006) High pressure neutron diffraction study of deuterated goethite (α-FeOOD). in preparation

Nelmes RJ, Loveday JS, Wilson RM, Besson JM, Klotz S, Hamel G, Hull S (1993a) Structure studies at high pressure using neutron powder diffraction. Trans Am Crystallogr Assoc 29:19-27

Nelmes RJ, Loveday JS, Wilson RM, Besson JM, Pruzan P, Klotz S, Hamel G, Hull S (1993b) Neutron-diffraction study of the structure of deuterated ice-VIII to 10 Gpa. Phys Rev Lett 71:1192-1195

Nelmes RJ, McMahon MI, Belmonte SA, Parise JB (1999) Structure of the high-pressure phase III of iron sulfide. Phys Rev B 59:9048-9052

Nguyen JH, Kruger MB, Jeanloz R (1994) Compression and pressure-induced amorphization of $Co(OH)_2$ characterized by infrared vibrational spectroscopy. Phys Rev B 49:3734-3738

Nguyen JH, Kruger MB, Jeanloz, R (1997) Evidence for "partial" (sublattice) amorphization in $Co(OH)_2$. Phys Rev Lett 49:1936-1939

Nield VM, Whitworth RW (1995) The structure of ice Ih from analysis of single-crystal neutron diffuse-scattering. J Phys Condens Matter 7:8259-8271

Parise JB, Antao SM, Michel FM, Martin CD, Chupas PJ, Shastri SD, Lee PL (2005) Quantitative high-pressure pair distribution function analysis. J Synch Rad 12:554-559

Parise JB, Brown GE (2006) New opportunities at emerging facilities. Elements 2:37-42

Parise JB, Cox H, Kagi H, Li R, Marshall W, Loveday J, Klotz S (1998a) Hydrogen bonding in $M(OD)_2$ compounds at high pressures. Rev High Press Sci Tech 7:211-216

Parise JB, Leinenweber K, Weidner DJ, Tan K, Von Dreele RB (1994) Pressure-induced hydrogen bonding: Neutron Diffraction study of brucite, $Mg(OD)_2$, to 9.3 GPa. Am Mineral 79:193-196

Parise JB, Locke DR, Tulk CA, Swainson I, Cranswick L (2006) The effect of pressure on the Morin transition in hematite (α-Fe_2O_3). Physica B in press

Parise JB, Loveday JS, Nelmes RJ, Kagi H (1999) Hydrogen repulsion "transition" in $Co(OD)_2$ at high pressure? Phys Rev Lett 83:328-331

Parise JB, Theroux B, Li R, Loveday J, Marshall W, Klotz S (1998b) Pressure dependence of hydrogen bonding in metal deuteroxides: a neutron powder diffraction study of $Mn(OD)_2$ and β-$Co(OD)_2$. Phys Chem Minerals 25:130-137

Paureau J, Vettier C (1975) New high-pressure cell for neutron-scattering at very low-temperatures. Rev Sci Instrum 46:1484-1488

Peter S, Parise JB, Smith RI, Lutz HD (1999) High-pressure neutron diffraction studies on laurionite-type Pb(OD)Br. J Phys Chem Solids 60:1859-1863

Prewitt CT, Parise JB (2000) Hydrous phases and hydrogen bonding at high pressure. Rev Mineral Geochem 41:309-333

Proffen T (2006) Analysis of disordered materials using total scattering and the atomic pair distribution function. Rev Mineral Geochem 63:255-274

Salzmann CG, Radaelli PG, Hallbrucker A, Mayer E, Finney JL (2006) The preparation and structures of hydrogen ordered phases of ice. Science 311:1758-1761

Searle CW (1967) On pressure dependence of low-temperature transition in hematite. Phys Lett A 25:256-259

Parise JB (2006) Introduction to neutron properties and applications. Rev Mineral Geochem 63:1-25

Shirane G (1959) A Note on the magnetic intensities of powder neutron diffraction. Acta Crystallogr 12:282-285

Shull CG, Strauser WA, Wollan EO (1951) Neutron diffraction by paramagnetic and antiferromagnetic substances. Phys Rev 83:333-345

Stone HJ, Tucker MG, Le Godec Y, Meducin FM, Cope ER, Hayward SA, Ferlat GPJ, Marshall WG, Manolopoulos S, Redfern SAT, Dove MT (2005a) Remote determination of sample temperature by neutron resonance spectroscopy. Nucl Instrum Methods Phys Res A547:601-615

Stone HJ, Tucker MG, Meducin FM, Dove MT, Redfern SAT, Le Godec Y, Marshall WG (2005b) Temperature measurement in a Paris-Edinburgh cell by neutron resonance spectroscopy. J Appl Phys 98:Art. No. 064905

Strassle, T, Saitta, AM, Le Godec, Y, Hamel, G, Klotz, S, Loveday, JS, and Nelmes, RJ (2006) Structure of dense liquid water by neutron scattering to 6.5 GPa and 670 K. Phys Rev Lett, 96: Art. No. 067801

Tse JS, Trouw F, Gutt C, Press W, Shpakov V, Belodludov W (2000) Structure and dynamics of clathrate hydrates: Neutron elastic and inelastic studies. Abstr Papers Am Chem Soc 220:U164-U164

Vogel SC, Priesmeyer H-G (2006) Neutron production, neutron facilities and neutron instrumentation. Rev Mineral Geochem 63:27-57

Von Dreele RB (2006) Neutron Rietveld refinement. Rev Mineral Geochem 63:81-98

Wenk H-R (2006) Neutron diffraction texture analysis. Rev Mineral Geochem 63:399-426

Werner SA, Arrott A, Atoji M (1968) Effects of pressure and a magnetic field on chromium studied by neutron diffraction. J Appl Phys 39:671-673

Wilding MC, Benmore CJ (2006) Structure of glasses and melts. Rev Mineral Geochem 63:275-311

Williams Q, Guenther L (1996) Pressure induced changes in bonding and orientation of hydrogen in FeOOH-goethite. Solid State Comm 100:105-109

Wilson RM, Loveday JS, Nelmes RJ, Klotz S, Marshall WG (1995) Attenuation corrections for the Paris-Edinburgh cell. Nucl Instrum Methods Phys Res A354:145-148

Wollan EO, Davidson WL, Shull CG (1949) Neutron diffraction study of the structure of ice. Phys Rev 75: 1348-1352

Worlton TG, Bennion RB, Brugger RM (1967) Pressure dependence of Morin transition in α-Fe_2O_3 to 26 Kbar. Phys Lett 24:653-655

Worlton TG, Brugger RM, Bennion RB (1968) Pressure dependence of Neel temperature of Cr_2O_3. J Phys Chem Solids 29:435-438

Xu JA, Mao HK, Hemley RJ, Hines E (2002) The moissanite anvil cell: a new tool for high-pressure research. J Phys Condens Matter 14:11543-11548

Xu JA, Mao HK, Hemley RJ, Hines E (2004) Large volume high-pressure cell with supported moissanite anvils. Rev Sci Instrum 75:1034-1038

Yan CS, Vohra YK, Mao HK, Hemley RJ (2002) Very high growth rate chemical vapor deposition of single-crystal diamond. Proc Nat Acad Sci USA 99:12523-12525

Zhao Y, He D, Qian J, Pantea C, Lokshin KA, Zhang J, Daemen LL (2005) Development of high P–T neutron diffraction at LANSCE – toroidal anvil press, TAP-98, in the HIPPO diffractometer. *In:* Advances in High-Pressure Technology for Geophysical Applications. Chen J, Wang Y, Duffy TS, Shen G, Dobrzhinetskaya LF (eds) Elsevier, p 461-474

Reviews in Mineralogy & Geochemistry
Vol. 63, pp. 233-254, 2006
Copyright © Mineralogical Society of America

10

Inelastic Scattering and Applications

Chun-Keung Loong

Intense Pulsed Neutron Source Division
Argonne National Laboratory
Argonne, Illinois, 60439-4814, U.S.A.
e-mail: ckloong@anl.gov

INTRODUCTION

Inelastic neutron scattering (INS) in general refers to scattering processes which involve energy and momentum exchange between the neutron and the scatterer. It is widely utilized for characterization of materials in basic and applied research across many disciplines including mineralogy. Obviously, the title here encompasses an area too vast to be adequately introduced in depth or in breadth in a single chapter. Thus the author hastens to confine the scope of this introduction to include mainly the basic concepts, some recent development in instrumentation, and a few illustrative examples. In the examples, we concentrate on showing the INS spectra as compared with the expected scattering cross section and pointing out the significant features rather than dwelling on the detailed interpretation. Further information regarding derivation of scattering formulation, discussion of techniques, and reviews of various subfields is referred to in the references.

Subjected to conservation of the total energy and momentum of the neutron+scatterer system, the energy and momentum transfer for a neutron which has undergone inelastic scattering are defined as:

$$E = E_0 - E_1 \tag{1}$$

and

$$E = E_0 - E_1 \tag{1}$$

where E is the energy and the product of the Planck's constant and the wavevector is the momentum. Subscripts 0 and 1 denote quantities pertaining to the incident and scattered neutrons, respectively. Positive and negative energy transfer E corresponds to, respectively, scattering with neutron energy loss and neutron energy gain, or, by analogy of Raman scattering, the Stokes and anti-Stokes lines. However, in case of light scattering, the wavevector $\mathbf{Q}$ is essentially zero hence Raman and IR spectroscopy measure only the Brillouin-zone-center modes of crystallite solids, where neutrons probe wave vectors spanning the entire Brillouin zone.

The goal of INS is to measure precisely the double differential cross section defined as:

$$\frac{d^2\sigma}{d\Omega dE} = \left(\frac{1}{N}\right)\frac{k_1}{k_0}\left(\frac{m_n}{2\pi\hbar^2}\right)^2 \left|\langle \mathbf{k}_1\xi_1|V|\mathbf{k}_0\xi_0\rangle\right|^2 \delta\left(E + E_{\xi_0} - E_{\xi_1}\right) \tag{3}$$

$$= \frac{k_1}{k_0} S(\mathbf{Q}, E)$$

where σ, Ω, N, m_n, are the scattering cross-section, solid angle, number of scattering units in the scatterer, and neutron mass, respectively. ξ_0 and ξ_1 denote the state of the scatterer before and after scattering, respectively. V is the scattering potential. In principle, the neutron

1529-6466/06/0063-0010$05.00   DOI: 10.2138/rmg.2006.63.10

may also undergo a change of its spin state after scattering but for simplicity we do not consider the scattering of a polarized neutron beam here. It can be seen from Equation (3) that INS effectively aims at the determination of the *scattering function* $S(\mathbf{Q},E)$ that contains $\langle \mathbf{k}_1\xi_1 | V | \mathbf{k}_0\xi_0 \rangle$, the matrix element characterized by the nature of many-body interactions within the scatterer. For example, it can be shown that the scattering function, depending on coherent or incoherent scattering, is related to respectively the inter-particle or self-particle space-time correlation functions of the scatterer under study. According to linear response theory, they are also linked to the Green functions and the dynamic susceptibilities of the scatterer, thereby enabling a direct connection between experimental result and theoretical prediction. Detailed description of the scattering functions for different systems and their corresponding connection to the Green functions and dynamic susceptibilities for studying dynamics and phase-transition behavior have been given in many texts and review articles (see for example (Squires 1978; Lovesey 1984, 1986) and will not be repeated here. Instead, we present a graphical illustration of Figure 1, showing the typical loci for significant result expected in the scattering function over the $(\mathbf{Q},E)$ space of interest to various systems.

EXPERIMENTAL CONSIDERATIONS

Neutron sources

It can be seen from Figure 1 that INS investigations of a specific system usually concentrate within a limited region of the $(\mathbf{Q},E)$ space. Therefore, experimenters have to choose a neutron-scattering facility that provides access of the desired region. Recalling the fact that 1 meV is equivalent to 11.6 K, neutron spectroscopy can then be subdivided into *cold, thermal,* and *epithermal* neutron scattering according the energy transfer E with loose respective boundaries of <8, 8-80, and >80 meV. Neutrons at modern facilities are either produced by steady-state reactors or pulsed, accelerator-based spallation sources. A pulsed reactor (e.g., IBR-2 at Dubna, Russia) and a steady-state-like spallation source (e.g., SINQ, Switzerland) are rare exceptions. Reactors generate primarily thermal neutrons near the core but such neutrons can be energized or slowed down to the epithermal or cold regime by hot or cold sources, respectively. Spallation sources always under-moderate the neutrons thus epithermal neutrons are plentiful, and cold neutrons can be generated by cold moderators. Therefore, in principle, both reactor and spallation neutron sources can accommodate INS experiments. However, more importantly, experimenters should choose the beamline/spectrometers carefully in order to match the scientific needs for the problem at hand.

Beamline and components

The goal of an INS instrument is to quantitatively measure the scattering events, in terms of counts of the scattered neutrons as a function of energy recorded by well-defined detectors (with known pixel resolution and position, efficiency $\eta(E_1)$, etc.), from which a scattering function, $S(\mathbf{Q},E)$ pertinent to the sample in the environment under study, can be accurately constructed. Figure 2 shows schematically the generic components, scattering kinematics, and the quantities involved. The scattering function, if well characterized over a wide enough range of $(\mathbf{Q},E)$, can permit the determination of the spatial-temporal correlation functions or the corresponding Green functions, of which the data represent a basic and powerful vehicle toward the understanding of a many-body system. For example, within the linear response theory, instabilities or phase transitions will manifest themselves near the poles (divergence) of the imaginary part of the dynamic susceptibility measurable by INS.

Given the vast complexity of scientific problems, as suggested in Figure 1, it is impossible to adequately capture all the information over the multiple decades in the $(\mathbf{Q},E)$ domain by one or two INS instruments. Analogously, that is to say, there is no single magic lens through which you can see the landscape of a whole mountain and the fine structure of a grain of

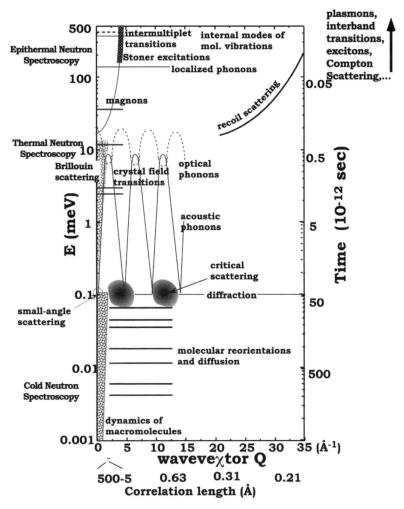

Figure 1. The loci of significant intensity expected in the scattering function of INS shown schematically for various systems. The wavevector-energy and space-time variable sets of the corresponding correlation functions are related reciprocally via a double Fourier transform (modified after Sinha 1985).

sand therein at the same time. The current state-of-the-art INS instruments were developed, to a large extent, through a combined effort of scientists and engineers from neutron centers and from multidisciplinary neutron user communities. The R&D for better instrumentation, software and hardware included, continues today as technologies advance and the scientific realm expands. Here, we present only a skeleton of some important design parameters and operation principles of INS instruments.

Figure 2 shows a schematic drawing of a generic INS instrument. The sample, whose orientation defined, for example, by its crystal orientation or magnetic quantization axis relative to the neutron beam, is seen to scatter the incident neutrons to a detector, characterized by a pixelized solid angle and neutron-detection efficiency, at a scattering angle ϕ. In order to differentiate the net energy change for a scattering event, an energy filter, which selects neutrons with a narrow distribution of energies and/or spins over a collimated solid angle, has

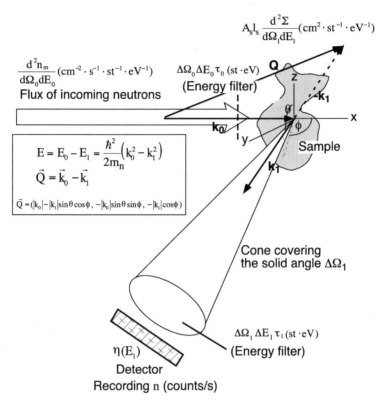

Figure 2. A drawing of a generic INS instrument, identifying the essential components with their parameters and measurement units. The inset outlines the neutron energy transfer and wavevectors governed by conservation of energy and momentum (modified after Windsor 1988).

to be inserted in the incident or scattered beam, sometimes in both places. An INS instrument operating in *direct* (*inverse*) geometry refers to the fixed incident energy + variable scattered energies (variable incident energy + fixed scattered energy) configuration. The upshot is, with proper calibration and performance of each component, one can derive the measured double differential cross section in absolute units from the recorded neutron counts of the detector.

The mechanism for neutron beam definition employed by an energy filter varies from transmission, diffraction, reflection, and refraction in materials as well as other means of neutron optics including manipulation of the neutron spins. Two most important energy filters, as illustrated in Figure 3, are crystal monochromators (M) for incident neutrons or analyzers (A) for scattered neutrons and choppers. In the case of chopper utilization, the transmitted beam yields narrow bursts of neutrons whose speeds (hence energies) are subsequently determined by the flight time over a known distance, i.e., by the *time-of-flight* (*TOF*) *method* that is particularly suitable to spallation sources owing to the inherent time structure of the neutron fluxes. The energy resolution, $\Delta E/E$, is governed, among many factors, mainly by the crystal Bragg angle, θ_X, $X = M$ or A, for the crystal case and by the rotational speed and slit-package configuration for the chopper case. From the Bragg law, we find $\Delta l/l \sim \cot\theta_X$. This fact accounts for the requirement of high monochromator "kick-off" angle or nearly "backscattering" geometry for high- resolution instruments. For high-resolution choppers, a rotation speed of 36,000 rotations per minute is routinely applied. Additionally, TOF instruments must deal with the effects of uncertainties and broadening (due to the finite size of

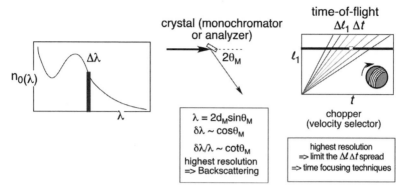

Figure 3. Crystals (middle) and choppers (right) used for selection neutrons with a narrow band of energies (left).

various components) in flight times and distances on the resolutions. The Q-resolution, often coupled with that of energy, is more complicated. *Time-focusing* is an important aspect in designing pulsed-source instruments.

Neutron spectrometers

INS instruments are highly complex. An attempt to dwell on the detailed design and operation of various machines is beyond the scope of this chapter. Three popular spectrometers, the triple-axis, chopper, and crystal-analyzer spectrometer, are briefly mentioned here so that readers may better appreciate the different kinds of experimental results given as examples in the following section.

The *triple-axis spectrometer*, pioneered by B. N. Brockhouse, provides ingenious flexibility in exploring the reciprocal lattice space, it is used mainly for studying elemental excitations (e.g., phonons and magnons) and dynamics associated with phase transitions in solid-state physics, see Figure 4. Full polarization operation and analysis of the incident and scattered neutrons are readily achievable. It operates in a steady-state mode and normally ignores any time structure of the neutron beam, see for example (Shirane et al. 2002). Thus triple-axis spectrometers are exclusively reactor instruments. Because it collects data point-by-point through the reciprocal space using a single or only a handful of detectors, data rate is relatively low. A reactor often is equipped with several triple-axis spectrometers in order to cover the cold and thermal neutron regimes.

A *chopper spectrometer* admits monochromatic bursts of neutrons incident on a sample and subsequently allows analysis of the scattered energy by TOF methods (see Fig. 5). For reactors, a first chopper has to be employed to "chop" the steady-state beam into polychromatic pulses, then subsequently monochromized by a second chopper at a downstream position. For pulsed sources, pulsation is inherently built in by the source so only one chopper is in principle needed. Furthermore, a pulsed-source chopper spectrometer enables the coverage of a very wide dynamic range from cold to thermal to epithermal neutrons on the same instrument without interrupting the sample environment, see for example (Loong et al. 1987). Chopper spectrometers are equipped with large detector banks, often with position-sensitive detectors, subtending a large solid angle. Thus voluminous data are collected in one setting, surveying a sizeable region of the (Q,E) space but sorting out data along specific **Q** directions is difficult because the fixed-detector configuration limits the flexibility in assessing **Q**. Therefore, traditionally the majority of samples used are polycrystalline or disordered materials. Nowadays, however, a state-of-the-art pulsed-source chopper spectrometer such as the MAP

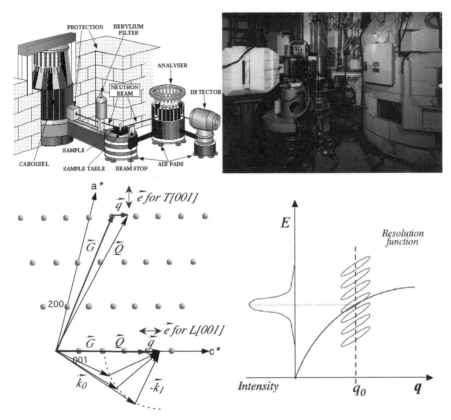

Figure 4. Top: The IN14 and T2 triple-axis spectrometers at ILL and LLB. Bottom: (left) Longitudinal and transverse scans for phonon dispersion measurements. The wavevector **Q** is the sum of a reciprocal vector **G** and a reduced wavevector **q**. **e** is the phonon polarization vector. (right) A constant-q scan showing the intersection of the resolution function of a triple-axis spectrometer with a phonon dispersion curve, giving rise to an intensity profile of a phonon group.

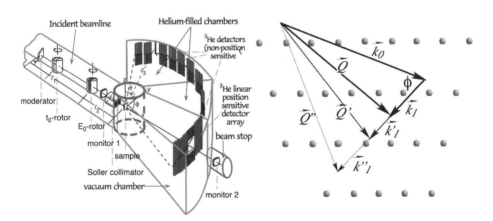

Figure 5. (left) A pulsed-source chopper spectrometer. (right) The constant-k_0 scans observed by a detector at a scattering angle ϕ.

instrument of ISIS is also capable of conducting single-crystal experiments for studying collective excitations, much like the triple-axis spectrometers at reactors.

A time-focused *crystal-analyzer spectrometer* operates in the mode of inverse geometry in which polychromatic neutron bursts are incident on the sample and only the scattered neutrons of a fixed final energy set by the Bragg reflection from a crystal are recorded as a function of the flight time. Since the total flight distance is known, the incident energy corresponding to each TOF channel can be calculated, giving rise to an energy spectrum. If scores of crystal analyzers are positioned at different scattering angles, data for a scattering function $S(\mathbf{Q},E)$ can be obtained. In general, the time-focusing method can only be applied to optimize a specific region of the energy transfer. If the energy resolution is optimized at the elastic-scattering region, the instrument is often referred to as a quasielastic scattering neutron spectrometer (QENS) (Fig. 6), otherwise, it is termed an inelastic crystal-analyzer spectrometer (Connatser et al. 2003). QENS excels in studying diffusive motions, particularly those of hydrogen atoms or molecules in soft matter. Additionally, a pulsed-source crystal-analyzer spectrometer may cover also the inelastic region up to about 200 meV with a reasonable resolution. Moreover, it is easy to install a few detectors (no analyzers) at high and low angles so that, concurrent with quasielastic to inelastic scattering measurements, powder diffraction and small-angle scattering data can also be obtained for the same sample.

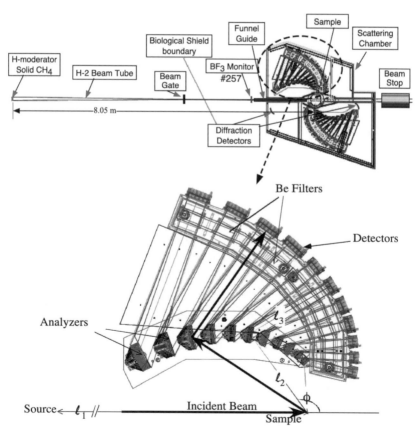

Figure 6. The layout of the pulsed-source crystal-analyzer spectrometer QENS of IPNS (Connatser et al. 2003).

Complementarity with other spectroscopic techniques

Neutron spectroscopy is one of many spectroscopic methods aiming to study the complex interactions in a many-body system at the atomic level. From a researcher's viewpoint, it is important to realize the complementarity of neutrons and other probes in fulfilling the goal of a scientific problem. A comparison of the various methods based on experimental techniques and data interpretation is available in a previous volume of MSA reviews (Hawthorne 1997). Here, we limit ourselves to a qualitative look at the complementary information obtainable from various spectroscopies over the (Q,E) space, which is shown in Figure 7, and comment on only the relationship of INS with some recently developed techniques.

It is noteworthy that INS techniques and other methods collectively do not fill the entire (Q,E) space. Therefore, there exist phenomena, especially for slow dynamics or low energy excitations, which presently are inaccessible for investigation by spectroscopic methods. Secondly, INS nevertheless encompass a major portion of the (Q,E) space that is essential to many scientific disciplines.

The advent of high-brilliance synchrotron radiation x-ray sources has brought about significant progress in the last decade especially on phonon measurements by means of inelastic X-ray scattering (IXS). Using crystal analyzers with the back-scattering geometry to realize the highest energy and spatial resolution and adopting the triple-axis spectrometer operation, nowadays phonon dispersion relations can be measured by IXS with energy resolutions (~meV) comparable to those of INS. Furthermore, nuclear inelastic absorption spectroscopy, which detects X-ray fluorescence from internally excited nuclei in a sample at resonance prompted by synchrotron radiation, can provide a measure of the partial phonon density of states of those nuclei. This method applies to only a handful of Mössbauer nuclei of which compounds containing [57]Fe are the mostly studied. For more details, reader may refer

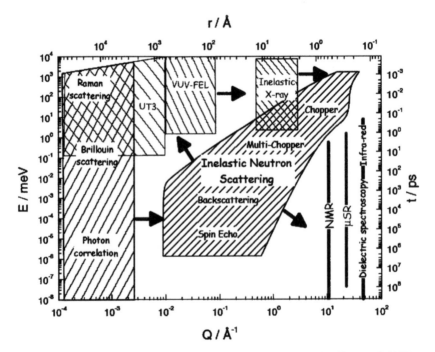

Figure 7. Coverage in the (Q,E) space by neutron and other spectroscopies (Boue et al. 2002).

to (Burkel 2000). In general, cross-fertilization among INS, IXS and other spectroscopies will lead to a fuller characterization of a complex science problem.

SCIENTIFIC EXAMPLES

Phonons and rare-earth crystal-field excitations in xenotime

Mixed natural rare-earth orthophosphates RPO_4 (R = rare earth elements) form the minerals monazite (R = La-Gd) and xenotime (R = Tb-Lu). Pure crystalline forms of these compounds can be synthesized by controlled precipitation techniques and single crystals can be grown by means of flux methods. Their outstanding properties, such as high melting point (above 2000 °C), structural and chemical stability, high corrosion resistance, rare-earth-activated luminescence etc., have been explored for technological applications as nuclear waste storage media, scintillators, phosphors, lasers, and magnetic refrigerants. Furthermore, thermodynamic properties, magnetic phase transitions, Jahn-Teller effects, and anomalous electron-phonon interactions associated with many of the RPO_4 compounds are of fundamental interest (Boatner 2002). A systematic INS investigation was carried out on the xenotime series which, compared with monazite, has a simpler crystal structure (zircon-type, body-centered tetragonal, space group $I4_1/amd$, four formula units per unit cell).

Phonon measurements of non-magnetic reference $LuPO_4$ The end member $LuPO_4$ is regarded as "non-magnetic" (weakly paramagnetic with no magnetic ordering at low temperatures) because Lu has a fully filled $4f$ shell. Therefore, phonons in $LuPO_4$ are expected to be the least influenced by magnetic or electronic interactions with the rare-earth ions, and a detailed measurement of the phonon dispersion curves of $LuPO_4$ would provide a base-line reference to distinguish effects due to rare-earth magnetism in other xenotime members. The primitive cell contains two formula units (12 atoms), implying 36 phonon branches along each direction. Such relatively simply structure affords an analysis of the data using a lattice-dynamics model from which the neutron double differential cross section and the phonon dispersion relations can be calculated through the diagonalization of the dynamical matrix in terms of parameters representing harmonic interatomic interactions between atoms. A group theoretical analysis of the phonon modes, which classifies the phonons according to the space-group symmetries in the Brillouin zone, enables a decomposition of the 36×36 dynamical matrix into smaller blocks according to the irreducible representation along a symmetry direction, thereby facilitating the assignment of observed data to phonon branches. Nowadays such an analysis scheme and set up of a lattice-dynamics model are readily incorporated in computer codes which are made accessible to researchers (Warren and Worlton 1972, 1974; Eckold et al. 1987).

INS measurements of phonons in materials rely on the *coherent* scattering from the nuclei; namely, it is the interference of the scattered wavefronts of the neutrons that gives rise to information regarding the collective motion of the atoms. The double differential cross section for a neutron that emits one phone of energy $\hbar\omega_i$ after scattering (neutron energy loss) is (Price and Sköld 1986)

$$\frac{d^2\sigma}{d\Omega dE} = \left(\frac{1}{N}\right)\frac{k_1}{k_0}\left(\frac{(2\pi)^3}{2V}\right)\sum_{G,q_j}\hbar\left|F\left(Q,q_j\right)\right|^2\frac{\langle n_j+1\rangle}{\omega_j(q)}\delta(Q-q-G)\delta\left(E-\hbar\omega_j(q)\right) \quad (4)$$

where the one-phonon structure factor

$$F\left(Q,q_j\right) = \sum_d \frac{b_d}{\sqrt{M_d}}e^{-W_d(Q)}e^{iQ\cdot d}Q\cdot e_d^j(q) \quad (5)$$

defines essentially the phonon intensity measured at Q which is the sum of the reduced wavevector q from the reciprocal vector G for phonon branch j. In Equations (4)-(5) V is the volume of the primitive cell, M is the mass of atom of index d at position **d** with a coherent scattering amplitude b_d, W is the Debye-Waller factor, and n is the population factor. **e** is the polarization vector.

The INS measurements and data analysis involve iterative processes of triple-axis spectroscopic measurements on large single-crystal samples and refinements of the model by fitting the data (phonon frequencies and intensities) to calculated results until self-consistency is reached. The experiments on $LuPO_4$ single crystals were performed at the Laboratoire Leon Brillouin (LLB, Scalay, France) using the 2T1 triple-axis spectrometer. A shell model using Born-von Karman-type axially symmetric forces and polarizable oxygen ions was constructed for data interpretation (Nipko et al. 1997).

Figure 8 shows the observed data (open circles) and calculated phonon dispersion curves (lines) of $LuPO_4$ at 10 K along the [100], [110] and [001] symmetry directions. The scarcity of data points at high energies reflects the fact that neutron flux above thermal energy is limited at reactor sources (unless a hot source is available). The Raman and IR data from the literature were added. As expected, they occur at the zone center only. The agreement between experimental data and model calculations is good.

The overall phonon frequency distribution was investigated by TOF INS using the HRMECS chopper spectrometer at the Intense Pulsed Neutron Source (IPNS, Argonne, U.S.A.). The data collected with $E_0 = 200$ meV were summed over scattering angles of 70° to 120° which correspond to Q varying between 11 to 20 Å^{-1}. Since in general Q is much larger than the dimension of the Brillouin zone or q, and the polycrystalline sample effectively provides an average over all crystallographic orientations, a *neutron-weighted* phonon density of states (DOS) can be obtained from the measured scattering function according to

$$G(E) = \frac{2\bar{M}}{\hbar^2}\left\langle \frac{e^{2W(Q)}}{Q^2}\frac{E}{n+1}S(Q,E)\right\rangle \approx \bar{M}\sum_i \frac{c_i\sigma_i}{M_i}F(E) \qquad (6)$$

where c_i, σ_i, M_i, and $F_i(E)$ are the concentration, scattering cross section, mass and partial phonon DOS, respectively, for the ith atomic species. $\bar{M}$ is the mean sample mass, and $\langle\cdots\rangle$ represents the average over all observed Q values. $2W(Q) \approx \alpha Q^2$ can be estimated from diffraction data. The validity of Equation (6) has been discussed elsewhere (Bredov et al. 1967; Oskotskii 1967; Taraskin and Eilliot 1997a,b).

Figure 9 displays the neutron-weighted one-phonon and two-phonon DOS calculated from the shell model, for which the sum compared well with the measured phonon spectrum. Knowledge of the phonon DOS of a material is essential in the understanding of its thermodynamic properties. For example, the lattice specific heat can be calculated. Figure 9 shows the calculated specific heat at constant volume for $LuPO_4$ in comparison with the measured specific heat at constant pressure for $ScPO_4$, an isostructural compound chosen because of the lack of experimental data on $LuPO_4$. The agreement is good.

We stress the importance of INS in the characterization of the atomic dynamics of minerals. In particular, the combination of epithermal neutron spectroscopy at pulsed sources, ancillary equipment capable of handling extreme sample environments, and advanced multiscale molecular-dynamics simulations or *ab initio* calculations is a very powerful method (Kieffer and Navrotsky 1985, Chaplot et al. 2002). INS is the only technique that is not restricted by selection rules or hampered by optical opacity and sample absorption, sensitive to excitations over the entire Brillouin zone, and relatively immune from impurities. On the other hand, the readers should be aware of the special isotopes that exhibit exceedingly large neutron absorption cross section, such as ^{113}Cd and ^{155}Gd. INS studies of materials containing elements dominated by such absorptive isotopes are very difficult.

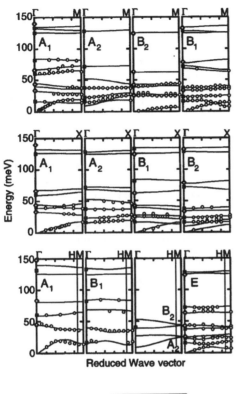

Figure 8. Phonon dispersion curves of $LuPO_4$ along the [100] (top), [110] (middle), and [001] (bottom) symmetry directions. The symbols, squares, diamonds, and circles correspond to Raman, infrared, and neutron data, respectively. The lines are calculated phonon dispersion curves using a lattice-dynamics shell model. Experimental errors are comparable to the size of the symbols. [Reprinted with permission from Nipko et al. (1997) Phys Rev B, Vol. 56, Figs. 4, 6, 7, p. 11584-11592. © 1997 by the American Physical Society. URL: *http://link.aps.org/abstract/PRB/v56/p11584*]

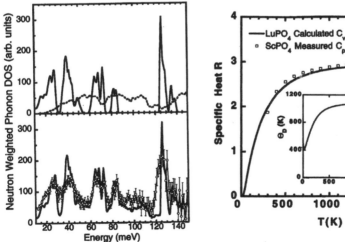

Figure 9. (LEFT) (top) The calculated one-phonon (solid line) and two-phonon (dashed line) neutron-weighted phonon DOS for $LuPO_4$. (bottom) The measured phonon spectrum (open circles) compared with the sum of the calculated one- and two-phonon neutron-weighted DOS (line). (RIGHT) The calculated lattice specific heat in terms of R per average atom (R is the universal gas constant) (solid line) for $LuPO_4$ and measured specific heat for $ScPO_4$. The calculated temperature dependence of the Debye temperature for $LuPO_4$ is plotted in the inset. [Reprinted with permission from Nipko et al. (1997) Phys Rev B, Vol. 56, Figs. 4, 6, 7, p. 11584-11592. © 1997 by the American Physical Society. URL: *http://link.aps.org/abstract/PRB/v56/p11584*]

Rare-earth crystal-field level structure of xenotime. The f-electrons of a rare-earth ion in a crystal are influenced by the crystalline electric fields of the surrounding ions, resulted in lifting the degeneracy of the Hund's-Rule ground multiplet into a level structure governed by the site symmetry. At temperatures comparable to the crystal-field (CF) splitting, all the levels are populated and the magnetic properties are relatively isotropic and weakly interactive. At low enough temperatures for the occupation of only the lowest states, the magnetization may be highly anisotropic thereby favoring strong coupling with the lattice and/or other rare-earth spins. In some cases, cooperative structural changes such as the Jahn-Teller transition or long-range magnetic ordering may occur. Therefore, it is paramount to characterize the rare-earth CF level structure for a better understanding of the magnetic and optical behavior of the materials. INS, in which neutrons are scattered via the interaction between the neutron spins and the magnetic moments of the rare-earth ions—neutron *magnetic scattering*—is the best method to fulfill such task.

Rare-earth ions in xenotime occupy the site of D_{2d} tetragonal symmetry and the quantization direction is defined by the four-fold c-axis of the crystal structure. $TbPO_4$, $DyPO_4$, and $HoPO_4$ exhibit spin-spin interactions at sub-helium temperatures and $TbPO_4$ at 2.3 K, and $YbPO_4$ over a wide range of temperatures up to 300 K shows evidence of strong spin-lattice coupling (Loong and Soderholm 1994). For INS determination of the CF level structure, polycrystalline samples are sufficient but for detailed studies of the spin-lattice or electron-phonon interactions, single crystal specimens, possibly also using polarized neutrons, are preferred.

The neutron-scattering cross section for CF transitions in N noninteracting ions is given in the dipole approximation by (Loong et al. 1993)

$$\frac{d^2\sigma}{d\Omega dE} = N \frac{k_1}{k_0} \frac{(\gamma r_0)^2}{4} g_J^2 S(\mathbf{Q},E) \tag{7}$$

where the scattering function is expressed as

$$S(\mathbf{Q},E) = f^2(\mathbf{Q}) e^{-2W(\mathbf{Q})} \sum_{n,m} \frac{\exp\left(-E_n/k_BT\right)}{Z} \left|\langle n|J_\perp|m\rangle\right|^2 \delta(E_n - E_m - E) \tag{8}$$

In the above equations, γ is the neutron magnetic moment in units of nuclear Bohr magnetons, r_0 is the electron classical radius, g_J is the Landé g factor, $f(\mathbf{Q})$ is the ionic magnetic form factor, and Z is the partition function. In general the ith CF state $\langle i|$ at an energy E_i is a mixture of $|J,M\rangle$ states including those from higher J multiplets, and $J_\perp$ is the component of the total angular-momentum operator perpendicular to $\mathbf{Q}$. For experiments using polycrystalline samples and unpolarized neutrons, the measured quantity is

$$\left|\langle|J_\perp|\rangle\right|^2 = \frac{2}{3}\left[\left|\langle|J_x + J_y|\rangle\right|^2 + \left|\langle|J_z|\rangle\right|^2\right]$$

and only the modulus of $\mathbf{Q}$ is retained. The CF states are calculated from the Schrödinger equation with the Hamiltonian:

$$H_{CF} = \sum_{k,q,i} B_q^k C_q^k(i) \tag{9}$$

where the $C_q^k(i)$ are spherical tensor operators of rank k, and depend on the coordinates of the ith electron. The summation of i is over all the f electrons of the ion, and the B_q^k are the CF parameters. The site symmetry at the ion of interest determines which B_q^k parameters are nonzero. For RPO_4 the rare-earth ions occupy lattice sites of tetragonal (D_{2d}) symmetry, so the CF term is characterized by five parameters of real numbers, B_0^2, B_0^4, B_4^4, B_0^6 and B_4^6 which are determined by fitting the neutron spectrum to the scattering cross section of CF transitions.

Figure 10a compares the measured and calculated CF spectrum for HoPO$_4$, which shows good agreement. The derived CF energy level structure of the Ho^{3+}:^{5}I$_8$ ground multiplet is shown in Figure 10b, where the arrows indicate the observed CF transitions by INS. The CF structure is the key to understand the magnetic properties at low temperatures. Figure 10c shows the calculated and measured paramagnetic susceptibility of HoPO$_4$ with the applied field directions perpendicular and parallel to the crystallographic c-axis. The very strong anisotropy, $\chi_\parallel \gg \chi_\perp$, suggests that the Ho magnetic ground state plays an important role as a "bootstrap" for the long-range ordering of the Ho moments at low temperatures. The ordered moments are parallel to the easy axis of the susceptibility (Loong et al. 1993).

Electron-phonon coupling in YbPO$_4$. The phonon measurements of non-magnetic LuPO$_4$ and the determination of the rare-earth CF level structure of the xenotime series, all of importance in their own right, allow, in addition, more involved investigations of the anomalous behavior caused by 4f-electron-phonon interactions in some members. For example, the Yb^{3+}:^{2}F$_{7/2}$ ground multiplet is lifted by CF into four Kramers doublets: Γ_6, Γ_7, Γ_6, and Γ_7 at 0, 12, 32, and ~43 meV, respectively, as illustrated in Figure 11. However, Raman data showed evidence of strong coupling of the two upper CF states with an E_g phonon over the 4-300 K temperature range (Becker et al. 1992), and Brillouin-scattering data showed a ~20% softening of the (C_{11}-

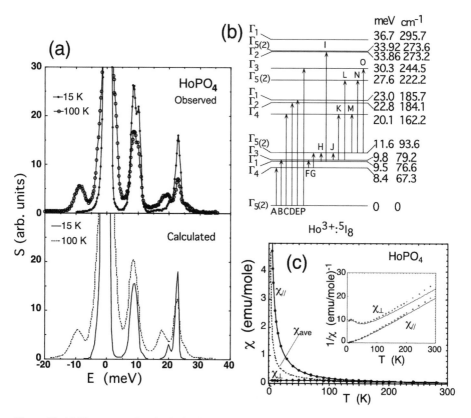

Figure 10. (a) The measured and calculated CF spectrum of HoPO$_4$ at 15 and 100 K. (b) A schematic diagram of the splitting of the Ho^{3+}:^{5}I$_8$ ground multiplet by the CF. The arrows refers to the observed transitions by INS. (c) The calculated (curves) and measured (symbols) paramagnetic susceptibility with the applied field directions perpendicular and parallel to the crystallographic c-axis. χ_{ave} is the powder-average susceptibility and the inset shows the inverse susceptibility of HoPO$_4$ (modified after Loong et al. 1993).

C_{12})/2 elastic constant with decreasing temperature from 300 to 10 K but no structural phase transition occurred (Nipko et al. 1996). Since rigorous theoretical model of such CF-state-phonon interaction is lacking, the INS experiments resorted to a comparison of the YbPO$_4$ excitation spectrum with that of LuPO$_4$—the normal reference, using single-crystal samples (Loong et al. 1999).

It was found that most of the phonons in YbPO$_4$ follow closely the corresponding phonon data of LuPO$_4$ and the predicted structure factors from a "phonon-only" lattice-dynamics model. Furthermore, the observed magnetic-scattering intensities of the CF transitions agree with those predicted by the matrix elements of the CF model. However, in the 30-50 meV region, as shown in Figure 11, there are anomalous features. First, the phonon intensities are smeared out by the superposition of the two CF peaks corresponding to transitions from the ground state to the upper Γ_6, and Γ_7, and the CF peaks are unusually broad. Second, the profile of the 32-meV peak is highly asymmetric. Third, all the CF transitions damp out with increasing temperature much more rapidly than the expected behavior based on the population factor estimate, and fourth, the origin of a small peak at ~24 meV, which was seen throughout the Brillouin zone, is not understood. A closer examination of this behavior tentatively suggested that a large fluctuating component associated with the monopole term of the 4f wavefunctions may give rise to a coupling of the upper Γ_6, and Γ_7 CF doublets with phonons of comparable strengths and energies.

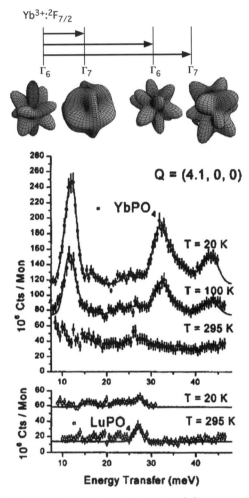

Figure 11. (top) The CF levels of the Yb^{3+}:^{2}F$_{7/2}$ ground multiplet in YbPO$_4$, the non-spherical features in the f-electron density of each state reflects the quadrupole nature of the CF potential. (bottom) An example of a constant-**Q** scan data for the excitation spectra of YbPO$_4$ and LuPO$_4$ [Reprinted with permission from Loong et al. (1999) Phys Rev B, Vol. 60, Fig. 1a, p. R12549-R12552. © 1999 by the American Physical Society. *URL: http://link.aps.org/abstract/PRBv60/pR12549*]

The monopole-type coupling does not require a compatible symmetry between specific phonon modes and the CF states and therefore is observed throughout the Brillouin zone.

Incoherent scattering from protonic species in minerals

The scattering of a neutron from an atomic nucleus depends on their relative spin directions, either parallel or antiparallel, so does the *scattering length*. If after scattering the neutron spin flips, the wavefunction then no longer interferes with the incident one, the scattering is *incoherent*. Incoherent scattering carries no information about the relative positions among atoms; its intensity does not depend on **Q**. In the case of scattering from a proton—the nucleus of a hydrogen atom, the spin of the neutron and the proton, both being ½, are combined to form a triplet state of total spin 1 and a singlet state of total spin 0. Experimentally, the scattering length of the triple and singlet state are found to be 1.04 and −4.74, respectively, in the units of 10^{-12} cm. Consequently, the coherent and incoherent scattering cross section of hydrogen (^{1}H), which correspond to the square of the mean scattering length and the mean-square deviation of the scattering lengths, respectively, multiplied by 4π, are 1.8 and 79.8, respectively, in the units of 10^{-24} cm (barn = bn). The *total* scattering cross section is the sum of the coherent and incoherent scattering cross section (Lovesey 1984).

Protonic species in minerals, incorporated naturally or synthetically, often play an important role in controlling the mechanical, chemical, and thermodynamic properties and functionality. Since hydrogen has an extraordinary large incoherent scattering cross section of ~80 bn as compared to other elements (e.g., the total cross section of 1.9, 2.17, 4.23, 11.6 bn for N, Si, O, and Fe, respectively), the observed INS spectra of hydrogen-containing minerals are dominant by features from single-particle motion of hydrogen atoms. Compared to optical spectroscopies, Raman and IR techniques only detect zone-center modes and the intensity of the hydrogen vibrations, amid comparable or stronger signals from other atoms, is difficult to quantify. X-ray scattering, relying on electromagnetic interaction with electrons in the sample, does not detect hydrogen with sufficient sensitivity.

If contributions from elements other than hydrogen are negligible, the incoherent-scattering double differential cross section for an assembly of protons can be expressed as (Price and Sköld 1986)

$$\left.\frac{d^2\sigma}{d\Omega dE}\right|_{inc} = \frac{k_1}{k_0}\left(\frac{\hbar Q^2}{2M_H}\right)\frac{\sigma_{inc}^H}{4\pi}e^{-2W(Q)}\frac{\langle n+1\rangle}{E}F\left(E/\hbar\right) \tag{10}$$

where M_H and σ_{inc}^H are the mass and incoherent cross section of hydrogen and $F(E/\hbar)$ is the vibrational density of states of hydrogen atoms. In general, protons are bonded to other elements to form different protonic species, thus INS is a good tool to detect molecular vibrations, rotation and diffusive motion associated with hydrogen atoms.

Hydrogen vibrational density of states in silicate hydrogarnet, alunite, and protonated Mn-spinel One of the fundamental goals of geology is to understand the thermodynamic properties of minerals and eventually to predict their structural stability under different conditions. Silicate hydrogarnet, with a general formula of $^{[8]}X_3^{[6]}Y_2(^{[4]}SiO_4)_{3-x}(O_4H_4)_x$ where the superscripts in brackets refer to the O coordination of the cations, allows chemical substitution as indicated by x, which produces a defect garnet structure comprising $(SiO_4)^{4-}$ and $(O_4H_4)^{4-}$ units. This substitution decreases the thermal stability and bulk modulus. Water stored in the garnets may be released in the Earth and act as a catalyst for certain types of geochemical reactions. Given this structural complexity, it is desirable to first characterize the hydrogen vibrations in an end-member compound with all the SiO_4 units replaced. Figure 12 shows neutron spectra of a synthetic hydrogarnet $Sr_3Al_2(O_4H_4)_3$ at 6 K over the 0-500 meV region, obtained by multiple runs with different incident neutron energies using a pulsed-source chopper spectrometer (Lager et al. 1998). The spectra are also compared with IR data. Some salient differences between the neutron and IR data include a much weaker band at about 60 meV for neutron than for IR because it involves motion of the Al sublattice and a narrow band at 76.7 meV, possibly due to Al-O-H bond bending that is observed by neutrons but not by IR. Furthermore, the O-H stretch vibrational band around 450 meV is of higher frequency and broader than that observed by IR, reflecting the fact that optical spectroscopy only senses the zone-center vibrations. From the neutron data, it was concluded that hydrogen bonding in $Sr_3Al_2(O_4H_4)_3$ is relatively weak.

Minerals of alunite-jarosite group have the general formula $MR_3(SO_4)_2OH)_6$, where M commonly refers H_3O^+, Na^+, K^+, Rb^+, Ag^+, Tl^+, NH_4^+, ½Pb^{2+}, or ½Hg^{2+}, and R to Al^{3+}

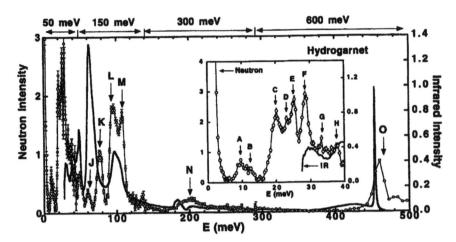

Figure 12. Comparison of the observed neutron (symbols) and IR (line) spectra of $Sr_3Al_2(O_4H_4)_3$. The neutron data collected at 6 K were composed of runs obtained with different incident energies as labeled on top. An enlarged portion of the low-energy spectrum is shown in the inset. The IR data were collected at ambient temperature. [Reprinted from Lager et al. (1998) with permission from Elsevier.]

(alunite) or Fe^{3+} (jarosite). Much attention has been directed at the nonstoichiometry and the questionable presence of oxonium ions (H_3O^+) in the materials. Both synthetic and natural samples are known to contain H_2O in excess of the stoichiometric amount. Some of the excess H_2O is generally believed to be present as H_3O^+ ions substituting for the K site. INS measurements were made on two synthetic samples: alunite and H_3O-anluite with the formula $K_{0.88}(H_3O)_{0.12}Al_{2.67}(SO_4)_2(OH)_{4.92}(H_2O)_{1.08}$ and $H_3OAl_{2.87}(SO_4)_2(OH)_{5.64}(H_2O)_{0.39}$, respectively, based on the results of chemical analysis. Figure 13 shows the INS spectra of these two compounds. There are significant differences in region around 20-100 meV. However, it is not clear if any of these features are related to the presence of oxonium in the H_3O-alunite but not in alunite (Lager et al. 2001).

Spinel is another family of important materials. For example, the phenomenon of lithium-ion extraction from aqueous solutions into a spinel-type λ-MnO_2 lattice is not well known although scavenging lithium from seawater using a λ-MnO_2-based adsorbent optimized for Li insertion was recently demonstrated (Koyanaka et al. 1997). This is achieved by an initial acid treatment whereby a highly active, protonated λ-MnO_2 or $H_xMn_2O_{4-y}$ powder is obtained. Li ions in seawater, nominally at a concentration of ~0.1 ppm (parts per million), can be extracted by dispersing the $H_xMn_2O_{4-y}$ nanoparticles in seawater through H^+-Li^+ ion exchange. Afterward, the adsorbent is filtered out. A wash by a 0.1 M HCl solution removes the Li and re-protonates the powder at the same time, rendering a fresh adsorbent ready for use again. In spite of much higher concentration of Na (11,000 ppm), Mg (1,200 ppm) and other ions in seawater, this adsorbent only extracts Li.

First-principles calculations of the electronic structure of pure $LiMn_2O_4$ spinel (space group $Fa3m$) were performed, followed by molecular-dynamics simulations of the phonon DOS, which agrees well with the INS data (Loong and Koyanaka 2005; Loong et al. 2006). The calculation was then extended to investigate the structural configuration of $(H_xLi_{1-x})Mn_2O_4$, namely, position of the protons substituting the Li. In $LiMn_2O_4$ spinel, the basic framework structure is formed by an array of MnO_6 octahedra and Li atoms occupying the 8a sites are coordinated by 4 O neighbors to form tetrahedral units. As Li atoms are replaced by protons, the underlying MnO_6 framework remains intact but the protons relax their positions from the Li site and move toward one of the nearby oxygen atoms, forming OH clusters in the lattice. As a result, the protons exhibit neither long-range ordering nor a local symmetry as the Li atoms do ($43m$) in $LiMn_2O_4$. This is in agreement with neutron-diffraction result. Figure 14 displays the INS spectrum of a protonated Li-spinel. The observed O-H librational and stretch-vibration frequencies are 121 and 416 meV, which are corroborated by the result (108 and 410 meV, respectively) from *ab initio* calculations.

Hydroxyl ions in nanostructured bone apatite crystals Bone is a nanocomposite containing principally collagen and mineral apatite. As compared to synthetic hydroxyapatite (HAp with chemical formula $Ca_{10}(PO_4)_6(OH)_2$) which normally has micrometer-size crystallites, bone apatite consists of platelet-shaped nanocrystals exhibiting rough surfaces, calcium deficiency, and possibly ion substitutions by CO_3 and HPO_4 groups. The crystal chemistry of bone apatite, such as the role of the bioactive ions presumably located on the surface or on modified mobile lattice sites, is important to biological functions which include bone remodeling and homeostasis. Due to the nanoscale dimensions and poor crystallinity, the content and location of

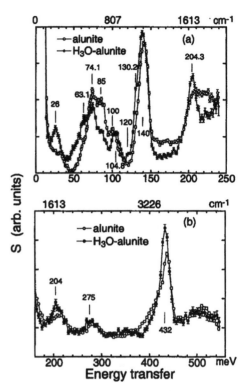

Figure 13. Comparison of the observed low- (a) and high-energy (b) INS spectra for alunite and H_3O-alunite at 20 K. Frequencies of major vibrational bands are labeled in units of meV. [Reprinted with permission of the Mineralogical Association of Canada, from Lager et al. (2001).]

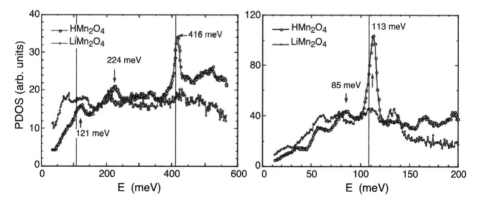

Figure 14. The observed INS spectra of $LiMn_2O_4$ and protonated spinel HMn_2O_4. The librational and stretch-vibrational frequencies of the OH species are 121 and 416 meV. The 224 meV peak is an overtone of the librational band and the origin of the small peak at 85 meV is unknown. The vertical lines denote the corresponding frequencies from *ab initio* calculations.

the hydroxyl ions in bone apatite have not been firmly established despite numerous diffraction and spectroscopic studies and the subject is still controversial today. Here, INS is advantageous due to the keen sensitivity to hydrogen content and mobility through the Q-dependence of the intensity. INS measurements find, as shown in Figure 15, no sharp excitations characteristic of the librational mode and stretch vibrations of OH ions around 80 and 450 meV (645 and 3630 cm^{-1}), respectively, in the bovine and rat bone apatites, whereas such salient features are clearly seen in micron- and nanometer-size crystals of pure hydroxyapatite powders (Loong et al. 2000). Thus the data provide additional definitive evidence for the lack of well-localized OH$^-$ ions in the crystals of bone apatite. Weak features at 160-180 and 376 meV, which are observed in the apatite crystals of rat bone and possibly also in adult mature bovine bone, but not in the synthetic hydroxyapatite, are assigned to the deformation and stretch modes of OH ions belonging to HPO$_4^-$ species.

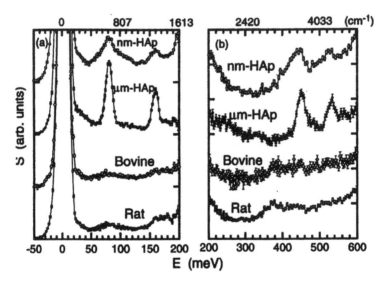

Figure 15. The low-energy (a) and high-energy (b) portion of the observed neutron spectra of micron- and nanometer-size HAp crystalline powders, and the bovine and rat bone crystals at 10 K. For clarity, the HAp and bovine bone spectra are shifted vertically by a constant interval [Reprinted from Loong et al. (2000) with permission from Elsevier.]

Quasielastic scattering

The aforementioned examples concern atoms on lattice sites or electrons in an orbital exhibiting a quantized energy level structure within which energy exchange with the neutrons results in sharp features at discrete energies in the INS spectrum corresponding to excitation or de-excitation processes. These spectral peaks in general occur only at certain wavevectors **Q** which reflects the spatial dependency of the excitations such as that expressed in the phonon structure factor in Equation (5). Therefore, **Q** in the Fourier space is analogous of the scope of a microscopic lens in the real space—a small Q corresponds to a small magnification power but over a large scope and, vice versa, a large Q covers a small scope with high resolution (see Fig. 1).

Mobile atoms or molecules in solids or soft matter undergo stochastic motions constantly, obeying thermodynamic laws characterized by macroscopic quantities such as temperature, pressure and volume. Microscopically, an atom may oscillate rapidly around a local site and then occasionally jumps to another nearby site, or a molecule may tumble about a bond and then flops over to associate with another bond. When neutrons are scattered by these atoms or molecules, no sharp inelastic feature appears but a diffuse profile of intensities very near the elastic ($E = 0$) region. This is called *quasielastic* scattering (QENS). Incoherent quasielastic neutron scattering can provide useful information regarding the microscopic diffusive motion of an atom or molecule in the following way. Imagine that we monitor the QENS intensity with a detector at certain Q, or conceptually equivalent to watching the atoms within a specific scope of a lens, a fast diffusing atom will wander outside the scope sooner than a slow moving atom does, which implies lower QENS intensity for the slow system and verse versa. Therefore, by careful measurements of the incoherent QENS intensities over a wide range of Q, we can in principle derive knowledge regarding the nature of the diffusion processes. If an atom migrates on the time average following a specific direction within the system, say, hydrogen jumping from site to site along the body-diagonal direction of a bcc lattice, then the QENS intensity will vary in a characteristic manner along the corresponding **Q** along the [111] direction.

The timescale of typical stochastic motions of macromolecules is of the order of ~1 ns or longer. The resolution of a quasielastic scattering spectrometer has to provide very high and well-defined energy resolution in the elastic region in order to measure the quasielastic component with minimum overlap with the elastic peak. This can be achieved by cold-neutron TOF chopper spectrometers, crystal-analyzer spectrometers at back-scattering geometry, and spin-echo spectrometers. The highest resolution attained to date is the spin-echo technique (Richter 2000). It makes novel use of the neutron intrinsic spin, which will precess in a magnetic field at a so-called Larmor frequency that is inversely proportional to the neutron speed. Thus by careful measurements of the fractional change of the Larmor precession angle over a flight distance before and after the scattering event, sensitivity of the order of 1 ppm of the neutron speed can be achieved. Furthermore, the data obtained are readily expressed as the *intermediate scattering function*, $I(\mathbf{Q},t)$, which relates to the scattering function $S(\mathbf{Q},E)$ through a time-energy Fourier transform. The intermediate scattering function in the time domain provides a more intuitive interpretation of the relaxation or diffusion processes.

One of the simplest models for diffusive motion is random, isotropic translational diffusion governed by the Fick's Law, which gives rise to an intermediate scattering function (Bée 1988)

$$I(Q,t) = e^{-DQ^2 t} \tag{11}$$

where D is the diffusion coefficient. A Fourier transform of $I(Q,t)$ yields

$$S(Q,E) = \frac{1}{\pi} \frac{D\hbar Q^2}{\left(D\hbar Q^2\right)^2 + E^2} \tag{12}$$

Therefore, we anticipate that in general, for each Q, results from spin-echo neutron spectrometry render a decay-like intermediate function in time and TOF QENS data a Lorentzian-like shape in energy. Different relaxation or diffusion mechanisms usually reveal themselves only with subtle differences in both scattering functions of QENS. As a result, quantitative analysis of QENS data often relies on sophisticated computer simulations or theoretical modeling.

Diffusion of water molecules in clay minerals Clay minerals, being common and abundant constituents of soil, are of importance to chemical and geological science and industrial applications, from water filtration and nutrient delivery in soil to clay-catalyzed RNA polymerization to isolation barrier for underground storage of nuclear wastes. Aluminosilicate clay crystals such as montmorillonite exhibit a layered structure comprising an Al-containing octahedron layer sandwiched between two tetrahedral sheets of silicate. The formation of negative charges on the layers occurs often due to substitution of other divalent or trivalent cations for Al^{3+} or Si^{4+}, respectively, and counter cations such as Ca^{2+}, Na^{2+} and Mg^{2+} fill the interlayer space and balance the charge on the layers. When water is absorbed into the space between the stacked layers, the volume can swell up to several hundred percent. Clearly, an understanding of water diffusion in clay is paramount.

Figure 16 shows the measured scattering functions in time and energy domain obtained from spin-echo and TOF measurements on a sodium montmorillonite at ambient temperature (Malikova 2005). As expected, the data do not fit the simple diffusion model expressed by Equations (11) and (12) because there exist multiple OH species that diffuse via different relaxation and translation-rotation characteristics. Instead, the data were fitted to an $I(Q,t)$ called the stretched potential function which assumes a distribution of local diffusivities D in the sample: (Colmenero et al. 1999)

$$I(Q,t) = \int_{-\infty}^{\infty} g(\ln D^{-1}) e^{-Q^2 t / D^{-1}} d(\ln D^{-1}) = e^{-(t/\tau)^{\beta}} \tag{13}$$

where β the stretched potential exponent, and τ the relaxation time, are the fit parameters. The corresponding Fourier-transformed $S(Q,E)$ does not have an analytic form and was computed numerically for the fitting. Figure 16 shows examples of the fits at typical Q values. After extracting τ as a function of Q, the diffusion coefficient can be estimated from the slope of plotting $(1/\tau)$ versus Q^2 at $Q \rightarrow 0$ because diffusion is defined as a phenomenon over long distances (Malikova et al. 2005).

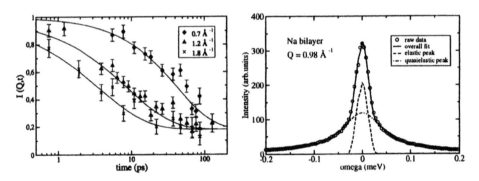

Figure 16. Examples of neutron TOF (left) and spin-echo (right) data (symbols) for a sodium montmorillonite at ambient temperature as compared with fitting the data to the stretched potential model (lines). [Used with permission of author from Malikova (2005).]

CONCLUSIONS

We have introduced the fundamental concepts of inelastic scattering of neutrons from a sample whereby the observed scattered intensity as a function of energy and wavevector yields important information regarding the spatial-temporal interactions of the atoms or electrons in the system. The practice of INS involves consideration of neutron sources—reactor-based or accelerator-based facilities, neutron spectrometers and their specialization, and data interpretation. In this short chapter, we focus the discussion of several instruments at reactor and spallation sources on mainly their basic operation and their capabilities and complementarities in the characterization of different problems. Examples are given to illustrate typical experimental data collected by a variety of INS instruments and their comparison with the expected scattering functions. In closing, the author adverts to the unprecedented investment in early 21st century in building and upgrading state-of-the-art neutron sources in the United States (e.g., the SNS at Oak Ridge National Laboratory), Japan (the Neutron Arena of the J-PARC Project), and the United Kingdom (the second target station of ISIS at Rutherford Appleton Laboratory). These high-flux sources will alleviate the low count-rate problem of INS and enable better data statistics, more systematic measurements, and utilization of smaller specimens. Furthermore, activities for users in the research and industrial communities to collaborate with neutron scientists and engineers on the R&D of advanced neutron spectroscopic techniques and data visualization/interpretation schemes will expand to uncharted territories (see Fig. 7) to meet new scientific demands. We look forward to the excitements of new discoveries by INS in coming years.

ACKNOWLEDGMENT

Work performed at Argonne National Laboratory is supported by the U. S. DOE-Basic Energy Science under the contract No. W-31-109-ENG-38.

REFERENCES

Becker PC, Williams GM, Edelstein NM, Koningstein JA, Boatner LA, Abraham MM (1992) Observation of strong electron-phonon coupling effects in $YbPO_4$. Phys Rev B 45:5027-5030

Bée M (1988) Quasielastic Neutron Scattering. Adam Hilger,

Boatner LA (2002) Synthesis, structure, and properties of monazite, pretulite, and xenotime. Rev Mineral Geochem 48:87-121

Boue F, Cywinski R, Furrer A, Glattli H, Kilcoyne S, McGreevy RL, McMorrow D, Myles D, Ott H, Rübhausen M, Weill G (2002) Neutron scattering and complementary experimental techniques. *In*: The ESS Project: New Science and Technology for the 21st Century. Vol. 2. The European Spallation Source Project, p 5-4

Bredov MM, Kotov BA, Okuneva NM, Oskotskii VS, Shakh-Budagov AL (1967) Possibility of measuring the thermal vibration spectrum g(ω) using coherent inelastic neutron scattering from a polycrystalline sample. Sov Phys Solid State 9:214-218

Burkel E (2000) Phonon spectroscopy by inelastic X-ray scattering. Rep Prog Phys 63:171-232

Chaplot SL, Choudhury N, Ghose S, Rao MN, Mittal R, Goel P (2002) Inelastic neutron scattering and lattice dynamics of minerals. Eur J Mineral 14: 291-329

Colmenero J, Arbe A, Alegria A, Monkenbusch M, Richter D (1999) On the origin of the non-exponential behavior of the a-relaxation in glass-forming polymer: incoherent neutron scattering and dielectric relaxation results. J Phys Condens Matter 11:A363-A370

Connatser RW Jr, Belch H, Jirik L, Leach DJ, Trouw FR, Zanotti J-M, Ren Y, Crawford RK, Carpenter JM, Price DL, Loong C-K, Hodges JP, Herwig KW (2003) The QuasiElastic Neutron Spectrometer (QENS) of IPNS: Recent upgrade and performance. Forschungszentrum Jülich GmbH, Jülich, Germany I: 279-288

Eckold G, Sten-Arsic M, Weber HJ (1987) UNISOFT - a program package for lattice-dynamical calculations. J Appl Crystallogr 20:134-139

Hawthorne FC (ed) (1997) Spectroscopic Methods in Mineralogy and Geology. Reviews in Mineralogy Vol. 18. Mineralogical Society of America

Kieffer SW, Navrotsky A (eds) (1985) Microscopic to Macroscopic Atomic Environments to Mineral Thermodynamics. Reviews in Mineralogy Vol. 14. Mineralogical Society of America

Koyanaka H, Koyanaka Y, Numata Y, Wakamatsu T (1997) Collection of lithium from sea water by an adsorption plate method. J Mining Mater Process Inst Japan 113:275-279

Lager GA, Nipko JC, Loong C-K (1998) Inelastic neutron scattering of the (O_4H_4) substitution in garnet. Physica B 241-243:406-408

Lager GA, Swayze GA, Loong C-K, Rotella FJ, Richardson JW Jr, Stoffregen RE (2001) Neutron spectroscopic study of synthetic alunite and oxonium-substituted alunite. Can Mineral 39:1131-1138

Loong C-K, Ikeda S, Carpenter JM (1987) The resolution function of a pulsed-source neutron chopper spectrometer. Nucl Instrum Methods A 260:381-402

Loong C-K, Kolesnikov AI, Koyanaka H, Takeuchi K, Fang C (2006) Extraction of metals from natural waters: a neutron characterization of the nanostructured manganese-oxide-based adsorbents. Physica B: in press

Loong C-K, Koyanaka H (2005) Harvesting precious metals and removing contaminants from natural waters - Can neutrons benefit industrial researchers in Japan? J Neut Res 13:15-19

Loong C-K, Loewenhaupt M, Nipko JC, Braden M, Boatner LA (1999) Dynamic coupling of crystal-field and phonon states in $YbPO_4$. Phys Rev B 60:R12549-R12552

Loong C-K, Rey C, Kuhn LT, Combes C, Wu Y, Chen S-H, Glimcher MJ (2000) Evidence of hydroxyl-ion deficiency in bone apatites: an inelastic neutron scattering study. Bone 26:599-602

Loong C-K, Soderholm L (1994) Rare earth crystal field spectroscopy by neutron magnetic scattering: from xenotime to High-Tc superconductors. J Alloys Compd 207-208:153-160

Loong C-K, Soderholm L, Hammonds JP, Abraham MM, Boatner LA, Edelstein NM (1993) Rare-earth energy levels and magnetic properties of $HoPO_4$ and $ErPO_4$. J Phys Condens Matter 5:5121-5140

Lovesey SW (1984) Theory of Neutron Scattering from Condensed Matter. Clarendon Press

Lovesey SW (1986) Condensed Matter Physics, Dynamic Correlations. Benjamin/Cummings Publishing

Malikova N (2005) Dynamique de l'eau et des ions dans des argiles de type montmorillonite par simulation microscopique et diffusion quasi-élastique des neutrons. PhD Thesis, Université Pierre et Marie Curie, Paris, France

Malikova N, Cadéne A, Marry V, Dubois E, Turq P, Zanotti J-M, Longeville S (2005) Diffusion of water in clays – microscopic simulation and neutron scattering. Chem Phys 317:226-235

Nipko J, Grimsditch M, Loong C-K, Kern S, Abraham MM, Boatner LA (1996) Elastic constant anomalies in $YbPO_4$. Phys Rev B 53:2286-2290

Nipko JC, Loong C-K, Loewenhaupt M, Braden M, Reichart W, Boatner LA (1997) Lattice dynamics of xenotime: the phonon dispersion relations and density of states of $LuPO_4$. Phys Rev B 56:11584-11592

Oskotskii VS (1967) Measurement of the phonon distribution function in polycrystalline materials using coherent scattering of slow neutrons into a solid angle. Sov Phys Solid State 9:420-422

Price DL, Sköld K (1986) Introduction to neutron scattering. *In*: Neutron Scattering. Part A. Sköld K, Price DL (ed)s Academic Press, p 1-97

Richter D (2000) Polymer dynamics by neutron spin echo spectroscopy. *In*: Scattering in Polymeric and Colloidal Systems. Brown W, Mortensen K (eds) Gordon and Breach, p 535-574

Shirane G, Shapiro SM, Tranquada JM (2002) Neutron Scattering with a Triple-Axis Spectrometer: Basic Techniques. Cambridge University Press

Sinha SK (1985) Introduction to electronic excitations. *In*: Proceedings of the 1984 Workshop on High-Energy Excitations in Condensed Matter. Vol. 2. Silver RN (ed) LANL, LA-10227-C, Los Alamos, New Mexico, p 346-357

Squires GL (1978) Introduction to the Theory of Thermal Neutron Scattering. Cambridge University Press

Taraskin SN, Eilliot SR (1997a) Connection between the true vibrational density of states and the derived from inelastic neutron scattering. Phys Rev B 55:117-123

Taraskin SN, Eilliot SR (1997b) How to calculate the true vibrational density of states from inelastic neutron scattering. Physica B 234-236:452-454

Warren L, Worlton TG (1972) Group-theoretical analysis of lattice vibrations. Comp Phys Commun 3:88-117

Warren L, Worlton TG (1974) Improved version of group-theoretical analysis of lattice dynamics. Comp Phys Commun 8:71-84

Windsor CG (1988) Neutron scattering instrumentation. *In*: Neutron Scattering at a Pulsed Source. Newport RJ, Rainford BD, Cywinski R (eds) Adam Hilger, p 111-126

Reviews in Mineralogy & Geochemistry
Vol. 63, pp. 255-274, 2006
Copyright © Mineralogical Society of America

11

Analysis of Disordered Materials Using Total Scattering and the Atomic Pair Distribution Function

Thomas Proffen

Lujan Neutron Scattering Center
Los Alamos National Laboratory
Los Alamos, New Mexico, 87545, U.S.A.
LA-UR 06-0622
e-mail: tproffen@lanl.gov

INTRODUCTION

Without a doubt, our ability to determine the atomic structure of complex materials has revolutionized our understanding of these materials over the last decades. Neutron scattering with its unique abilities has contributed greatly to this success. The range of applications of neutron scattering in geosciences are apparent when reading through this volume of Reviews in Mineralogy & Geochemistry. Usual structure determination, the realm of crystallography, is based on the analysis of Bragg intensities. In this article we want to look beyond or underneath the Bragg peaks and discuss the atomic pair distribution function (PDF) method as an approach to more fully understand the atomic structure of complex materials, from the local to the medium range all the way to the long range structure of the material.

As mentioned above, structure determination is commonly based on the analysis of Bragg peaks either using single crystal diffraction (Ross and Hoffman 2006, this volume) or powder diffraction (Harrison 2006, this volume). However, one needs to keep in mind, that the structure resulting from Bragg scattering is only the *long range average structure* of the material. Many materials owe their interesting properties to defects within the material and obviously a complete structural picture and understanding of the properties will require knowledge of the "true" atomic structure. Geological samples are no exception and in fact are among the most complex systems known. Deviations from the average structure, e.g., in form of defects, manifest themselves as diffuse scattering. This is illustrated in an example shown in Figure 1. Two 2D structures have been simulated; both have the same lattice parameters, contain one atomic site in the unit cell as well as contain a total of 30% vacant sites. The difference is the vacancy ordering. Figure 1a shows the structure with a random distribution of vacancies. The structure shown in Figure 1b shows chemical short range order (SRO), in this case a preferred ordering along the x- and y-direction. Figures 1c and 1d show the calculated single crystal scattering pattern for both structures. If one would integrate the Bragg intensities for both cases, the result would be identical as one can verify by inspecting the powder diffraction pattern shown in Figure 1e. This is not all surprising, since the average structure is given by the lattice parameters, the atomic site positions within the unit cell as well as its occupancy. All these parameters are the same and contain no information about the chemical SRO. The scattering between the Bragg peaks (Fig. 1c,d) on the other hand is quite different. In the case of a random vacancy distribution, one observes a flat background (it is not zero, however), whereas in the case of vacancy SRO, streaks of diffuse scattering along the h and k-directions can be seen. Analysis of this diffuse scattering pattern would allow one to construct a model of the chemical SRO. The analysis of single crystal diffuse scattering is beyond the scope of this paper and the reader might refer to these books (Nield and Keen 2001; Welberry 2004). Obviously, in powder diffraction directional information is lost, however, in many cases single

1529-6466/06/0063-0011$05.00

DOI: 10.2138/rmg.2006.63.11

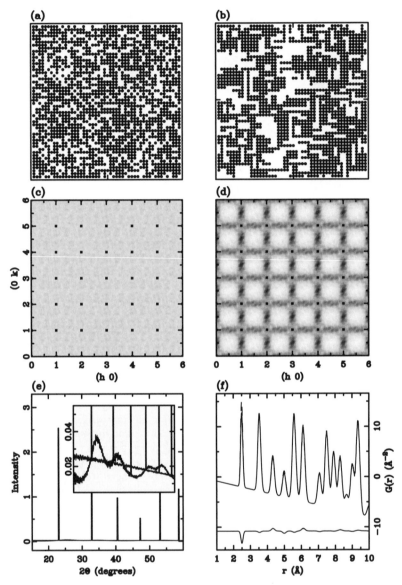

Figure 1. Panel (a) shows a simulated structure with a random distribution of vacancies and (b) a structure with preferred ordering along the *x*- and *y*-direction. (c) and (d) show the corresponding neutron scattering patterns, (e) the calculated powder diffraction pattern for both structures with the insert showing an enlarged section of the background. (f) shows the PDFs calculated for both structures. The difference is shown below.

crystals are not available. Figure 1e shows the powder diffraction pattern corresponding to the two example structures. The Bragg scattering is identical, but the diffuse scattering shows up as modulation in the powder pattern as shown in the insert. Finally we make the connection to the PDF. The PDF is basically obtained via a Fourier transform of the total (Bragg + diffuse) scattering pattern. The resulting PDFs for our example are shown in Figure 1f. The difference curve drawn below the PDFs shows clearly that the chemical SRO changes the PDF and this information can in fact be extracted from the PDF (Proffen 2000; Proffen et al. 2002).

From this very simple example, it is clear that one needs to analyze the total scattering pattern to obtain a realistic picture of the atomic structure of a material. In general diffuse scattering contains information about two-body correlations in the structure including chemical SRO, distortions, correlated motion or orientation SRO of molecules to name a few examples. In the next sections, we will discuss the details of the PDF approach from measurements to modeling as well as present a few examples selected from our own work. An in depth discussion of the PDF method, its formalism and many more applications can be found in a recent book (Egami and Billinge 2003).

THE ATOMIC PAIR DISTRIBUTION FUNCTION (PDF)

What is a PDF?

The PDF gives the probability of finding an atom at a given distance r from another atom. In other words it can be understood as a weighted bond length distribution. Looking back at Figure 1f, it is immediately apparent that the nearest neighbor distance in our example structure is 2.5 Å. In fact the PDF method originated from the study of liquids and glasses (see Wilding and Benmore 2006, this volume; and Cole et al. 2006, this volume) which have no long range order at all. The strength of the PDF approach is the fact that it can be applied to a wide range of materials from amorphous materials to crystals. This is illustrated in Figure 2 showing the experimental PDFs for three samples: (a) silica glass rod, (b) crystalline quartz powder and (c) gold nano-particles with an average diameter of ~3.5 nm. In all cases the PDF is shown out to atom-atom distances of 40 Å. One can see immediately that the glass has no long range order, since there are no PDF peaks at larger atom-atom distances. However, the PDF shows two sharp clearly defined nearest neighbor peaks corresponding to Si-O and O-O nearest neighbors. In addition some broader medium range order peaks can be observed before the PDF goes to zero. The quartz on the other hand, being a crystal, shows PDF peaks over the full range shown and in fact the peaks persist to even higher distances until dampened by the resolution of the instrument as we discuss in the next section. The insert in Figure 2a shows the near neighbor region of the PDF for the silica glass as well as the quartz powder, showing a remarkable agreement. This again is no surprise, since the basic building block, the SiO_4 tetrahedron is the same in both samples and just their medium and long range arrangement is different. One of the examples discussed later will give more details about the mysteries of SiO_2. The third example (Fig. 2c) is the PDF of gold nanoparticles. These particles have an average diameter of ~35 Å and no atom-atom correlations should exist for distances larger than the diameter. In fact the number of pairs will start decreasing for smaller distances resulting in a dampening of the PDF peaks as function of r. This can be seen in Figure 2c and we will discuss these nanoparticles in more detail in one of the examples given later.

MEASURING A PDF

The first step in a PDF measurement is to obtain the *total scattering structure function*, $S(Q)$, from the measured intensities using

$$S(Q) = \frac{1}{\langle b \rangle^2} \left(\frac{d\sigma_c}{d\Omega} \right) + \left(\frac{\langle b \rangle^2 - \langle b^2 \rangle}{\langle b \rangle^2} \right) \tag{1}$$

Here b is the scattering length, the angle brackets denote the average over the sample and $d\sigma_c/d\Omega$ is the coherent single scattering cross section of the sample which is related to the observed and normalized intensity (Egami and Billinge 2003). We will skip further details on the derivation of the structure function and move on to the calculation of the PDF itself. The

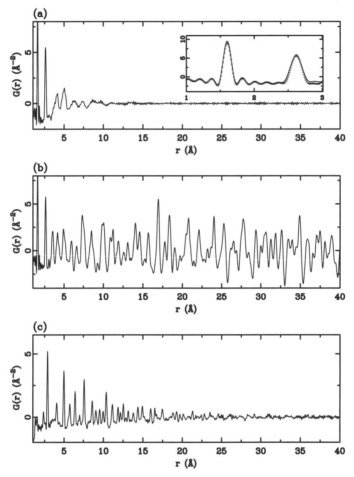

Figure 2. Panels (a) and (b) show the experimental PDF of a fused silica rod (glass) and quartz (crystalline), respectively. For comparison, the near neighbor region of both PDFs is shown in the insert on the top panel. Panel (c) shows the experimental PDF from gold nanoparticles with an average diameter of ~35 Å. All data were collected on the NPDF instrument and the diffraction data were terminated at $Q_{max} = 35$ Å^{-1}.

PDF, $G(r)$, is obtained via a Fourier transform of the structure function $S(Q)$

$$G(r) = 4\pi r \left[\rho(r) - \rho_0\right] = \frac{2}{\pi} \int_0^\infty Q\left[S(Q) - 1\right]\sin(Qr)dQ \qquad (2)$$

here $\rho(r)$ is the microscopic pair density, ρ_0 is the average number density and Q is the magnitude of the scattering vector. For elastic scattering $Q = 4\pi\sin(\theta)/\lambda$ with 2θ being the scattering angle and λ the wavelength of the radiation used. In the following sections, we will refer frequently to the "experimental PDF" and one should keep in mind that the PDF, $G(r)$, is in fact a quantity extracted from the experimental scattering data. Details about this step are discussed a little later in this section. So far we have talked about the total PDF, where all atoms in the sample contribute. By changing the scattering power of one or more elements it is possible to extract a chemically specific PDF. In a binary alloy two measurements contrasting atom B would yield the differential PDF containing contributions from A-A and A-B pairs. Using three independent measurements, all partial PDFs A-A, A-B and B-B can be extracted.

For X-ray scattering one utilizes the fact that close to the absorption edge of a specific element the scattering power changes due to anomalous scattering. An example of the chemically specific PDF of the semiconductor alloy $In_{0.5}Ga_{0.5}As$ is given in (Petkov et al. 2000). In case of neutrons one can use the fact that different isotopes of an element have different neutron scattering length. Isotope substitution studies are applied to glasses and liquids but are also used to study disordered crystals, e.g., in a study by (Louca et al. 1999) on high temperature superconducting cuprates.

Data processing. At first glance, a powder diffraction experiment carried out to obtain a PDF is not any different from an experiment designed to yield data for Rietveld refinements. In fact some programs such as PDFgetN (Peterson et al. 2000) use the same data input file format as the Rietveld refinement program GSAS (see Knorr and Depmeier 2006, this volume). In many cases one wants to carry out Rietveld refinements to obtain the average structure of the sample as well as refinements of the PDF obtaining information about disorder in the material. However, there are some differences relating to the experiment. Since data are usually not refined out to very high values of Q in the case of Rietveld analysis, one needs to take care to obtain sufficient statistics at large values of Q required for PDF analysis. In addition to the sample measurement itself, the following characterization runs are needed to obtain a PDF: a vanadium or null scatterer measurement to characterize the incident neutron spectrum as well as a background and empty sample container measurement to account for the parasitic scattering not related to the sample. These data files as well as information about the sample and geometry of the experiment are used to generate the function $S(Q)$ as well as the desired PDF, $G(r)$ (see Fig. 3). Neutrons are used for quite some time for PDF studies of glasses and liquids and the two main data processing packages are the GLASS package (Price undated) and the ATLAS package (Soper et al. 1989). These packages are also used to process data obtained from crystalline materials. PDFgetN is a graphical user interface to the GLASS package. It has been developed to simplify data processing and make PDF analysis available beyond the circle of specialists. It can be difficult, especially for the non-specialist, to judge the quality of a resulting PDF or to maintain conditions when processing a large number of PDFs which were collected e.g., as a function of temperature. The program PDFgetN also offers a set of quality criteria to aid in these questions (Peterson et al. 2003). In fact for well behaved samples, the PDF processing can now be done completely automatic using PDFgetN. Other program packages to process PDF from neutron diffraction data are GUDRUN and total scattering processing has been added to the ISAW package. Many times a certain package will be the default data processing choice for a particular neutron diffractometer. For a list of programs and download locations, refer to Table 1.

Selecting an instrument. Next we want to discuss the question how to select a suitable instrument for a certain experiment. As mentioned earlier, the PDF technique can be used to study a variety of samples from liquids and glasses to disordered crystals and nano materials. The trouble with obtaining the PDF using a Fourier transform is that of course we are not

Table 1. List of data processing and modeling software mentioned in this article.

Program	URL
PDFgetN	*http://pdfgetn.sourceforge.net/*
ISAW	*http://www.pns.anl.gov/computing/isaw/*
GUDRUN	*http://www.isis.rl.ac.uk/disordered/Manuals/gudrun/gudrun_GEM.htm*
DISCUS PDFFIT	*http://discus.sourceforge.net/*

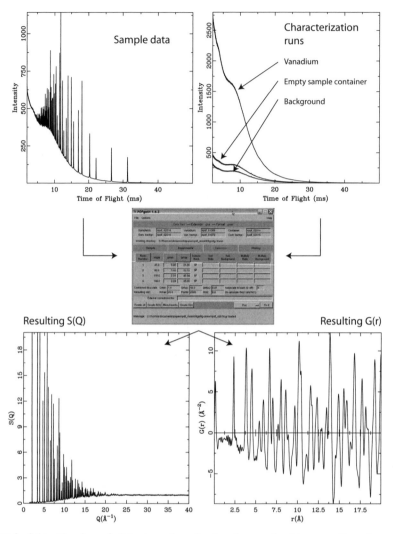

Figure 3. Schematic view of data processing showing the sample data as well as the required characterization runs in the top row. This input is used to generate the final $S(Q)$ and $G(r)$ shown in the bottom row. In this example, the data of the backscattering bank of silicon powder are shown.

able to measure to infinite momentum transfer Q. We also have to take into account the finite resolution of the measurement. First, the cutoff at finite Q decreases the real-space resolution of the PDF and causes so-called termination ripples. This effect is illustrated in Figure 4c. Here we see the nearest neighbor peak of the semi-conductor alloy $ZnSe_{0.5}Te_{0.5}$. The interesting feature in these alloys is the difference in Zn-Se and Zn-Te nearest neighbor bond lengths. The PDF obtained using the data out to $Q_{max} = 40$ Å^{-1} clearly shows the two distinct bond lengths. The same data, however, terminated at $Q_{max} = 17$ Å^{-1} show only one broad peak shown with the cross symbols in Figure 4c. The required real-space resolution of the PDF will determine the value of Q_{max} that is required for the experiment.

Next we will investigate the role of the resolution of the diffraction experiment. In Figures 4a and 4b the PDF of nickel powder measured on two different neutron powder

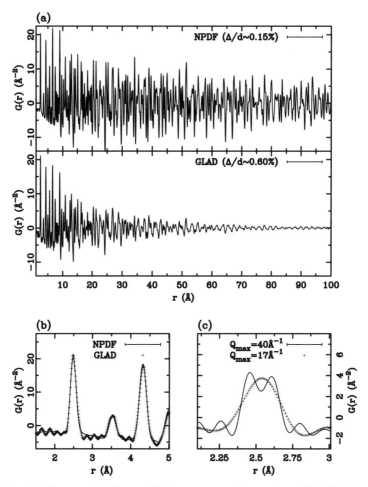

Figure 4. Panel (a) illustrates the influence of the instrument resolution on the resulting PDF, here of nickel powder. The PDF shown on top was taken on the high resolution instrument NPDF at the Lujan Neutron Scattering Center and the PDF below was taken on the lower resolution instrument GLAD at the Intense Pulse Neutron Source. Both PDFs were terminated at $Q_{max} = 35$ Å^{-1}. Panel (b) show a comparison of both data sets in the near neighbor region. Panel (c) shows the PDF of the near neighbor peaks of ZnSe$_{0.5}$Te$_{0.5}$ for different values of Q_{max} (details see text).

diffractometers is shown. The top panel shows the data taken on the high resolution instrument NPDF (Fig. 5) at the Lujan Neutron Scattering Center at Los Alamos National Laboratory and the bottom panel shows data taken on the lower resolution instrument GLAD at the Intense Pulsed Neutron Source at Argonne National Laboratory. GLAD has been designed for measurements of glasses and liquids which do not require high resolution. Note that the PDFs are plotted up to atom-atom distances of $r_{max} = 100$ Å. The Q resolution causes an exponential dampening of the PDF peaks (Toby and Egami 1992) as function of distance r. As a result, the PDF obtained from the high resolution instrument NPDF allows one to extract atom-atom correlations out to distances in excess of 200 Å (Chung et al. 2005) in contrast to lower resolution measurements where no structural information beyond distances of 50 Å or so can be extracted. The near neighbor region of the PDF is not affected by the instrument resolution as can be seen in Figure 4b comparing the PDFs from NPDF and GLAD.

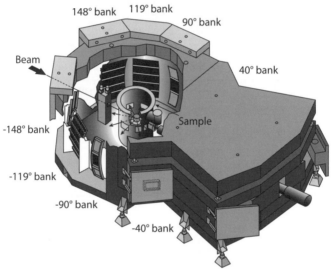

Figure 5. The top panel shows the high resolution Neutron Powder DiFfractometer NPDF located at flightpath 1 at the Lujan Neutron Scattering Center at Los Alamos National Laboratory. The bottom panel shows a schematic view of the instrument showing the sample chamber and the detector panels.

Having discussed the principles of a total scattering experiment, we now move to some of the more practical aspects. First a high value of Q_{max} requires the use of high energy X-rays from a synchrotron source or high energy neutrons as can be found at a spallation neutron source (Vogel and Priesmeyer 2006, this volume). By considering the real space resolution of $G(r)$ that is needed for a particular problem, the required value of Q_{max} can be estimated. Since PDF peaks have a minimum width determined by thermal vibrations, and a value of Q_{max} > 50 Å^{-1} will usually not add any structural information. Considering the desired maximum value r_{max} of the PDF allows one to determine the required resolution of the instrument. As we have seen earlier, if only the near neighbor region is of interest, a lower resolution but higher intensity instrument should be chosen. In cases where one wants to study extended defects such as nano-domains, a high resolution instrument is needed. The ability to extract structural information over such a large length scale using high resolution instruments such as NPDF opens a new territory for the total scattering or PDF approach, e.g., in the study of materials with inhomogeneities on the nanometer scale or nanomaterials to name just a few examples.

A list of existing neutron diffractometers suitable for total scattering analysis is given in Table 2. In addition, two new instruments for total scattering studies, NIMROD and NOMAD, are currently designed at the new target station at ISIS and the SNS at Oak Ridge National Laboratory, respectively.

Table 2. List of existing neutron diffractometers at selected spallation neutron sources. The information listed was taken from the respective instrument web sites.

Instrument	Resolution $\Delta d/d$	Q range
Lujan Neutron Scattering Center *http://www.lansce.lanl.gov/*		
NPDF	0.15 – 0.8 %	0.85 – 51.1 Å^{-1}
HIPPO	0.40 – 5.0 %	0.13 – 52.4 Å^{-1}
HIPD	0.30 – 3.0 %	0.20 – 60.0 Å^{-1}
Intense Pulsed Neutron Source *http://www.pns.anl.gov/*		
GPPD	0.26 – 0.8 %	1.10 – 60.4 Å^{-1}
SEPD	0.34 – 4.7 %	0.60 – 47.3 Å^{-1}
GLAD	0.60 – 5.0 %	0.20 – 45.0 Å^{-1}
ISIS *http://www.isis.rl.ac.uk/*		
GEM	0.34 – 4.7 %	0.60 – 100 Å^{-1}
POLARIS	0.50 – 1.0 %	0.30 – 31.5 Å^{-1}
SANDALS	2.0 – 3.0%	0.05 – 50.0 Å^{-1}

Modeling a PDF

After obtaining an experimental PDF from neutron or X-ray diffraction data, the question arises how to analyze it. As we have seen in the previous sections, a lot of information can be extracted by just "looking" at the PDF itself. So, the simplest analysis is to obtain bond length information and coordination numbers by fitting the near neighbor peaks of the PDF or to study the evolution of a particular PDF peak as function of e.g., temperature. The r dependence of the PDF peak width contains information about correlated motion. In principle, motion of the atoms will influence the width of the observed PDF peaks. Imagine two atoms being strongly bonded, they will tend to move in phase, causing the corresponding PDF peak to sharpen. A simple analysis of the peak width can reveal information about correlated motion (Jeong et al. 2003). However, in most cases a comparison or refinement of a structural model will be desired. The PDF is simply the bond length distribution of the material weighted by the respective scattering powers of the contributing atoms. The PDF can be calculated from a structural model using:

$$G_c(r) = \frac{1}{r} \sum_{ij} \left[\frac{b_i b_j}{\langle b \rangle^2} \delta(r - r_{ij}) \right] - 4\pi r \rho_0 \qquad (3)$$

here the sum goes over all pairs of atoms i and j within the model crystal separated by r_{ij}. The scattering power of atom i is given by b_i, and $\langle$ b $\rangle$ is the average scattering power of the sample. To account for the limited range in Q, the calculated function $G(r)$ is then convoluted with a termination function: $S(r) = \sin(Q_{max}r)/r$. There is a number of techniques to refine an experimental PDF (Proffen and Page 2004). Using a model made up of relatively few atoms, full profile refinement of the PDF based on a structural model can be carried out using the program PDFFIT (Proffen and Billinge 1999). The program allows one to refine structural parameters such as lattice parameters, anisotropic atomic displacement parameters, position and site occupancies. Even though this is similar to the results of a Rietveld refinement, one needs to realize that the structural model obtained is strictly only valid for length scales corresponding to the *r*-range used for the refinement. This opens up the possibility to study structural parameters as a function of length scale by varying the *r*-range refined as we will see in more detail in the examples section.

In other cases a model structure made up of a large number of atoms (typically tens of thousands) is needed. This approach is usually used if a system shows disorder that cannot be captured in a few unit cells, e.g., chemical SRO. Once such a starting structure is generated, there are a number of techniques to refine the structural parameters to fit the experimental PDF. One approach is the so-called Reverse Monte Carlo technique (Tucker et al. 2001). The technique basically works as follows: First, the PDF is calculated from the chosen crystal starting configuration and a goodness-of-fit parameter χ^2 is computed as

$$\chi^2 = \sum_{i=1}^{N} \frac{\left[G_{\text{exp}}(r) - G_{\text{calc}(r)} \right]^2}{\sigma^2} \tag{4}$$

The sum is over all measured data points r_i, G_{exp} stands for the experimental and G_{calc} for the calculated PDF. The RMC simulation proceeds with the selection of a random site within the crystal. The system variables associated with this site, such as occupancy or displacement, are changed by a random amount, and then the model PDF and the goodness-of-fit parameter χ^2 are recalculated. The change $\Delta\chi^2$ of the goodness-of-fit χ^2 before and after the generated move is computed. Every move which improves the fit ($\Delta\chi^2 < 0$) is accepted. "Bad" moves worsening the agreement between the observed and calculated PDF are accepted with a probability $P = \exp(-\Delta\chi^2/2)$. Obviously in a large model structure, the number of parameters is much larger than the number of PDF peaks observed and one needs to take care to constrain the refinement in an appropriate fashion, such as applying geometric constraints of bond lengths and bond angles as well as refining the total scattering and the integrated Bragg intensities (Tucker et al. 2001). An alternative approach is to generate the model structure based on a few model or interaction parameters rather than relaxing individual atoms. The parameters chosen will be model dependent and reflect the physics and chemistry of the system. The refinement of these parameters can be done using a least-square algorithm (Welberry et al. 1998) or evolutionary algorithms (Weber 2005). We have just begun to explore extended models that can capture the details of a PDF over a wide range in *r* obtained from complex materials.

EXAMPLES

In following section we present a very subjective list of applications of the PDF technique. It should be stressed that there are a number of other studies and the application to glasses and liquids are discussed by Wilding and Benmore (2006, this volume) and Cole et al. (2006, this volume). The examples chosen are designed to give an overview of the types of problems that can be addressed using the PDF.

Strain in semiconductor alloys

The first example of the PDF technique is the study of local atomic strain in semiconductor alloys, $ZnSe_{1-x}Te_x$ (Peterson et al. 2001). The technological interest in these materials is given by the fact that the band gap can be tuned as function of the composition *x*. The so-called II-IV semiconductor alloy $ZnSe_{1-x}Te_x$ can be made over the entire range of compositions. Looking at the crystallographic structure, we find that both end members ZnTe and ZnSe have the zincblende structure (*Fm3m*) with Se/Te occupying (0,0,0) and Zn occupying (¼,¼,¼). The lattice parameter of the alloys interpolates linearly between the end member values as function of the composition x. Conventional structure analysis also results in nearest neighbor bond lengths interpolating between the respective end member values. However, X-ray absorption fine structure (XAFS) experiments (Boyce and Mikkelsen 1989) show that the atomic NN distances stay close to their natural lengths found in the end-member compounds: $L^0_{Zn-Te} = 2.643(2)$ Å and $L^0_{Zn-Se} = 2.452(2)$ Å. XAFS also yields local structural information, however in contrast to PDF, it is limited to the nearest and second nearest neighbor shells. Time of flight neutron powder diffraction data were measured on the GEM diffractometer at ISIS. The data were processed using PDFgetN. The reduced structure functions, $Q[S(Q) - 1]$, obtained from the ZnSe$_{1-x}$Te$_x$ samples are shown in Figure 6a and the corresponding PDFs are shown in Figure 6b. Some conclusions can be draw from simple inspection of the PDFs shown in Figure 6b. Focusing on the area between the vertical dotted lines labeled (L) in the figure. Starting from the bottom ($x = 0$) we see the ZnTe NN peak decrease in height and the SeTe NN peak increase as a function of composition. It can also be clearly seen that the corresponding bond lengths stay close to their respective end member values. The region labeled (L) reflects the local structure. The situation is different in the area between the vertical dotted lines labeled (A) in Figure 6b. Here the two peaks around $r \sim 7$ Å shift to larger distances as the value of x increases. Peaks are not appearing and disappearing as observed for the NN peaks. In this region the behavior corresponds to the one of the average structure. However, it is also clear that the PDF peaks broaden for intermediate compositions, which indicates large static displacements caused by strain in the alloys.

The NN peaks were characterized by profile fitting. The resulting NN bond length and their 2σ error bars are shown as filled circles in Figure 6c. The values determined by the XAFS study (Boyce and Mikkelsen 1989) are shown as empty circles in the same figure. There is excellent agreement between the results. Next the data were refined using a model consisting of a single unit cell. With occupancies of Se/Te set to their nominal values, only the average structural model could be refined. This can be seen in Figure 6d as dashed line. Clearly this model is not describing the local structure, since it only predicts a single NN peak whereas the data show a peak split. Next the model was modified in two ways, two sites were fully occupied by Te and two sites within the unit cell by Se. Additionally displacements along the <111> direction were allowed on all atomic sites. Only the magnitude of the displacement was refined. The result can be seen as solid line in Figure 6d, the PDF data are shown as + symbols. Clearly the distorted model shows the NN peak split observed in the data. The resulting distortion were $\xi_{Zn} = 0.06$ Å and $\xi_{Te,Se} = 0.03$ Å (Jeong et al. 2001; Peterson et al. 2001). This example has shown that simple inspection of the PDFs, here as a function of composition, allows one to draw conclusions about the local versus medium range structure. In this case we observe the NN peak split and a broadening of PDF peaks going out in distance r. This is an indication of strain present in the sample. More refined modeling allows to extract more details about the strain, true NN bond lengths as well as to answer the question, if there is any indication of chemical SRO in this sample.

Domain structures

In this example, we will explore refinements where the refinement range is varied. This allows one to e.g., explore structures containing domains. The PDF calculated from a simulated domain structure will be used as "experimental" data. In our case, the matrix consists of a cubic lattice of atoms M with a lattice parameter of $a = 3$ Å. The model contains spherical domains containing atoms D on the same lattice, coherently imbedded in the matrix structure. The diameter of the domains is $d = 15$ Å. The overall composition of the model structure is 85% M and 15% D atoms. A cross-section of the simulated structure is shown in

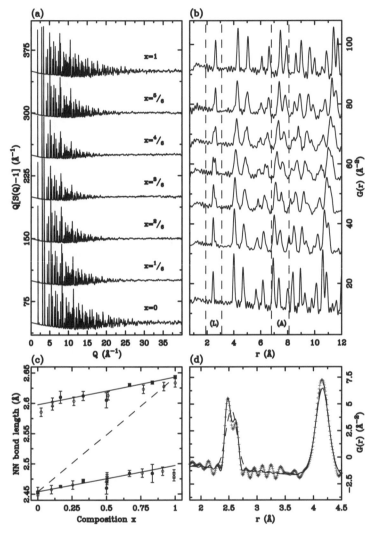

Figure 6. Measurements of the semi-conductor alloy $ZnSe_{1-x}Te_x$ as function of doping x. Panel (a) shows the reduced structure function, panel (b) shows the corresponding PDFs. Panel (c) shows the extracted nearest neighbor bond length as a function of composition. Panel (d) shows a refinement of the average structural and a disordered model. All data were collected on the instrument GEM at ISIS at the Rutherford Appleton Laboratory. For details see text.

Figure 7b. The domain structure as well as the resulting PDF were created using the program DISCUS (Proffen and Neder 1997). All subsequent refinements of the simulated data were carried out with the program PDFFIT, used previously. The structural model is the same as the matrix structure, only each site is occupied by atoms M and D and the occupancy parameter, o_D, is refined. The first refinement was carried out over a range of $2.5 < r < 27.5$ Å. The refined occupancy is $o_D = 0.85$, just as expected. The refinement is shown in Figure 7a in the top panel as solid line. The data are represented as + symbols and the difference between data and calculated PDF is plotted below. The overall R-value for this refinement is $R = 0.21\%$. Please note, that in the context of this paper the R-value is used only as relative measure to

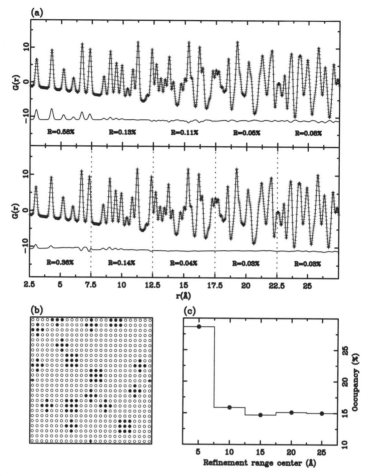

Figure 7. (a) PDF refinement of a simulated structure containing domains. The top panel shows the result of a refinement of the complete range. The bottom panel shows refinement results of individual regions, with a width of $\Delta r = 5$ Å. The R-values are shown for each region. For comparison R-values for the same regions are calculated and displayed in the top panel. In both cases, the "experimental" data are shown as + symbols and the calculated result as solid line. The difference between observed and calculated values multiplied by a factor of 50 is shown below the curves in both panels. (b) Cross-section of simulated structure containing domains (filled circles). (c) Refined site occupancy as function of refinement range. Details see text.

compare refinements. No attempts have been made to compute sensible standard deviations for the simulated data. Inspection of the top panel of Figure 7a shows residuals between calculated PDF and the data, mainly at distances $r < 7.5$ Å. Note, that the difference curves in the Figure are enlarged by a factor of fifty for clarity. Below the difference curve, R-values for individual sections with a width of $\Delta r = 5$ Å are listed. As is already obvious from the difference curve, the agreement up to $r = 7.5$ Å is worse. This can be understood considering that at distances smaller than the radius of the domains ($d/2 = 7.5$ Å) more D-D pairs are sampled than in the random case. At distances larger than $r = 7.5$ Å, only M-M and M-D pairs are sampled. We neglect at this point correlations D-D originating from different domains. Since the refinement range chosen extends to $r = 27.5$ Å, the model assumes the correct overall composition at the expense of worse agreement at low r.

However, the refinement range can be freely chosen, and rather than using the complete range of the PDF in a single refinement, sections of the PDF are refined individually. A set of five refinements over a range of $\Delta r = 5$ Å each is shown in the bottom panel of Figure 7a. The refinement ranges are indicated by the vertical dotted lines. Again, the data are represented as + symbols and the refinement result as solid line. The difference and individual R-values shown below the data indicate a much better overall agreement. For the first range up to $r = 7.5$ Å we see the biggest improvement. The refined occupancy is higher, $o_D = 0.29$ and the R-value has significantly improved. Observing a higher occupancy of atom type D is consistent with the view that sampling with distances smaller than the domain radius yields more D-D pairs. For all other refinement ranges we find the overall occupancy of $o_D = 0.15$. The occupancy o_D, as a function of refinement range, is shown in Figure 7c and one can immediately see that for ranges larger than the domain size, the occupancy is that of the average structure. This way r range dependent refinements can be used to estimate the domain size.

As we have seen, refining the PDF in sections, structural information can be obtained as a function of length scale. In this simple example, it was evident from the agreement of the data with a simple model, that there were structural differences when sampling with distances below and above $r = 7.5$ Å. This is of course exactly the radius of the domains used to create the input data. This is a very simple demonstration and more complex models could be constructed using the multi phase refinement capabilities of PDFFIT. This is a real strength of the PDF approach applied to disordered materials and a more complicated example investigating inhomogeneities in $LaMnO_3$ using this approach can be found in (Qiu et al. 2005).

Local structure of gold nanoparticles

Another group of materials, which lends itself to PDF analysis, are nanoparticles. In this example, we discuss a PDF study of gold nanoparticles (Page et al. 2004a). The obvious difference between nanoparticles and their bulk counterparts is their size. Standard diffraction techniques are based on an infinitely periodic system and as nanoparticles become smaller and smaller, this assumption is no longer valid and standard techniques often fail in determining the structure of nanoparticles. The PDF method on the other hand has no such limitation, since no periodicity is required from the start. Another difference is the significance of structural relaxation at the surface. For a bulk material the number of atoms on the surface can be neglected compared to the number of atoms in the core. For nanoparticles, however, depending on their size, the fraction of atoms belonging to the surface can be as large as 20%. Again, since the PDF does not rely on periodicity, it is the ideal method to study the difference between surface and core structure. Fluorothiol-capped gold nanoparticles as well as a bulk gold sample were measured on the instrument NPDF. A transmission electron micrograph of the nanoparticles is shown in Figure 8c. The individual particles can easily be identified. The inset on the top right shows the particle size distribution. The average size is approximately 3.6 nm. One of the complications doing neutron scattering experiments, is the usually large sample amount required. For this experiment, we used 1.5 g of nanoparticles. The measuring time was approximately 15 hours. The resulting PDFs, $G(r)$, are shown in Figure 8a. The top panel shows the PDF for a bulk gold sample. Even for the bulk sample, a decrease in PDF intensities as function of r is visible. In this case this falloff is determined by the finite resolution of the instrument NPDF as discussed earlier. However, atom-atom correlations are clearly visible out to the largest distances. For the nanoparticles sample on the other hand (bottom panel), it is quite obvious that the PDF peaks diminish at much lower distances. This is expected, since there are no atom-atom pairs at distances larger than the particle diameter which is marked as dashed line in Figure 8a. The next step is to refine the experimental PDFs using a structural model. A representative result obtained from the data measured at $T = 15$ K is shown in Figure 8b. The resulting structural parameters for the bulk sample as well as the gold nanoparticles are listed in Table 3.

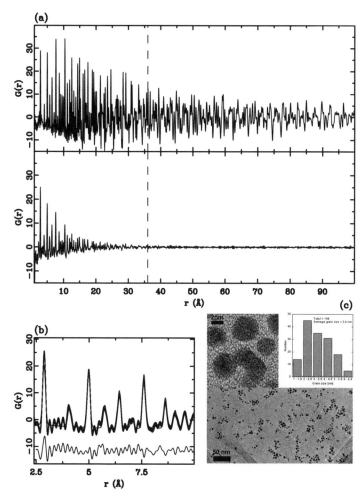

Figure 8. (a) Top: Experimental PDF of bulk gold. Bottom: Experimental PDF of gold nanoparticles. Data were collected on NPDF. The dashed line marks the average diameter of the nanoparticles. (b) Refinement of the gold nanoparticles data, the difference curve is shown below. (c) Transmission electron micrograph showing the nanoparticles. The insets show an enlarged particle and the size distribution.

Table 3. Refinement results for bulk gold and gold nanoparticles measured at $T = 15$ K and $T = 300$ K (see text for details). The units for lattice parameters are Å and the values of $<u^2>$ are given in Å^2 and multiplied by 1000. Standard deviations on the last digit are given in parentheses.

	Bulk sample $T = 300$ K	Nano sample $T = 300$ K	Bulk sample $T = 15$ K	Nano sample $T = 300$ K
a	4.0833(3)	4.0757(3)	4.0697(2)	4.0650(1)
$<u^2>$	8.833(2)	6.551(4)	2.172(2)	2.495(1)
Q	0.012(9)	0.1565(5)	0.027(3)	0.1415(4)

A number of conclusions can be drawn from these results. The lattice parameters confirm previous findings that gold nanoparticles have slightly smaller unit cells than bulk samples. In addition, the coefficient of expansion of the nanoparticles seems to be slightly smaller than that of bulk gold. On the other hand there are other conclusions that can be drawn from the $G(r)$ patterns that are very interesting. For example, the sharpness of the shortest Au–Au vector at $r = 2.8$ Å shows that the lattice parameter remains essentially constant across the whole nanoparticles, i.e., there is no significant relaxation of the structure near the surface of the nanoparticle, which would lead to an asymmetrical peak at $r = 2.8$ Å. It is interesting to note, that the diffraction pattern of the nanoparticle could not be satisfactorily analyzed using the Rietveld technique.

The many faces of SiO_2

In this example we present a quantitative analysis of PDF data of Fontainebleau sandstone, where an unexpected amorphous phase was discovered (Page et al. 2004b; Proffen et al. 2005). The average structure of many consolidated sandstones appears to be mainly quartz. However, peculiar nonlinear properties have been observed in sandstone (Darling et al. 2004), that are not found in pure quartz. Neutron PDF measurements were carried out on NPDF, in order to find a local structural explanation for these properties. The sample was a cylindrical piece of Fontainebleau sandstone, approximately 50 mm high with a diameter of 9 mm, loaded in a vanadium can. The measuring time was 24 h. Rietveld analysis confirmed earlier X-ray measurements, the Fontainebleau sandstone was indeed more than 99% crystalline quartz. Next the data were processed to obtain a reduced structure function $F(Q) = Q[S(Q) - 1]$ (Fig. 9a). The high quality or low noise level of the data is easily visible. In addition a significant modulation of the scattering data at high Q can be observed. The data were terminated at $Q_{max} = 40$ Å^{-1}, to obtain the corresponding PDF shown as crosses in Figure 10a together with the results of refinements of the quartz structural model over three different ranges. The top panel shows the refinement over the complete range shown. Two observations can be made: First the agreement between model and experimental PDF at distances with $r > 3$ Å is excellent and secondly the agreement for $r < 3$ Å is very poor. This is also reflected in the weighted R-value, R_{wp}, listed in Table 4 for the different ranges.

Next identical refinements for two additional refinement ranges of $3 < r < 20$ Å and $1 < r < 3$ Å were carried out. The results are shown in the center and bottom panel of Figure 10a.

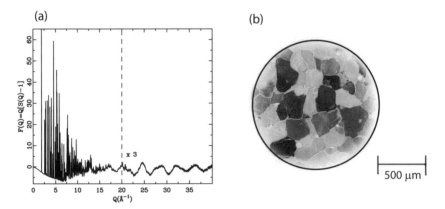

Figure 9. (a) Reduced structure function, $F(Q)=Q[S(Q)-1]$, for a Fontainebleau sandstone sample measured on NPDF. Data above $Q = 20$ Å^{-1} (dashed line) are enlarged by a factor of three. The strongest peak is cut off for clarity. (b) Thin section of the Fontainebleau sandstone sample showing the individual quartz crystallites.

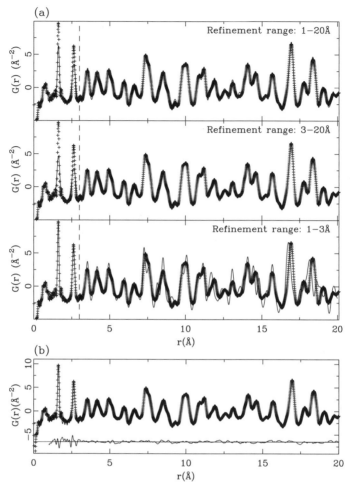

Figure 10. (a) PDF refinements of a single quartz structural phase over three different refinement ranges indicated on the graphs. (b) Result of a two-phase refinement of the PDF over the complete range shown. Details are discussed in the text.

Table 4. PDF refinement results for different refinement ranges r. The units for lattice parameters are Å and the values of $<u^2>$ are given in Å^2 and multiplied by 1000. The numbers in parentheses are the standard deviation of the last digit.

Refinement range	$1 < r < 20$ Å	$3 < r < 20$ Å	$1 < r < 3$ Å
a	4.9198(1)	4.9176(1)	4.9320(7)
c	5.4044(2)	5.4069(1)	5.315(2)
$<u^2>$ (Si)	3.666(3)	4.800(4)	3.908(2)
$<u^2>$ (O)	7.500(3)	10.303(4)	4.777(1)
R_{wp} (1-20 Å)	20.6%	23.7%	64.8%
R_{wp} (3-20 Å)	14.5%	9.6%	72.5%
R_{wp} (1-3 Å)	35.1%	48.2%	13.3%
Scale	1.009(1)	1.118(1)	0.999(3)

In this case, the agreement between model and data is good over the refined range, but poor outside it. Structural parameters as well as R-values for all three r-ranges are listed in Table 4. Obviously, the crystalline quartz model cannot describe the observed PDF of the sandstone sample over the complete distance range. However, the differences manifest themselves in an excess intensity of the first two PDF peaks, corresponding to the nearest Si-O and O-O distances. This is also confirmed when studying the scale factors for the different refinement ranges. If just the region containing the first two peaks is refined, the scale factor is one (Table 4) as expected for properly normalized data. However, for the range $3 < r < 20$ Å the scale factor needed to match the observed data is 1.1, and a good agreement between observed and calculated data is achieved over both regions. This in fact rules out local distortions or thermal diffuse scattering (TDS) as explanation since this type of disorder influences the shape and position of PDF peaks, but the total number of contributing atom pairs is conserved with respect to the perfect crystal. In other words no change in scale factor would be seen. The effect of TDS manifests itself in significantly smaller atomic displacement parameters, $<u^2>$, when refining only over the range $1 < r < 3$ Å, accounting for correlated motion as discussed earlier. A possible explanation is offered, by assuming that approximately 10% of the SiO_4 tetrahedra do not participate in the crystalline long range ordered phase and consequently are not present in the observed PDF peaks above $r > 3$ Å. This is also consistent with the modulation observed at high Q. The nearest Si-O distance of about 1.6 Å results in the modulation in Q of $2\pi/1.6 = 3.9$ Å^{-1}, that can be observed in Figure 9a.

The next step is the refinement of a two phase structural model to account for the extra PDF intensities at low r values. The simplest approach is to construct each phase from the crystalline structure of quartz. However, to account for the missing long range order, one of the phases creates correlations only out to a cutoff value, in this case $r_{cut} = 3$ Å. For the crystalline phase with correlations over the complete fitting range, we used the PDF refinement result obtained for $3 < r < 20$ Å. Only a scale factor and the correlated motion parameter were allowed to refine. For the amorphous phase all structural parameters were refined, except the atomic positions. The result is shown in Figure 10b. First we observe that for the range $3 < r < 20$ Å the refinements yields the same agreement of $R_{wp} = 9.6$% as for the single phase refinement over the same range. The significant improvement is seen in the low r-range and overall the refinement has improved from $R_{wp} = 20.6$% for the single phase model to $R_{wp} = 12.6$% for the two phase model. From the scale factors result in a phase fraction of the amorphous phase of 9.3%. However, closer inspection of Figure 10b shows that despite the large improvement by adding a second phase, the low r-region of the experimental PDF is still not modeled as well as the data at $r > 3$ Å. This is most likely due to the lack of the model to account for the more complicated correlated motion present in SiO_4 tetrahedra. Several tests with different models for the correlated motion in the amorphous phase gave very similar results. However, the corresponding scale factors yielded phase fractions between 5 to 10%. Regardless, these two phase refinements provide clear evidence for amorphous silica phase present in Fontainebleau sandstone. Obtaining more detailed quantitative information regarding phase fractions as well as structure requires more development of our modeling techniques, specifically related to the modeling of correlated motion.

SUMMARY

The total scattering approach and the PDF technique allow one to obtain a picture of the "true" atomic structure from a variety of materials from glasses to crystalline disordered materials, multi phase systems containing amorphous and crystalline components as well as nanoparticles. In all of these cases, the study of the PDF adds important structural information or makes a structure determination possible at all. Great advances have been made to allow the rapid and straight forward collection of total scattering data. In case of neutron scattering, the instrument NPDF at the Lujan Neutron Scattering Center is at the forefront of these efforts. For a standard sample, the structure function, $S(Q)$, as well as the PDF, $G(r)$, are automatically generated and accessible to the instrument user via a web site. The true challenge is the understanding and modeling of the structure of complex systems. A very simple approach was demonstrated in the section about domain structures. The ultimate goal would be to construct a sufficiently large model structure consisting of the matrix structure, the domains as well as a structural description of the domain boundary. Although in principle possible, this is currently still very much at the limit of what is possible with today's programs and available computing power. However, current efforts focus on the development of better and more user friendly modeling software. There is no doubt, at least in the author's mind, that the PDF technique will continue to grow and become the structural tool of choice to study complex materials.

ACKNOWLEDGMENTS

This work has benefited from the use of NPDF at the Lujan Center at Los Alamos Neutron Science Center, funded by DOE Office of Basic Energy Sciences and Los Alamos National Laboratory funded by Department of Energy under contract DE-AC52-06NA25396. The upgrade of NPDF has been funded by NSF through grant DMR 00-76488.

REFERENCES

Boyce JB, Mikkelsen JC (1989) Local structure of pseudobinary semiconductor alloys: an X-ray absorption fine structure study. J Cryst Growth 98:37-43

Chung JH, Proffen T, Shamoto S, Ghorayeb AM, Croguennec L, Tian W, Sales BC, Jin R, Mandrus D, Egami T (2005) Local structure of $LiNiO_2$ studied by neutron diffraction. Phys Rev B 71:064410

Cole DR, Herwig KW, Mamontov E, Larese J (2006) Neutron scattering and diffraction studies of fluids and fluid-solid interactions. Rev Mineral Geochem 63:313-362

Darling TW, TenCate JA, Brown DW, Clausen B, Vogel SC (2004) Neutron diffraction study of the contribution of grain contacts to nonlinear stress-strain behavior. Geophys Res Lett 31:L16604-1–L16604-4

Egami T, Billinge SJL (2003) Underneath the Bragg Peaks: Structural Analysis f Complex Materials. 1st Edition. Pergamon

Harrison RJ (2006) Neutron diffraction of magnetic materials. Rev Mineral Geochem 63:113-143

Jeong IK, Heffner RH, Graf MJ, Billinge SJL (2003) Lattice dynamics and correlated atomic motion from the atomic pair distribution function - art. no. 104301. Phys Rev B 67:104301-104301

Jeong IK, Mohiuddin-Jacobs F, Petkov V, Billinge SJL, Kycia S (2001) Local structure of $In_xGa_{1-x}As$ semiconductor alloys by high-energy synchrotron X-ray diffraction. Phys Rev B 63:205202

Knorr K, Depmeier W (2006) Application of neutron powder-diffraction to mineral structures. Rev Mineral Geochem 63:99-111

Louca D, Kwei GH, Dabrowski B, Bukowski Z (1999) Lattice effects observed by the isotope-difference pair density function of the $(YBa_2Cu_3O_{6.92})$-Cu-63/65 superconductor. Phys Rev B 60:7558-7564

Nield VM, Keen DA (2001) Diffuse Neutron Scattering from Crystalline Materials. Oxford University Press

Page K, Proffen T, Terrones H, Terrones M, Lee L, Yang Y, Stemmer S, Seshadri R, Cheetham AK (2004a) Direct observation of the structure of gold nanoparticles by total scattering powder neutron diffraction. Chem Phys Lett 393:385-388

Page KL, Proffen T, McLain SE, Darling TW, TenCate JA (2004b) Local atomic structure of Fontainebleau sandstone: Evidence for an amorphous phase? Geophys Res Lett 31:L24606-1–L24606-4

Peterson PF, Bozin ES, Proffen T, Billinge SJL (2003) Improved measures of quality for the atomic pair distribution function. J Appl Crystallogr 36:53-64

Peterson PF, Gutmann M, Proffen T, Billinge SJL (2000) PDFgetN: a user-friendly program to extract the total scattering structure factor and the pair distribution function from neutron powder diffraction data. J Appl Crystallogr 33:1192

Peterson PF, Proffen T, Jeong IK, Billinge SJL, Choi KS, Kanatzidis MG, Radaelli PG (2001) Local atomic strain in $ZnSe_{1-x}Te_x$ from high real-space resolution neutron pair distribution function measurements. Phys Rev B 63:165211-1–165211-7

Petkov V, Jeong IK, Mohiuddin-Jacobs F, Proffen T, Billinge SJL, Dmowski W (2000) Local structure of $In_{0.5}Ga_{0.5}As$ from joint high-resolution and differential pair distribution function analysis. J Appl Phys 88:665-672

Price DL (undated) GLASS package. Intense Pulsed Neutron Source, Internal Report No. 14.

Proffen T (2000) Analysis of occupational and displacive disorder using the atomic pair distribution function: a systematic investigation. Z Kristallogr 215:661-668

Proffen T, Billinge SJL (1999) PDFFIT, a program for full profile structural refinement of the atomic pair distribution function. J Appl Crystallogr 32:572-575

Proffen T, Neder RB (1997) DISCUS: A program for diffuse scattering and defect-structure simulation. J Appl Crystallogr 30:171-175

Proffen T, Page KL (2004) Obtaining structural information from the atomic pair distribution function. Z Kristallogr 219:130-135

Proffen T, Page KL, McLain SE, Clausen B, Darling TW, TenCate JA, Lee SY, Ustundag E (2005) Atomic pair distribution function analysis of materials containing crystalline and amorphous phases. Z Kristallogr 220:1002-1008

Proffen T, Petkov V, Billinge SJL, Vogt T (2002) Chemical short range order obtained from the atomic pair distribution function. Z Kristallogr 217:47-50

Qiu XY, Proffen T, Mitchell JF, Billinge SJL (2005) Orbital correlations in the pseudocubic O and rhombohedral R phases of LaMnO$_3$. Phys Rev Lett 94:177203

Ross NL, Hoffman C (2006) Single-crystal neutron diffraction: present and future applications. Rev Mineral Geochem 63:59-80

Soper AK, Howells WS, Hannon AC (1989) ATLAS - Analysis of Time-of-Flight Diffraction Data from Liquid and Amorphous Samples. ISIS Facility, Rutherford Appleton Laboratory. *http://www.isis.rl.ac.uk/disordered/Manuals/ATLAS/ATLAS_manual.htm*

Toby BH, Egami T (1992) Accuracy of pair distribution function analysis applied to crystalline and non-crystalline materials. Acta Crystallogr Sect A 48:336

Tucker MG, Dove MT, Keen DA (2001) Application of the reverse Monte Carlo method to crystalline materials. J Appl Crystallogr 34:630-638

Vogel SC, Priesmeyer H-G (2006) Neutron production, neutron facilities and neutron instrumentation. Rev Mineral Geochem 63:27-57

Weber T (2005) Cooperative Evolution - a new algorithm for the investigation of disordered structures via Monte Carlo modelling. Z Kristallogr 220:1099-1107

Welberry TR (2004) Diffuse X-Ray Scattering and Models of Disorder. Oxford University Press

Welberry TR, Proffen T, Bown M (1998) Analysis of single-crystal diffuse X-ray scattering via automatic refinement of a Monte Carlo model. Acta Crystallogr Sect A 54:661-674

Wilding MC, Benmore CJ (2006) Structure of glasses and melts. Rev Mineral Geochem 63:275-311

Reviews in Mineralogy & Geochemistry
Vol. 63, pp. 275-311, 2006
Copyright © Mineralogical Society of America

12

Structure of Glasses and Melts

Martin C. Wilding

Institute of Mathematical and Physical Sciences
University of Wales
Aberystwyth
Ceredigion, SY23 3BZ, United Kingdom
e-mail: martin.wilding@aber.ac.uk

Chris J. Benmore

Intense Pulsed Neutron Source
Argonne National Laboratory
9700 Cass Avenue
Argonne, Illinois, 60439, U.S.A.
e-mail: benmore@anl.gov

INTRODUCTION

Melting

The liquid state dominates terrestrial and planetary processes. The history of the early solar system for example involved the accretion of primarily molten bodies. In cooler temperature regimes, interplanetary dust, comets and the moons and planets in the outer part of the solar system are dominated by ices which may be present in liquid or amorphous forms (Cernicharo and Crovisier 2005; Porco et al. 2005a,b). The interior of the Earth and terrestrial planets are dominated by magnesium silicate minerals, a reflection of separation of iron dominated liquids from chondrite composition planetesimals (Poirier 2000). Subsequently planets evolved through segregation, crystallization and volcanic activity; all dominated by liquid state processes. The oceanic and continental crust, while compositionally distinct form the mantle is dominated by liquid state processes, the oceanic crust resulting from basaltic volcanism. Hydrothermal processes are important agents for geochemical processes in the Earth's crust and also, according to recent surveys on Mars (Neukum et al. 2004; Bullock 2005; Newsom 2005).

Phase diagrams at constant pressure maps out stability domains of various crystalline and liquid phases as a function of composition and temperature. For example simple binary oxides MgO and SiO_2 (Wilding et al. 2004a) (Fig. 1) can show a variety of crystalline and liquid state structures which maybe further restricted in terms of important stoichiometric end-members such as Mg_2SiO_4 and $MgSiO_3$.

The thermodynamic and transport properties of liquids reflect the atomic or molecular scale structure of the liquid. So probes of the liquid structure are essential to understanding physical and chemical processes. Rarely can high pressure and temperature liquid state processes be observed *in situ* and alternative routes to structural investigation are sometimes taken, including the study of glasses and amorphous state materials. The glasses themselves can be chemically and structurally complex and are not necessarily direct representatives of equivalent liquids.

Thermodynamics and structure

The three main states of matter are shown in Figure 2, together with the *triple point* at which all three states co-exist in equilibrium, and the *critical point* where the densities of the

 DOI: 10.2138/rmg.2006.63.12

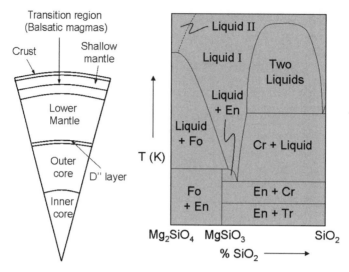

Figure 1. Section of the earth and phase diagram of forsterite-SiO₂ system at ambient pressure (adapted from Levin 1956). Fo; fosterite, En; enstatite, Cr; cristobalite, Tr; tridymite.

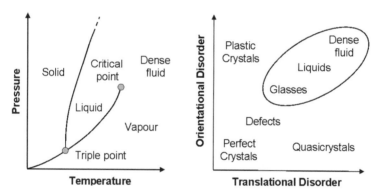

Figure 2. Thermodynamic state and structural disorder of materials. Adapted from Egelstaff (1967) and Price et al. (1991), respectively.

liquid and vapor become equal. Upon cooling or compression a will condense to form a liquid, with well-defined shells of nearest neighbor atoms. Solid crystalline or glassy forms may form upon cooling the liquid, where as solid amorphous forms are commonly formed by vapor deposition (Egelstaff 1967; Stanley 1971; Barrat and Hansen 2003).

Liquids lack the *long range* orientational and translational atomic order that characterizes the crystalline state but do exhibit *short range order*. Glasses and amorphous solids also have short range order and may possess *intermediate range order*. Short range order can be described using a probability function i.e., the probability of finding an atom at a distance, r, from an atom at the origin. This *radial distribution function* $\rho(r)$ is often referred to as a *pair distribution function* as it shows the sum of the correlations between different pairs of atoms (Fig. 3).

Glasses formed upon supercooling the liquid exhibit a *glass transition* (Zallen 1983; Debenedetti 1996), where properties such as heat capacity change abruptly. Consequently glasses are metastable i.e., not in thermodynamic equilibrium, and reflect the continued

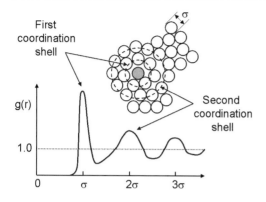

First
coordination
shell

g(r)

1.0

Second
coordination
shell

0   σ   2σ   3σ

Figure 3. Atomic configurations for a liquid of hard spheres of diameter σ showing the first and second coordination shells (Egelstaff 1967; Barrat and Hansen 2003). The PDF shows peaks corresponding to the first and second coordination shells.

rearrangement of structural configurations (relaxation) in the supercooled liquid regime, below the melting temperature. Glasses are not therefore snapshots of the stable liquid but neither are they disordered forms of crystalline phases. In contrast, amorphous solids do not exhibit a glass transition but are also metastable.

Amorphization may occur through a number of different routes including chemical synthesis, deposition of the vapor onto a cold substrate, radiation damage and pressure induced amorphization. Glassy and amorphous structures produced by various methods are usually different (Price et al. 1988) and heavily dependent on their precise chemical, thermal and pressure history, making them inherently *polyamorphic* (Grimsditch 1984, 1986). This has led to some confusion in recent years as the term "polyamorphism" has been associated with a first order phase transition between different glassy, amorphous or liquid states. First order phase transitions are more commonly associated with the reversible liquid-solid, vapor-solid or vapor-liquid transitions denoted by the solid lines in Figure 2(a).

The role of neutron diffraction

Although of fundamental importance to physics, chemistry, geology and technology, the structure of glass, liquid and amorphous materials remains one of the least understood states of matter. The local chemical bonding and intermediate range order interactions that are seen in the pair distribution function play an important role in determining structure-property relationships. For example, the local coordination number of liquid silicon is associated with a metal/insulator transition. The pair distribution function is obtained by Fourier transformation of the total structure factor, $S(\mathbf{Q})$, obtained from diffraction data from liquids, glasses and amorphous materials probed by radiation with wavelengths comparable to the inter-atomic separation. Total neutron scattering is one of the few experimental probes that can provide a direct measure of nuclear arrangements over the wide range of length scales illustrated in Figure 2(a) (Egelstaff 1965, 1967; Price 1985, 1986). Thermal and high energy neutrons are highly penetrating and a powerful bulk probe, which can provide high resolution information at the atomic level which is needed for the study of liquid structures. In addition, neutrons are sensitive to light elements, for example hydrogen (deuterium) and oxygen, which varies from isotope to isotope depending on their nuclear spin. Isotopic substitution is therefore a powerful probe in determining detailed atomic structure of specific atom pairs.

In this chapter we will outline the application of neutron diffraction to the study of liquids and amorphous materials, including those of interest to the earth science community. First we will outline the structure of liquids and glasses and we will discuss some of the recent models for liquid behavior and the relationship between microscopic structure and bulk, macroscopic behavior. Next we will discuss in detail how elastic neutron scattering measurements are made, outlining basic scattering theory, data correction and transformation procedures. Included in this section will be discussion of instruments available for neutron diffraction and some of the strategies available for *in situ* investigation of high temperature and high pressure processes. We will provide some examples of how neutron diffraction measurements have been applied to the geosciences and finally we will outline future developments including the design of new instruments and neutron diffraction techniques.

THE STRUCTURE RELATED PROPERTIES OF LIQUIDS, GLASSES AND AMORPHOUS MATERIALS

The pair distribution function and the pair potential

We have introduced the concept of the pair distribution function, describing the averaged separation of atom pairs in a liquid or amorphous solid (see also Proffen 2006, this volume). Close to the origin of the central atom in Figure 2 there is a region where the pair distribution function is zero, which denotes the hardcore repulsion of two neighboring atoms. There is a pronounced first peak in the pair distribution function which in a monatomic liquid approximates two atomic diameters, σ. A series of smaller peaks at higher-r represents next neighbor shells at increasing distances from the central atom, until the mean density of the system is reached at $\rho(r) = 1$. The separation of atoms in the $\rho(r)$ reflects the interaction of atom pairs and is therefore related to the atomic pair potential (Egelstaff 1967). At small-r the potential is dominated by a large positive repulsive energy term that results from overlapping electron shells, therefore limiting the closest approach of atoms. At longer distances attractive potential energy terms are small and negative. At a separation slightly greater than σ, the energy is at a minimum and the atom pair is stable. A simple form of the pair-potential is the Lennard-Jones potential which can be used to illustrate the relationship between $\rho(r)$ and the pair potential $V_{ij}(r)$ as shown in Figure 3.

Glasses and supercooled liquids

Many liquids however fail to crystallize when cooled below the stable melting temperature, these liquids are metastable and have a Gibbs free energy higher than that of the corresponding crystal. In this supercooled liquid regime the structure of liquids may remain unchanged for relatively long observation times, but the liquids are considered "non ergodic" or not in equilibrium on an infinite timescale. Supercooled liquids but can be thought however of as being "locally ergodic" or in a localized equilibrium (Wales et al. 2000; Wales 2003). As the temperature of a liquid is decreased the viscosity increases. In macroscopic terms this is associated with an increase in the relaxation time of structurally dependent properties. The combined lowering of temperature and increase in relaxation time means fewer structural configurations are explored (non-ergodicity) and structures are eventually trapped in a non crystalline configuration.

From a classical viewpoint vitrification is regarded as a competition between two competing timescales (Debenedetti 2003). The time taken to crystallize a given volume of a crystal, that is the thermodynamic driving force for nucleation, and the structural relaxation time, which is the kinetic control on crystal growth.

The formation of a glass is a continuous process and properties such as volume and enthalpy change over a relatively large temperature interval within the supercooled regime. The vitrification process is therefore is not an equilibrium one and the properties quenched into the glass depend very much on quench rate. The formation of a glass though is not simply a result of kinetic arrest. As was first pointed out in 1948 by Kauzmann (1948) the isobaric heat capacity (C_p) of a liquid exceeds that of the corresponding crystal and one consequence of this is that the entropy of the liquid decreases more rapidly with temperature(Angell and Moynihan 2000;

Angell et al. 2000a; Moynihan and Angell 2000). In the metastable, supercooled regime there is a temperature at which the curves for liquid and crystal entropy intersect. This temperature is referred to as the Kauzmann temperature, T_k. In practice the Kauzmann temperature is never reached because the calorimetric glass transition is intercepted. Continued extrapolation of the supercooled entropy curve would result in negative values of entropy as $T \to 0$ K, inconsistent with the Boltzmann formula for entropy and unphysical. This is referred to in the literature as the Kauzmann Paradox (Debenedetti 1996).

Configurational entropy and fragility

The structure of a supercooled liquid is a reflection of its potential energy surface, or energy landscape (Fig. 4). The energy landscape model of a supercooled liquid implies that constituent atoms vibrate within potential minima and that there are infrequent jumps between different minima that represent different structural configurations (Angell et al. 1986, 2000b; Debenedetti 1996; Angell 1997a,b; Angell and Moynihan 2000; Wales et al. 2000; Stillinger and Debenedetti 2002, 2003; Debenedetti et al. 2003; Wales 2003; Debenedetti and Torquato 2004). In these models vibrational and configurational contributions to the liquid properties are separable. The entropy loss on cooling in the energy landscape view is a progressive loss of structural configurations and inevitably influences the transport properties of the supercooled liquid. As the liquid is cooled the size of regions in which molecules can rearrange themselves (cooperative regions) decreases. The decreasing size of cooperative regions (i.e., non ergodicity) and decrease in configurational entropy (S_{conf}) is the basis for the Adam-Gibbs model of structural relaxation.

Supercooled liquids often deviate from Arrhenius viscosity temperature relations and a useful classification has been adopted, introduced by Angell and following from earlier ideas developed by Laughlin and Uhlmann (1972). Angell et al. (1986, 1994) scaled viscosity data to the temperature at which the viscosity reached 10^{12} Pa·s (the accepted value of T_g). Liquids that have close-to-Arrhenius viscosity over the entire temperature range (e.g., SiO_2, GeO_2) are termed strong, while fragile liquids deviate (Fig. 5).

The structural stability of liquids of different fragility is reflected in the change in heat capacity at the glass transition and also the extent at which the changes in heat capacity in the glass transition interval reflect changes in structural relaxation time; this is an important

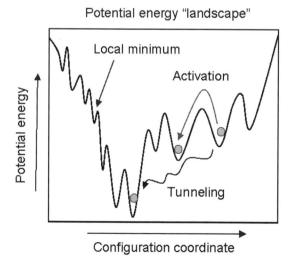

Potential energy "landscape"

Local minimum

Activation

Tunneling

Potential energy

Configuration coordinate

Figure 4. Representation of a two dimensional potential energy hypersurface (energy landscape). There are a series of minima separated by potential energy barriers; these represent different structural configurations in the super-cooled liquid. Jumps between minima (different structural configurations) are by thermal activation. Tunneling between local minima would reflect a transition between two different metastable amorphous forms for example.

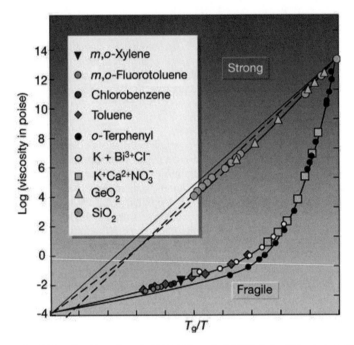

Figure 5. The liquid fragility (Angell et al. 1986; Debenedetti 1996; Saika-Voivod et al. 2001; Saika-Voivod et al. 2002). The viscosity and hence relaxation behavior of several glass forming liquids is plotted as a function of temperature scaled to the glass transition (T_g). Strong network forming liquids have an Arrhenius viscosity-temperature relation while fragile liquids show dramatic changes in viscosity as a function of temperature. The fragility of liquids is also reflected in the jump in heat capacity at the glass transition, i.e., the configurational entropy.

link between liquid fragility and the Adam Gibbs model of structural relaxation but there is no apparent direct link between the liquid or glass structure and fragility. Strong liquids are characterized by small increased (ΔC_p) in heat capacity at the glass transition indicating persistence of intermediate and long-range structural ordering the supercooled regime. More fragile liquids have a greater ΔC_p and, because the relaxation times are non-Arrhenius, relatively narrow glass transition intervals (Moynihan 1993; Moynihan et al. 1996).

Pressure induced amorphization

Cooling a liquid until it vitrifies is not the only way of producing a disordered material. When solids are compressed they can become amorphous, a process known as pressure-induced amorphization; this process is interpreted as a logical extension of the observations made originally by Bridgeman (1937). The dP/dT slopes of melting curves may be negative or overturn with pressure, these negative dP/dT slopes indicate an increase in the density of the substance on melting, and an overturn of the melting curve will represent a change in liquid volume with pressure. If the negative dP/dT curve is extended into a metastable regime then this curve may be intercepted when low pressure crystalline phases are compressed (Richet and Gillet 1997).

Abrupt transitions between structurally different amorphous forms of the same substance and overturns of the melting curve with pressure are used as arguments to support first order liquid-liquid or amorphous-amorphous transitions, i.e., "polyamorphism" (Poole et al. 1997; Brazhkin et al. 2002). Such first order liquid-liquid transitions would, when followed to the end of the transition line, result in a second critical point. These types of transition have suggested in a number of systems, including Si (Deb et al. 2001; Sastry and Angell 2003), Ge, SiO_2 (Lacks 2000; Mukherjee et al. 2001)and GeO_2 (Wolf et al. 1992)and importantly in amorphous forms of ice (Mishima et al. 1984, 1985; Mishima 1994; Mishima and Stanley 1998a, b). Moreover, the existence of first order transitions has been contested in the cases of vitreous GeO_2 (Guthrie) and amorphous ice (Tulk et al. 2002), largely on the basis of diffraction data. Rather in these studies it has been suggested that the transitions occur continuously through a continuum of intermediate forms. No second critical point has yet been found experimentally in any of these systems.

Clearly then liquids and amorphous materials show potential changes in structure and associated bulk macroscopic properties as a function of composition, pressure and temperature. In many cases, liquids that are of interest are fragile and are not good glass-formers. Structural changes as a function of extremes in pressure and temperature are also difficult to probe experimentally, not least the proposed second critical points that lie at the heart of discussions about polyamorphic transitions in liquids and amorphous materials. Neutrons offer the opportunity to investigate the microscopic structure of stable and metastable liquids and also amorphous materials, and, because neutrons are more sensitive to light elements, particularly hydrogen, aqueous solutions and ices can also be studied. Experimental strategies for neutron diffraction studies of amorphous materials include examination of glasses or liquids as a function of composition and also as a function of pressure or temperature. The latter requires specialized sample environments and novel synthesis methods.

NEUTRON DIFFRACTION THEORY AND MEASUREMENT

Scattering theory

Structural studies of liquid or amorphous states generally involve elastic or inelastic scattering of electromagnetic radiation (Bacon 1963, 1975, 1987; Egelstaff 1965; Wright et al. 1991). As with crystalline (elastic) diffraction, a beam of radiation is directed at the sample and the intensity of the signal measured as a function of scattering angle 2θ and the wavelength of the incident radiation. The intensity is expressed as a function of the momentum transfer of the scattered particle, $\mathbf{Q}$, where $\mathbf{Q}$ is defined as $\mathbf{Q} = \mathbf{k}_{incident} - \mathbf{k}_{scattered}$, with $\mathbf{k}_{incident}$ and $\mathbf{k}_{scattered}$ the incident and scattered wave vectors respectively. The scattered intensity contains information on the structure of the material in reciprocal space and the extracted structure factor is related to the pair distribution function by a Fourier transform (Egelstaff 1965; Bacon 1975; Lovesey 1984).

Various types of measurement can be made by scattering neutrons from a sample. In a neutron diffraction experiment, the diffraction pattern is obtained from the counts per second measured by a detector placed at a solid angle $d\Omega$ and expressed as the differential cross section. If the energy exchange between incident and scattered neutrons is small, a static approximation is made. This approximation assumes that scatted neutrons are counted in a detector regardless of their energies and that the measurements are an integration of the double differential cross section. A comprehensive theoretical basis leading up to this point has been described by several authors (Squires 1978; Lovesey 1984; Bée 1988)and more recently for the case of liquids and glasses by Fisher et al (Fischer et al. 2003), this will not be covered here.

The double differential cross section is defined by the equation ($d^2\sigma/d\Omega dE$). Given that I_0 is the flux of incident neutrons this can be related to the observed intensity I through

$$I = I_0 \frac{d^2\sigma}{d\Omega dE_1} d\Omega dE_1 \tag{1}$$

If the energy of the scattered neutrons is not analyzed and the neutron counts are integrated at a given solid angle we obtain the differential cross section and total cross section respectively,

$$\frac{d\sigma}{d\Omega} = \int_0^\infty \frac{d^2\sigma}{d\Omega dE_1} dE_1 \quad \text{and} \quad \sigma = \int_\Omega \frac{d\sigma}{d\Omega} d\Omega \tag{2}$$

The scattered neutron intensity can be expressed as a function of the momentum transfer $\hbar Q = \mathbf{k}_{incident} - \mathbf{k}_{scattered}$ and energy transfer $\hbar\omega = E_{incident} - E_{scattered}$.

The scattering length, b, characterizes the strength of the neutron-nucleus interaction. The quantity b can be complex and its' value depends on the particular isotope and the spin state of the neutron-nucleus system. Scattering lengths can be positive or negative and fluctuate erratically from one isotope to another. In general the absorption cross section is less than the scattering cross section and $\sigma_{absorption} \propto \upsilon^{-1}$, where υ is the neutron velocity. The coherent and incoherent cross sections are given by $\sigma \equiv 4\pi b^2$. This allows us to define the coherent scattering law as

$$S(\mathbf{Q},\omega) = \frac{1}{N} \frac{\mathbf{k}_{incident}}{\mathbf{k}_{scattered}} \frac{4\pi}{\sigma_{coh}} \frac{d^2\sigma}{d\Omega dE_1}\Big|_{coh} \tag{3}$$

Where $S(\mathbf{Q},\omega)$ is the dynamic structure factor. $S(\mathbf{Q},\omega)$ is related to the van Hove correlation functions $G(\underline{r},t)$ through a Fourier transformation. The static structure factor $S(\mathbf{Q})$ is obtained by integrating $S(\mathbf{Q},\omega)$ with respect to ω at constant $\mathbf{Q}$.

However in a typical reactor based neutron diffraction experiment the detector performs an integration over the final energies at constant angle rather than constant $\mathbf{Q}$. Since

$$\mathbf{Q}^2 = \mathbf{k}_{incident}^2 + \mathbf{k}_{scattered}^2 - 2\mathbf{k}_{incident}\mathbf{k}_{scattered}\cos 2\theta \tag{4}$$

and if we make the static approximation where the incident neutron energy is assumed to be much greater than the energy transfers involved in the neutron-nucleus interaction such that $\mathbf{k}_{scattered} \approx \mathbf{k}_{incident}$ we can write

$$\mathbf{Q}^2 \approx 2\mathbf{k}_{incident}^2(1 - \cos 2\theta) \quad or \quad \mathbf{Q} = \frac{4\pi}{\lambda}\sin\theta \tag{5}$$

Experimentally an effective cross section is measured because the detector is not black, i.e., it has a wavelength dependence efficiency, resulting in two different integration paths in $\mathbf{Q}$-ω space between the defined $S(\mathbf{Q})$ and the experimentally measured $S(\mathbf{Q})$. The static approximation holds for thermal neutrons scattering from heavy nuclei but breaks down for scattering from light nuclei. The inelastic correction to liquid neutron data will be discussed in the next section.

Multicomponent systems

In a polyatomic system, the total scattering comprises of the sum of several partial atom-pair contributions (Fig. 6). The measured differential cross section has two parts, the self scattering part which is the incoherent scattering from individual scattering centers and the distinct part, the coherent interference term from different atom pairs. This is written as:

$$\frac{1}{N}\left[\frac{d\sigma}{d\Omega}(\mathbf{Q})\right] = F(\mathbf{Q}) + \sum_\alpha^n c_\alpha \bar{b}_{\alpha,inc}^2 \tag{6}$$

where $F(\mathbf{Q})$ is the total interference function, i.e., the distinct scattering, and c_α the concentration of chemical species α. It is convenient to use the convention of Faber-Ziman (1965) to express the total interference function $F(\mathbf{Q})$, in terms of a dimensionless static structure factor $S(\mathbf{Q})$:

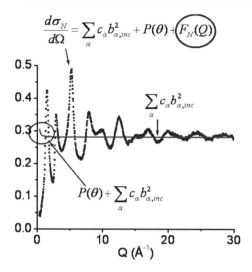

$$\frac{d\sigma_N}{d\Omega} = \sum_\alpha c_\alpha b_{\alpha,inc}^2 + P(\theta) + \boxed{F_N(Q)}$$

$$\sum_\alpha c_\alpha b_{\alpha,inc}^2$$

$$P(\theta) + \sum_\alpha c_\alpha b_{\alpha,inc}^2$$

Figure 6. Differential cross section for a time of flight spallation neutron diffraction experiment on glassy SiO_2, showing the self-scattering component for a polyatomic sample and the distinct scattering (the interference function). Also shown is the magnitude of the correction for inelastic (Plazeck) scattering, $P(\theta)$.

$$F(\mathbf{Q}) = \sum_{\alpha,\beta}^{n} c_\alpha c_\beta \bar{b}_\alpha \bar{b}_\beta \left[S_{\alpha,\beta}(\mathbf{Q}) - 1 \right] \tag{7}$$

Where for n components there are $n(n + 1)/2$ partial structure factors, $S_{\alpha\beta}(\mathbf{Q})$, labeled species α and β. Other formalisms also exist and have recently been reviewed by (Dove et al. 2002). Bhatia-Thornton formalism has been used extensively to describe the topology of liquid and glassy systems (Salmon 2002, 2005).

For a liquid at $\mathbf{Q} = 0$, $S(\mathbf{Q})$ is proportional to the isothermal compressibility of the fluid. For example, under normal conditions a liquid is usually incompressible so $S(\mathbf{Q}) \rightarrow 0$, but if at the critical point $S(\mathbf{Q}) \rightarrow \infty$ and a dramatic rise in low-$\mathbf{Q}$ signal is observed. In practice there is a small gap between $\mathbf{Q} = 0$ and the minimum experimentally measure $\mathbf{Q}$-value. This region is usually extrapolated with minimal effects on Sine Fourier transformation used to obtain the radial distribution function $G(r)$. A much more significant factor affecting the transformation process is the truncation of the data at a maximum $\mathbf{Q}$-value, which is equivalent to multiplying the data by a step function. The resulting real space correlation function has a strong oscillatory component in $G(r)$ known as termination ripples (which decreases with increasing r) and may affect the position of the first real space peak. This problem can be minimized by multiplying by a smoothly decaying modification function, e.g., a Lorch function (Lorch 1970), which reduces the effect of the step function before inversion. In practice the maximum measured $\mathbf{Q}$ value does not always yield the most accurate transform, noisy data can also lead to spurious oscillations in $G(r)$ and truncating at a positive node can reduce the step function effect. It is usually good practice to try Fourier transforming the data varying these parameters to ensure small features observed in $G(r)$ are not Fourier artifacts.

The Sine Fourier Transform of $S_{\alpha\beta}(\mathbf{Q})$ itself leads to the partial pair distribution function through:

$$g_{\alpha,\beta}(r) - 1 = \frac{1}{2\pi^2 r \rho_0} \int_0^\infty \mathbf{Q} \left[S_{\alpha,\beta}(\mathbf{Q}) - 1 \right] \sin(\mathbf{Q}r) d\mathbf{Q} \tag{8}$$

The total number density of atoms is ρ_0 and $g_{\alpha\beta}(r)$ is the probability of finding an atom β at a distance r from an atom α. The coordination number can be found by integrating the partial radial distribution function; this gives the average number of β atoms in a spherical shell around α between distances of r_1 and r_2.

$$\bar{n}_\alpha^\beta = 4\pi\rho_0 c_\beta \int_{r_1}^{r_2} g_{\alpha\beta}(r) r^2 dr \tag{9}$$

A Fourier transform of the total multi-component $F(\mathbf{Q})$ defines the total pair distribution function $G(r)$, which is the weighted sum of all partial values for neutron diffraction data,

$$G(r) = \frac{1}{2\pi^2 r \rho_0} \int_0^\infty \mathbf{Q} F(\mathbf{Q}) \sin(\mathbf{Q}r) d\mathbf{Q} = \sum_{\alpha,\beta}^n c_\alpha c_\beta \bar{b}_\alpha \bar{b}_\beta \left[g_{\alpha,\beta}(r) - 1 \right] \tag{10}$$

For neutron and X-ray studies of glasses and amorphous materials the total correlation function, $T(r)$, or differential distribution function, $D(r)$, are often used rather than the G(r) which is commonly calculated for liquids in molecular simulations for example :

$$T(r) = 4\pi r \rho_0 \left[G(r) + \sum_{\alpha,\beta}^n c_\alpha c_\beta \bar{b}_\alpha \bar{b}_\beta \right] \tag{11}$$

$$D(r) = 4\pi r \rho_0 \left[G(r) + \sum_{\alpha,\beta}^n c_\alpha c_\beta \bar{b}_\alpha \bar{b}_\beta \right] - 1$$

The reason for this is that $T(r)$ highlights correlations at higher r, while the $G(r)$, is favored by those studying liquids, as it emphasizes the local structure. The differential distribution function $D(r)$ has the advantage that the bulk density is removed.

Total scattering techniques may also be applied to polycrystalline materials as well as glasses and liquids. The strength of this technique is that it can be applied to any system. This includes multiphase materials that have both crystalline and amorphous components (see Proffen 2006, this volume).

Neutron instrumentation

Although the principles of diffraction from a liquid or amorphous material are the same as that for a polycrystalline sample, the requirements for liquid and amorphous instruments are different. In amorphous materials, the intensity of the scattered signal is considerably weaker and instruments must be designed to have a high count rate and minimal background contribution to the scattered signal. High counts come at the expense of the instrumental resolution that is more critical for studies of crystalline samples. Another important requirement of instruments for liquids and amorphous materials is high real space resolution. Glasses and liquids exhibit short-range order and the resolution increases by increasing the maximum $\mathbf{Q}$ value of the data used in the Fourier Transform.

From the equation $\mathbf{Q} = (4\pi/\lambda)\sin\theta$, it can be seen there are two ways of measuring the scattered intensity as a function of scattered vector, $\mathbf{Q}$. At a steady state source the variation with $\mathbf{Q}$ can be measured as a function of the scattering angle 2θ at fixed wavelength. Whereas in a time of flight measurement λ is varied at constant 2θ.

Reactor sources use a conventional twin axis instrument with a flux of neutrons coming from the moderator (Fig. 7). Thermal neutrons have a Maxwellian distribution of velocities that is time-independent. The two axes are the monochromator and the sample, with a detector either moved through 2θ, or more commonly an array of position sensitive detectors covering a large solid angle. Diffractometers at reactor sources are usually extremely stable, have low backgrounds and the sample dependent corrections to the data can usually be calculated (more) precisely, which is suited to very accurate measurements. Individual experiments are

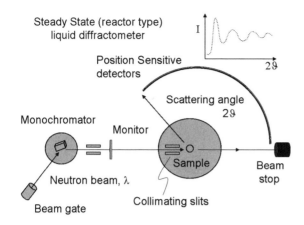

Figure 7. Schematic of a reactor type neutron diffraction instrument. In this twin axis experiment, a beam of thermal neutrons from a reactor; has a Maxwellian distribution of velocities and is time-independent. A beam of wavelength λ is scattered into a detector through angle 2θ.

however limited in **Q**-range. For example, if 0.7 Å incident wavelength neutrons are used the maximum **Q** value that can be reached is 17.9 Å^{-1}, which although ideal for most liquid scattering experiments (where $S(\mathbf{Q}) \approx 1$ by this value), it is not ideal for the study of glasses and amorphous materials which generally have strong oscillations in $S(\mathbf{Q})$ out to much higher **Q**-values. Alternatively if a wavelength of 0.35 Å is used (usually at a significant loss in flux), this extends the maximum **Q**-range to 35.9 Å^{-1} but it also pushes the minimum **Q**-value from 0.47 to 0.94 Å^{-1} (assuming a practical minimum scattering angle of $2\theta = 3°$) which may or may not be a problem depending on the position of the first peak or features in $S(\mathbf{Q})$ for a particular sample. If experiments on the same sample are performed at the two different wavelengths, the data can be merged to obtain a wide **Q**-range. Current neutron diffractometers at reactors routinely used for structural studies of liquids and glasses include D4C (ILL, France) and 7C2 (Saclay, France).

Time-of-flight neutrons for glass diffraction measurements (Fig. 8) use neutrons with epithermal and thermal energies, most commonly available at spallation sources. A single pulse of neutrons therefore contains a range of wavelengths e.g., the most useful being in the range from 0.1 to ~4 Å ,which are recorded by the arrival time at a detector (Soper 1982; Ellison et al. 1993; Howells and Hannon 1999). It is possible in a time of flight experiment to measure several diffraction patterns simultaneously using an array of position sensitive detectors at fixed angles. At each fixed angle the scattered time of flight intensity covers a slightly different **Q**-range, when combined, can range from 0.02 to 50 Å^{-1}. This enables significantly higher count rates

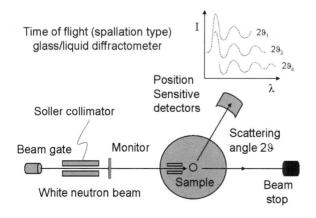

Figure 8. Schematic of a time of flight neutron diffraction instrument at a spallation source. In this configuration a series of detectors, each at fixed angle 2θ, record the scattered neutrons as a function of the time-of-flight. The data recorded at different angles cover different λ (or **Q**) ranges and are merged together in the final analysis.

to be obtained, however the corrections to the data are more significantly complex compared to single wavelength measurements and the backgrounds are generally higher. Current neutron diffractometers routinely used for structural studies of liquids and glasses at spallation sources include the Glass, Liquid and Amorphous materials Diffractometer (GLAD, IPNS), the Small Angle Neutron Diffractometer for Amorphous and Liquid Samples (SANDALS) and GEneral Materials Diffractometer (GEM, both at ISIS), and HIT-II (KENS, Japan).

DATA ANALYSIS PROCEDURE AND SAMPLE ENVIRONMENT

Corrections

In a neutron diffraction experiment the measured property is the intensity as a function of momentum transfer $\hbar\mathbf{Q}$. To obtain the differential cross section for the sample from this intensity data there are several instrument and sample dependent corrections that need to be made (Fig. 9). Firstly, a correction for the detector efficiency is performed and the conversion made to $\mathbf{Q}$. Secondly, the contribution to the scattered intensity from the sample container and instrument background must be subtracted. There are also contributions to the diffraction pattern from the attenuation of neutrons by the sample and multiple scattering effects, which must be accounted for. Because of the static approximation, corrections for inelastic scattering are also made and these can be large if the elements have low atomic number. The corrected sample data has to be normalized (divided by) a known standard, most commonly this is a vanadium sample of the same dimensions of the sample, to eliminate any additional geometric corrections. Vanadium is used because the cross section is largely incoherent (99%) and isotropic (Mayers 1984).

Attenuation corrections are generally based on simple sample geometries using the Paalman and Pings formalism (Paalman and Pings 1962). The correction uses in series of coefficients of the type $A_{S,S}(\lambda,\theta)$ that depend on both wavelength and scattering angle. Multiple scattering corrections, M_S, are generally more complicated and depend on the sample and sample environment geometries. In a typical diffraction experiment from a liquid the sample thickness is chosen such that only 10% of the incident beam is scattered, since the multiple/primary scattering ratio increases significantly if more of the beam is scattered, making it more difficult to extract the single scattering events. Numerical solutions for multiple scattering are based on a series of individual elastic, isotropic scattering events and have been evaluated for both

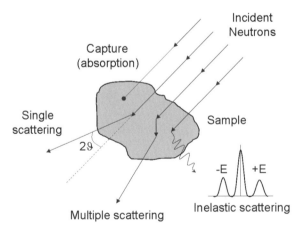

Figure 9. Schematic diagram of neutron processes which occur in the sample. Elastic single scattering events contain the useful structural information required in a liquid diffraction experiment The other processes are calculated and removed from the measured spectra.

cylindrical (Soper and Egelstaff 1980) and flat plate (Soper 1983) geometries. Attenuation and multiple scattering corrections are applied to the measured intensity in the following manner,

$$I_{corrected} = \left(\frac{I_S - B}{A_{S,SC}} - \frac{A_{C,SC}(I_C - B)}{A_{S,SC}A_{C,C}} \right) - \frac{\sigma_{SC}}{4\pi b^2} M_{SC} \tag{12}$$

Where I_S, I_C and B are the measured intensities of the sample plus container, container and background respectively. $A_{S,SC}$ denotes the attenuation from the sample in the presence of the sample plus container etc., and M_{SC} is the ratio of multiple to primary scattered neutrons (Fig. 9).

Inelastic (Placzek) corrections are necessary when the scattering nuclei move due to recoil effects or thermal motion. The correction is normally negligible for heavy elements, but is problematic for samples containing hydrogen or deuterium. In a reactor experiment this correction can be made by subtraction of a polynomial in powers of Q^2 from the differential cross section in order to obtain the distinct part of scattering after correction (Placzek 1952; Yarnell et al. 1973). The correction manifests itself as a "droop" at high angles i.e., high Q values, and is most noticeable in hydrogen containing samples. Conversely, at spallation sources the inelastic effects occur most prominently at the lowest Q-values which correspond to slower neutrons with the largest wavelengths. The effect can be minimized by using longer wavelength neutrons scattered at smaller scattering angles. Due to the wide range of energies used in a spallation type experiment the correction is considerably more complex for H or D containing samples. Parameterization methods have been attempted (Zetterstrom 1996) but a more common and practical approach is to fit a low order Chebyshev polynomial to remove the broad inelastic scattering component from the measured cross section (Soper and Luzar 1992).

Neutron absorption resonances are anomalous changes in an isotope absorption cross section at a particular wavelength. In some cases there is a corresponding change in the scattering cross section also. These normally occur at short wavelengths and are not a problem at thermal energy reactor sources, but do present problems at spallation sources where a significant number of higher energy neutrons are used. In practice the data collected around the region of a resonance is removed and discarded.

Data analysis software

Total neutron diffraction has been used to study the structure of liquid and glassy materials for fifty years. The methodology was first developed in the 1950's using reactor sources (Egelstaff 1967), but it was the advent of time-of-flight instrumentation at spallation neutron sources in the 1980's which led to a dramatic increase in the number of detectors and the need for more sophisticated data analysis software. At the current time, pulsed neutron scattering is undergoing a renaissance with the development of very high power spallation sources around the world and with these changes significant progress in data analysis and visualization software can be expected.

For reactor single wavelength experiments the corrections are relatively straight forward to implement in most software packages or programming languages and several research groups developed their own versions of the formalism outlined previously, e.g., *FOURIER* at Bristol University. The suite of software most well-known for spallation time of flight experiments was a package written by Soper et al. (1989) and entitled *Analysis of Time of flight data for Liquid and Amorphous Samples* (*ATLAS*). Another software package for total scattering data analysis called *PDFgetN* (Billinge and Kanatzidis 2004) was derived from an older *GLASS* package written at Argonne but is limited to cylindrical geometry samples. Most recently the *ATLAS* package has been upgraded at ISIS (*GUDRUN*) and IPNS (*ISAW*) (Tao et al. 2005). Figure 10 shows some examples of time of flight data and the corrections used in *ISAW* to obtain the interference function $F(\mathbf{Q})$. Note the comparative scales of the y-axis for the Plazeck correction for the two samples.

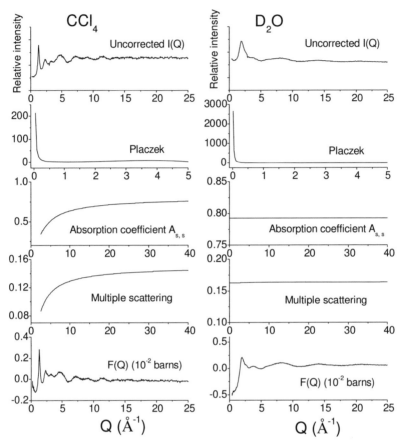

Figure 10. Examples of the measured intensity and correction factors associated with analyzing data from liquid CCl$_4$ and D$_2$O. (from top to bottom) normalized but uncorrected sample scattering intensity vs. **Q**; calculated Placzek effect in terms of a dimensionless ratio between the Placzek corrected scattering cross section and the uncorrected one (note the difference in *y*-scale); calculated coefficient of neutrons attenuated by the sample; calculated multiple scattering in terms of multiple/primary scattering ratio; corrected sample scattering intensity vs. *Q*. Corrections were calculated at the scattering angle $2\theta = 113.6°$ for GLAD diffractometer at IPNS.

Sample environments

Sample cans. Containers with crystalline diffraction peaks introduce difficulties in accurately extracting the weak diffuse scattering signal of a liquid or glass. Therefore sample containers are usually made of a mechanically strong, incoherent scattering material such as vanadium, which yields a constant scattering signal vs. **Q**. In the case of "null scattering" TiZr containers, the fact that Ti has a negative coherent scattering length is used to produce an alloy (of 47.8% wt Ti) such that the sum of the coherent scattering lengths are zero and produces no Bragg scatter. This is a common container for aqueous samples as water reacts with vanadium.

The sample volume of glass or liquid is usually relatively large in a neutron scattering experiment, generally several grams, if reasonable counting statistics are to be obtained. Cylindrical geometry cans are usually preferred because the attenuation and multiple scattering corrections are more accurate than for flat plate cans. However, in the case of highly absorbing materials or liquids with a large incoherent component such as hydrogen flat plate containers of 1 mm or less (together with a large size incident neutron beam) are generally used to minimize multiple scattering effects.

Furnaces. Changes in the structure of melts at elevated temperatures can be studied using a variety of furnace types. These range from furnaces with cartridge heaters situated above and below the sample that can be used to heat samples from room temperature to ~350 °C. To furnaces based on vanadium resistive elements with several aluminum heat shields, which can be used to heat samples to temperatures of ~1100 °C at which point the vanadium starts to soften. For these experiments the samples are normally contained in a silica tube. In both cases experiments can suffer from significant temperature gradients across the sample, in the attempt to keep a minimal amount of (crystalline) material in the beam.

Refrigerators, baths and cryostats. A closed cycle refrigerator or displex based on a cold finger design is generally used to cool a sample down to ~20 K or so, and adequate for most liquid, glass sample studies. Oil baths provide greater stability around room temperature reaching ~−20 °C and have been used for clathrate studies. For more accurate temperature control at lower temperatures an "Orange" cryostat with a helium and nitrogen reservoir is required, e.g., for studying abrupt transitions in amorphous ices.

Pressure cells. Gas cells can be used to study fluids up to pressures of ~0.6 GPa. TiZr or Al alloys are most commonly used. Gas pressure cells have been used to study clathrates and clays using a gas such as He as the pressure medium.

High pressure neutron diffraction studies can be carried out using the Paris-Edinburgh press (Loveday et al. 1996; Le Godec et al. 2004). These pressure cells can be used to generate pressures of up to 25 GPa. Pressure is usually generated by two opposed torroidal anvils, made of tungsten carbide or sintered diamond, that deform a metal gasket (usually TiZr). Incident neutrons can be directed down the compression axis with neutrons detected in the plane of the gasket, a configuration used for example on the instrument PEARL at ISIS, or neutrons can be directed through the gasket and detected in the same plane, as on GLAD (IPNS). The anvils themselves can be coated with boron nitride to act as collimators in this latter configuration although there can still be contributions form the anvils, which are often hard to subtract because the anvils deform when compressed. Since the anvils close on compression, typically from 1.6 mm to 0.8 mm at 5 GPa for a TiZr gasket, the scattered signal decreases significantly with increasing pressure. Several noticeable studies have been completed including studies of amorphous ices and GeO_2. More recently studies have been completed on Mg-silicate glasses and vitreous B_2O_3.

Containerless levitation. Studies of supercooled liquids using furnaces can be difficult or impossible because the container induces nucleation. In the containerless leviator technique (Weber et al. 1994; Landron et al. 2000) a bead of refractory sample material is placed on a water cooled conical vanadium nozzle and levitated by a gas jet, the bead can be laser heated to temperatures of up to 3050 K (Fig. 11)(Weber et al. 2003a). The absence of heterogeneous nucleate sites means that liquids can be supercooled up to a few hundred degrees below the stable liquidus curve. Although there is no container to subtract the droplet only levitates a few hundred microns, such that its' mid-point is level with the top of the nozzle, attenuating the lower half of the scattered neutrons in the vertical plane, so accurate backgrounds are crucial. Nonetheless some studies of liquid Al_2O_3 and $CaO-Al_2O_3$ have been performed (Weber et al. 2003a) and with the advent of new spallation sources of high flux and with developments in focusing optics, there are future opportunities for the *in situ* study of super cooled liquids by containerless levitation.

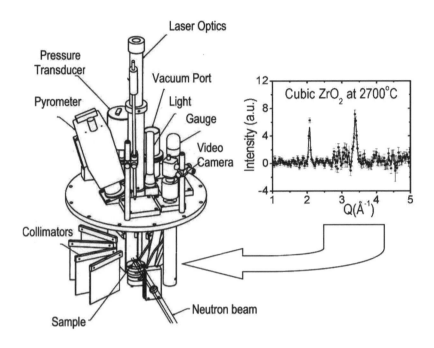

Figure 11. Containerless levitator with laser heating as used for *in situ* liquid diffraction measurements at GLAD at IPNS. The diffraction pattern for ZrO_2 at 2700 °C is shown (inset).

INTERPRETATION OF NEUTRON DIFFRACTION DATA

Glass and liquid structure

Consider a liquid such as SiO_2, frequently used as a starting point for discussions of geological or commercially important glasses. The basic structural units forming pure SiO_2 glass are SiO_4 tetrahedra, that corner-share to form a three dimensional framework. If additional oxide components are added, for example CaO or Na_2O, that modify the silicate framework by forming non-bridging oxygens, and the glass-forming ability is often improved. Good glass formers, including those in metallic alloys, tend to be formed close to stable or metastable eutectics where there are no stoichiometric crystalline phases.

The simplest structural model of a glass, such as the continuous random network (CRN) (Zachariasen 1932) indicates that a simple glass such as an alkali silicate would consist of a framework of SiO_2 tetrahedra and randomly arrange network modifying cations. However, the local atomic arrangements (and intermediate range) are far from random and governed by strong inter-atomic forces. Several studies using diffraction techniques have shown that glass structures are more ordered than the CRN model and in the case of alkali silicates, domains rich in the modifying cations exist. Such modified random network models form a good basis for interpreting diffraction data but the interpretation of the full glass structure is difficult, based solely on the pair distribution function measured in a neutron diffraction experiment. It is common therefore to combine neutron results with other data such as X-ray spectroscopy, NMR and thermodynamic property data. Diffraction modeling techniques such as Reverse Monte Carlo (McGreevy 2001) and Empirical Potential Structure Refinement (Soper 2005) have been developed in recent years to fit model structures of glasses and liquid diffraction data, in an attempt to provide an analogy to the modeling crystalline powder patterns with Rietveld refinement methods.

Raman spectroscopy is often used to study glasses and is very sensitive to changes in local structure; the results are only qualitative however. Nuclear magnetic resonance spectroscopy has long been used in combination with diffraction to interpret silicate glass structure, usually in terms of the modification of the large polymerized silicate, aluminosilicate and aluminate frameworks (Q speciation). NMR is isotope specific and can be used for example to probe [11]B, [29]Si, [27]Al and [18]O environments providing information on bond angle distribution (Grandinetti 2003) and ring statistics as well as the modification of network. Other spectroscopic techniques such as Extended X-ray Absorption Fine Structure (Filipponi 2001) are also element specific and act as a local structural probe, EXAFS is restricted however to high Q values although is sensitive to low concentrations and can be very effectively combined with diffraction techniques (Filipponi et al. 2001). Generally, one of the goals of the study of liquid and glass structures is an attempt to link the bulk macroscopic properties to the microscopic structure (Wright et al. 2005) and structure is usually related to both density and T_g.

With the development of third generation synchrotron sources, there has recently been huge progress in producing instrumentation for using highly penetrating "neutron" like X-rays of ~100 keV (Neuefeind 2002) for the study of liquid and glass structure. These high energy X-rays act as a bulk probe and cover a wide Q-range, comparable to neutron instruments at spallation sources. Neutron and high energy X-ray diffraction can be viewed as complementary "sister" techniques and are particularly useful for studying oxide or hydrogenous systems, as while neutron scattering lengths vary erratically across the periodic table, X-ray form factors vary as a function of atomic number.

Ranges of order in glass and liquid structure

The structure factor and associated real space distribution function for a multi-component system can be complex and sometimes any peaks past the nearest neighbor can be difficult to interpret, but the data provide hard constraints for any structural model. The use of techniques such as isotopic substitution, isomorphic substitution or combined neutron and X-ray data sets usually enable more detailed information to be obtained, than any single data set can yield.

A useful technique, used to interpret diffraction data of amorphous and liquid structures is to delimit characteristic distance ranges, a methodology developed by Wright (Wright 1979; Wright et al. 1980) and Price (1996), see Figure 12. Range I relates to the nearest neighbor distances corresponding to the basic structural unit, i.e., the short-range order. Range II is associated with the connectivity of these structural units. Range III relates to the presence of larger correlations of several structural units and usually referred to as intermediate range order and is associated with the presence of a first sharp diffraction peak in the neutron spectra. It is usually in range III that there is most disagreement between experiment and most structural models.

Short range order and connectivity

Peaks reflecting range I are the sharpest in $G(r)$ and contain information on the first neighbor coordination number and bond length. For SiO_2, short range order refers to the Si-O and O-O correlations that reflect average size and shape of the basic 'intra-molecular' tetrahedron. Some chemical knowledge is usually assumed to deduce likely atom pair bond lengths, through comparison with similar known crystal structures or comparison with bond valence theory. There are various methods to calculate the coordination number once the atom pair is identified. If there is a clear minimum after the first peak and only a single atom pair type is expected, the coordination number can be calculated directly by integration to the first minimum. If there are overlapping peaks Gaussians can usually be fitted centered on the mean bond length (convoluted with the Q-space step function) to decompose the contributing atom pair correlations.

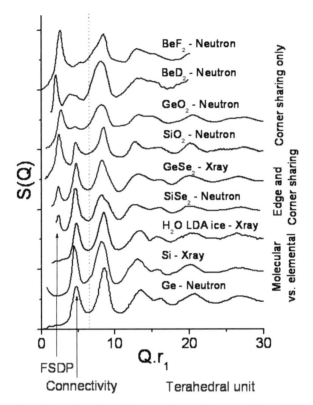

Figure 12. Comparison of several tetrahedral glasses and amorphous materials, scaled by the first peak in $G(r)$, denoted r_1. The three structural ranges are denoted (I) tetrahedral unit, (II) connectivity and (III) first sharp diffraction peak (FSDP) (Benmore et al. 2005).

The next distance range, Range II order reflects the connectivity of the main structural units and overlaps with Range I, for example may be associated with the ratio bridging and non-bridging oxygens, or degree of edge to corner sharing units. Range II distances are characteristic of next nearest neighbor Si-Si distance in the case of SiO_2, reflecting inter-tetrahedral distances. The peaks are broader than the range I peaks and reflect the distribution of Si-O-Si inter-tetrahedral and torsional bond angles. Normally these angles cannot be determined directly from a single neutron measurement, but need to be combined with (for example) an X-ray dataset on the same sample.

Intermediate range order

Range III is characteristic of intermediate range order, the formation of cages, rings, layers, chains or other structures through the connection of the basic structural units. In real space this topology is usually characterized by a distance range that extends up to 10-20 Å, although order may also extend beyond that (Salmon 2004). The most prominent signature of intermediate range order in the measured structure factor $S(Q)$ is the presence of a first sharp diffraction peak (FSDP), that is a peak or feature below $Q \cdot r_1 < 3$ Å^{-1} for tetrahedral materials as shown in Figure 12. The FSDP in network glasses can be attributed to structural correlations on a "length scale" in real space of periodicity $2\pi/Q_P$, which decreases in magnitude with increasing r. Where Q_P is the position of the FSDP. The FSDP peak height reflects both the degree of periodicity and is also a function of the packing of the structural elements.

It has been stated that it is futile for researchers performing structural studies on glasses to search for a universal theory of the first sharp diffraction peak (FSDP) (Wright et al. 2004). Nonetheless there have been several suggestions for the origin of the FSDP which are listed below :

(1) *Crystalline models*. e.g., P.H. Gaskell and D.J. Wallis (Gaskell and Wallis 1996). This model works well for layered structures such as $ZnCl_2$, but since the FSDP is also observed in some liquids this cannot be a universal explanation for most glass structures (Madden and Wilson 1996).

(2) *A range of structural elements can give rise to the FSDP*. A low **Q** peak can arise from the structure factor for the molecular centers although it's detailed shape is modified by the simultaneous rapid fall-off of the molecular form factor (Wright et al. 1985). That is to say, the FSDP is dominated by inter-molecular cation-cation interactions, but affected by the intra-molecular shape, as illustrated in Figure 13.

(3) *The FSDP is a chemical-order pre-peak due to interstitial volume (voids)* around cation-centered structural units (Elliott 1991a, 1991b). This is a similar argument to (2) but whereas the FSDP may arise from interactions between atoms across a void in case (2), in case (3) it is argued that it arises from void-void interactions. The void model explains a difference in the neutron FSDP height for alkali silicate glasses which has contributions from both positive (Na, K) and negative (Li) neutron scattering lengths.

Since both (2) and (3) correspond to the same structure either could be true, although we note that neutrons only scatter from atomic nuclei.

If intermediate range structure is to be observed directly in real space, correlations such as the differential distribution function $D(r)$ are required, or even $r \cdot D(r)$. The example of amorphous ices provides a good example of the variety of different structures that can be observed in this regime, which is dominated by oxygen-oxygen correlations at high densities

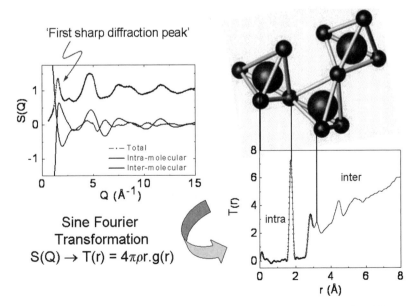

Figure 13. A deconvolution of the $S(\mathbf{Q})$ and $G(r)$ functions into "intra-molecular" and "inter-molecular" parts for the case of vitreous GeO_2 (Sampath et al. 2003; Guthrie et al. 2004).

as illustrated in Figure 14. The intermediate range order is greatly affected by the number and position of interstitial molecules which lie just outside the first coordination shell defined by the tetrahedron. Very high density ice has on average two interstitial molecules and high density amorphous ice one.

There are long range fluctuations in glassy and amorphous solids that reflect long range fluctuations in density. These cannot easily be discerned by wide angle diffraction but require small angle techniques (Wright et al. 1980). A description of long range order from PDF techniques is given elsewhere in this book (Proffen 2006, this volume).

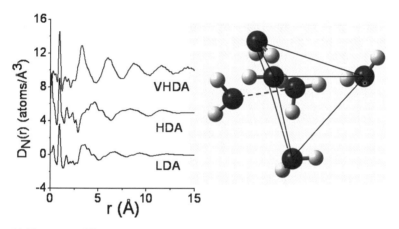

Figure 14. The neutron differential distribution function for Low density amorphous (LDA), high density amorphous (HDA) and very high density amorphous (VHDA) ice. Right : A schematic diagram of the local structure of HDA ice.

PARTIAL STRUCTURE FACTOR DETERMINATION

Considerable insight into the structure of glassy and liquid materials can be achieved by the separation of the total diffraction pattern $F(\mathbf{Q})$ into it's contributions from the individual partial structure factors $S_{\alpha\beta}(\mathbf{Q})$. Experimental methods for partial structure factor determination include, neutron isotope substitution, isomorphic substitution, anomalous X-ray scattering and combining neutron and X-ray data. A review of these methods has been given by Suck and Maier (1993).

Isotopic substitution

The neutron scattering length of elements may vary for different isotopes. If the contrast in scattering length is sufficiently high i.e., practically $\Delta b < 3$ fm, then samples with different isotopic composition can be used to determine unequivocally the partial structure factors for each atom pair. Neutron diffraction with isotopic substitution consists of measuring the differential cross section for several identical samples with different isotopic composition. For a binary sample this will require three separate measurements if all the partial structure factors are to be determined independently.

Following the seminal paper of Enderby et al. (1966) inventing the technique of isotopic substitution in neutron diffraction, the technique has been applied to several systems (Enderby et al. 1967, 1973) making use of pronounced difference in neutron scattering lengths between different isotopes of the same element, e.g., for Nickel $b(^{58}\text{Ni}) = 14.4$ fm and $b(^{62}\text{Ni}) = -8.7$ fm. For a multi-component system of n elements there are $n(n + 1)/2$ partial structure factors. If neutron diffraction experiments are made on two structurally identical samples, except for the isotopic composition of one of the components α, a *first order difference function* can be extracted which contains information on the interactions only involving the substituted species i.e., α-*X*, where *X* refers to all the species in the system. If three experiments are made and the isotopic composition of the same element is again different a *second order difference function* may be obtained to obtain the partial structure factor for the α-α interactions. The middle isotopic composition is usually a 50:50 mixture of the two extremes to maximize contrast. Of course the experimental statistics become progressively worse as more differences are taken and sufficient isotopic contrast is required to obtain meaningful data.

One of the largest isotopic neutron scattering length contrasts is between hydrogen and deuterium, which is ideal for the study of aqueous solutions, and has been successfully implemented on a range of systems using the SANDALS instrument at ISIS for many years. We give the example of the second order difference technique applied to pure water. In this case the measured radial distribution function can be written :

$$G_{H_2O}(r) = \frac{c_H^2 b_H^2}{A} g_{HH}(r) + 2\frac{c_H c_O b_H b_O}{A} g_{OH}(r) + \frac{c_O^2 b_O^2}{A} g_{OO}(r) \qquad (13)$$

where $A = c_H^2 b_H^2 + 2c_H c_O b_H b_O + c_O^2 b_O^2$.

If three samples H_2O, D_2O and a 50:50 mixture (denoted HDO to maximize the contrast) are measured. Given that $b_H = -3.74$ fm, $b_D = 6.67$ fm, $b_{MIX} = (b_H + b_D)/2$ and $b_O = 5.80$ fm we can write three simultaneous equations,

$$G_{H_2O}(r) = 0.318\, g_{HH}(r) - 0.492\, g_{OH}(r) + 0.190\, g_{OO}(r) \qquad (14)$$

$$G_{D_2O}(r) = 0.486\, g_{HH}(r) + 0.423\, g_{OH}(r) + 0.091\, g_{OO}(r)$$

$$G_{HDO}(r) = 0.113\, g_{HH}(r) + 0.446\, g_{OH}(r) + 0.441\, g_{OO}(r)$$

These equations can be solved to yield the partial pair distribution functions $g_{HH}(r)$, $g_{OH}(r)$ and $g_{OO}(r)$ and the solutions are shown in Figure 15. It should be noted that the $g_{OH}(r)$ partial in

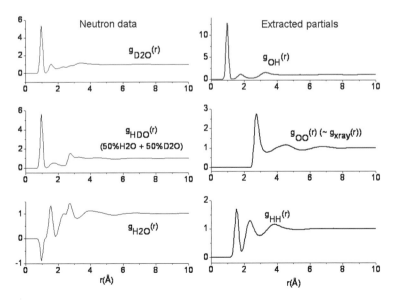

Figure 15. The second order difference H/D substitution technique applied to liquid water, see Equation (14).

$G_{H_2O}(r)$ has a negative Faber-Ziman weighting factor because of the negative scattering length of H. This means that the $g_{OH}(r)$ partial is inverted in $G_{H_2O}(r)$, giving negative peaks.

The advantage of this technique is that systematic errors are small, provided the experiments are all conducted on the same instrument under the same experimental conditions. The limitation is that all the different isotope samples need to have the same structure, which can sometimes be problematic for glasses or amorphous materials out of equilibrium, or if some isotopes contain significant impurities which may effect the structure.

Combining neutron and high energy X-ray data

A more recent trend is combining neutron and high energy X-ray data. This has the advantage that both measurements can be performed on the same sample. However the two different experimental set-ups, with different instrumental resolutions mean that the systematic errors are much greater than the isotopic substitution method. However this method is ideally suited to the study of oxide glasses where the contrast is large, with neutrons most sensitive to the oxygen interactions and the X-ray signal is dominated more by the signal associated with the cations. Usually a first order difference function is obtained, eliminating one species in order to deconvolute two overlapping peaks in real space. This means that specific partial structure factors are not extracted, but rather removed from the spectra. The elimination of a partial structure factor is similar but no quite as straight forward as for isotopic substitution since the weighting factors are not Q-independent, but are governed by the electronic form factors $f_\alpha(Q)$.

For the case of Germania (Sampath et al. 2003; Guthrie et al. 2004) we can write a general diffraction equation for both neutron and X-ray scattering from the same sample,

$$S_{GeO_2}(r) = W_{GeGe}\,S_{GeGe}(Q) + W_{GeO}\,S_{OH}(Q) + W_{OO}\,S_{OO}(Q) \qquad (15)$$

Where $W_{\alpha\beta}(Q)$ are the normalized Faber-Ziman weighting factors. For neutrons we use the same format we previously described for water, but for X-rays the neutron scattering length b_α is replaced by the electronic form factor for that element, $f_\alpha(Q)$. These weighting factors are shown graphically for the case of GeO_2 in Figure 16.

We can see that for the neutron case the most weighting goes to the oxygen-oxygen partial structure factor, whereas this has the least weighting in the X-ray function. The two techniques are therefore sensitive to different aspects of the same structure.

Reverse Monte Carlo and empirical potential structure refinement

Given that the pair correlation function is a one-dimensional representation of the three dimensional structure, a more complete structural picture invariably requires some form of modeling. In the case of a system for which the interaction is purely pairwise additive there is theoretical justification for the determination of a three dimensional structure from $g(r)$. This is not generally the case for a real system where triplet and higher body correlations usually exist. However additional "chemical" constraints may help provide a realistic model of the system.

In powder diffraction studies of crystalline system Rietveld refinement is used to model the crystal structure. Analogous techniques are available for modeling amorphous and liquid structures. There are two models that are most commonly used; Reverse Monte Carlo (RMC) modeling and Empirical Potential Structural Refinement.

Reverse Monte Carlo modeling (McGreevy 2001, 2003) uses a three-dimensional arrangement of atoms as a starting configuration. This may be based on a crystal structure or random distribution of atoms. Atoms are moved randomly one at a time and the structure factor calculated and compared to experimental data. If the random move improves the agreement with the experimental data then the move is accepted with some probability. The random movement of atoms is continued until the difference between the modeled and experimental data is

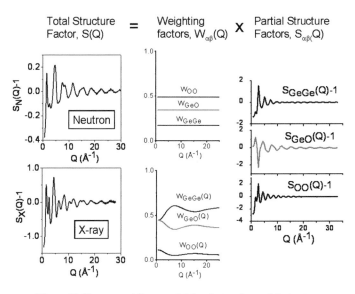

Figure 16. Neutron and X-ray weighting factors for partial structure factors in the case of vitreous GeO_2, see Equation (15).

minimized. A Reverse Monte Carlo model can be used with different data sets, for example neutron and X-ray diffraction data and the structure can be fitted to both simultaneously e.g., see Figure 17. Reverse Monte Carlo has proven to be effective in analyzing glassy or amorphous materials particularly if independently derived constraints, coordination numbers form NMR, X-ray or EXAFS, for example, are used. Reverse Monte Carlo models can be adapted to refine models with pre-determined topology and they can also be used to study the diffuse scattering in disordered crystalline structures (Keen et al. 2005).

The Reverse Monte Carlo technique does not use inter-atomic potentials. However similar method to RMC, Empirical Potential Structural Refinement does use interatomic potentials

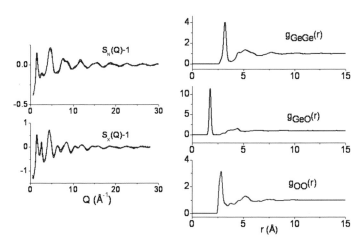

Figure 17. RMC modeling of neutron and X-ray data. Left: The measured neutron and X-ray structure factors for vitreous GeO_2 and RMC fits calculated from a 3D model. Right: The partial pair distribution functions calculated from the model.

and these are determined from the difference between the measured and calculated partial pair correlation functions (Soper 2005). The Empirical Potential Structural Refinement method is generally applied to molecular liquids which easily become stuck in local minima using RMC. Firstly, a three dimensional model with the correct density is constructed and a reference potential chosen. An empirical potential is then determined from each partial pair correlation function and these are then added as correction terms to the reference potential. The modeling proceeds iteratively until there are no significant differences between the model and the data. Empirical potential structural refinement has been used to model complex molecular liquids and although recent applications also include amorphous ices (Finney et al. 2002a,b) and water at high temperature and pressure (Soper 2000, 2001).

One disadvantage of both types of refinement method is that they can result in non-unique solutions, different structures that fit the experimental data equally well. However these structures may differ only very slightly. At best, if several additional constraints are made and a good fit obtained to the diffraction data, a realistic 3D model of the glass may be obtained.

STUDIES OF GLASSES, LIQUIDS AND AMORPHOUS MATERIALS

There are a large number of elastic neutron scattering studies of liquids and amorphous materials, representing in excess of fifty years of studies. These range from studies of elemental and molecular liquids, to molten salts, oxide glasses and clays of interest to the materials and earth science communities. Below we give some examples of the diversity of applications of liquid and amorphous state diffraction studies in areas that overlap materials and earth science. Rather than an exhaustive survey we illustrate the experimental strategies and progress that has been made in linking the microscopic structure to observable properties.

Water and aqueous solutions

Water is a liquid that has been studied extensively by neutron diffraction (see also Cole et al. 2006, this volume). Neutron diffraction in many ways is an ideal probe for aqueous solution structures, as alternative techniques, such as X-ray diffraction are only indirectly sensitive to hydrogen bonding through the measurement of oxygen-oxygen and oxygen-ion correlations. Neutrons are very sensitive to the presence of hydrogen (deuterium) and hydrogen-ion interactions. Neutrons also enable the $g_{OO}(r)$ $g_{OH}(r)$ and $g_{HH}(r)$ contributions to the pair distribution functions to be extracted (Narten and Hahn 1982; Narten et al. 1982). Water at room temperature has been studied by neutron diffraction using deuterium-enriched samples to reduce the incoherent scatter from ^{1}H. Water has been shown to consist of tetrahedrally coordinated oxygen ions at room temperature (Narten and Hahn 1982). However the possibility of larger ring structures has often been argued, most recently by Nilsson and co-workers (Nilsson). The water molecule is non-spherical and EPSR has been used (Soper and Silver 1982; Soper 2000, 2001, 2002; Soper and Rossky 2000) to determine the most probable configurations of water molecules and provides a test for the inter-atomic potentials that have been used to simulate liquid water structure as a function of temperature and pressure.

Neutron diffraction with isotopic substitution can be used to understand the distribution of ions surrounding water molecules in electrolyte solutions and solutions containing large molecules such as DNA or proteins (Neilson et al. 2001). Neutron scattering lengths are isotope dependent and if difference techniques are used the partial contributions to the total structure factor and pair distribution function can be established. Studies of electrolyte solutions using different isotopes of Li and Ni have been performed (deJong et al. 1996; Howell and Neilson 1996, 1997; Ansell and Neilson 2000); different isotopes of Cl can be used to study the hydration shell around the Cl anions and the relative degrees of interaction with the cations, for example the hydration shell is more strongly influenced by the cations in Li-Cl solutions than in Ni-Cl solutions. In trying to understand the structures of aqueous solutions it is important to establish how solute and solvent molecules interact.

The study of aqueous solutions has been extended to include studies of water in the interlayer regions of clays (Sposito et al. 1999; Skipper et al. 2000). The general view has been that the structure of interlayer fluids is similar to that of a concentrated ionic solution. However, isotopic substitution has been used to investigate the liquid structure more thoroughly, establishing that the dominantly octahedral coordination of alkali ions, Na and K, in the interlayers of Wyoming montmorillonite are bonded to the siloxene clay surfaces through the complex interaction of hydrophobic molecules,

Simple oxide liquids and glasses

Natural silicate liquids are compositionally complex. It has been the convention therefore for experimental studies to be confined to systems that are simple analogies of natural compositions in order to understand their behavior. For neutron diffraction studies unravelling the structure is further complicated because oxygen may form up to 60-70% of the sample. Because neutrons are sensitive to the presence of oxygen, a large fraction of $S(Q)$ and the real space transform can result from oxygen-related correlations. In pair distribution functions of oxides, the O-O peak often overlaps cation-cation interactions. Unequivocal contributions to the total structure factor cannot be made using isotopic difference techniques but even with simple binary oxide systems, careful experiments as a function of composition, temperature or pressure can provide considerable insight into the salient changes in liquid and glass structure.

Vitreous SiO₂. Vitreous SiO_2 has been studied by both neutron and X-ray diffraction. An early model of its structure is that of Zachariasen (1932) and supposes that the bonding nature of the glassy phase is similar to that of crystalline SiO_2. The CRN SiO_2 glass differs from crystalline SiO_2 in that the individual SiO_4 tetrahedra are connected randomly to give an disordered structure lacking long range order (Wright and Leadbetter 1976; Wright 1985a; Wright and Sinclair 1985; Wright et al. 1991). This continuous random network contrasts with earlier models that noted that the diffuse scattering from SiO_2 was coincident with the Bragg peaks from crystalline phases and was used to suggest that the glass was formed of crystallites, small crystal units. Warren (1969) and Mozzi and Warren (1969) rejected the crystallite model on the basis that there was no additional scattering at low values of **Q**. Modified crystallite models have been proposed in which the network is not considered completely random but in which there are spatial fluctuations indicative of more ordered regions. Both the continuous random network and modified crystallite models produce three dimensional networks but these differ in the way in which defects are distributed. The continuous random network would have defects randomly distributed throughout the network, while there would be defects concentrated in less ordered regions, in modified crystallite models, with higher ring size distributions.

Neutron diffraction studies of SiO_2 provide data of up to 45 Å^{-1} and have been in interpreted using the characteristic distance ranges favored by Wright (Wright et al. 1980; Wright 1985b). The main structural unit, the SiO_4 tetrahedra is assumed to have a symmetric Si-O peak with a mean inter-atomic distance of 1.605-1.608 Å for neutrons. This value is slightly higher for X-rays (Wright 1994). The connectivity of the tetrahedra involves an inter-tetrahedral bond angle and two torsion angles, the distribution of which would be uniform if the network were completely random. However, recent NMR studies have shown that in SiO_2 glass the shorter Si-O bond lengths correlate with smaller Si-O-Si angles (Grandinetti 2003). The topology of amorphous materials is investigated using modeling techniques and the models are compared to the diffraction data. Invariably 3D models include a distribution of different ring sizes of linked tetrahedra which maybe compared to Raman spectroscopy to identify the presence of three and four-membered planar rings. Many studies of SiO_2 using spectroscopic techniques compare the structure with a crystalline phase, usually β-cristobalite (e.g., Phillips 1982,

1986), however these models are generally inconsistent with the diffraction data (Galeener and Wright 1986). Longer range fluctuations in density can in principle be studied by small angle techniques; such studies show no evidence for crystallite or modified crystallite structure (Wright et al. 2005).

K_2O-SiO_2 glasses. The addition of alkalis to a silicate system modifies the network by formation of non-bridging oxygens. In Q-speciation terms this is described by an equilibrium of the form, $2Q^n = Q^{n+1} + Q^{n-1}$, recorded in the quenched glass structure by NMR and Raman spectroscopy. Temperature-dependent configurational changes are associated with an anomalous temperature-dependence of the heat capacity. Neutron diffraction measurements provide data on the short-range order of glasses and liquids, i.e., the SiO_4 units which are not expected to change significantly with temperature, other than through the small effects of thermal broadening and slight changes in due to different non-bonded oxygen/bonded oxygen distances and changes in intermediate range order with increasing K_2O content.

In the $S(\mathbf{Q})$, the first peak is attributed to changes on the medium length scale (Cormier 2003; Majerus et al. 2004). For K_2O-SiO_2 glasses, the first two peaks in the $S(\mathbf{Q})$ occur at 0.97 and 2.1 Å^{-1}. These two features replace the single sharp peak in pure SiO_2 at 1.60 Å^{-1} and result form the changes in intermediate range order as potassium is introduced. When glasses are heated form 300 to 1273 K, the intensity of the first peak increases, while the peak at 2.1 Å^{-1} decreases in intensity. In real space, peaks at 1.62 and 2.67 Å correspond to Si-O and O-O correlations respectively. The K-O peak overlaps with that of O-O and is therefore difficult to resolve. The most noticeable change in the pair-correlation function with temperature is the change in the appearance of a shoulder at 3.7 Å on the low-r side of a peak attributed to K-O and K-Si at 4.0 Å. As temperature is increased the shoulder at 3.7 Å decreases in intensity and this is interpreted as a change in the geometry of the potassium environment. Potassium is coordinated by 7-12 oxygen atoms with distances (Cormier 2003; Majerus et al. 2004) of 2.6 to 3.6 Å. The potassium ions occupy segregations consistent with the modified random network model of alkali silicate glasses (Greaves and Davis 1974; Greaves et al. 1981; Greaves 1985). With temperature increase, the alkali channels are expanded, consistent with more rapid alkali diffusion and decoupling from the network Si-O diffusion. Models of K_2O-SiO_2 glasses, based on the diffraction data suggest that there are changes in the Si-O-Si bond angles that may reflect differences in the ring statistics of the corner-shared SiO_4 tetrahedra, but it is not clear how these change with Q-speciation.

Na_2O-SiO_2 glasses. The structure of Na_2O-SiO_2 glasses have been studies by neutron diffraction and combined with Raman spectroscopy and Reverse Monte Carlo modeling (Zotov et al. 1998; Zotov and Keppler 1998). The Reverse Monte Carlo model is used to determine the partial structure factors bond-angle distribution and ring statistics which are then compared with the spectroscopy data. Whilst the Reverse Monte Carlo model is not a unique fit to the diffraction data it reproduces feature such as the first peak in the $S(\mathbf{Q})$ very well and also yields short range order information consistent with the non-random distribution of sodium and non-bridging oxygen ions suggested by other studies of alkali silicate glasses.

The Si-O-Si bond angle is the simplest structural feature that can be obtained form the Reverse Monte Carlo fit and reflects the distribution of Q^n species in the glass. The asymmetric distribution of Si-O-Si bond angles reported in this study (Zotov et al. 1998; Zotov and Keppler 1998), reflect a clustering of Q^4 units and not a random Q^3-O-Q^4 distribution. The model is further interpreted as reflecting a large number of 3- and 4-membered Q^4 rings, again consistent with the non-random distribution of non-bridging oxygens expected in the modified random network models of alkali silicate glasses.

When heated to the temperature of the glass transition and also at temperatures slightly above, neutron diffraction studies of this sodium silicate glass do not show any significant changes in either short- or medium-range order. The position or shape of the first diffraction peak in the $S(\mathbf{Q})$ does not change and emphasizes that there is no apparent direct connection between the structural relaxation process and mid-range order. The only changes with temperature reported are minor changes in the Si-O distance and a broadening of the Si-O correlation in the PDF (Zotov et al. 1998; Zotov and Keppler 1998), these are interpreted as thermally activated distortion of the SiO_4 tetrahedron.

CaO-SiO₂ glasses. The conventional view of binary silicate glasses is one of a silicate network with interstices of modifying ions distributed in a non-random way. The role of neutron and X-ray diffraction is to establish the positions of atom pairs with sufficient precision to test structural models of the distribution of modifying cations. Calcium-bearing systems lend themselves to isotopic and once separated the $\Delta S(\mathbf{Q})$ differences can be Fourier transformed to isolate $G_{CaX}(r)$ to yield Ca-O distances and coordination numbers. In addition, and more importantly, Ca-Ca and Ca-Si correlations can be distinguished and these provide information on the mid-range order. Compositions close to $CaSiO_3$ have been studied by Gaskell (Gaskell et al. 1991; Cormier et al. 1998, 2001) and compared with the equivalent pyroxenoid structures. The crystal structures in this case consist of Si-O chains with a close-packed oxygen sub-lattice, Ca cations occupy octahedral interstices. In some of the models of $CaSiO_3$ glass produced by Gaskell (Gaskell et al. 1991; Cormier et al. 1998, 2001), the Ca-O polyhedra (with a mean coordination number of 4.67) form planar sheets similar to those in the crystalline phases. Were these sheets to be buckled, Gaskell notes that the second Ca-O distances would be at 5.3 Å, and these distances are absent form the measured pair distribution function. The Ca-O interstices are therefore planar and indicate a high degree of ordering in the first neighbor coordination shell and beyond.

CaO-Al₂O₃-SiO₂ glasses. Calcium aluminosilicate glasses of low silica content can be produced by conventional melt-quenching techniques and have small amounts of SiO_2 added to assist in glass formation. The addition of SiO_2 to the $CaO-Al_2O_3$ composition influences physical behavior, the glass-forming ability and also the glass transition temperature (T_g). That is the rheological properties. There is a maximum in the glass transition temperature at 10-20 mol% SiO_2 and on the basis of several structural models it has been suggested that as SiO_2 is added, SiO_4 tetrahedra are incorporated in to the aluminate framework and non-bridging oxygens become more randomly distributed, the net polymerization of the system increases and viscosity increases.

Combined neutron and X-ray measurements (Cormier et al. 2004, 2005) allow structural changes to be followed according to composition and by using isotope difference techniques the Ca-Ca and Ca-tetrahedra correlations can be isolated. These results show that aluminum is dominantly fourfold coordinated by oxygen in the low silica compositions and forms a Q^3 species. As SiO_2 is added, Si is localized to highly polymerized Q^4 species that increase net polymerization. The addition of further SiO_2 results in a more even distribution of SiO_2 and continuous structural modifications occur, to the point where the amount of octahedrally coordinated Al-O also increase, and a dominantly tetrahedral aluminosilicate network changes to one in which SiO_4 dominates.

MgO-SiO₂ glasses. As outlined above, many structural studies of glass are limited to compositions where glasses are easily formed. More fragile liquids are therefore not generally studied. This is particularly critical for neutron diffraction where relatively large amounts of sample are required. With specialized synthesis techniques however it is possible to collect diffraction data for exotic glass forming systems. A study of $MgO-SiO_2$ glasses, made by containerless levitation techniques has recently been completed (Wilding et al. 2004b, 2004a). This study has shown that there are significant changes in glass structure over the compositional interval 50-33% SiO_2, corresponding to the mineral compositions enstatite to forsterite. In this study the partial structure factors were extracted by combining neutron diffraction data with high energy X-ray data and the weighted difference $\Delta S(\mathbf{Q})$ used to eliminate Si-O contributions to the $S(\mathbf{Q})$ and real space transform.

The combined data sets (Fig. 18) show a non-linear change in Mg-O coordination number. For glasses with SiO_2 contents between 50 and 38% the mean Mg-O coordination number is 4.5, interpreted as a mixture of MgO_4 and MgO_5 polyhdera. The Mg-O coordination number increases to 5.0 at the forsterite composition (33% SiO_2) and the glass is dominated by MgO_5 polyhedra. Kohara et al. (2004) have interpreted this glass structure as reflecting the limit to formation of a silicate network glass, there being insufficient SiO_4 tetrahedra to form a silicate network. The change in Mg-O coordination number coincides with increased density, a change in the glass transition temperature and the rheological properties (Wilding et al. 2004a,b).

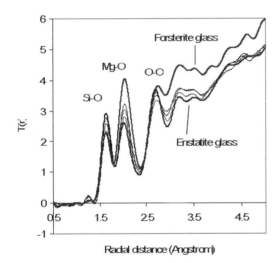

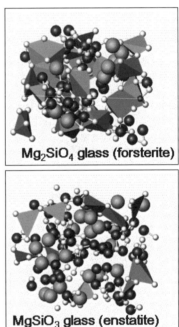

Figure 18. Pair distribution function ($T(r)$) from neutron diffraction studies of magnesium silicate glasses ranging from forsterite to enstatite composition (Wilding et al. 2004a,b). There is a jump in the Mg-O coordination number between 38 and 33% SiO_2. The modeled (RMC) structures are shown for forsterite and enstatite composition glasses.

In situ diffraction studies of oxide liquids

The structure of liquids with low melting points can be carried out in silica tubes in conventional furnaces. For refractory liquids diffraction studies are more challenging. *In situ* studies of aluminate systems have been carried out using containerless levitation techniques. There are two published studies which, though hampered by small sample size nevertheless show that *in situ* studies of refractory liquids are feasible. An *in situ* neutron diffraction study of levitated Al_2O_3 (Landron et al. 2001, 2003) shows that the coordination number of Al-O in the liquid contrasts with that in the crystalline phase. At 2500 K liquid Al_2O_3 has up to 60% of the aluminum ions four-coordinated by oxygen with 20% in five-fold coordination. These coordination numbers are based on EPSR modeling of the $S(\mathbf{Q})$ from the neutron diffraction experiments.

A neutron diffraction study of CaO-Al_2O_3 and SrO-Al_2O_3 liquids between 1950 and 2000 K, using containerless levitation, (Weber et al. 2003b) again shows a liquid dominated by four-coordinate aluminum. Real space transforms of the diffraction data show that the Ca-O and Sr-O coordination numbers are between 6 and 8 and that the intermediate range order is decreased relative to the glassy state.

In both studies the data are noisy and are only available to a limited value of Q. As new instruments are developed and smaller, more intense (possibly focused) neutron beams can be used, the quality of data can be improved and the collection time reduced. This will provide future opportunities for studying the structure of refractory liquids in the supercooled regime, so critical in understanding fragile and liquid-liquid transition behavior.

Amorphous ices

It is believed that most of the water in the solar system is in the form of amorphous ice (Cernicharo and Crovisier 2005). Much of this amorphous ice will occur at high pressures, for example in the interiors of large icy bodies such as the moons of Saturn. One intriguing feature of amorphous ices is that they show an apparent amorphous-amorphous transition, a feature observed originally by Mishima (Mishima et al. 1984, 1985; Mishima 1994) and the subject of much controversy (see also Kuhs and Hansen 2006, this volume). It is only relatively recently, however, that *in situ* diffraction studies have been carried out. In H_2O there is an apparent "first order-like" transition between a low density amorphous ice (LDA) and a high density form (HDA). Thermodynamic models (Moynihan 1997) have been used to argue for the presence of a second critical point, below which a first order transition is possible.

The HDA phase of water can be made by piston cylinder techniques and the recovered samples studied *ex situ*. A comparative study of LDA and HDA, using neutron diffraction with isotopic substitution and combined with EPSR (Finney et al. 2002b) has been used to ascertain the differences in the pair-distribution function of the two forms. Both forms of amorphous ice are fully hydrogen-bonded tetrahedral networks. The structure of HDA resembles that of liquid water at high pressure (Soper and Ricci 2000) while LDA is similar to ice 1h (Soper 1986, 2002). The pair distribution functions for the two forms differ most notably because of the presence of an interstitial water molecule in the HDA form, which lies just beyond the first O-O coordination shell.

The structures of amorphous ice are further complicated by the presence of a further recoverable form VHDA (Very High Density Amorphous ice) (Finney et al. 2002a). This form is distinguished from HDA by a very intense, sharp first peak in the $S(Q)$, that is shifted to higher Q. This amorphous form is associated with a second interstitial molecule and leads to a huge increase in the degree of intermediate range ordering (Guthrie 2004).

In situ neutron diffraction studies of the abrupt transition between the low and high density amorphous ice forms have been performed using the Paris-Edinburgh pressure cell. The high pressure phase (HDA) has been studied to pressures of 2.2GPa (Klotz et al. 2002, 2005b) Studies of the transition itself, far from clarifying the nature of the LDA-HDA transition suggest two possible routes for the transition; continuous and discontinuous. A discontinuous transition is consistent with Raman spectroscopic data. The diffraction data for the transition shows the growth of a peak at 2.25 Å^{-1} in the $S(Q)$, attributed to HDA and a decrease in a peak at 1.7 Å^{-1} (LDA) as LDA is compressed from 0 to 0.5 GPa (Klotz et al. 2002, 2005b). At pressures of 0.3 GPa, the diffraction data is interpreted as a mixture of two forms, consisting of different amounts of HDA and LDA and implying the nucleation and growth of one amorphous phase in the matrix of another (Klotz et al. 2005a). However an arbitrary shift parameter is introduced to account for composition effects, which assumes crystal like behavior. As discussed in the previous section, the change in the first peak in the diffraction pattern is associated with both the degree of periodicity and packing of the structural units. Studies of HDA within the transition region (Tulk et al. 2002, 2003; Guthrie et al. 2003; Urquidi et al. 2004) indicate therefore that the average position of the interstitial molecule changes continuously until an LDA structure is adopted (Fig. 19) i.e., that there is a distinct structural relaxation process associated with a

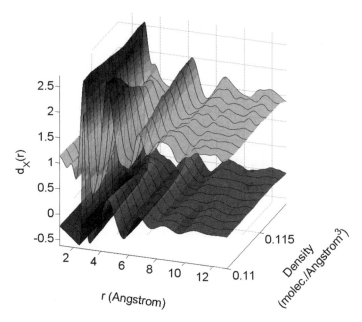

Figure 19. Oxygen-oxygen partial differential distribution function for amorphous ice. The diffraction data (X-ray) is shown at the top) while the results from a molecular dynamics simulation are shown at the bottom. The collapse of the second shell can be clearly observed as the density increases and the interstitial molecules are pushed into the first O-O shell.

continuum of intermediate forms. This may be explained as an abrupt change in short range order, but a continuous change in intermediate range order, unlike a crystal-crystal transition.

In situ high pressure studies of GeO₂ and other amorphous materials

In a network glass such as GeO_2, corner shared tetrahedra form an open three-dimensional network comprised of cages. At high pressures, Ge can be six-fold coordinated by oxygen, and this system provides the first opportunity to observe a purely octahedral glass network. *In situ* neutron diffraction studies of GeO_2 (combined with high energy X-ray diffraction studies and molecular dynamics simulations) have been used to investigate the nature of the change in short- and intermediate-range order (Sampath et al. 2003; Guthrie et al. 2004). It has also been suggested that vitreous GeO_2 may undergo a first order amorphous-amorphous transition. As GeO_2 glass is compressed the height and position of the first peak in the $S(Q)$ changes and this indicates an decrease in intermediate range order (Sampath et al. 2003; Guthrie et al. 2004) through the shrinkage and collapse of the cage structures (Fig. 20). Prior to a coordination change there are changes in the O-O correlations as oxygen atoms move closer to central germanium atoms. Between 6 and 10 GPa the nearest neighbor coordination number increases and a mixture of 4, 5 and 6 Ge polyhedra co-exist. This is again an intermediate state and not simply a mixture of 4 and 6 coordinate Ge. As the pressure is increased to above 15 GPa a high pressure octahedral glass forms, which is not recoverable. This network comprises of a mixture of edge- and face-shared Ge-O octahedral units.

SUMMARY AND FUTURE DIRECTIONS

Neutrons offer the opportunity to determine the structure of liquid and amorphous materials directly by diffraction. The total structure factor, $S(\mathbf{Q})$, determined in the diffraction

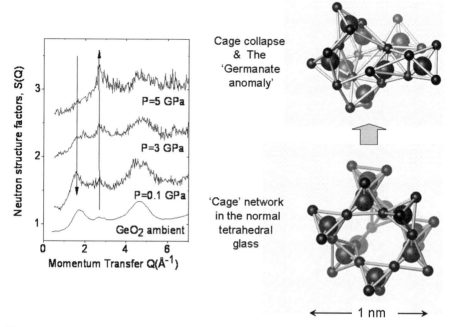

Figure 20. The measured neutron diffraction signal from germania at high pressure showing the disappearance of the FSDP (collapse of cages in the network) and rise of the second connectivity peak just prior to the start of formation of GeO$_5$ units at 6GPa (Sampath et al. 2003; Guthrie et al. 2004).

experiment is directly related to the real space pair distribution function of the liquid or glass by Fourier transform. Neutrons are in many ways ideal probes for liquid and amorphous materials, they are very penetrating and generally equally sensitive to light and heavy elements. It has however been the case that application of neutron diffraction to amorphous materials of interest to the geoscience community is limited because the samples of interest are chemically complex, large samples are required for neutron diffraction and *in situ* studies at pressures and temperatures pertinent to the earth's interior are not easily achievable. Nevertheless, there is an extensive literature devoted to studies of glass and liquids. Chemically complex liquids and glasses can be studied if the partial structure factors can be determined either by using isotopic substitution techniques, or more cheaply, by combining data from neutron and X-ray diffraction experiments. In oxide materials, the scattering of neutrons from oxygen in the sample may account for up to 60% of the scattered signal which means that in the real space transform, oxygen-related correlations can mask metal-metal correlations that reflect clustering and connectivity of coordination polyhedra, that potentially can be used to link amorphous structure with macroscopic properties. Combined neutron and high energy X-ray measurements can be used to identify different partial contributions to the PDF; this is particularly effective for the oxygen-bearing systems that dominate the interior of the Earth and other planets. The PDF is merely the starting point for interpreting amorphous structure, some form of model is invariably used to interpret the data and the quality of these models is judged on how well the short- and intermediate-range structure is reproduced in the $S(\mathbf{Q})$ or $G(r.)$

Developments in the specialized sample environments for use in combination with neutron diffraction mean that the change in liquid or glass structure with pressure and temperature can now be ascertained. Although few studies on the changes in amorphous structure with pressure have been made, they generally show large changes in both intermediate and short range order.

The nature of these changes remains controversial with regard to polyamorphism and the high pressure liquid regime is as yet largely unexplored. Similarly, high temperature sample environments provide opportunities to examine the structure of metastable, supercooled liquids where most of the changes in liquid transport properties (i.e., viscosity) occur but in which a direct link between structure, rheology and structural relaxation has yet to be made. As sample environments become developed there are opportunities to probe extremes of temperature and pressure offered by the advent of new neutron sources and instruments.

The Spallation Neutron Source (SNS) and new second target station at ISIS offer high neutron fluxes and there is the opportunity to examine small samples such as beads of levitated liquid, exotic glasses and samples in pressure cells. Instruments such as SNAP and NOMAD (SNS) are being commissioned (2008 and 2010 respectively) to examine disordered materials, SNAP is dedicated to examining materials under extremes of pressure, and new cells for high pressure experiments built, micro focusing and supermirror guides will result in better counting statistics and opportunities to examine liquid and amorphous materials at high pressure *and* temperature (Parise 2006, this volume). The NOMAD instrument will be dedicated to the study of atomic scale structure of liquids, glasses and disordered crystals. Following the philosophy of the SANDALS instrument at ISIS and high energy diffraction at 3rd generation synchrotron sources, the combination of high energies and low angles of scatter will be used to minimize the corrections and enable sophisticated sample environment apparatus to be used. Elliptical ^{3}He tubes will increase detector efficiency of the higher energy neutrons and continuous angular coverage will be maintained. Focusing devices will be used for longer wavelength neutrons. The large increase in neutron intensity will therefore enable higher resolution, experiments on smaller samples and smaller contrast isotope substitution experiments than are currently feasible.

REFERENCES

Angell CA (1997a) Glass forming liquids with microscopic to macroscopic two-state complexity. Prog Theor Phys Supp 126:1-8
Angell CA (1997b) Landscapes with megabasins - polyamorphism in liquids and biopolymers and the role of nucleation in folding and folding diseases. Physica D 107:122-42
Angell CA, Goldstein M (eds) (1986) Dynamic Aspects of Structural Change in Liquids and Glasses. New York Academy of Sciences
Angell CA, Moynihan CT (2000) Ideal and cooperative bond-lattice representations of excitations in glass-forming liquids: Excitation profiles, fragilities, and phase transitions. Metall Mater Trans B 31:587-96
Angell CA, Moynihan CT, Hemmati M (2000a) "Strong" and "superstrong" liquids, and an approach to the perfect glass state via phase transition. J Non-Cryst Solids 274:319-331
Angell CA, Ngai KL, McKenna GB, McMillan PF, Martin SW (2000b) Relaxation in glass forming liquids and amorphous solids. J Appl Phys 88:3113-3157
Angell CA, Poole PH, Shao J (1994) Glass-forming liquids, anomalous liquids, and polyamorphism in liquids and biopolymers. Nuovo Cimento Soc Ital Fis D 16:993-1025
Ansell S, Neilson GW (2000) Anion-anion pairing in concentrated aqueous lithium chloride solution. J Chem Phys 112:3942-3944
Bacon GE (1963) Applications of Neutron Diffraction in Chemistry. Pergamon Press [distributed in the Western Hemisphere by Macmillan]
Bacon GE (1975) Neutron Diffraction. Clarendon Press
Bacon GE (ed) (1987) Fifty Years of Neutron Diffraction: The Advent of Neutron Scattering. Institute of Physics Publishing
Barrat J-L, Hansen JP (2003) Basic Concepts for Simple and Complex Liquids. Cambridge University Press
Bée M (1988) Quasielastic Neutron Scattering: Principles and Applications in Solid State Chemistry, Biology, and Materials Science. Institute of Physics Publishing
Benmore CJ, Hart RT, Mei Q, Price DL, Yarger J, Tulk CA, Klug DD (2005) Intermediate range chemical ordering in amorphous and liquid water, Si, and Ge. Phys Rev B 72:22011-4
Billinge SJL, Kanatzidis MG (2004) Beyond crystallography: The study of disorder, nanocrystallinity and crystallographically challenged materials with pair distribution functions. Chem Commun 7:749-60

Brazhkin VV, Buldyrev SV, Rhzhov VN, Stanley HE (2002) New Kinds of Phase Transitions: Transformations in Disordered Substances. Kluwer Academic Publishers

Bullock MA (2005) Mars - the flow and ebb of water. Nature 438:1087-1088

Cernicharo J, Crovisier J (2005) Water in space: The water world of Iso. Space Sci Rev 119:29-69

Cole DR, Herwig KW, Mamontov E, Larese J (2006) Neutron scattering and diffraction studies of fluids and fluid-solid interactions. Rev Mineral Geochem 63:313-362

Cormier L (2003) Neutron diffraction analysis of the structure of glasses. J Phys IV 111:187-210

Cormier L, Calas G, Gaskell PH (2001) Cationic environment in silicate glasses studied by neutron diffraction with isotopic substitution. Chem Geol 174:349-363

Cormier L, Gaskell PH, Calas G, Soper AK (1998) Medium-range order around titanium in a silicate glass studied by neutron diffraction with isotopic substitution. Phys Rev B 58:11322-11330

Cormier L, Neuville DR, Calas G (2005) Relationship between structure and glass transition temperature in low-silica calcium aluminosilicate glasses: The origin of the anomaly at low silica content. J Am Ceramic Soc 88:2292-2299

Cormier L, Neuville DR, Massiot D (2004) Structure of tectosilicate and peraluminous glasses in the cao-al2o3-sio2 system. Geochim Cosmochim Acta 68:A650-A

Deb SK, Wilding M, Somayazulu M, McMillan PF (2001) Pressure-induced amorphization and an amorphous-amorphous transition in densified porous silicon. Nature 414:528-530

Debenedetti PG (1996) Metastable Liquids: Concepts and Principles. Princeton University Press

Debenedetti PG (2003) Supercooled and glassy water. J Phys Condens Matter 15:R1669-R1726

Debenedetti PG, Stillinger FH, Shell MS (2003) Model energy landscapes. J Phys Chem B 107:14434-14442

Debenedetti PG, Torquato S (2004) Frank H. Stillinger, theoretical chemist: A tribute. J Phys Chem B 108: 19569-1956970

deJong PHK, Neilson GW, Bellissent Funel MC (1996) Hydration of Ni^{2+} and Cl in a concentrated nickel chloride solution at 100 °C and 300 °C. J Chem Phys 105:5155-5159

Egelstaff PA (1965) Thermal Neutron Scattering. Academic Press

Egelstaff PA (1967) An Introduction to the Liquid State. Academic Press

Elliott SR (1991a) Medium-range structural order in covalent amorphous solids. Nature 354:445-452

Elliott SR (1991b) Origin of the 1st-sharp diffraction peak in the structure factor of covalent glasses. Phys Rev Lett 67:711-714

Ellison AJG, Crawford RK, Montague DJ, Volin KJ, Price DL (1993) The new glass, liquids and amorphous materials diffractometer (GLAD) at IPNS. J Neutron Res 1:61-70

Enderby JE, Howells WS, Howe RA (1973) Structure of aqueous-solutions. Chem Phys Lett 21:109-112

Enderby JE, North DM, Egelstaf Pa (1966) Partial structure factors of liquid Cu-Sn. Philos Mag 14:961-970

Enderby JE, North DM, Egelstaf Pa (1967) Partial structure factors of liquid alloys. Adv Phys 16:171-175

Faber TE, Ziman JM (1965) A theory of electrical properties of liquid metals. 3. Resistivity of binary alloys. Philos Mag 11:153

Filipponi A (2001) EXAFS for liquids. J Phys Condens Matter 13:R23-R60

Filipponi A, Di Cicco A, De Panfilis S, Trapananti A, Itie JP, Borowski M, Ansell S (2001) Investigation of undercooled liquid metals using XAFS, temperature scans and diffraction. J Synchrotron Rad 8:81-86

Finney JL, Bowron DT, Soper AK, Loerting T, Mayer E, Hallbrucker A (2002a) Structure of a new dense amorphous ice. Phys Rev Lett 89:5503-7

Finney JL, Hallbrucker A, Kohl I, Soper AK, Bowron DT (2002b) Structures of high and low density amorphous ice by neutron diffraction - art. No. 225503. Phys Rev Lett 8822:5503

Fischer HE, Salmon PS, Barnes AC (2003) Neutron and X-ray diffraction for structural analysis of liquid and glass materials. J Phys IV 103:359-390

Galeener FL, Wright AC (1986) The Phillips, J.C. model for vitreous SiO_2 - a critical-appraisal. Solid State Commun 57:677-682

Gaskell PH, Eckersley MC, Barnes AC, Chieux P (1991) Medium-range order in the cation distribution of a calcium silicate glass. Nature 350:675-677

Gaskell PH, Wallis DJ (1996) Medium-range order in silica, the canonical network glass. Phys Rev Lett 76: 66-69

Grandinetti PJ (2003) Does phase cycling work for nuclei experiencing strong quadrupolar couplings? Solid State Nucl Mag Resonance 23:1-13

Greaves GN (1985) EXAFS and the structure of glass. J Non-Cryst Solids 71:203-217

Greaves GN, Davis EA (1974) Continuous random network model with 3-fold coordination. Philos Mag 29: 1201-1206

Greaves GN, Fontaine A, Lagarde P, Raoux D, Gurman SJ (1981) Local-structure of silicate-glasses. Nature 293:611-616

Grimsditch M (1984) Polymorphism in amorphous SiO_2. Phys Rev Lett 52:2379-2381

Grimsditch M (1986) Annealing and relaxation in the high-pressure phase of amorphous SiO_2. Phys Rev B 34:4372-4373

Guthrie M, Tulk CA, Benmore CJ, Xu J, Yarger JL, Klug DD, Tse JS, Mao HK, Hemley RJ (2004) Formation and structure of a dense octahedral glass. Phys Rev Lett 93:art. no.-115502

Guthrie M, Urquidi J, Tulk CA, Benmore CJ, Klug DD, Neuefeind J (2003) Direct structural measurements of relaxation processes during transformations in amorphous ice. Phys Rev B 68:art. no.-184110

Howell I, Neilson GW (1996) Li+ hydration in concentrated aqueous solution. J Phys Condens Matter 8:4455-4463

Howell I, Neilson GW (1997) Ni²⁺ coordination in concentrated aqueous solutions. J Molec Liquids 73-4: 337-348

Howells WS, Hannon AC (1999) LAD, 1982-1998: The first ISIS diffractometer. J Phys Condens Matter 11: 9127-38

Kauzmann W (1948) The nature of the glassy state and the behavior of liquids at low temperatures. Chem Rev 43:219-56

Keen DA, Tucker MG, Dove MT (2005) Reverse Monte Carlo modelling of crystalline disorder. J Phys Condens Matter 17:S15-S22

Klotz S, Hamel G, Loveday JS, Nelmes RJ, Guthrie M, Soper AK (2002) Structure of high-density amorphous ice under pressure. Phys Rev Lett 89:2582202-5

Klotz S, Strassle T, Nelmes RJ, Loveday JS, Hamel G, Rousse G, Canny B, Chervin JC, Saitta AM (2005a) Nature of the polyamorphic transition in ice under pressure. Phys Rev Lett 94:025506-1–025506-7

Klotz S, Strassle T, Saitta AM, Rousse G, Hamel G, Nelmes RJ, Loveday JS, Guthrie M (2005b) *In situ* neutron diffraction studies of high density amorphous ice under pressure. J Phys Condens Matter 17:S967-S74

Kohara S, Suzuya K, Takeuchi K, Loong CK, Grimsditch M, Weber JKR, Tangeman JA, Key TS (2004) Glass formation at the limit of insufficient network formers. Science 303:1649-1652

Kuhs WF, Hansen TC (2006) Time-resolved neutron diffraction studies with emphasis on water ices and gas hydrates. Rev Mineral Geochem 63: 171-204

Lacks DJ (2000) First-order amorphous-amorphous transformation in silica. Phys Rev Lett 84:4629-4632

Landron C, Hennet L, Coutures JP, Jenkins T, Aletru C, Greaves N, Soper A, Derbyshire G (2000) Aerodynamic laser-heated contactless furnace for neutron scattering experiments at elevated temperatures. Rev Sci Instrum 71:1745-1751

Landron C, Hennet L, Thiaudiere D, Price DL, Greaves GN (2003) Structure of liquid oxides at very high temperatures. Nucl Instrum Methods Phys Res Sect B 199:481-488

Landron C, Soper AK, Jenkins TE, Greaves GN, Hennet L, Coutures JP (2001) Measuring neutron scattering structure factor for liquid alumina and analysing the radial distribution function by empirical potential structural refinement. J Non-Cryst Solids 293:453-457

Laughlin WT, Uhlmann DR (1972) Viscous flow in simple organic liquids. J Phys Chem 76:2317-2325

Le Godec Y, Strassle T, Hamel G, Nelmes RJ, Loveday JS, Marshall WG, Klotz S (2004) Techniques for structural studies of liquids and amorphous materials by neutron diffraction at high pressures and temperatures. High Pressure Res 24:205-217

Levin EM (1956) Phase Diagrams for Ceramists. American Ceramic Society

Lorch E (1970) Conventional and elastic neutron diffraction from vitreous silica. J Phys C 3:1314-1320

Loveday JS, Nelmes RJ, Marshall WG, Besson JM, Klotz S, Hamel G, Hull S (1996) High pressure neutron diffraction studies using the paris-edinburgh cell. High Pressure Res 14:303-309

Lovesey SW (1984) Theory of Neutron Scattering from Condensed Matter. Clarendon Press

Madden PA, Wilson M (1996) Covalent effects in ionic systems. Chem Soc Rev 25:339-350

Majerus O, Cormier L, Calas G, Beuneu B (2004) A neutron diffraction study of temperature-induced structural changes in potassium disilicate glass and melt. Chem Geol 213:89-102

Mayers J (1984) The use of vanadium as a scattering standard for pulsed-source neutron spectrometers. Nucl Instrum Methods Phys Res Sect A 221:609-618

McGreevy RL (2001) Reverse Monte Carlo modelling. J Phys Condens Matter 13:R877-R913

McGreevy RL (2003) Reverse Monte Carlo modeling. J Phys IV 111:347-371

Mishima O (1994) Reversible 1st-order transition between two H_2O amorphs at ~0.2 GPa and ~135 K. J Chem Phys 100:5910-5912

Mishima O, Calvert LD, Whalley E (1984) Melting ice-i at 77 K and 10 kbar - a new method of making amorphous solids. Nature 310:393-395

Mishima O, Calvert LD, Whalley E (1985) An apparently 1st-order transition between two amorphous phases of ice induced by pressure. Nature 314:76-78

Mishima O, Stanley HE (1998a) Decompression-induced melting of ice iv and the liquid-liquid transition in water. Nature 392:164-168

Mishima O, Stanley HE (1998b) The relationship between liquid, supercooled and glassy water. Nature 396: 329-335

Moynihan CT (1993) Correlation between the width of the glass transition region and the temperature dependence of the viscosity of high-tg glasses. J Am Ceramic Soc 76:1081-1087

Moynihan CT (1997) Two species/non ideal solution model for amorphous/amorphous phase transitions. Mater Res Soc Proc 455:411-425

Moynihan CT, Angell CA (2000) Bond lattice or excitation model analysis of the configurational entropy of molecular liquids. J Non-Cryst Solids 274:131-138

Moynihan CT, Lee SK, Tatsumisago M, Minami T (1996) Estimation of activation energies for structural relaxation and viscous flow from dta and dsc experiments. Thermochim Acta 280:153-62

Mozzi RL, Warren BE (1969) Structure of vitreous silica. J Appl Crystallogr 2:164

Mukherjee GD, Vaidya SN, Sugandhi V (2001) Direct observation of amorphous to amorphous apparently first-order phase transition in fused quartz - art. No. 195501. Phys Rev Lett 8719:5501

Narten AH, Hahn RL (1982) Direct determination of ionic solvation from neutron-diffraction. Science 217: 1249-1250

Narten AH, Thiessen WE, Blum L (1982) Atom pair distribution-functions of liquid water at 25-degrees-c from neutron-diffraction. Science 217:1033-1034

Neilson GW, Mason PE, Ramos S, Sullivan D (2001) Neutron and X-ray scattering studies of hydration in aqueous solutions. Philos Trans Royal Soc London Ser A 359:1575-1591

Neuefeind J (2002) High energy XRD investigations of liquids. J Molec Liquids 98-9:87-95

Neukum G, Jaumann R, Hoffmann H, Hauber E, Head JW, Basilevsky AT, Ivanov BA, Werner SC, van Gasselt S, Murray JB, McCord T (2004) Recent and episodic volcanic and glacial activity on mars revealed by the high resolution stereo camera. Nature 432:971-979

Newsom H (2005) Planetary science - clays in the history of Mars. Nature 438:570-571

Parise JB (2006) High pressure studies. Rev Mineral Geochem 63:205-231

Phillips JC (1982) Spectroscopic and morphological structure of tetrahedral oxide glasses. Solid State Phys Adv Res Appl 37:93-171

Phillips JC (1986) The Phillips J.C. model for vitreous SiO_2 - a critical-appraisal - comments. Solid State Commun 60:299-300

Placzek G (1952) The scattering of neutrons by systems of heavy nuclei. Phys Rev 86:377-388

Poirier JP (2000) Introduction to the Physics of the Earth's Interior. Cambridge University Press

Poole PH, Grande T, Angell CA, McMillan PF (1997) Polymorphic phase transitions in liquids and glasses. Science 275:322-323

Porco CC, Baker E, Barbara J, Beurle K, Brahic A, Burns JA, Charnoz S, Cooper N, Dawson DD, Del Genio AD, Denk T, Dones L, Dyudina U, Evans MW, Fussner S, Giese B, Grazier K, Helfenstein P, Ingersoll AP, Jacobson RA, Johnson TV, McEwen A, Murray CD, Neukum G, Owen WM, Perry J, Roatsch T, Spitale J, Squyres S, Thomas P, Tiscareno M, Turtle EP, Vasavada AR, Veverka J, Wagner R, West R (2005a) Imaging of Titan from the Cassini spacecraft. Nature 434:159-168

Porco CC, Baker E, Barbara J, Beurle K, Brahic A, Burns JA, Charnoz S, Cooper N, Dawson DD, Del Genio AD, Denk T, Dones L, Dyudina U, Evans MW, Giese B, Grazier K, Heifenstein P, Ingersoll AP, Jacobson RA, Johnson TV, McEwen A, Murray CD, Neukum G, Owen WM, Perry J, Roatsch T, Spitale J, Squyres S, Thomas PC, Tiscareno M, Turtle E, Vasavada AR, Veverka J, Wagner R, West R (2005b) Cassini imaging science: Initial results on Phoebe and Iapetus. Science 307:1237-1242

Price DL (1985) Proceedings of the workshop on research opportunities in amorphous solids with pulsed neutron sources, Argonne, Illinois, April 17-19, 1985 - preface. J Non-Cryst Solids 76:R7-R8

Price DL (1986) Inelastic-scattering from amorphous solids. Physica B & C 136:25-29

Price DL (1996) Intermediate-range order in glasses. Curr Opin Solid State Mater Sci 1:572-577

Price DL, Moss SC, Reijers R, Saboungi ML, Susman S (1988) Intermediate-range order in glasses and liquids. J Phys C 21:L1069-L1072

Price DL, Saboungi ML, Susman S, Volin KJ, Wright AC (1991) Neutron-scattering function of vitreous and molten zinc-chloride. J Phys Condens Matter 3:9835-9842

Proffen T (2006) Analysis of disordered materials using total scattering and the atomic pair distribution function. Rev Mineral Geochem 63:255-274

Richet P, Gillet P (1997) Pressure-induced amorphization of minerals: A review. Eur J Mineral 9:907-933

Saika-Voivod I, Poole PH, Sciortino F (2001) Fragile-to-strong transition and polyamorphism in the energy landscape of liquid silica. Nature 412:514-517

Saika-Voivod I, Poole PH, Sciortino F (2002) Interrelationship of polyamorphism and the fragile-to-strong transition in liquid silica. *In:* New Kinds of Phase Transitions: Transformations in Disordered Substances. Brazhkin VV, Buldyrev SV, Rhzhov VN, Stanley HE (eds) Kluwer Academic Publishers, p 169-78

Salmon PS (2002) Amorphous materials - order within disorder. Nature Materials 1:87-88

Salmon PS (2005) Moments of the bhatia-thornton partial pair-distribution functions. J Phys Condens Matter 17:S3537-S3542

Sampath S, Benmore CJ, Lantzky KM, Neuefeind J, Leinenweber K, Price DL, Yarger JL (2003) Intermediate-range order in permanently densified GeO_2 glass. Phys Rev Lett 90:55021-4

Sastry S, Angell CA (2003) Liquid-liquid phase transition in supercooled silicon. Nature Mater 2:739-743

Skipper NT, Williams GD, de Siqueira AVC, Lobban C, Soper AK (2000) Time-of-flight neutron diffraction studies of clay-fluid interactions under basin conditions. Clay Minerals 35:283-290

Soper AK (1982) An instrument for liquids, amorphous and powder diffraction. AIP Conf Proc 89:23-34

Soper AK (1983) Multiple-scattering from an infinite-plane slab. Nucl Instrum Methods Phys Res 212:337-347

Soper AK (1986) How well can the structure of water be determined by neutron-diffraction. Physica B & C 136:322-324

Soper AK (2000) The radial distribution functions of water and ice from 220 to 673 K and at pressures up to 400 mpa. Chem Phys 258:121-137

Soper AK (2001) Tests of the empirical potential structure refinement method and a new method of application to neutron diffraction data on water. Molec Phys 99:1503-1516

Soper AK (2002) Water and ice. Science 297:1288-1289

Soper AK (2005) Partial structure factors from disordered materials diffraction data: An approach using empirical potential structure refinement. Phys Rev B 72

Soper AK, Egelstaff PA (1980) Multiple-scattering and attenuation of neutrons in concentric cylinders.1. Isotropic first scattering. Nucl Instrum Methods Phys Res 178:415-425

Soper AK, Howells WS, Hannon AC (1989) Atlas-analysis of time of flight diffraction data from liquid and amorphous samples. Technical report RAL 89-046

Soper AK, Luzar A (1992) A neutron-diffraction study of dimethyl-sulfoxide water mixtures. J Chem Phys 97:1320-1331

Soper AK, Ricci MA (2000) Structures of high-density and low-density water. Phys Rev Lett 84:2881-2884

Soper AK, Rossky PJ (2000) Liquid water and aqueous solutions - preface. Chem Phys 258:107-108

Soper AK, Silver RN (1982) Hydrogen-hydrogen pair correlation-function in liquid water. Phys Rev Lett 49:471-474

Sposito G, Skipper NT, Sutton R, Park SH, Soper AK, Greathouse JA (1999) Surface geochemistry of the clay minerals. Proc Nat Acad Sci USA 96:3358-3364

Squires GL (1978) Introduction to the Theory of Thermal Neutron Scattering. Cambridge University Press

Stanley HE (1971) Introduction to Phase Transitions and Critical Phenomena. Clarendon Press

Stillinger FH, Debenedetti PG (2002) Energy landscape diversity and supercooled liquid properties. J Chem Phys 116:3353-3361

Stillinger FH, Debenedetti PG (2003) Phase transitions, kauzmann curves, and inverse melting. Biophys Chem 105:211-220

Suck J-B, Maier B (eds) (1993) Methods in the Determination of Partial Structure Factors of Disordered Matter by Neutron and Anomalous X-ray Diffraction. World Scientific

Tulk CA, Benmore CJ, Klug DD, Urquidi J, Neuefeind J, Tomberli B (2003) Another look at water and ice - response. Science 299:45

Tulk CA, Benmore CJ, Urquidi J, Klug DD, Neuefeind J, Tomberli B, Egelstaff PA (2002) Structural studies of several distinct metastable forms of amorphous ice. Science 297:1320-1323

Urquidi J, Benmore CJ, Egelstaff PA, Guthrie M, McLain SE, Tulk CA, Klug DD, Turner JFC (2004) A structural comparison of supercooled water and intermediate density amorphous ices. Mol Phys 102:2007-2014

Wales DJ (2003) Energy Landscapes. Cambridge University Press

Wales DJ, Doye JPK, Miller MA, Mortenson PN, Walsh TR (2000) Energy landscapes: From clusters to biomolecules. Adv Chem Phys 115:1-111

Warren BE (1969) X-ray studies of glass structure. Am Ceram Soc Bull 48:872

Weber JKR, Benmore CJ, Tangeman JA, Siewenie J, Hiera KJ (2003a) Structure of binary $CaO-Al_2O_3$ and $SrO-Al_2O_3$ liquids by combined levitation-neutron diffraction. J Neut Res 11:113-121

Weber JKR, Hampton DS, Merkley DR, Rey CA, Zatarski MM, Nordine PC (1994) Aeroacoustic levitation - a method for containerless liquid-phase processing at high-temperatures. Rev Sci Instrum 65:456-465

Weber JKR, Tangeman JA, Key TS, Nordine PC (2003b) Investigation of liquid-liquid phase transitions in molten aluminates under containerless conditions. J Thermophy Heat Transfer 17:182-185

Wilding MC, Benmore CJ, Tangeman JA, Sampath S (2004a) Coordination changes in magnesium silicate glasses. Europhy Lett 67:212-218

Wilding MC, Benmore CJ, Tangeman JA, Sampath S (2004b) Evidence of different structures in magnesium silicate liquids: Coordination changes in forsterite- to enstatite-composition glasses. Chem Geol 213:281-291

Wolf GH, Wang S, Herbst CA, Durben DJ, Oliver WF, Kang ZC, Halvorson K (1992) Pressure induced structural collapse of the tetrahedral framework in crystalline and amorphous GeO_2. *In:* High-pressure Research: Application to Earth and Planetary Sciences. Syono Y, Manghnani MH, (eds) Terra Scientific Pub. Co., American Geophysical Union, p. 503-17

Wright AC (1979) Modern techniques for glass structure determination. Am Ceramic Soc Bull 58:380-90

Wright AC (1985a) How much do we really know about the structure of amorphous solids. J Non-Cryst Solids 75:15-27

Wright AC (1985b) Scientific opportunities for the study of amorphous solids using pulsed neutron sources. J Non-Cryst Solids 76:187-210

Wright AC (1994) Neutron-scattering from vitreous silica. 5. The structure of vitreous silica - what have we learned from 60 years of diffraction studies. J Non-Cryst Solids 179:84-115

Wright AC, Connell GAN, Allen JW (1980) Amorphography and the modeling of amorphous solid structures by geometric transformations. J Non-Cryst Solids 42:69-86

Wright AC, Hulme RA, Grimley DI, Sinclair RN, Martin SW, Price DL, Galeener FL (1991) The structure of some simple amorphous network solids revisited. J Non-Cryst Solids 129:213-232

Wright AC, Hulme RA, Sinclair RN (2005) A small angle neutron scattering study of long range density fluctuations in vitreous silica. Phys Chem Glasses 46:59-66

Wright AC, Leadbetter AJ (1976) Diffraction studies of glass structure. Phys Chem Glasses 17:122-145

Wright AC, Shaw JL, Sinclair RN, Vedishcheva NM, Shakhmatkin BA, Scales CR (2004) The use of crystallographic data in interpreting the correlation function for complex glasses. J Non-Cryst Solids 345-46:24-33

Wright AC, Sinclair RN (1985) Neutron-scattering from vitreous silica. 3. Elastic diffraction. J Non-Cryst Solids 76:351-368

Wright AC, Sinclair RN, Leadbetter AJ (1985) Effect of preparation method on the structure of amorphous solids in the system As-S. J Non-Cryst Solids 71:295-302

Yarnell JL, Katz MJ, Wenzel RG, Koenig SH (1973) Structure factor and radial-distribution function for liquid argon at 85 K. Phys Rev A 7:2130-2144

Zachariasen WH (1932) The atomic arrangement in glass. J Am Ceram Soc 54:3841-3851

Zallen R (1983) The Physics of Amorphous Solids. Wiley

Zetterstrom P (1996) Parameterization of the Van Hove dynamic self-scattering law $S_s(\mathbf{Q}, \omega)$. Molec Phys 88: 1621-1634

Zotov N, Delaplane RG, Keppler H (1998) Structural changes in sodium tetrasilicate glass around the liquid-glass transition: A neutron diffraction study. Phys Chem Minerals 26:107-110

Zotov N, Keppler H (1998) The structure of sodium tetrasilicate glass from neutron diffraction, reverse Monte Carlo simulations and Raman spectroscopy. Phys Chem Minerals 25:259-267

Reviews in Mineralogy & Geochemistry
Vol. 63, pp. 313-362, 2006
Copyright © Mineralogical Society of America

13

Neutron Scattering and Diffraction Studies of Fluids and Fluid-Solid Interactions

David R. Cole

Chemical Sciences Division
Oak Ridge National Laboratory
Oak Ridge, Tennessee 37831-6110, U.S.A.
e-mail: coledr@ornl.gov

Kenneth W. Herwig and Eugene Mamontov

Spallation Neutron Source
Oak Ridge National Laboratory
Oak Ridge, Tennessee 37831-6475, U.S.A.

John Z. Larese

Department of Chemistry
University of Tennessee
Knoxville, Tennessee, 37996, U.S.A.

INTRODUCTION

Role of fluids in earth systems with the focus on H_2O

Geologic fluids (defined liberally as gases, liquids, and supercritical solutions) act as reaction media, reactants, and carriers of energy and matter in the natural environment. Among the many different types of geologic fluids, those containing volatile C-O-H-N-S species and those enriched in chloride salts (brines) are of particular interest. They occur widely in varied geochemical settings, commonly contain significant quantities of dissolved and suspended compounds (complex hydrocarbons, organic macromolecules, colloids/nanoparticles), play a crucial role as primary reaction media, and are important sources and sinks of greenhouse gases. The consequences of coupled reactive-transport processes common to most geological environments depend on the properties and reactivities of these fluids over broad ranges of temperature, pressure and fluid composition. The relative strengths of complex molecular-scale interactions in geologic fluids, and the changes in those interactions with temperature, pressure, and fluid composition, are the fundamental basis for observed fluid properties. Understanding these solvent-mediated interactions for broad classes of solutes and suspensions in natural systems over the range of conditions typical of geologic fluids will greatly improve our capability to model and predict fluid behavior, reactivity, and the partitioning of elements and isotopes between coexisting species and phases.

Complex intermolecular interactions of C-O-H-N-S fluids (H_2O, CO_2, H_2, H_2S, N_2, CH_4) result in their unique thermophysical properties, including large deviations in the volumetric properties from ideality, vapor-liquid equilibria, and critical phenomena. Water is one of the best general solvents for inorganic materials due to its molecular structure and the distribution of electric charge (Neilson et al. 2002). In aqueous fluids containing various solutes (electrolytes, metals, organic/bio-molecules), numerous solute-solute and solute-solvent reactions lead to specific interactions, including complexation, binding, local ordering, and clustering. Indeed, a key goal in geochemistry is to develop a comprehensive understanding of the thermophysical

1529-6466/06/0063-0013$10.00   DOI: 10.2138/rmg.2006.63.13

properties, structures, dynamics, and reactivities of complex geologic fluids and molecules (water and other C-O-H-N-S fluids, electrolytes, and organic-biological molecules) at multiple length scales (molecular to macroscopic) over wide ranges of temperature, pressure, and composition. This knowledge is foundational to advances in the understanding of other geochemical processes involving mineral-fluid interfaces and reactions. It is also becoming increasingly clear that organic molecules in aqueous and mixed-volatile fluids—ranging from simple hydrocarbons and carboxylic acids to branched and cyclic compounds, to proteins and humic substances—play major roles in controlling geochemical processes, not just at the Earth's surface, but also deep within the crust. The origin of life may be partly attributable to the properties of such molecules in complex fluids under extreme conditions, as they appear to play an important role in mineral reactivity and templating of mineral precipitates.

Fluid/fluid and fluid/solid interfaces are regions across which all elemental transfers take place. The complex structural and dynamic changes that occur at interfaces can profoundly affect hydrodynamics, reaction rates and reaction mechanisms. Recently, there has been considerable effort devoted to behavior at 'stable' interfaces, involving the juxtaposition of phases that largely retain their bulk characteristics at some distance from the interface. This enables us to isolate the unique features of the interface, which are controlled in part by changes in the compositions and state conditions of the bulk phases, whose structural and dynamic properties are often better understood (e.g., Brown et al. 1999; Wesolowski et al. 2004).

Nanoscale porosity can be generated from structural modifications (Radlinski 2006, this volume) in a reaction zone, and this process can be understood by correlating the extent and rate of change of chemical and isotopic signatures with contemporaneous structural modifications. Confinement of fluids in the porous reaction (leached) zones can impact fluid-phase behavior, chemical transport across the zone, and reactivity. A molecular-level understanding of the reactive interface between the product and parent phases is starting to yield a more predictive view of time- and state-dependent reaction progress (e.g., Cole et al. 2004).

Finally, the importance of water in deep earth and planetary processes even at trace levels, cannot be overstated (Williams and Hemley 2001). Water affects physical and chemical processes such as melting, mantle convection and the chemistry of melts generated from the earth's mantle. How water is stored in the earth, what the water content of mineral phases is, and how it is mobilized are key issues addressed by numerous methodologies, not the least of which involves the use of neutron scattering and diffraction.

Why neutrons – hydrogen (and deuterium) is the key?

Given the complexity of natural geo-fluids and their role in mediating surface interactions and reactivity with mineral phases, there can be no doubt that a quantitative understanding is needed of molecular-level fluid properties and fluid interactions with solids. There is a wide spectrum of analytical approaches that can be brought to bear, including, but certainly not limited to dynamic light scattering, IR, microscopy (e.g., electron; force), NMR, synchrotron-based X-rays, and neutron scattering and diffraction. When coupled with molecular simulation, this wide array of methods provides the means by which we can interrogate the structure and dynamics of fluids and their interactions with solids. Each of these provides a unique window into the properties and behavior of fluids and their reactivity.

Neutron and X-ray scattering are two of the most important tools available to probe the atomic structure of fluids, including aqueous solutions and complex liquids. Scattering patterns obtained experimentally and the distribution function, $g(r)$, are related through Fourier transformation (e.g., Nielson et al. 2002). Interatomic distances, coordination numbers and the extent of local order around a particular atom can be delineated through a quantitative understanding of distribution functions, either individually, or as combinations ($G_\alpha(r)$) specific to a particular atom (or ion; α or β), or as total fluctuations ($G(r)$), of all species in the fluid.

The local structure can be divided into several parts such as the contact distance, next nearest neighbor distance, etc., and eventually the end of short-range order (see Fig. 3 in Wilding and Benmore 2006, this volume). Furthermore, from an examination of the shapes of the correlation in $g(r)$, one can obtain a qualitative guide to the degree of complexation (Nielson and Adya 1997). Experimental structural information can be used to assess the interatomic correlations in all types of fluids, and to test the robustness of simulations based on model potentials for specific components.

There are a number of distinct advantages in using neutrons and X-rays to determine the structure of fluids, glasses and other amorphous materials (see also Wilding and Benmore 2006). The principal advantage of neutron methods stems from the fact that neutrons interact directly with the atomic nuclei within a molecule through the strong force and the interaction is isotropic. Consequently, structural information in the form of $g_{\alpha\beta}(r)$ is accessible directly from the experimental diffraction data. The strength of this interaction varies irregularly with atomic number, so that even isotopes of the same element do not have the same neutron scattering cross-section or scattering length. For example, as shown in Table 1, the most significant isotopic variation occurs for hydrogen, which has a coherent scattering length of −3.74 barn, whereas for deuterium the scattering length is 6.67 barn. Therefore, two fluids with the same composition of atoms but containing isotopically different nuclei (e.g., CH_4 vs. CD_4; $H^{35}Cl$ vs. $H^{37}Cl$, etc.) will yield different neutron scattering patterns. The fact that neutrons are sensitive to hydrogen and differences between its isotopes permits observation and measurement of the hydrogen structural correlations in water (and other hydrogeneous phases) that are not easily obtainable by X-rays. In contrast to X-rays, neutrons are coherently scattered equally strongly by light and heavy elements—i.e., hydrogen scatters coherently just as effectively as manganese. Hence, neutrons can be used to probe the structure of molecules containing lighter atoms. We will see later in this chapter that isotopic substitution will play a pivotal role in determination of structural features of both simple fluids (e.g., H_2O) as well as more complex solutions such as electrolyte-bearing waters.

One final aspect of neutron interaction with fluids is worth discussing. The isotopic compositions of molecular species can be tailored such that their scattering or diffraction properties differ from one another and from solid matrices with which they may be interacting.

Table 1. Coherent scattering lengths (fm[a]) of important elements common to fluids found on and within the Earth (from Enderby et al. 1987).

Element or Isotope	b	Element or Isotope	b	Element or Isotope	b
H	−3.74	^{40}Ca	4.9	^{65}Cu	11.1
D	6.67	^{44}Ca	1.8	Zn	5.686
^6Li	1.87	Fe	9.51	^{64}Zn	5.5
^7Li	−2.2	^{54}Fe	4.2	^{68}Zn	6.7
N	9.36	^{56}Fe	10.1	Ag	5.97
^{14}N	9.37	^{57}Fe	2.3	^{107}Ag	7.64
^{15}N	6.44	Ni	10.3	^{109}Ag	4.19
K	3.67	^{58}Ni	14.4	^{113}In	5.39
^{41}K	2.58	^{60}Ni	2.82	^{115}In	4.00
Cl	9.58	^{62}Ni	−8.7	Ba	5.07
^{35}Cl	11.7	^{64}Ni	−0.37	^{130}Ba	−3.6
^{37}Cl	3.1	Cu	7.718	^{137}Ba	6.82
Ca	4.9	^{63}Cu	6.7		

[a] 1 fm = 10^{-15} m

Hence, neutron scattering and diffraction permits interrogation of the structure and dynamics of fluids at interfaces and within confined geometries. The behavior of fluids at interfaces or in confinement commonly differs from their bulk behavior in interesting, and not necessarily intuitive, ways. For example, how water (and other fluid species) orients and moves near and on a mineral (or biomolecular) surface or within a nanopore can be addressed by neutron techniques which are non-invasive and nondestructive (e.g., Mamontov 2004, 2005). Indeed, a number of neutron methods exist that allow the study of fluid structural properties involving translational as well as orientational ordering of fluids on surfaces or within micro- or mesoporous materials (Idziak and Li 1998). These topics will be discussed in detail later in this chapter.

Important molecular-level behavior amenable to neutron scattering and diffraction

While it is clear that no one method can sufficiently resolve structure at the required level around all species in bulk complex fluids or fluids interacting with an interface, neutrons do provide a unique probe useful in approaching certain aspects of a number of key molecular issues. One prime example involves the phenomenon of *hydrogen bonding* which plays a fundamental role in the structure, function and properties of many fluids of geochemical and biogeochemical importance. The interaction manifests itself as a pronounced, directional correlation between specific sites on neighboring molecules (Bruni et al. 1996). It also produces a highly characteristic vibrational density of states in hydrogen bonded molecules, and dramatically affects the thermodynamic properties of a fluid compared to what they would be if the molecules interacted simply by van der Waals interactions. In crystalline materials the hydrogen-bond direction will relate to some aspect of the crystal symmetry, but in disordered fluids and materials the ordering effect of the orientational correlations is influenced by the inherent disorder of the molecular arrangement (Dore 1991). The complexity of hydrogen-bonded networks poses many problems that have still not been resolved and arguments persist over the true cause of the hydrogen bond interaction. In fact, the question of how many bonds each H_2O molecule makes with its nearest neighbor was identified as one of the top 125 fundamental science questions (Kennedy and Norman 2005).

Perturbation of hydrogen-bonding and associated molecular fluid structures may arise from either the presence of dissolved constituents (e.g., solutes in aqueous solution) or interfacial interactions between a fluid and a solid such as a colloid or nanoparticle, a 2D mineral surface, or within a porous or highly fractured network. A molecular-level understanding of the nature of speciation of solutes in aqueous solution is a critical question with direct impact on the quantities such as solid solubilities and the alteration of geological materials. Neutron diffraction with isotopic substitution (NDIS) permits the direct determination of radial distribution functions $g_{\alpha\beta}(r)$, providing the appropriate level of molecular-level structural detail. Small-angle neutron scattering and diffraction can also be employed in real-time studies of the nucleation and assembly of macromolecules (e.g., biopolymers, including proteins) and low-dimensional building blocks that give rise to crystalline nanoparticles. *In situ*, non-invasive neutron methods permit interrogation of poorly constrained mechanisms of transformation of a mixture of solid and fluid components into a complex extended solid (Walton et al. 1999, 2002). Neutron scattering techniques are ideally suited for the study of the structure and dynamics of atoms and molecules physisorbed on surfaces (e.g., Larese 1997, 1999) including processes such as multi-layer development, melting and rotational tunneling on unreactive stable surfaces. Furthermore, structural and dynamical features associated with chemisorption involving the formation of surface hydroxyl groups that saturate the coordination of surface cations and complete the interrupted bulk crystalline network can be "imaged" by neutron methods. The dynamical behavior of fluids and gases contained within porous solids is controlled by processes occurring at the interface as well as the rates of supply and removal of mobile components. The richness and complexity of fluid behavior (e.g., phase transitions, molecular orientation and relaxation, diffusion, adsorption, wetting, capillary condensation, etc.) in confined geometries has been, and continues to be, the focus of numerous applications of neutron scattering.

Objectives

The focus of this chapter is the application of neutron scattering and diffraction methods to the study of fluids and their interaction with solid matrices. By way of numerous examples, emphasis is placed on what neutrons can tell us about the molecular-level properties and behavior of geo-fluids and the processes attendant with their interaction with solid surfaces. Discussion focuses on two main themes, homogeneous fluids with a major emphasis on both water and aqueous solutions containing dissolved constituents, and the structure and dynamics of fluids interacting with either solid surfaces or within confined geometries, again with an emphasis on water. While we recognize the increased importance of inorganic-organic-biologic interactions, particularly at surface and near-surface earth conditions, the review of the complex molecular interplay of these chemical systems is unfortunately beyond the scope of this review.

HOMOGENEOUS FLUIDS

Background

The structure of pure liquids, such as water, and aqueous solutions is conveniently interrogated using scattering techniques. The most common of these includes the use of either X-rays or neutrons. Differences arise due to the variations in the scattering properties of single atoms for each kind of radiation. In either case, radiation penetrates the liquid and is scattered through an angle θ. Analytical information is obtained by assessing the intensity of the scattered radiation as a function of this angle. The scattering of radiation by condensed matter may be related to the distribution of atomic positions if the wavelength λ of the radiation is of the order of magnitude of the interatomic spacing r. The distribution of scattering intensity contains information on the distribution of atoms, and the measured spectrum of energy transfer contains information on single and collective particle motion. A major goal in the study of fluids is to identify and test the relationship between intermolecular forces and the microscopic structure and dynamics of the fluids. Understanding the structure and dynamics of bulk fluids is a prerequisite to studying these for fluids interacting with surfaces or confined to porous matrices. The following discussion of the basic principles of neutron scattering in liquids is intended to complement the description provided in Loong (2006, this volume) and Wilding and Benmore (2006, this volume) with emphasis placed here on a description of structure and dynamics of water and aqueous solutions containing inorganic solutes.

The neutrons used in a neutron scattering experiment penetrate matter rather easily. Unlike X-rays which are scattered by electromagnetic interaction with the atomic electron distribution, the neutron is scattered by the nucleus which behaves as a point scatter since its size ($\sim 10^{-15}$ m) is much smaller than the wavelength of the incident neutron ($\sim 10^{-10}$ m). The essential features of the scattering pattern are related to a structure factor that characterizes the basic arrangement of scattering centers for the fluid under investigation. In ordered materials, such as crystals, the regular arrangement of the lattice leads to a series of well-defined peaks governed by the Laue conditions. By contrast, scattering in partially disordered materials such as fluids or amorphous solids produces patterns composed of broad oscillatory structures which still contain structural information through the phase differences arising from interference of the scattered waves. The formalism for describing the scattering of neutrons is much simpler than that for X-rays because there is no need to introduce an atomic scattering factor to represent the spatial distribution for each individual scattering unit. There are, however, other characteristics arising from the mechanism of nuclear interaction that must be accounted for and have a significant bearing on the design of experiments and interpretation of results.

Scattering experiments fall into two categories (Winter 1993). The first involves experiments where only the angular distribution of the scattering intensity is determined, and the second considers both the angular distribution for the intensity and the energy transfer. In the former, the intensity is related to the static structure factor $S(Q)$, and in the latter the intensity is related to the dynamic structure factor $S(Q,E)$, where $\mathbf{Q}$ and E are the momentum and energy transferred to the particle in the scattering process, respectively. The direction of a wave vector is the direction of propagation of the plane wave and its magnitude is $2\pi/\lambda$. The incoming radiation in the experiment is characterized by its wavelength, λ and by its intensity, I_o. The λ is chosen so that it is commensurate with the average distance between atoms, but much smaller than the sample dimensions.

Scattering theory

In the context of water, Wilding and Benmore (2006, this volume) provide a fairly complete description of scattering theory, experimental methodologies and various corrections attendant with neutron diffraction. Our intent here is only to amplify and where necessary, expand a bit on some of the concepts that were introduced. Neutron diffraction concerns the measurement of the coherent interference of scattered waves. The experimental method may be monochromatic angle-dispersive (typically reactor based) or time-of-flight energy-dispersive (typically based at a pulsed source). Diffraction experiments probe the differential cross (DC) section of neutrons (Egelstaff 1992), $d\sigma/d\Omega$, defined as the ratio of the scattering cross (SC) section $d\sigma$ scattered into the solid angle $d\Omega$ about the scattering angle θ. For an assembly of fixed nuclei (forming molecules or not) the DC is described by the static approximation:

$$\frac{d\sigma}{d\Omega} = \frac{1}{N}\sum_{\alpha,\beta}\bar{b}_\alpha\bar{b}_\beta\left\langle e^{-i\mathbf{Q}\cdot\mathbf{r}_\alpha}\,e^{-i\mathbf{Q}\cdot\mathbf{r}_\beta}\right\rangle \tag{1}$$

where $\bar{b}_i$ is the coherent scattering length associated with the nucleus with position $\mathbf{r}_i$. The operators $\mathbf{r}_\alpha$ and $\mathbf{r}_\beta$ are assumed to both be evaluated at $t = 0$ according to the Heisenberg convention inside the thermal expectation bracket. For a complete evaluation of Equation (1) we assume that all relevant energy transfers have been integrated over the momentum transfer for the elastic process for the given $\mathbf{Q}$, and that N is the total number of scatterers in the sample. A variety of corrections to experimental data are required accounting for effects such as incoherent and multiple scattering, inelastic effects, container absorption, etc (Head-Gordon and Hura 2002; Wilding and Benmore 2006, this volume). This cross section may be split into a self ($\alpha = \beta$) and distinct (coherent interference) part:

$$\frac{d\sigma}{d\Omega} = \left(\frac{d\sigma}{d\Omega}\right)^{self} + \left(\frac{d\sigma}{d\Omega}\right)^{distinct} \tag{2}$$

The advantage of this division of the SC is to isolate the unwanted effect of incoherent scattering into the self term. The self part can be further split into self-coherent and an incoherent part:

$$\left(\frac{d\sigma}{d\Omega}\right)^{self} = \left(\frac{d\sigma}{d\Omega}\right)^{self}_{coh} + \left(\frac{d\sigma}{d\Omega}\right)_{incoh} \tag{3}$$

In the case of water, the contribution of incoherent scattering comes from either hydrogen or deuterium since the incoherent cross-section of oxygen is virtually zero. Using D_2O in scattering experiments avoids the large contribution of incoherent scattering from hydrogen to the unwanted background of structural measurements. In the case of pure D_2O

$$\left(\frac{d\sigma}{d\Omega}\right)^{self} = \left(b_O^2 + 2b_D^2\right) + 2\left(\frac{\sigma_{D,incoh}}{4\pi}\right) \tag{4}$$

where b_O and b_D are the coherent scattering lengths for oxygen and deuterium atoms, respectively, b_O = 5.85 barn, b_D = 6.674 barn, and where $\sigma_{D,incoh}$ = 2.032 barn is the incoherent scattering cross section due to deuterium atoms D. The distinct part can be separated into an intra- and inter-molecular part corresponding to correlations between atoms in the same molecule or atoms belonging to different molecules, respectively:

$$\left(\frac{d\sigma}{d\Omega}\right)^{distinct} = \left(\frac{d\sigma}{d\Omega}\right)^{intra} + \left(\frac{d\sigma}{d\Omega}\right)^{inter} \quad (5)$$

Thus, the DC may now be expressed by the relation:

$$\left(\frac{d\sigma}{d\Omega}\right) = \left(\frac{d\sigma}{d\Omega}\right)^{self} + \left(\frac{d\sigma}{d\Omega}\right)^{intra} + \left(\frac{d\sigma}{d\Omega}\right)^{inter} \quad (6)$$

The intensity of scattering for molecular systems is converted to a normalized molecular structure factor, $S_M(Q)$ where **Q** is the elastic scattering vector (Fig. 1). The $S(Q)$ is directly related to the distinct part of the DC which for D_2O is given as:

$$S_M(\mathbf{Q}) = \frac{(d\sigma/d\Omega)^{distinct} + (b_O^2 + 2b_D^2)}{(b_O + 2b_D)^2} \quad (7)$$

For a molecular fluid like D_2O or H_2O, it is convenient to separate terms into those arising from correlations in the same molecule (intramolecular terms) and those from correlations in different molecules (intermolecular terms). Thus the structure factor may be split into two parts as:

$$S_M(Q) = f_1(Q) + D_M(Q) \quad (8)$$

where $f_1(Q)$ is the molecular form factor and the $D_M(Q)$ function contains all of the intermolecular correlations. The $f_1(Q)$ function is equivalent to the

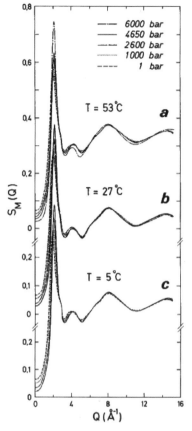

Figure 1. Examples of $S_M(Q)$ for D_2O obtained at three constant temperatures and pressures. (a) 53 °C, (b) 27 °C, (c) 5 °C. [Used by permission of American Physical Society, from Bellissent-Funel and Bosio (1995), J Chem Phys, Vol. 102, Fig. 4, p.3721]

diffraction pattern that would be observed for individual molecules in the low-density regime and it is solely dependent on the molecular conformation and coherent scattering lengths of the constituents. For D_2O molecules this becomes (Dore 1985):

$$\left[f_1(Q)\right]^{D_2O} = \frac{1}{(b_O + 2b_D)^2}\left[b_O^2 + 2b_D^2 + 4b_Ob_DF(Q,r_{OD}) + 2b_D^2F(Q,r_{DD})\right] \quad (9a)$$

where r_{OD} is the bond length and $r_{DD} = 2r_{OD}\sin\theta/2$ for an intramolecular bend angle, θ, and:

$$F(Qr_{\alpha\beta}) = \frac{\sin Qr_{\alpha\beta}}{Qr_{\alpha\beta}}\exp(-Q^2u_{\alpha\beta}^2/2) \quad (9b)$$

is the interference term; $u_{\alpha\beta}^2$ is the square amplitude of vibration for the distance $r_{\alpha\beta}$. The exponential term is an effective Debye-Waller factor that allows for variation of the time-averaged bond distance arising from thermal vibrations. The resulting r-parameter is an ensemble average over the thermally excited states of the molecule. The values of r_{OD} and r_{DD} may be used to define the mean bond-angle D···O···D for the water molecule in the liquid state (Dore et al. 2004).

In principle, at lower Q-values the difference term, $D_M(Q)$ contributes to the overall diffraction pattern and may be formally expressed as:

$$D_{\mathrm{M}}(Q) = \frac{1}{N_M \left[\sum b_n\right]^2} \left\langle \sum_{\alpha \neq \beta} \exp\left(\mathbf{Q} \cdot \mathbf{r}_{c\alpha\beta}\right) \sum_{n_\alpha n_\beta} b_{n_\alpha} b_{n_\beta} \exp\left(i\mathbf{Q} \cdot \left(\mathbf{r}_{c_{n,\alpha}} - \mathbf{r}_{c_{n,\beta}}\right)\right) \right\rangle \quad (10)$$

where n labels the nucleus within the molecule; α and β refer to different molecules in which the molecular center is situated; $\mathbf{r}_{cn}$ is the intramolecular distance of nucleus n from the molecular center and $\mathbf{r}_{c\alpha\beta}$ is the intermolecular distance between centers of molecules designated α and β. This function contains the required information on the spatial arrangement of the molecules in the fluid or glassy assembly as distinct from that of the molecule. The term "structure" corresponds to the ensemble average for all molecules in the system at a specific time. Since the fluid is composed of a large number of molecules in constant motion it is more convenient to think of this as a long-time average of the various changing configurations around a single molecule. The structure factor as defined in Equation (7) (and in Wilding and Benmore 2006, this volume) arises from an ensemble of snap shot images with a "shutter-speed" that is short compared to the characteristic motion of the scattering centers.

The most useful quantity for expressing the static structural characteristics of fluid order is the total pair correlation function, $g(r)$ (Proffen 2006, this volume). This is related to the Fourier transform of $S(Q)$ by the relation:

$$4\pi r \rho_M \left(g(r) - 1\right) = \frac{2}{\pi} \int_0^\infty Q(S_M(Q) - S_M(\infty)) \sin Qr dQ \quad (11)$$

where $S_M(\infty) = (b_O^2 + 2b_D^2)/(b_O + 2b_D)^2$ is the asymptotic value of $f_1(Q)$ at large Q and ρ_M is the molecular density. The function $g(r)$ is a combination of different partial (weighted) correlation functions and includes both inter- and intramolecular distances. In the case of water it is possible to remove the intramolecular terms by subtracting the molecular form factor from $S_M(Q)$ to obtain $D_M(Q)$, which may be Fourier transformed in order to obtain the pair correlation function $g_L(r)$ for the intermolecular term only:

$$d_{\mathrm{L}}(r) = 4\pi r \rho_M \left(g_{\mathrm{L}}(r) - 1\right) = \frac{2}{\pi} \int_0^\infty Q D_M(Q) \sin(Qr) dQ \quad (12)$$

Composite functions of this form will in general contain contributions from all the partial correlation functions weighted according to the b-values and relative concentrations, i.e.,

$$g_{\mathrm{L}}(r) = \sum c_\alpha^2 b_\alpha^2 g_{\alpha\alpha}(r) + 2 \sum_{\alpha \neq \beta} c_\alpha c_\beta b_\alpha b_\beta g_{\alpha\beta}(r) \quad (13)$$

For neutron scattering by D_2O with two components we need to account for $n(n + 1)/2$ correlations, so the composite function is weighted with the following proportions:

$$g_{\mathrm{L}}(r) = 0.489 \, g_{DD}(r) + 0.421 g_{OD}(r) + 0.090 g_{OO}(r) \quad (14)$$

This result indicates that the signal is dominated by D-D and O-D correlations where D atoms make a significant contribution to the total scattering pattern in contrast to X-ray diffraction studies where it is the O atoms the predominate (Bellissent-Funel and Bosio 1995).

The measurement of the total coherent scattering intensity by means of a two- or three-axis diffractometer (reactor source) or a time-of-flight spectrometer (pulsed source) enables the diffraction pattern to be obtained but careful data reduction incorporating experimental and analytical corrections is needed to normalize $S_M(Q)$ and $D_M(Q)$. Even for this simplest of cases, D_2O neutron diffraction, there is a need to account for a number of corrections. The major errors arise from inelasticity (Placzek correction), sample attenuation, multiple scattering, and absorption resonance (Wilding and Benmore 2006, this volume).

Difference methods

The structural information obtained from a single measurement does not necessarily provide much quantitative insight into the properties of the fluid although it can prove useful in checking the validity of simulation predictions from molecular dynamics or Monte Carlo methods. A more detailed study of a sample type such as H_2O can be achieved by isotopic substitution. In Wilding and Benmore (2006, this volume) the use of isotopic substitution to achieve either a *first-order* or *second-order difference function* from neutron diffraction was discussed. We will address these in more detail below as they apply to the interrogation of the structure of aqueous solutions containing dissolved ions. However, let us consider neutron diffraction isotope substitution (NDIS) in the context of our discussion of water. For H_2O, a similar expression to Equation (9a) can be used for the $f_1(Q)$ function with b_H instead of b_D. For isotopic mixtures, the effective coherent scattering is represented by the average value, $\langle b_{HD} \rangle = \alpha_D b_D + \alpha_H b_H$ where α_D is the mole fraction of deuterium [$\alpha_D + \alpha_H = 1$), provided it is assumed that the H and D atoms are in equivalent positions. The efficacy of the iso-structural assumption between light and heavy water has been documented extensively—e.g., Hart et al. (2005) and references therein.

From Table 2 we see that since $b_D = 6.67$ fm and $b_H = -3.74$ fm it is possible to change $\langle b_{HD} \rangle$ over a wide range and even to choose the special case of $\alpha_D = 0.359$ for which $\langle b_{HD} \rangle = 0$. The coherent and incoherent contributions for various mixtures of H_2O and D_2O are also given in Table 2. Soper and Egelstaff (1981) and Soper and Silver (1982) demonstrated that the technique of H-D substitution could be used to extract information on water structure. By performing three measurements on pure H_2O, pure D_2O, and a mixture (or more, as described by Dore et al. 2004) of H_2O and D_2O, one can exploit the scattering contrast of hydrogen

Table 2. Neutron parameters for various isotopes of hydrogen and oxygen, and the coherent and incoherent contributions for various H_2O/D_2O mixtures (Dore et al. 2004).

Neutron parameters for various isotopes of hydrogen and oxygen.

	b/fm	σ_{coh}/barns	σ_{incoh}/barns	$\sigma_c\sigma_i$
H	−3.74	1.76	79.9	0.022
D	6.67	6.00	2.0	3.0
^{16}O	5.80	4.24	0	-
^{17}O	5.78	4.2	0.2	-
^{18}O	5.84	4.3	0	-

Coherent and incoherent contributions for various H_2O/D_2O mixtures.

α_H	0	0.05	0.10	0.20	0.50	0.641	1.00
$\langle b_{HD} \rangle$/fm	6.67	6.15	5.63	4.59	1.47	0	−3.74
σ_{coh}	15.3	13.7	12.2	9.5	4.8	4.2	7.6
σ_{incoh}[H/D]	4.0	11.8	19.6	35.2	82	104	160
σ_{incoh}[mix]	0	1.3	2.4	4.2	6.6	7.6	0
$\Sigma\sigma_{incoh}$/mol	4.0	13.1	22.0	39.4	88.6	112	160
$\sigma_c\sigma_i$	3.8	1.05	0.47	0.24	0.054	0.037	0.048

and deuterium to isolate the three site-site correlation functions directly. If a 50:50 H/D mixture is used the inelastic contribution practically cancels in the data reduction process. This constitutes a *second-order difference* method which yields a total pair correlation function for each species as given by Wilding and Benmore (2006, this volume). These functions form the basis for calculating the partial pair distribution functions $g_{HH}(r)$, $g_{OH}(r)$, and $g_{OO}(r)$ examples of which are provided in Figure 2 for ambient conditions (Head-Gordon and Hura 2002). Modeling is required to provide a more complete 3D structural reconstruction of the pair correlation function which is a 1D representation. For fluids like water, the Empirical Potential Structural Refinement (EPSR) method summarized most recently by Soper (2005) provides one possible path to accomplish this.

Structure of water at ambient and high temperature-pressure conditions

Ambient water. Despite the difficulties with the NDIS approach arising from the large uncertainty in the inelastic correction, numerous studies have been reported describing the use of this method to study water structure at ambient (e.g., Thiessen and Narten 1982; Soper and Philips 1986; Bruni et al. 1996; Soper et al. 1997; Ricci and Soper 2002) and elevated temperature and/ or pressure conditions (e.g., Neilson et al. 1979; Gaballa and Neilson 1983; Ichikawa et al. 1991; Postorino et al. 1993; Tromp et al. 1994; Tassaing et al. 1998; Yamaguchi 1998; Soper 2000; Bellissent-Funel 2001). There are also numerous papers dealing with the subject of supercooled water which we have elected not to discuss in this section (e.g., Dore 1984, 1994; Bellissent-Funel et al. 1989; Bellissent-Funel 1998; Dore et al. 2000).

It is worth devoting some space to a discussion of the structure of bulk water to set the stage for perturbations produced

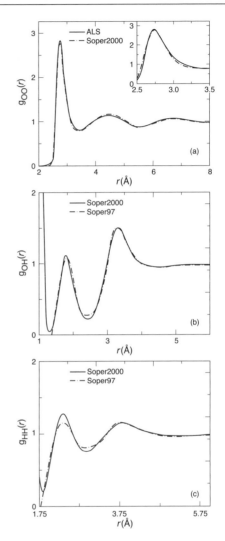

Figure 2. Comparison of neutron data (as $g(r)$) on pure water at 25 °C and 1 atm. (a) Comparison of ALS X-ray experimental $g_{OO}(r)$ from Hura et al. (2000), Sorenson et al. (2000) – solid line with reanalysis of Soper (2000) NDIS data (dashed line). Comparison of neutron data on pure water at 25 °C, 1 atm: Soper et al. (1997) – dot-dash line; Soper (2000) – solid line; for (b) $g_{OH}(r)$ and (c) $g_{HH}(r)$. [Used by permission of the American Chemical Society, from Head-Gordon and Hura (2002), Chem Rev, Vol. 102, Fig. 4a,b,c, p. 2661]

by dissolved ions and the interactions with various types of solid interfaces to be described in later sections. Let us first consider the scattering amplitude for oxygen, which is smaller than that for hydrogen. Because the $g_{OO}(r)$ distribution is the most difficult to determine it is common for structural data from X-ray scattering to be used for comparison and incorporated into simulation studies. Indeed, Soper (1996, 2000) reported new neutron analysis results based on previously obtained neutron diffraction data using the EPSR method where the $g_{OO}(r)$ is in very good agreement with X-ray scattering data reported by Hura et al. (2000) and Sorenson et al. (2000) as shown in Figure 2a. Head-Gordon and Hura (2002) point out that the use of the SPC/E water model for the reference potential by Soper probably helped improve this agreement. As Figure 2a shows the first peak gives us the first neighbor oxygen-oxygen distance which occurs at about 2.8 Å. Of special interest, however, is the estimate of the oxygen-oxygen coordination number which requires consideration of the distribution density over the whole region of the maximum in $g_{OO}(r)$. To do this, one can integrate the distribution function up to the first minimum, considering spherical shells of volume $4\pi r^2 dr$. The coordination number, n_C can then be described by:

$$n_c = 4\pi\rho_O \int_o^{r_{min}} g_{OO}(r)r^2 dr \tag{15}$$

where ρ_O is the density of oxygen atoms in the fluid and r_{min} is the location of the first minimum in $g_{OO}(r)$, which is roughly 3.5 Å. Depending on the NDIS data set, the n_C values can range from 4 to very close to 5. This coordination number depends critically upon the upper cut-off distance used in performing the interrogation under the peak (Finney and Soper 1994). Beyond the first minimum, $g_{OO}(r)$ oscillates with a second maximum at about 4.5 Å which corresponds to an O-O-O angle of about 110°. This is close to the tetrahedral angle, implying that the local water molecule geometry is on average tetrahedral, though the broad nature of the second nearest neighbor peak shows there is considerable variation around this average. Head-Gordon and Hura (2002) note the strict adherence to hydrogen-bonded hexagons of ice water give way to greater translational and rotational motion of waters and a broader distribution of hydrogen-bonded configurations, including a variety of polygons of varying sizes and degrees of puckering, or distortion, all of which result in a more compact arrangement of water molecules. Recent work by Wernet et al. (2004) based on X-ray scattering suggest that water exists in mixed form—two-hydrogen bonded (60-75%) and tetrahedral (25-40%) species.

A much different diffraction pattern is observed for the oxygen-hydrogen distribution function (Fig. 2b) where a very strong peak occurs at about 1 Å. This peak corresponds to the intramolecular O-H distance, and is a diagnostic indicator that the experiment is reporting correct results because it is located at the known OH distance with an area predicting two hydrogen atoms (Finney and Soper 1994). The second maximum is centered on about 1.85 Å and is associated with the hydrogen atoms in surrounding molecules that are hydrogen bonded to the oxygen atom in a central molecule—i.e., the intermolecular O-H. Integration out to the first minimum gives a coordination number of about 1.7 whereas this number becomes approximately 3.3 for the second peak integrated out to the second minimum. The third peak centered on about 3.25 Å occurs at the first minimum of the $g_{OO}(r)$ distribution curve and refers to non-hydrogen-bonded oxygen-hydrogen distances on neighboring molecules. At this point, the total number of hydrogen atoms around the central oxygen is close to eight which is two times the number of oxygens in the same volume of solution.

Finally, the distribution of hydrogen atoms with respect to a central hydrogen atom, the hydrogen-hydrogen $g_{HH}(r)$ correlation, is shown in Figure 2c. This function is the easiest to determine experimentally because of the large difference in scattering length for hydrogen and deuterium. The first peak (not shown) is at the intramolecular H-H distance of 1.55 Å and again confirms the molecular geometry of the water molecule. The second peak is just above 2.4 Å and represents the closest H-H distances between hydrogen-bonded neighbors. The coordination number calculated by integration out to the second minimum at 3.1 Å is roughly 5.8. The third broad, asymmetric peak occurs around 3.7-3.8 Å and refers to the distant H-H distances on neighboring water molecules. Comparing $g_{OH}(r)$ and $g_{OO}(r)$, we see that the maxima on the H-H distribution function are located further from the central atom by 0.6 Å with respect to those on the O-H distribution (Fawcett 2004). This is entirely reasonable because of the O-H distance in water.

High temperature and/or high pressure water. We can assess the extent by which temperature and/or pressure influences the structure of water through examination of how the correlation functions vary as these two state variables are either raised or lowered. Qualitatively one can expect that as we lower temperature, structural order will increase as exemplified by results obtained on supercooled water (e.g., Dore et al. 2000; Botti et al. 2002). Conversely as we raise temperature we can anticipate a loss in ordering which will be reflected in a broadening of the various correlation peaks as well as some shift in peak positions. The changes in the structure of the three-dimensional hydrogen-bonded network of water are interrogated relative to the reference ambient temperature liquid using NDIS as described above.

At normal pressure, the structure of liquid water is fragile as manifested by its unusually large temperature dependence. Numerous neutron and X-ray scattering studies indicate that when water is isothermally compressed, the number of hydrogen bonds per water molecule is not altered by any appreciable amount relative to what is observed at ambient conditions (e.g., Bellissent-Funel and Bosio 1995; Yamanaka et al. 1994; Gorbaty and Demianets 1983; Gorbaty and Okhulkov 1994). However, the hydrogen bonds do become bent out of their ideal orientation and are correspondingly weaker energetically (Head-Gordon and Hura 2002). The most significant effects occur in the change of O-O separation with increasing pressure (into the 100's of MPa range) where the second peak in the $g_{OO}(r)$, indicative of local tetrahedral structure, is diminished with increasing pressure. With increasing pressure, the effect of temperature on liquid structure becomes smaller until the pressure dependency is nearly independent of temperature variation.

This is not the case when water is heated above ambient conditions up to and beyond the supercritical state points. By way of example we show the results of Soper et al. (1997) who provided a reassessment of both old and new diffraction data leading to improved site-site pair correlation functions of water from 25 to 400 °C and pressures up to 280 MPa (Fig. 3a-c). This composite plot (with the intramolecular portions removed) shows the changes in $g_{HH}(r)$, $g_{OH}(r)$, and $g_{OO}(r)$ as a function of increasing temperature and decreasing density. Despite some difficulties with small residual oscillations in these functions arising from non-trivial truncation and statistical uncertainties, qualitative trends are apparent. In the case of the O-H correlations (Fig. 3a) the first peak becomes significantly broader as the critical conditions are approached and appears to shift to larger radius values—i.e., from 1.85 Å at ambient conditions to about 2.15 Å at 573 K on the coexistence curve. The number of hydrogen bonds tends to decrease from 3.3 at ambient conditions to values more on the order of 2.2-2.4 at 573 K based on integration under the first O-H peak. In the supercritical state, at 673 K, this O-H peak no longer appears as a distinct peak and has been washed out into a broad shoulder. Based on these results it can be concluded that above the critical point, the hydrogen bonding in water has been significantly modified to the point where no distinct O-H site-site correlation peak is preserved. From an examination of the O-O correlation functions (Fig. 3b), we observe a similar trend wherein there is a gradual shift of the first peak to larger radius values with increasing temperature to a value of about 3.1 Å (X-ray diffraction data show this peak position to be closer to 3.5 Å). As temperature is increased to the critical point, the H-H pair correlations (Fig. 3c) exhibit a decided degradation of the pattern with considerable shoulder development even at 423 K to such an extent that it is difficult to tell how much the first ambient peak at 2.4 Å shifts with increasing temperature.

There can be no doubt that based on both neutron and X-ray diffraction, the molecular arrangement of water molecules is greatly perturbed at elevated temperatures compared to ambient water. The question that has been hotly debated in the past is whether or not there is

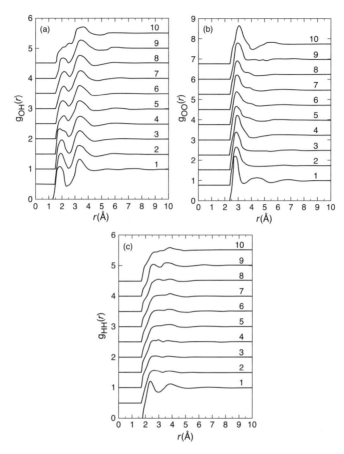

Figure 3. Radial distribution functions for H_2O: (a) $g_{OH}(r)$, (b) $g_{OO}(r)$ and (c) $g_{HH}(r)$ as a function of temperature and density from NDIS experiments. Temperature-pressure-density conditions: 1: 298 K, 0.1 MPa, 0.0334 molecules/Å³; 2: 423 K, 190 MPa; 0.0334; 3: 423 K, 100 MPa, 0.0308; 4: 573 K, 280 MPa, 0.308; 5: 573 K, 197 MPa, 0.0296; 6: 573 K, 110 MPa, 0.0278; 7: 573 K, 5.0 MPa, 0.0260, 8: 573, 10 MPa, 0.0240; 9: 573 K, 9.5 MPa, 0.0240; 10: 673 K, 80 MPa, 0.0211. [Used by permission of the American Physical Society, from Soper et al. (1997), J Chem Phys, Vol. 106, Figs. 2, 3, 4, p. 251]

a temperature at or above which there is total breakdown of hydrogen bonding. According to the work of Soper et al. (1997) highlighted above, calculated bond angle distributions tend to suggest that the degree of true hydrogen bonding is greatly reduced in the supercritical state compared to ambient conditions. Complementary experimental studies using Raman (Walrafen et al. 1999), X-ray scattering (Gorbaty and Kalinichev 1995) and IR (Bondarenko and Gorbaty 1973) suggest that tetrahedral hydrogen bonding persists to at least 650 K and perhaps even up to 700 K at 100 MPa (Head-Gordon and Hura 2002). Bolstered by simulation studies using improved water models in concert with various scattering data, a consensus has been established that local hydrogen bonding is still present near the critical temperature and density, but that the space-filled percolating hydrogen-bonded network dominant at ambient conditions collapses.

Structure of aqueous solutions of geochemical relevance

Understanding the speciation of solutes (e.g., cations, anions, non-polar species, etc.) in aqueous solutions is a critical issue with direct impact on important geochemical processes such mineral solubilities, metal mobilization and transport, and precipitation of potential contaminant phases. Of particular importance is the nature of ion hydration, rates of exchange of coordinated ions with water, and interaction energies between ions and water molecules. Given the current state of molecular-based experimental and modeling techniques, many of these aspects of complex aqueous solutions have been largely speculative and commonly model-dependent. The effect of dissolved ions on the average structure of water may range from negligible in very dilute solution to significant in concentrated solutions. The structure of water in aqueous solutions has commonly been described relative to the structure of pure water and terms such as "structure maker and "structure breaker" are intended to indicate deviations from the pure liquid (Soper and Turner 1993). These terms originated from the comparison of the correlation times of water molecules in aqueous solution with those in pure water (Ohtaki and Radnai 1993). Early on it was recognized that rather than focusing on single parameters such as the hydration number, the bond length, or the diffusion coefficient, that what really mattered were correlation functions which reflected the disordered and dynamic nature of aqueous solutions (e.g., Enderby 1983, 1995; Neilson and Enderby 1989).

In the context of an aqueous solution, let's revisit the concept of the radial distribution function, $g_{\alpha\beta}(r)$ related to the static structure. The function measures the probability of finding a β-particle at a distance r from an α-type particle placed at the origin. The average number of β-type particles that occupy a spherical shell of radius r and thickness dr at the same instance of time is given by:

$$dn_{\alpha\beta} = 4\pi\rho_\beta g_{\alpha\beta}(r)r^2 dr \tag{16}$$

where $\rho_\beta = N_\beta/V$ and N_β is the number of β species contained in the sample of volume, V (Neilson and Enderby 1996). The integrated form of this expression was given in Equation (15). A multicomponent solution in the form of a salt (MX_n) in H_2O can be described by 10 pair radial distribution functions grouped into three sub-systems: three representing the solvent structure, $g_{HH}(r)$, $g_{OH}(r)$, and $g_{OO}(r)$; three characterizing the solute structure, $g_{MM}(r)$, $g_{MX}(r)$, and $g_{XX}(r)$; and four which describe the solute-solvent structure (hydration phenomena), $g_{MO}(r)$, $g_{MH}(r)$, $g_{XO}(r)$, and $g_{XH}(r)$. The difficulty is that a single diffraction pattern contains a weighted average of all 10 partial structure factors, and to disentangle them is far from trivial. To fully appreciate this complexity, refer to Equation (10) in Wilding and Benmore (2006, this volume) which describes the formalism for the total pair distribution function, $G_{\alpha\beta}(r)$.

In Figure 4a we show examples of two typical $g(r)$s which represent (a) $g_{MO}(r)$ and (b) $g_{MH}(r)$ for an aqueous solution. The ratio Δ/Δ' gives a measure of the stability of the aqua-ion hydration complex. From measurements of r_{MO} and r_{XO} and knowledge of the interatomic distances of the H_2O molecule, it is possible to determine the ion-water conformation and to calculate an average tilt, as depicted in Figure 4c. In principle by comparison of n_{MO} and n_{MH} one can evaluate the extent to which an aqua ion is acidic, since the dissociation given as:

$$[M(H_2O)_n]^{p+} = [MH_2O)_{n-1}(OH)]^{(p-1)+} + H^+ \tag{17}$$

will lead to $n_{MH} < 2n_{MO}$. In practice, however, only strong acid behavior (e.g., Fe^{3+}) has been detected because of the errors in determination of n (Neilson et al. 1993).

The difference methods of NDIS referred to above (and in Wilding and Benmore 2006, this volume) are ideally suited to the determination of structure in terms of the individual $g_{\alpha\beta}(r)$ or as their linear combinations of the form $G_{\alpha\beta}(r)$, which is specific to the isotopically substituted species α. The *first-order difference* method applied to cations or anions by isotopic exchange M^* for M or X^* for X can be used to obtain information concerning aqua-ion structure in terms of the function $G_M(r)$ or $G_X(r)$. For $G_M(r)$ we can show (from Enderby et al. 1987):

$$G_M(r) = A[g_{MO}(r) - 1] + B[g_{MH}(r) - 1] + C[g_{MX}(r) - 1] + D[g_{MM}(r) - 1] \tag{18}$$

where $A = 2c_M c_O b_O \Delta b_M$, $B = 2c_M c_H b_H \Delta b_M$, $C = 2c_M c_X b_X \Delta b_M$, $D = c^2_M(b^2_M - (b^*_M)^2)$, $\Delta b_M =$

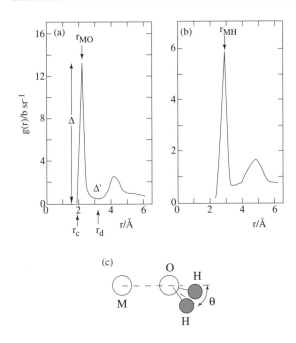

Figure 4. Two typical $g(r)$'s which represent (a) $g_{MO}(r)$ and (b) $g_{MH}(r)$ for a solution MX_n in H_2O. Δ/Δ' gives an estimate of the stability of the aqua-ion hydration complex. r_c is the nearest distance of approach of the two particles, r_{MO} and r_{MH} are the most probable nearest-neighbor separations between particles, r_d is the probable end of the first coordination shell. (c) Possible ion-water conformation and tilt angle, θ. [Used by permission of The Royal Society of Chemistry, from Neilson et al. (1993), J Chem Soc Faraday Trans, Vol. 89, Fig. 1, p. 2928]

$b_M - b_{M^*}$, and c is the atomic concentration of either M, X, O, or H (D), whose neutron coherent scattering length is b_i. In practice only the first two terms matter because c_M and $c_X \ll c_H$ and c_O such that the $G_M(r)$ reflects almost totally the structure of the coordination complex. The strength of the interactions dictate to a large extent the range of concentrations used in scattering experiments and in many cases molal quantities are required to achieve satisfactory results. Of course the key prerequisite here is having M cations (e.g., Li$^+$, K$^+$, Mg^{2+}, Ca^{2+}, Ni^{2+}, Zn^{2+}, Sr^{2+}, etc.) and X anions (Cl$^-$, I$^-$, NO$_3^-$, etc.) with isotopes exhibiting sufficient differences in their coherent scattering length (refer to Table 1 for a partial list). From our discussion of NDIS in water we know that H-D exchange occurs in all aqueous solutions such that by measuring the $G_M(r)$ for both heavy and light water, the $g_{MO}(r)$ and $g_{MH}(r)$ can be determined separately. This assumes that H and D can be treated as isomorphic—i.e., $g_{MH}(r) = g_{MD}(r)$. In an analogous manner, the hydration of anions can be studied, the derived quantities being, $G_X(r)$, $g_{XH}(r)$ [$= g_{XD}(r)$], and $g_{XO}(r)$. D$_2$O was used in the early applications of NDIS to salt solutions to avoid problems that could arise from the large incoherent scattering of H$_2$O. However, improvements in scattering technology made it possible to also work with salt solutions in H$_2$O.

The *second-order difference* experiments involving isotopic exchanges of M^* for M, X^* for X, and deuterium (D) for hydrogen (H) can be used to obtain individual pair functions for the solute [$g_{MM}(r)$, $g_{MX}(r)$, $g_{XX}(r)$] and the solvent [$g_{HH}(r)$, $g^*_{OH}(r)$]. Here the asterisk on $g_{OH}(r)$ denotes that it is only an approximation to the true $g_{OH}(r)$, except in the limit of greatly increased dilution where it will approach the exact function, as in pure H$_2$O. This approach makes it possible to improve resolution of ionic hydration structure especially with regard to the second hydration shell. It is worth repeating that because the scattering length of the oxygen isotopes, ^{16}O, ^{17}O, and ^{18}O, are almost identical, determination of $g_{OO}(r)$ in pure water is subject to large uncertainties, and is not readily accessible in solution. For a determination of either $G_M(r)$ or $G_X(r)$, it is preferential to work with solutions of heavy water, D$_2$O, because of the large incoherent scattering of the proton (H) spins and the large inelastic scattering of the proton nuclei in H_2O. For the most part it is sufficient to only describe the total distribution, $G_M(r)$ or $G_X(r)$ unless one really needs to resolve the hydration at the pair distribution function level (Neilson et al., 2001). The hydration number, n_I for an ion, M or X, in solution can be defined from Equation 15 where the $g(r)$'s can be $g_{MO}(r)$, $g_{MH}(r)$, $g_{XO}(r)$, or $g_{XH}(r)$, and concentration of O as c_O or H as c_H appear outside the integral.

Rather complete reviews on complex solutions including aqueous electrolytes of interest to geochemists studied by neutron and/or X-ray scattering have been provided by Enderby (1995), Neilson and Adya (1997) and Neilson et al. (2002) so we will only highlight some relevant results to illustrate the power of neutron scattering.

Solvent structure. How the radial distribution functions of water change with an added salt are of considerable interest to geochemists. Manifestations of changes in water structure on the dissolution of ions are evident in many macroscopic experiments, e.g., changes in compressibility, solvation enthalpies and entropies, etc. H/D isotopic substitution of the water molecules in concentrated aqueous solutions (e.g., common chloride salts of LiCl, KCl, NaCl, etc.) is used to calculate $g_{HH}(r)$ and $g_{OH}(r)$, and estimate $g_{OO}(r)$, whose accuracy is appreciably less than the other two functions. For example, the increase in concentration of LiCl on $g_{HH}(r)$ is shown in Figure 5 for two solutions, 1 and 10 molal (mol/kg H_2O), compared against pure H_2O (Tromp et al. 1992). They concluded that both the intermolecular and intramolecular structure of water are relatively unaffected by the presence of salt at the 1 molal level. Clearly, at higher concentrations, a significant effect takes place, and in this case, at 10 molal, the number of hydrogen bonds decreases by about 30% from that characteristically of pure H_2O. This type of work was extended by Leberman and Soper (1995) to include other salts, NaCl, Na_2SO_4, and NH_4SO_4, and also more recently by Bruni et al. (2001), who examined effects of high concentrations (to 10 molar) of hydroxide, as NaOH, on water structure. Leberman and Soper (1995) concluded that ions effectively produce a change in water structure that is equivalent to an increase in pressure—i.e., ionic concentrations of a few molal have equivalent pressures that exceed 100 MPa. This is not entirely correct since pressure on a solution is isotropic whereas this effect is "localized" and solvent-mediated due to electrostriction about the ions. Interestingly, cations commonly found in nature, such as Na^+ and K^+, as chlorides, will also not significantly alter the structural properties of water solvent at concentrations less than about the 1 molal range (Horita and Cole 2004).

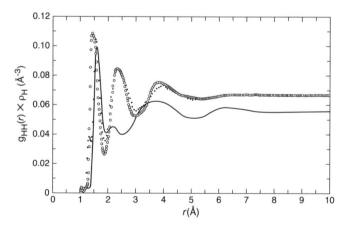

Figure 5. Partial pair radial distribution function $g_{HH}(r)$ multiplied by ρ_H; pure water – open symbols; 1 molal LiCl – dots; 10 molal LiCl – solid curve. [Used by permission of the American Physical Society, Tromp et al. (1992), J Chem Phys, Vol. 96, Fig. 7, p. 8467]

Ion-water structure. A tremendous amount of effort has gone into studies of ion-water interactions in aqueous solution. As noted above, recent reviews by Neilson and Adya (1997), Neilson et al. (2002), and most recently by Danielewicz-Ferchmin and Ferchmin (2004) provide very nice overviews on behavior of a wide range of cation and anion hydration behavior as interrogated by neutron, X-ray and computational methods. The NDIS method has contributed to the several key questions that are central to understanding this behavior including; (i) what are the length scales over which the primary hydration shell, secondary shell, intermediate zone and transition to bulk water range?; (ii) how sharply differentiated are these zones?; and (iii) how stable is the inner most hydration shell to changes in temperature, pressure, ion concentration and counter ion? We will only touch upon a few examples here to highlight how neutron scattering has contributed to answering these questions.

The alkali and alkaline earth metal ions (Li^+, Na^+ [isomorphic with Ag^+], K^+, Rb^+, Mg^{2+}, Ca^{2+}, Sr^{2+}) have probably been the most extensively studied of all the cations (e.g., Howell et al. 1991; de Jong and Neilson 1996, 1997). Example diffraction patterns, as $G_M(r)$ vs. r are shown in Figure 6. In general, as the ion size increases and the charge density is reduced there is a progressive weakening in the ion-water structure—i.e., the hydration shell becomes less well resolved. For example, the ionic hydration of Li^+ and Mg^{2+} (which is isomorphic with Ni^{2+}) is more clearly defined compare to that for K^+ or Sr^{2+}, respectively, and Na^+ and Ca^{2+} occupy

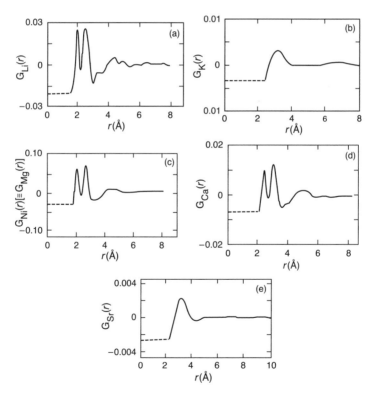

Figure 6. Examples of cation radial distribution functions $G_M(r)$ at 25 °C for (a) Li^+ in 3.6 M D_2O solution (Newsome et al. 1980); (b) K^+ in 4 M KCl D_2O solution (Neilson and Skipper 1985); (c) Ni^{2+} in a 2 M $NiCl_2$ D_2O solution (Powell et al. 1989), which is isomorphic with $MgCl_2$ in aqueous solution (Skipper et al. 1989); (d) Ca^{2+} in a 2.8 M $CaCl_2$ D_2O solution (Hewish et al. 1982); and (e) Sr^{2+} in a 3 M $Sr(ClO_4)_2$ D_2O solution (Neilson and Broadbent 1990). [Used by permission of Elsevier Ltd, Howell et al. (1991), J Molec Struct, Vol. 250, Figs. 2, 3, p. 286]

intermediate ion-water conformational structures. In particular, Sr^{2+} has a relatively weak hydration shell and no longer-range structure. Interestingly, it appears that the coordination number for the +1 valence alkali cations does not change appreciably from a value of between 5 and 7. As studies have shown that with the exception of Li^+, the extent of the perturbation to the structure of water is limited primarily to the range of the first hydration shell for alkali ions.

Other ions of interest to geochemists that have received considerable attention include transition metals (e.g., Cr^{3+}, Fe^{2+}, Fe^{3+}, Ni^{2+} and Cu^{2+}) and the lanthanides (Nd^{3+}, Dy^{3+}, Yb^{3+}). Characteristically, the transition metals have a first hydration shell usually containing 6 water molecules, a distinguishing second hydration shell, and considerable short-range ordering of water around them (Neilson et al. 2001). Of the transition metals, the most extensive work on aqua ion structure has been carried out on Ni^{2+} using NDIS, and is commonly used as a reference system to test the accuracy of other methods. This is particularly true when exploring the effects of increased concentration, temperature and pressure. NDIS experiments on aqueous solutions have been conducted at temperatures and pressures up to and beyond their critical points. It has been documented that the hydration structures of Li^+, Ni^{2+} and Cl^- are highly susceptible to changes in temperature, and there is a general reduction in hydration number with temperature presumably accompanied by an increase in anion-cation contact (Neilson and Adya 1997). Recently, Badyal and Simonson (2003) interrogated the effects of temperature (90, 175, 230 °C) on the hydration structure around Ni^{2+} in concentrated chloride solutions. They observed a gradual weakening of the hydration structure and a steady reduction in the average water coordination number with increasing temperature. Structural properties of the lanthanides have been less well investigated, but NDIS–based studies have suggested that the hydration number decreases approximately one unit across this series, from 9 to about 8 (e.g., Hahn et al. 1986; Helm et al. 1994; David et al. 2001).

The NDIS method has been used to interrogate the hydration behavior of important anions (Cl^-, Br^-, NO_3^-, ClO_4^-, etc.) but is particularly well suited to the study of chloride because of its favorable isotopes, ^{35}Cl and ^{37}Cl. It is considered as an ideal probe of structural perturbations which are induced when aqueous electrolyte solutions contain large polyions (Wilson et al. 1997; Tromp and Neilson 1996). In general, anion hydration is weaker than that for cations particularly for complex anions where the low charge density inhibits the formation of long-lived aqua-ion species (Herdman and Neilson 1990). Figure 7 shows example results (Botti et al. 2004) based on the *second-order difference* method for $g_{ClO}(r)$, $g_{ClH}(r)$ and $g_{ClCl}(r)$ for HCl (1:9 solution of HCl in H_2O at 298 K). The pronounced first peak in $g_{ClOw}(r)$ and $g_{ClHw}(r)$, along with a low first minimum indicate a well-defined first solvation shell and suggests that the Cl^- ions are likely substituting a water molecule in the H-bond network. There are 4.4±1.4 water molecules around each Cl^- ion with a radius of 3.5 Å from a central negative ion. Its structure is remarkably independent of counterion, concentration changes to ~1 molal, and moderate changes in temperature and pressure (Powell et al. 1993). At elevated concentrations above 1 molal, and in the presence of counterions such as Zn^{2+} or Cu^{2+}, there is an appreciable decrease in the coordination number from 6 to about 4. However, the general conformation of the $Cl – H_2O$ interaction does not change with concentration. The absence of a conformational change with concentration in nickel chloride solutions is probably the result of effective self-screening by the relatively strong Ni^{2+} cation (e.g., Tromp and Neilson 1996). The effect of temperature on Cl^- hydration appears to be more pronounced for relatively weaker counterions such as in LiCl solutions where the hydration shell is more strongly affected when compared to a solution such as $NiCl_2$.

Solute structure. Neilson et al. (2002) point out that determination of the pair radial correlation functions among ions in solution (e.g., Ni-Ni, Cl-Cl, Ni-Cl) represents one of the most challenging experimental aspects of the molecular-level characterization of aqueous solutions. *Second-order difference* experiments have been conducted at very high concentrations of LiCl (Ansell and Neilson 2000), copper chloride (Ansell et al. 1995), and nickel chloride and nickel sulfate (Howell and Neilson 1997). However, results from these

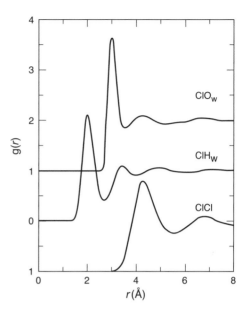

Figure 7. Radial distribution functions for $g_{ClHw}(r)$, $g_{ClOw}(r)$ and $g_{ClCl}(r)$ for HCl_{aq} at 298 K conditions (from Botti et al. 2004) around a central chlorine ion. The $g_{ClOw}(r)$ and $g_{ClCl}(r)$ have been shifted vertically for clarity. [Used by permission of the American Physical Society, Botti et al. (2004), J Chem Phys, Vol. 121, Figs. 3, p. 7843]

studies are not conclusive due to the difficulties inherent with this type of NDIS experiment. Enhancements in neutron intensity and instrument stability at facilities such as SNS offer the potential for more definitive *second-order difference* experiments on solute structure.

Dynamics in water and electrolyte solutions

Thus far we have restricted our discussion on bulk solutions to the structural aspects and have not addressed molecular dynamics behavior. Of particular interest are microscopic dynamical properties such as translation and rotation, diffusion, and the lifetime of hydrogen bonding in both pure water as well as aqueous solutions. At the microscopic level, liquid water may be viewed as a "transient gel" or a network of hydrogen bonds with a local tetrahedral symmetry. This picture was originally developed from the percolation model of Stanley and Teixeira (1980) and the connectivity properties have since been predicted by molecular dynamics (MD) simulations (e.g., Kalinichev 2001). It appears that the hydrogen bonded network includes small spatially-correlated "clusters" of four-bonded molecules and that the local density near a "cluster" is lower than the global density. The dimensions of the clusters have been evaluated by estimating their radius of gyration. The value of the mean characteristic length is approximately 20 Å whereas the correlation length of the density fluctuation has been measured by both X-ray (Michielsen et al. 1988) and neutron scattering (Bosio et al. 1989). It turns out that because of the differences between scattering cross sections of H_2O and D_2O, inelastic neutron scattering is a useful tool to probe the individual and collective motions of liquid water and water containing dissolved ions. Specifically, the diffusive motions (translation and rotation) can be interrogated by quasi-elastic neutron scattering (QENS), whereas the vibrational density of states, including both the inter- and intra-molecular parts, are accessible through the use of inelastic neutron scattering (INS). Many of these properties are strongly dependent on both temperature and pressure. The theoretical background for these two approaches is presented later in this chapter in the sections of fluid confinement and fluid-surface interactions, and to a limited extent in Loong (2006, this volume).

QENS is generally applied to the study of nonpropagating relaxation modes in liquids (e.g., Chen and Teixearia 1986). When applied to molecules containing hydrogen atoms, it measures the dominant incoherent scattering from them. In this case, the spectra are dominated by the Fourier transform of the van Hove self-correlation function of the H atoms. In liquid H_2O, the principle contributions to the quasi-elastic line are from the self-diffusion of the two equivalent H atoms and from the short-time reorientation of the molecule. QENS studies have been conducted on water (as either H_2O or D_2O) at ambient to supercooled conditions (e.g., Chen et al. 1982, 1999; Teixeira et al. 1985; Bellissent-Funel and Teixeira 1991; Petrillo et al. 2000), in the supercritical range (e.g., Tassaing and Bellissent-Funel 2000; Uffindell et al. 2000; Beta et al. 2003), and at high pressure (Cunsolo et al. 2006). Parameters obtained from QENS include the residence time, τ_0 (in ps) which has the strongest temperature dependence, a reorientational (hindered rotation) time, τ_1 (in ps), which exhibits Arrhenius behavior ($E_A = 1.85$ kcal mol^{-1}), and the characteristic mean jump diffusion length, L in Å. As seen in Figure 11 below (Confinement section), both τ values increase with decreasing temperature, with τ_0 displaying a relative steep non-linear slope (1.25 ps at 20 °C to >10 ps at −20 °C). We know that water molecules are instantaneously connected with most of their neighbors through hydrogen bonds. The breaking of hydrogen bonds is due to the libration motions which have large amplitudes and occur when the vibrational angle exceeds the critical value of 28°. When at least three bonds are broken simultaneously diffusion is possible where the proton jumps to the nearest site which is at an average distance L of about 1.6 Å. The value τ_0 is a direct measure of the average duration which is required for a water molecule to break loose from its hydrogen bondage by rotational excitation. The τ_0 is a time constant that no other technique except neutron scattering can access. The time associated with hindered rotations defined by τ_1 is interpreted as the typical hydrogen-bond lifetime. Because the number of hydrogen bonds increases with decreasing temperature, the diffusion process is strongly temperature dependent. At supercritical conditions for water (400-450 °C), τ_0 remains relatively constant at 0.29-0.3 ps. The self-diffusion coefficient increases significantly with increasing temperature and decreasing density with values of 3.71-5.21 × 10^{-4} cm^2 s^{-1} for the T interval of 400-450 °C (Beta et al. 2003). Based on QENS results, it appears that even at supercritical conditions, motion in water is still constrained by the presence of a non-trivial degree of hydrogen bonding.

QENS has also provided the means by which ions (cations in particular) can be characterized according to the lifetimes of water molecules in cationic hydration structures on time scales of the order of 10^{-9} sec—i.e., smaller than those accessible by nuclear magnetic relaxation techniques where the lowest limit is ~10^{-7} s (e.g., Herdman and Neilson 1990). The QENS method allows one to differentiate between those cations which hydrate water molecules on timescales, τ_b much longer than 5×10^{-9} s or much shorter than 10^{-10} s; it is also possible to identify cations such as Cu^{2+} and Zn^{2+} with hydration shells which exchange water molecules with the bulk on a timescale of $5 \times 10^{-9} \geq \tau_b /s \geq 10^{-10}$. Figure 8 summarizes the residence times of water molecules in the neighborhood of aqua ions and ranges of experimental techniques that be used to interrogate specific timescales (Neilson and Enderby 1989). For a more complete, although not very recent summary of the translational self-diffusion coefficients for ions, residence times of water in the first hydration shell, reorientational time of hydrated water molecules, and rates of water substitution reactions involving ions, refer to the nice compendium provided by Ohtaki and Radnai (1993). Inspection of Figure 8 indicates that ions (usually strong cations such as Ni^{2+} or Cr^{2+}) which can coordinate with protons for a time $>5 \times 10^{-9}$ s are said to possess hydrated water molecules that exchange slowly with the bulk water. Conversely, water molecules coordinated to weak cations (e.g., Na^+, Li^+, Cs^+) and all anions (e.g., Cl^-, I^-, F^-) exchange with bulk water on a timescale $<10^{-10}$ s which is classified as fast exchange. There are also several cations with hydrated water molecules which exchange with bulk water in an intermediate time domain of ~10^{-9} s such as some of the lanthanide cations.

Role of molecular-based simulations in neutron diffraction studies

A comparison between experimental scattering data and simulations lies at the heart of the validation of any molecular-based modeling effort. In recent years there has been a

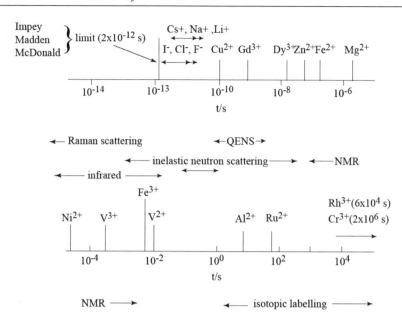

Figure 8. Residence times of water molecules in the neighborhood of aqua ions, and ranges of experimental techniques available to interrogate these time scales. [Used by permission of Elsevier Ltd., from Neilson and Enderby (1989), Adv Inorg Chem, Vol. 34, Fig. 2, p. 197]

constructive interplay between various types of simulation methodologies (e.g., reverse Monte Carlo, EPSR, classical and *ab initio* MD, density functional theory, etc.) and neutron scattering experiments over the full range of temperature and pressure, but particularly so on supercritical water (e.g., Soper 1996, 2005; Chialvo et al. 1998). A full discussion of this important component of neutron scattering studies is beyond the scope of this review, but a few comments are worth making that emphasize the importance of making the synergistic link between theory, modeling and simulation (TMS) and experimental results. For recent examples of these links, the reader is encouraged to examine the very nice summary provided by Kalinichev (2001) who described the molecular simulations of liquid and supercritical water, and Soper (2005) who presented in more detail the utility of the EPSR method. An interesting new result (Chialvo and Simonson 2006) demonstrates that MD simulation in concert with *first-order difference* experiments involving null and heavy water can be used to assess the magnitude of M^{v+}-X^{v-} ion-pair formation for a salt $M^{v+}X_n^{v-}$ in aqueous solution.

It was recognized in the mid 1990's that existing potential models for water did not adequately predict the features of the partial pair distribution functions, $g(r)$ for water measured by neutron scattering (e.g., Chialvo and Cummings 1998). It was unclear whether the disagreement between predicted and measured correlation functions was a reflection of unrealistic intermolecular models, inadequate processing of the raw NDIS data, or a combination of both. Suspicion focused on the inadequacies in the proper accounting of the inelasticity (Placzek errors) corrections associated with the NDIS experiments. These corrections become a large source of uncertainty at high temperature and remarkably difficult to estimate for the light isotopes. Molecular models pinpointed the inaccuracies in the scattering data, while at the same time revealing that improvements in how to constrain the hard-core diameters used in models helped avoid unphysical behavior over short length scales (Chialvo et al. 1998). Several conclusions were reached that are important for future work on the molecular features of water: (a) There is incentive to build new neutron instruments that use 0.1 Å wavelengths to help minimize problems associated with the inelastic correction; (b)

There is a need for an independent method to verify NDIS data because of its experimental complexity and the likelihood of undesirable numerical artifacts in the raw data; (c) It would be desirable to develop alternative experimental methods to determine microstructure, to avoid use of isotopic substitution and its troublesome inelasticity correction; and (d) We need alternative or complementary molecular-based methods to efficiently check the reliability of experimental structural results, before they are used in model parameterization.

FLUIDS CONFINED IN NANOPOROUS REGIMES

Confining matrices of interest to earth sciences

The behavior of liquids in confinement typically differs from their bulk behavior in many ways. Some important factors influencing the structure and dynamics of confined liquids include the average pore size and pore size distribution, the degree of pore interconnection, and the character of the liquid-surface interaction (e.g., hydrophilic vs. hydrophobic pore surfaces). While confinement of liquids in hydrophobic matrices, such as carbon nanotubes, or near the surfaces of mixed character, such as many proteins, had been the area of rapidly growing interest, the confining matrices of interest to earth sciences usually contain oxide structural units and thus are characterized by hydrophilic pore surfaces. This is because, unless dehydroxylation procedures are applied, there is always chemisorbed water on an oxide surface in the form of hydroxyl groups. The pore size distribution and the degree of interconnection vary greatly amongst porous matrices. At one end of the spectrum there are materials with irregular porous structure, such as porous glasses (e.g., Vycor), xerogels, aerogels, and rocks. At the other end, there are materials characterized by regular porous structure, such as mesoporous silicas (e.g., SBA-15, MCM-41, MCM-48), zeolites, and layered systems, for instance, clays. In many matrices, the pore size may be tailored by means of adjusting the synthesis regimen. In clays, the interlayer distance may depend on the level of hydration. Although studied less frequently, matrices such as artificial opals and chrysotile asbestos represent other interesting examples of ordered porous structures. In this section, we provide a brief review of research performed on liquids confined in materials of interest to the earth sciences (silicas, aluminas, zeolites, clays, rocks, etc.) using various neutron scattering techniques with special emphasis on dynamical behavior.

Neutron scattering as a probe of confinement processes

The properties of neutrons make them an advantageous probe for studying liquids in confinement compared to X-rays. Neutrons can be scattered either coherently or incoherently, thus providing opportunity for various kinds of analysis of both structural and dynamic properties of confined liquids. Such analysis is possible due to the fact that wavelengths of thermal and cold neutrons are comparable with intermolecular distances in condensed phases, while the neutron energy can be tailored to probe both high- (collective and single-particle vibrational) and low-frequency (single-particle diffusive) motions in the system. Importantly, the large incoherent scattering cross-section of hydrogen compared to other elements allows acquisition of spectra dominated by the scattering from H-containing species (see recent review article by Neumann 2006), whereas the X-ray scattering from such systems, which is virtually insensitive to hydrogen, would be dominated by the signal from the confining matrix. Last but not least, the large difference in the coherent and incoherent neutron scattering cross-sections of hydrogen and deuterium allows preferential selection of atoms to dominate the scattering signal by means of selective deuteration of the fragments of liquid molecules or the confining matrix.

Application of various techniques: neutron diffraction, small-angle scattering, inelastic spectroscopy

The structural properties of confined liquids can be assessed using coherent scattering techniques, neutron diffraction and small-angle neutron scattering (SANS). The former allows one to measure the static structure factor, $S(Q)$, which can be then Fourier-transformed to obtain the radial pair-distribution function, $g(r)$, that describes the distribution of the distances between the coherently scattering nuclei in the liquid. For hydrogen containing species, deuteration is required in order to obtain diffraction patterns from coherently scattering D nuclei.

Perhaps the most important application of neutron diffraction from confined liquids is detection of freezing-melting transitions. These transitions are known to be depressed to lower temperatures compared to the corresponding bulk liquids and to exhibit a substantial hysteresis. Unlike, for example, calorimetric measurements, a neutron diffraction experiment can not only detect a freezing transition in the confined liquid, but also determine the crystal structure of the resultant solid phase. This phase may be either crystalline but different from that of the bulk solid phase, or amorphous. Information on the average size of the crystalline particles in confinement can be obtained from broadening of the Bragg peaks.

Most of the neutron diffraction studies of freezing of the supercooled confined liquids involved water-ice transition in mesoporous silicas, silica sol-gels, or Vycor glass (e.g., Steytler et al. 1983; Bellissent-Funel et al. 1992; Takamuku et al. 1997; Baker et al. 1997; Dore et al. 2002; Venuti et al. 2004). A review by Dore covers a number of structural studies of supercooled confined water (Dore 2000). It was found by several researchers that in sufficiently small pores formation of a metastable cubic ice phase takes place instead of a bulk hexagonal ice phase. In even smaller pores, where the development of hydrogen-bonded network is strongly suppressed, formation of amorphous ice was observed. Even though the temperatures in experiments with supercooled confined water are outside the range which is typically of interest to earth sciences, such experiments are important because they elucidate the effect of confinement on hydrogen-bonded network. The structure of liquid water confined in Vycor glass at ambient temperature was also addressed (e.g., Bruni et al. 1998).

In general, studies involving porous silicas as confining matrices typically emphasize the nature of the hydrogen-bonded network of confined water molecules, whereas water-matrix interactions receive less attention. These interactions, however, are of central importance in another active area of structural studies that concerns liquid-like interlayer water in clays (e.g., Skipper et al. 1990, 1994, 1995; Williams et al. 1997, 1998; Powell et al. 1998; de Siqueira et al. 1999; Pitteloud et al. 2000, 2001, 2003). In these studies, H/D isotopic substitution was typically employed in order to separate the diffuse diffraction pattern due to interlayer water from the Bragg diffraction due to the crystalline structure of the clay. The emphasis in the clay experiments is usually on the interaction of water molecules with the inter-layer counterions and the hydration structure of these counterions. These experiments are typically carried out at room temperatures. We note in passing that the neutron diffraction technique in combination with standard methods of crystallographic analysis such as Rietveld structural refinement is well suited to address the structural position of D_2O or deuterated organic molecules adsorbed in structurally complex crystalline materials such as zeolites (e.g., Goyal et al. 2000; Floquet et al. 2004).

While diffraction measurements of liquids in confinement probe structural correlations not exceeding a few molecular diameters, SANS measurements provide coverage over much broader range in the real space (Radlinski 2006, this volume). This is because SANS involves measuring neutron intensities at very low values of the scattering vector, Q (i.e., at small angles). For a two-phase system, the SANS intensity is proportional to the scattering contrast, i.e., the square of the difference in the scattering length densities between the two phases. Therefore, information on both the magnitude and the spatial distribution of the scattering density variations can be obtained from SANS data. The molecular-level structure of the constituencies is unimportant because it is averaged out over large distances. This is exploited to a good advantage in contrast variation methods (e.g., deuteration of hydrogenous phases), when the pores are filled, or partially filled with an adsorbate to bring the difference in the scattering length densities (and, concurrently, the SANS intensity) to zero. For instance, in a contrast variation experiment open and closed pores could be distinguished since the latter would be inaccessible to the adsorbate. The contrast technique can be used to study the structures of both the empty pores and the adsorbate. In general, these structures do not have to be the same, especially in matrices with fractal pores, since an adsorbate may condense in the most strongly curved pore regions rather than cover the surface with a wetting film of uniform thickness (e.g., capillary condensation, Broseta et al. 2001).

SANS provides considerable flexibility in the choice of an adsorbate because the scattering length density can be tuned through partial deuteration of the adsorbate molecules. For a reader interested in a brief introduction to the SANS technique as well as a survey of the experimental results the review paper by Ramsay (1998) can be recommended. Fluids in various confining environments have been studied by SANS. Examples include water in Vycor glass and other porous silicas (Li et al. 1994; Agamalian et al. 1997), water in a synthetic clay and a natural rock (Broseta et al. 2001; Knudsen et al. 2003), and binary mixtures in various porous matrices (Dierker and Wiltzius 1991; Lin et al. 1994; Frisken et al. 1995), to name a few. Investigation of the liquid-gas critical phenomena in single-component fluids in confinement is another interesting area of SANS research (e.g., studies of cyclohexane in porous silica by Webber et al. (1996) and of carbon dioxide in a silica aerogel by Melnichenko et al. (2004, 2005, 2006)). Typically, SANS experiments are carried out at ambient or elevated temperatures (usually a few tens of degrees).

Dynamics of confined liquids is investigated by neutron spectroscopy that relies mainly on the incoherent scattering of neutrons from hydrogen nuclei. Even though the actual motions of nuclei in the molecules of a condensed phase are very complex, they usually can be separated into fast vibrational and librational and slow rotational and translational motions. The more energetic vibrational and librational modes are probed using dedicated neutron spectrometers with moderate energy resolution and reasonably high incident neutron energies. On the time scale of such spectrometers, rotational and translational motions are very slow and can be neglected. This type of measurement is known as inelastic neutron spectroscopy (INS). Compared to infrared spectroscopy, INS benefits from the absence of optical selection rules and the large incoherent scattering cross-section of hydrogen.

As in the structural studies, water is the medium whose dynamics in confinement has been studied most extensively. Traditionally, the field was dominated by studies of water (or hydroxyl groups) dynamics in zeolites (e.g., Jobic et al. 1992, 1996; Mitchell et al. 1993; Beta et al. 2004; Corsaro et al. 2005). More recently, INS was used to probe dynamics of hydroxyl groups in mesoporous silica (Geidel et al. 2003). In such studies, the fundamental vibrational modes of water molecules and hydroxyl groups adsorbed at well defined sites of the host framework are usually probed at low (helium) temperatures with neutron energy loss in order to sharpen the inelastic peaks, which suffer loss in quality due to multiphonon scattering in ambient temperature measurements with neutron energy gain. It should be noted that measuring the high-frequency stretching modes of OH and H_2O species (with an energy is excess of 400 meV or 3200 cm^{-1}) with neutron energy loss requires an inelastic neutron spectrometer built either at a spallation neutron source or a hot reactor neutron source in order to have access to sufficiently high incident neutron energies. Another recent development in the area of INS studies of confined liquids concerned measuring low-frequency (at about 100 meV and below) dynamics in supercooled water confined in Gelsil and Vycor glasses (Venturini et al. 2001; Crupi et al. 2002a,d). The emphasis in this area of research is on the properties of confined water, as opposed to studies of zeolites, where measured vibrational frequencies are typically used to characterize the adsorption sites.

There have been numerous INS studies of molecules other than water adsorbed in zeolites; e.g., alcohols (Schenkel et al. 2004a,b), chloroform (Mellot et al. 1998; Davidson et al. 2000), ethane and ethene (Henson et al. 2000), furan (Beta et al. 2001), ferrocene (Kemner et al. 2002), also see Jobic (1992, 2000b) for surveys of earlier and more recent zeolite studies. A study by Rosi et al. (2003) that has attracted much interest involved investigation of hydrogen adsorbed in metal organic framework MOF-5. The INS spectra indicated the presence of two well-defined binding sites in this material, which is characterized by significant hydrogen uptake. As in the case of confined water, INS studies of organic molecules in confining matrices with a well-defined structure are mainly aimed at characterization of the adsorption sites. A study of propylene glycol confined in a porous glass by Melnichenko et al. (1995) represents an example of research where the properties of the confined medium were of the main interest.

Basics of quasielastic neutron scattering (QENS)

Quasielastic neutron scattering (QENS) can be used to assess the mobility of confined liquids, which is the property affected the most by a confinement: a change by one to two orders of magnitude in the mobility of a confined liquid is common. Here we provide the most basic description of the technique; for a detailed discussion see Bée (1988), Roe (2000), and Hempelmann (2000).

The incoherent scattering function $S_{inc}(Q,E)$ measured in an experiment is the space and time Fourier-transform of the self-correlation function $G_{self}(r,t)$, which is the probability that a particle that was at $r = 0$ at $t = 0$ is at position r at time t:

$$S_{inc}(Q,E) = \int I_{inc}(Q,t) e^{i\frac{E}{\hbar}t} dt = \int\int (G_{self}(r,t) e^{-iQr} dr) e^{i\frac{E}{\hbar}t} dt \tag{19}$$

where the intermediate scattering function, $I_{inc}(Q,t)$, is the space Fourier transform of $G_{self}(r,t)$. For the simplest case of Fickian diffusion with a diffusion coefficient D:

$$G_{self}(r,t) = \frac{e^{\left(-\frac{r^2}{4Dt}\right)}}{(4\pi Dt)^{3/2}} \tag{20}$$

$$I_{inc}(Q,t) = e^{-tDQ^2} \tag{21}$$

$$S_{inc}(Q,E) = \frac{1}{\pi} \frac{\hbar DQ^2}{(\hbar DQ^2) + E^2} \tag{22}$$

That is, the signal measured by a spectrometer as a function of neutron energy transfer, E, is a Lorentzian with a half width at half maximum (HWHM) $\Gamma = \hbar DQ^2$. In general, QENS probes rotational and translational diffusive motions of molecules that result in the broadening of the elastic peak. In QENS measurements, the effects of faster vibrational and librational motions manifest themselves in the overall reduction of scattering intensities (Debye-Waller factor). It should be noted that knowledge of the resolution function and an extremely good energy resolution is of paramount importance in QENS. For this reason, time-of-flight and backscattering spectrometers built at cold neutron sources are frequently employed in this type of experiment. Though conceptually different from QENS, neutron spin-echo technique, which has very high energy resolution and yields the intermediate scattering function in the time space, $I(Q,t)$, can be also used for studying slow diffusive motions.

Diffusion in confined and bulk liquids cannot be completely described by Equations (20-22). A simple model that captures the essential features of the diffusion process is called the "jump diffusion" model. It assumes that the diffusing particles rest for a time τ_T between jumps at a distance r. In this model, the scattering function in the energy space is still a Lorentzian for a translational diffusion process, but the Q-dependence of its broadening becomes:

$$\Gamma(Q) = \frac{\hbar}{\tau_T}\left(1 - \frac{\sin(Qr)}{Qr}\right) \qquad (23)$$

(Bée 1988). The residence time between jumps and the jump length are related through the diffusion coefficient $D = r^2/6\tau_T$. A fixed jump-length jump diffusion model describes translational diffusion of interstitial species in some solids. In liquids, which are characterized by a distribution of jump lengths, $P(r)$, the HWHM averaged over the jump lengths becomes:

$$\Gamma(Q) = \frac{\hbar}{\tau_T} \frac{\displaystyle\int_0^\infty \left(1 - \frac{\sin(Qr)}{Qr}\right)P(r)dr}{\displaystyle\int_0^\infty P(r)dr} \qquad (24)$$

The choice of $P(r)$, which is not unique, defines the Q-dependence of the HWHM. Two models are typically employed. For a Gaussian distribution of jump lengths, the half width at half maximum derived by Hall and Ross (1981) is:

$$\Gamma(Q) = \frac{\hbar}{\tau_T}\left[1 - \exp(-DQ^2\tau_T)\right] \qquad (25)$$

where the three-dimensional diffusion coefficient $D = \langle r^2\rangle/6\tau_T$. Alternatively, an exponential distribution of jump lengths (Singwi and Sjölander 1960; Egelstaff 1967) yields:

$$\Gamma(Q) = \frac{\hbar}{\tau_T}\left[1 - \frac{1}{1 + DQ^2\tau_T}\right] \qquad (26)$$

For both types of distribution, $\Gamma(Q \to 0) = \hbar DQ^2$ and $\Gamma(Q \to \infty) = \hbar/\tau_T$. Therefore, for a translational diffusion process the residence time between jumps and the diffusion coefficient can be obtained from the high- and low-Q limits of the QENS broadening, respectively.

To simplify analysis, faster rotational and slower translational diffusion motions are usually assumed to be independent (decoupling approximation). Then in the time space one can write the intermediate scattering function as a product of the rotational and translational components. Consequently, the scattering function in the energy space becomes a convolution of the rotational and translational components. A number of model scattering functions describing various types of rotational or other localized motions are discussed by Bée (1988). A common feature, which distinguishes them from the translational scattering functions such as the one described by Equation (22), is the existence of an elastic term due to a finite probability of finding a diffusing particle at the initial position at time t. This elastic term is frequently called elastic incoherent structure factor (EISF). For a rotational diffusion process, both the characteristic time between rotational jumps and the jump geometry can be derived from the Q-dependences of QENS broadening and the EISF. Refer to the section on *Fluid Interaction with Surfaces* for a more detailed discussion of scattering functions characterizing rotational diffusive motion.

Application of QENS to fluid-confined systems

The landmark study of supercooled bulk water by Teixeira et al. (1985) has established the framework for interpretation of QENS data from bulk water and water in various confining matrices. Examples of the QENS spectra obtained from water at various values of Q are shown in Figure 9. The Q-dependence of the Lorentzian broadening describing the translational diffusion component is shown in Figure 10. Fits of good quality could be obtained for the data through the entire temperature range using Equation (26). In this study, the water had to be supercooled in order to separate characteristic times between the rotational and translational

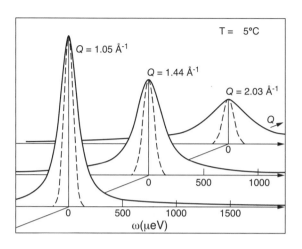

Figure 9. QENS data from supercooled water. [Reprinted with permission from Teixeira et al. (1985), Phys Rev A, Vol. 31, Fig. 1, p. 1915. © 1985 by the American Physical Society. URL: *http://link.aps.org/abstract/PRA/v31/p1913*]

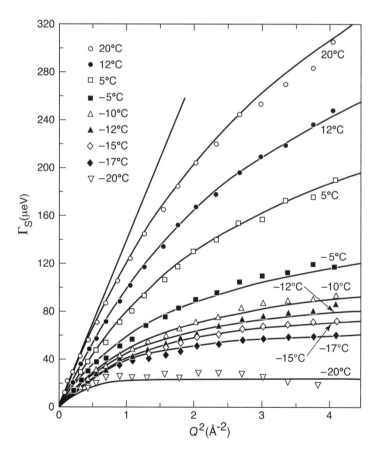

Figure 10. Linewidth of the translational diffusion component of supercooled water fitted with a model assuming exponential distribution of diffusion jump lengths. [Reprinted with permission from Teixeira et al. (1985), Phys Rev A, Vol. 31, Fig. 2, p. 1915. © 1985 by the American Physical Society. URL: *http://link.aps.org/abstract/PRA/v31/p1913*]

diffusion jumps. In bulk water, these characteristic times are both close to a picosecond at ambient temperature. Upon supercooling down to 253 K, the translational jumps slow down by an order of magnitude, whereas the time between rotational jumps increases by just about a factor of two, as one can see in Figure 11. Remarkably, confined water shares many of the same features observed in supercooled bulk water. In particular, the rotational diffusion jumps of water molecules in confinement tend to slow down much less compared to the translational diffusion jumps. Thus, the translational and rotational diffusion components are usually well separated at ambient temperature.

Because of the similarity between supercooled bulk and confined water, numerous studies of supercooled water in mesoporous silicas (Takamuku et al. 1997; Takahara et al. 1999; Mansour et al. 2002; Faraone et al. 2003b,c; Liu et al. 2004; Takahara et al. 2005), Vycor glass (Bellissent-Funel et al. 1993, 1995; Zanotti et al. 1999), and Gelsil glass (Crupi et al. 2002b,c) are relevant to understanding the mobility of water in confinement through a wide temperature range. As a result of these studies, two conceptually different approaches have evolved for interpretations of the QENS data on confined water. One is the "confined diffusion" model, where the shape and size of the confining pore manifest itself through the form-factor, which defines the Q-dependence of the EISF. This model assumes that water molecules perform unrestricted diffusion jumps within a pore until they encounter the pore boundaries. The second concept, commonly called "relaxing cage" model (Chen et al. 1995; Zanotti et al. 1999), suggests that on the time scale of translational jumps of a water molecule, the relevant confinement size is not the size of the pore, but that of the "cage" formed by the neighboring water molecules, from where the molecule can escape through the structural relaxation. In the

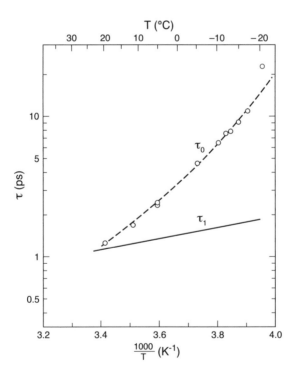

Figure 11. Temperature dependence of residence times for the translational (denoted τ_0) and rotational (denoted τ_1) diffusion components of supercooled water [Reprinted with permission from Teixeira et al. (1985), Phys Rev A, Vol. 31, Fig. 3, p. 1916. © 1985 by the American Physical Society. URL: *http://link.aps.org/ abstract/PRA/v31/p1913*]

"relaxing cage" model, the self-correlation function is not a Gaussian, and the functional form of the intermediate scattering function in the time space is no longer described by the simple exponential decay as in Equation (21). Instead, it is represented by a stretched exponential decay, $\exp[-(t/\tau)^\beta]$. A stretched exponential decay has the same functional form one can obtain assuming a distribution of local diffusivities in the confined medium (Colmenero et al. 1999).

QENS studies of water confined in materials other than silica tend to focus on the influence of the matrix parameters (such as cation charge and channel size in zeolites) on the water dynamics. Various examples of matrices used to confine water include zeolites (Paoli et al. 2002; Crupi et al. 2004a,b,c), porous alumina (Mitra et al. 1998, 2001), tuff (Maddox et al. 2002), layered silicate AMH-3 (Nair et al. 2005), aluminosilicate glasses (Indris et al. 2005), and a silica-alumina-NaO$_2$ molecular sieve (Swenson et al. 2005). In these studies, ambient and elevated temperatures were of primary interest, even though it was possible to supercool water confined in small pores. In an interesting series of experiments, Fratini et al. (2001, 2002) and Faraone et al. (2002) investigated the dynamics of water in cements (hydrated dicalcium and tricalcium silicates) as a function of aging (i.e., cement curing time) at ambient temperature. Water mobility in clays is an actively developing area (e.g., Gay-Duchosal et al. 2000; Swenson et al. 2000; Swenson et al. 2001a,b; Malikova et al. 2005, 2006). Because oriented clay platelets are readily available, it is possible to differentiate between in-plane and out-of-plane interlayer water dynamics by means of properly orienting the sample with respect to the scattering vector. Both ambient and supercooled temperature regimes of interlayer water diffusion in clays have been investigated. Examples of the less common systems interrogated by QENS to study the dynamics of confined water include a layered material V$_2$O$_5$ (Takahara et al. 2000; Kittaka et al. 2005) and chrysotile asbestos (Mg$_3$Si$_2$O$_5$(OH)$_4$) fibers with macroscopically aligned one-dimensional channels (Mamontov et al. 2005b).

Mobility of confined materials other than water has been also investigated in a number of QENS studies. Examples include hydrogen (Glanville et al. 2003), methane (Benes et al. 2001), toluene (Alba-Simionesco et al. 2003), and cyclohexane (Jobic et al. 1995) in micro- and mesoporous silicas, n-hexane (Stepanov et al. 2003; Jobic et al. 2003), H$_2$/D$_2$ (Fu et al. 1999; Bär et al. 1999; Jobic et al. 1999), N$_2$ and CO$_2$ (Papadopoulos et al. 2004), benzene (Jobic et al. 2000b) and propane (Mitra and Mukhopadhyay 2003) in zeolites, and benzene in chrysotile asbestos fibers (Mamontov et al. 2005a). In these experiments, the temperature range varied greatly, depending on the properties of the confined material. It should be noted that the signal from coherently scattering nuclei is dependent on collective properties (such as transport diffusivity), whereas the signal from incoherently scattering nuclei such as H, which is measured in the overwhelming majority of QENS experiments, describes self-diffusion. While both QENS and neutron spin-echo techniques can measure data from coherently and incoherently scattering samples, spin-echo is intrinsically more effective for coherent scattering measurements. This is why most spin-echo measurements involve deuterated samples and probe transport rather than self-correlation properties. Studies by Swenson et al. (2001) and Jobic et al. (2000a) provide examples of spin-echo experiments on hydrogenated (water) and deuterated (benzene) liquids in confinement, respectively.

Dynamics of complex hydrocarbon molecules in confinement has been also addressed by QENS. Jobic (2000a) studied linear and branched alkanes (for chains up to C-14) confined in ZSM-5 zeolite. Because of the relatively large size of confined molecules, the diffusion could be observed within the time window of a backscattering spectrometer only at high temperatures. Branched alkanes were found to diffuse much more slowly than linear alkanes: for example, the diffusion of CH$_3$(CH$_2$)$_6$CH$_3$ at 400 K was significantly faster than that of CH(CH$_3$)$_3$ at 570 K.

Role of molecular-based simulation

In many experiments, simple structural and dynamic analytical models can provide only limited insight into the system's properties. For such cases, computer simulations become necessary in order to understand the system's behavior and interpret the experimental results. While a large number of simulation studies of liquids in confinement have been performed, instead of presenting numerous examples we would like to outline some approaches that can be taken when using simulated data.

The data obtained in the course of a simulation (frequently, molecular dynamics) can be used in several ways. The simulated data are often compared directly with experimental data. A typical example is a study by Ricci et al. (2000), who used MD to calculate the radial correlation functions for bulk water and water confined in Vycor glass and compare them with the experimental radial correlation functions. Single-particle diffusional dynamics of water confined in Vycor glass was also calculated in this work. Another very recent example of simulation-experiment synergy was described by Skipper et al. (2006) for behavior of water in clays.

An alternative approach is to analyze simulated data without directly comparing them with an experiment, especially when the experiment would be difficult to perform. For example, Gallo and Rovere (2003) modeled dynamics of water in Vycor glass at low hydration levels. They reported a profound difference between the water molecules near the surface of cylindrical pores and the rest of the confined water molecules. The former exhibited an anomalous diffusion, whereas the latter showed ordinary Brownian motion. Such a distinction may be difficult to assess in an experiment, where constant exchange between molecules near and far from the surface takes place. Another example of an interesting model system which would be difficult to investigate experimentally is a study of water confined to a slab geometry (Zangi 2004).

Some computational studies are used to set up a new framework for interpretation of the whole class of experimental data. For example, molecular dynamics studies by Chen et al. (1999) and Liu et al. (2002) were used to establish the relaxing cage model for water that we have discussed in the previous section. Another example is a discussion of the limits of applicability of the decoupling approximation for rotational and translational diffusion components in water by Faraone et al. (2003a). Similarly, the recent work by Wang et al. (2006) provides MD descriptions of water behavior in a variety of layered minerals including brucite, gibbsite, muscovite, and talc which should be experimentally verified with neutron scattering and other complementary methods such as NMR.

Last but not least, simulation can be used to link results obtained using different experimental techniques and discuss their validity. Malikova et al. (2005, 2006) recently compared computer simulations with QENS and spin-echo neutron measurements of diffusion of water in clays to demonstrate that underestimation of the relaxation times (that is, overestimation of water dynamics) may occur if the limitations of experiment or simulation preclude assessment of sufficiently long relaxation times.

FLUID INTERACTION WITH SURFACES

Many of the chapters included in this volume describe how neutrons can contribute to our understanding of bulk-condensed matter in cases of geological importance and interest. Despite the fact that neutron methods are traditionally viewed as a bulk technique (since they scatter from the nucleus and because of the lower fluxes available at even the most intense neutron sources) they can also be successfully applied to study surfaces and interfaces (e.g., gas-solid, fluid-solid, fluid-fluid and buried interfaces). In an earlier volume of the Reviews in Mineralogy and Geochemistry series Fenter (2002) and Bedzyk and Cheng (2002) discuss how X-ray reflectivity and X-ray standing wave techniques can be used to probe mineral surfaces and interfaces. Here we illustrate how neutron scattering techniques could and have been used to investigate problems of similar types by presenting several prototypical examples of how diffraction and inelastic scattering techniques have been applied to the study of gas-solid and liquid-solid interfaces and within porous media. The main advantage that neutron techniques offer in this arena is the possibility to study BOTH the structure and the dynamics of the system of interest at the microscopic level. We will discuss how neutron diffraction and inelastic scattering can be used to study the interaction of gases and fluids with lamellar compounds (i.e., graphite and clays), metal oxides and within porous media (such as Vycor glass and nanometer scale tubular arrays). Our objective here is to illustrate the power of the technique and to stimulate the use of neutrons to study of problems relevant to the interest of geochemists.

Structure based on neutron diffraction from layered and two-dimensional systems

While single crystal surfaces and interfaces are readily studied using X-ray and electron diffraction techniques, neutrons can only be applied to such interfacial studies if high surface to volume materials can be used or where large beam footprints are employed such as in reflectometry studies. As noted above, these studies typically involve materials like clays, mica, graphite, and silica. One of the earliest studies relevant to the discussion here is that of Warren (1941) who explained how the diffraction from crystalline powders of random layer materials (like graphite and carbon blacks) exhibits diffraction patterns with characteristic lineshapes that are shaped like a sawtooth. Warren considered a model system composed of a random layer of lattice structures arranged parallel and equidistant from one another, but a random translation parallel to a layer and rotation about the normal. Warren explained his results by noting that the calculated lineshape was a direct result of the powder averaging of a perfect two-dimensional (2D) reciprocal lattice of rods that correspond to the real space "sheets" of carbon that form the basal plane of graphite. Figure 12 (left panel) shows schematically how the asymmetric sawtooth line shape arises while Figure 12 (right panel) is a plot of the diffraction pattern from a randomly oriented material that is a collection of 2D planes. This diffraction profile is often referred to as the "Warren" lineshape. His argument holds true for other lamellar materials as well, so it is relevant to our discussion here.

Since we are considering a 2D system, we note that this is a monomolecular system and all of the centers of mass of the molecules are in the same plane (even if individual nuclei in the molecule are not). $S(Q)$, introduced earlier in this chapter, therefore only depends on the component of Q projected onto the scattering plane, $Q_{parallel}$. Just as Warren did we now

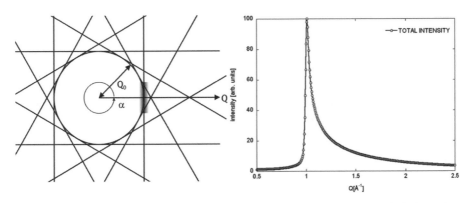

Figure 12. Schematic representation of how the projection of Bragg rods (left panel) leads to the asymmetric shape shown in the Warren lineshape figure illustrated in the right hand panel. The origin of this asymmetry is discussed in the text. This diffraction pattern uses a Lorentzian profile for the structure factor of the Bragg rod.

consider the case of lamellar systems like graphite and mica and assume that the scattering system has random orientations of the crystallites about an axis normal to the basal (or 2D) plane. Because the magnitude of the parallel component of the structure factor $S(Q)$ is the relevant quantity we note that:

$$S(Q_{\text{parallel}}) = \frac{1}{2\pi}\int_0^{2\pi} S(Q_{\text{parallel}},\xi)\,d\xi \tag{27}$$

where ξ is the angle made by Q_{parallel} with respect to an arbitrary reference direction in the scattering plane. An average must be made over the distribution of the surface normal to the individual crystallites that make up the powdered sample. Figure 13 illustrates the scattering geometry for a single crystallite. This figure shows a crystallite (labeled as the primed coordinate system) where the surface normal (z' axis) is oriented at an angle θ with respect to the unprimed laboratory frame. The primed frame is chosen so that both the primed and unprimed y axes lie in the same plane and the z' axis lies in the yz plane of the lab frame. For this illustration we chose the **Q** vector to be along the y axis. Hence Q_{parallel} can be given by:

$$Q_{parallel} = \left|Q\hat{\mathbf{y}} - (\mathbf{Q}\cdot\hat{\mathbf{z}}')\hat{\mathbf{z}}\right| = Q(1-\sin^2\theta\sin^2\phi)^{\frac{1}{2}} \tag{28}$$

where $\hat{\mathbf{y}}$, $\hat{\mathbf{z}}$ and $\hat{\mathbf{z}}'$ are unit vectors pointing along the respective primed and unprimed axes. We use some straightforward vector geometry to express the unit vector $\hat{\mathbf{z}}'$ in terms of the laboratory (unprimed) frame. If the probability distribution of the crystallite plane normals per unit solid angle is $P(\theta)$ then the fraction of plane normals within the solid angle $\sin\theta d\theta d\phi$ is given by their product $[P(\theta)\sin\theta d\theta d\phi]$. By combining our results from above and assuming that $F(Q)$ (see Eqn. 9a,b) depends only on the magnitude of Q, we find that the cross section can be written as:

$$\frac{d\sigma}{d\Omega} = B^2\left|F(Q)\right|^2\int_0^{2\pi} d\phi\int_0^{\pi} d\theta\sin\theta P(\theta)S[Q(1-\sin^2\theta\sin^2\phi)^{\frac{1}{2}}] \tag{29}$$

While this is the essence of what is required to calculate the cross section it is noteworthy to indicate that Warren deduced that in the vicinity of the two dimensional reciprocal lattice vector τ_{hk} for a finite sized 2D crystallite of coherence length L the structure factor can be approximated by:

$$S(Q) = \rho_0(L/a)^2 e^{-[|Q_{parallel}-\tau_{hk}|^2 L^2/4\pi]} \tag{30}$$

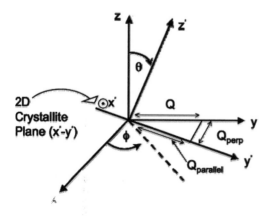

Figure 13. Scattering geometry for a 2D crystallite with the surface normal (z') tilted with respect to the fixed laboratory z axis. Note that the x' axis comes out of the plane of the paper.

Numerous authors have more recently expanded the expression for the cross section given in Equation (29) in various ways. Most of this work is for determining the structure of adsorbed films on solid substrates in particular graphite. The details of these expansions can be found in Stephens et al (1984), Dutta et al. (1980) and Dimon et al. (1985). The work of Dimon et al. (1985) is particularly noteworthy in that it finds that a Lorentzian distribution gives a more satisfactory fit to the lineshape. Our recent experimental evidence suggests that the Lorentzian lineshape can be successfully used to describe the scattering profile of adsorbed molecular films on substrates like graphite foam and MgO where a truly random distribution of the crystallites seems to be realized. Figure 14 illustrates the overall quality of the Lorentzian lineshape description for an a monolayer butane film adsorbed on MgO(100) surfaces. We have focused our attention on describing the lineshape for 2D solid systems, however in cases where the range of spatial correlations is restricted to near neighbor and next near-neighbor coordination rings and where translational mobility also is present (i.e., 2D liquid/fluids), we assume that the reader understands that these lineshapes would broaden in response to the decay of the spatial correlations. Later in this discussion we will indicate how INS and QENS can be used in conjunction with diffraction to gain further microscopic information about the details of the dynamics in interfacial solids and fluids.

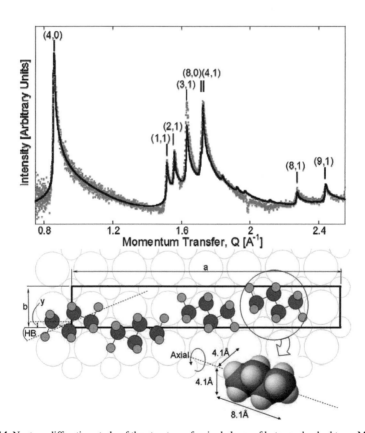

Figure 14. Neutron diffraction study of the structure of a single layer of butane adsorbed to an MgO (100) surface. The sawtooth shaped peaks in the diffraction pattern (upper panel) indicate that the molecular film is two-dimensional. The lower panel illustrates how the four molecules in the unit cell (a = 29.5 Å; b = 4.21 Å) are arranged. The molecules are found to "register" with the MgO lattice with a $2\sqrt{2} \times 7\sqrt{2}$ cell configuration. (Arnold et al. 2006a).

It is of some interest to consider the situation where one of the idealized 2D systems that we have addressed can be followed in a layer by layer growth mode from a strictly 2D plane to one that is more 3D like. Such is the situation in the formation of multilayer molecular films adsorbed to uniform substrates or where epitaxial metal or soft matter growth is realized in chemical vapor deposition, molecular beam epitaxy or polymeric deposition systems. The "lineshape" discussion above has to be modified to account for the development of the third dimension of order in the system. Conceptually this is rather straightforward. Instead of considering, as Warren (1941) did, an ideal 2D reciprocal lattice composed of an ordered array of uniform rods; the reciprocal lattice for an idealized multilayer (e.g., 2-5 individual layer) system is characterized by an array of modulated rods, where the functional form of the modulation of those rods represents the details of the spatial distribution of the scatters in the direction perpendicular to the original 2D plane. Stephens et al. (1984) considered the multilayer diffraction situation in X-ray scattering studies while Larese and his collaborators considered how neutron scattering could be used to follow the growth and melting of atomic and molecular systems on graphite. Figure 15 illustrates how neutron diffraction patterns evolve with film thickness for the growth of ideal two and three layer close-packed films on the graphite basal plane.

Dynamics derived from inelastic neutron scattering

The general concept of using inelastic neutron scattering to obtain information concerning the dynamical properties of condensed matter systems has been covered in numerous places in this volume and for the case of QENS applied to the study of translational diffusion in confined liquids. In the following sections we discuss how rotational and vibrational motions might also be incorporated into these interfacial investigations. For a more detailed discussion of scattering functions characterizing translational diffusion refer to the earlier section on QENS studies of confinement.

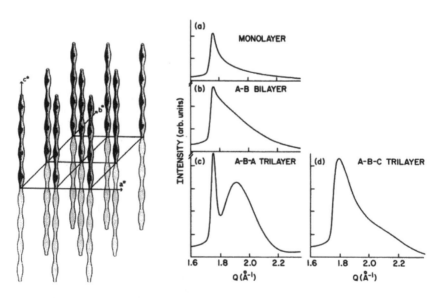

Figure 15. Schematic view (left panel) of a modulated lattice of rods characteristic of a spatially finite bilayer structure. Model line-shape calculations (right panel) for a powder-averaged diffraction profile resulting from an *A-B* stacking sequence for (a) monolayer, (b) bilayer, (c) trilayer thick film, (d) illustrates the diffraction profile for an *A-B-C* trilayer. A triangular, commensurate, close-packed structure was used in each case. The in-plane lattice spacing was the same in all cases. The interplane distance was held fixed between all layers. [Used with permission from Larese et al. (1988a) Phys Rev B, Vol. 37, Figs. 1,2, p. 4736. © by the American Physical Society. URL: *http://link.aps.org/abstract/PRB/v37/p4735*]

Neutron scattering can be applied effectively in the study of the vibrational and rotational motion of hydrogen bearing molecules or molecular species. The large incoherent cross section for hydrogen means that incoherent neutron scattering can also be used for vibrational spectroscopy. The major advantage that neutrons offer as a tool for vibrational spectroscopy over the more common optical techniques of infrared and Raman scattering is that there are no symmetry-dependent selection rules that cause certain modes to be unobservable. A very recent monograph by Mitchell et al. (2005) is an excellent and practical resource for gaining a foundation and understanding of neutron vibrational spectroscopy as it has been applied in chemistry, biology, materials science and catalysis. While it is clear that we need to once again consider the neutron scattering cross section, we will simplify this discussion here and point out that information about the normal modes of vibration of a molecule or of molecules attached to a surface or chemically bound to an interface can be realized using the incoherent scattering (we will not discuss the measurements of phonons in crystals or collective vibrations here but naturally these can be measured). The intensity of the ith molecular vibrational transition is proportional to:

$$I_i \propto Q^2 U_i^2 e^{-Q^2 U_{total}^2} \sigma \qquad (31)$$

where U_i is the vibrational amplitude of the atoms within the particular mode. The exponential term in Equation (31) is the Debye-Waller factor and U_{total} is the mean square displacement of the molecule. The magnitude of this is partially determined by the thermal motion of the molecule. A very good example of how this type of spectroscopy can be used in the mineralogical context is shown in Figure 16 (Dove 2002).

Another important dynamical process which neutrons can probe is the characterization of rotational motion. If a molecule rotates randomly about its center of mass (COM), the positions of the hydrogen atoms $\mathbf{r_n}(t)$ move over a sphere of constant radius $|\mathbf{r_n}|$. This type of motion is referred to as rotational diffusion. For illustrative purposes, consider the simple case of the diffusion of a single-atom-bearing molecule that is rotationally diffusing. Such a model would be a reasonably good choice for a system like crystalline methane at temperatures just below the

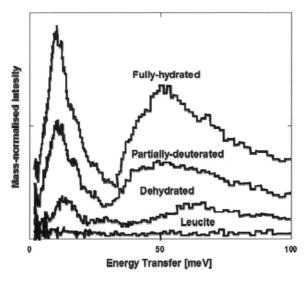

Figure 16. Incoherent inelastic neutron scattering from samples of analcime with various degrees of hydration and compared with scattering from leucite. [Used by permission of Schweizerbart Science Publishers (*http://www.schweizerbart.de*), from Dove (2002), Euro J Min, Vol. 14, Fig. 12, p.218]

melting point. Here the molecular COM is fixed on a lattice but the molecules are freely reorienting. One finds that a rate law similar to the one applicable to translational diffusion exists:

$$\frac{\partial G}{\partial t} = D_R \nabla_\Omega^2 G \tag{32}$$

where G is the time dependant angular probability function and D_R is the rotational diffusion constant, the inverse of which gives the time constant for rotational motion. It is convenient to express the operator ∇^2 in spherical coordinates. While the expression seems rather straightforward, the complexity of the actual solution is significant. As with the expression for translation diffusion the result can found in the monograph by Bee (1988). The solution is:

$$G[\Omega(t)\Omega(0)] = 4\pi \sum_{t=0}^{\infty} e^{-(D_R l(l+1)t)} \sum_{m=-l}^{+l} Y_m^l[\Omega(t)] Y_m^{l*}[\Omega(0)] \tag{33}$$

where the $Y_m^l(\Omega)$ are the spherical harmonics and $G[\Omega(t),\Omega(0)]$ is the probability of finding the bond orientation at $\Omega(t)$ at time t; when it is at $\Omega(0)$ at time $t = 0$. The intermediate scattering function takes the form:

$$F_{\text{inc}}(Q,t) \propto j_0^2(Qr) + \frac{1}{\pi} \sum_{l=1}^{\infty} (2l + j_l^2(Qr) e^{-[D_R l(l+1)]} \tag{34}$$

where j_l's are the regular Bessel functions. The incoherent scattering function is derived immediately from the Fourier transform as:

$$S(Q,\omega) \propto j_0^2(Qr)\delta(\omega) + \frac{1}{\pi} \sum_{l=1}^{\infty} (2l+1) j_l^2(Qr) \frac{D_R l(l+1)}{[D_R l(l+1)]^2 + \omega^2} \tag{35}$$

where $\delta(\omega)$ is the Dirac delta function. Notice that the expression has two parts, the ($l = 0$) component which is the elastic line centered at $\omega = 0$ and the second one which is a summation of Lorentzian peaks with widths independent of Q. The lowest order term in the latter component ($L = 1$) has a width of $2D_R$, whereas higher order terms have greater widths. The amplitudes of the quasielastic components are Q dependent via the Bessel functions. The temperature dependence of the QENS component often follows an Arrhenius relation:

$$\tau^{-1} = D = D_0 e^{(-E_R/RT)} \tag{36}$$

where E_R is the activation energy related to the rotational motion (see Fig. 17).

While we have shown that the quasi-elastic component of the incoherent scattering is useful for determining the time constant for rotational diffusion, the elastic incoherent component can also be used to obtain useful information concerning the position of the hydrogen atoms. The elastic component can be used to define the Elastic Incoherent Structure Factor (EISF) or $I_{\text{EISF}}(Q)$ which is:

$$I_{\text{EISF}}(Q) = \frac{S_{el}(Q)}{S_{el}(Q) + S_{qe}(Q)} \tag{37}$$

For isotropic rotational motion of molecules with radius r, the EISF has the simple form:

$$I_{\text{EISF}}(Q) = \frac{\sin(Qr)}{Qr} \tag{38}$$

The EISF has been used to better understand the nature of the rotational dynamics in a wide range of interfacial systems. In Figure 18 we show one such study of EISF reported by Simon et al. (1990) on nitric acid intercalated in graphite.

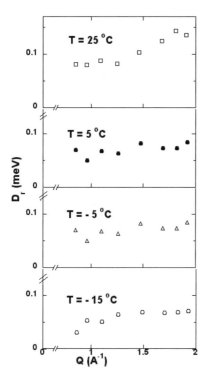

Figure 17. The rotational linewidth as a function of Q for four temperatures for the rotational diffusion of water in Vycor glass. [Reprinted with permission from Bellissent-Funel et al. (1995), Phys Rev E Vol. 51, Fig. 7, p.4565. © 1995 American Physical Society. URL: *http://link.aps.org/abstract/PRE/v51/p4558*]

It is commonly found that at lower temperatures or when the barrier to rotation is high (i.e., the rotational motions are significantly hindered) neutrons have been used to study the discrete transitions between rotational levels, librational motion and quantum (rotational) tunneling. A recent review article by Johnson and Kearley (2000) and the monograph by Press (1981) should be consulted for an introduction to this topic. The extreme sensitivity of the tunnel spectrum to the potential energy surface (PES) demonstrates that high resolution neutron spectroscopy provides a unique probe of intermolecular interactions that is unrivalled in its sensitivity to the potential energy landscape. Molecular groups that exhibit rotational tunneling provide a direct measure of the symmetry and magnitude of intermolecular and weak intramolecular interactions (comparable to the thermal energy kT at room temperature). Tunneling (translational) is purely a quantum phenomenon familiar to all beginning students of quantum mechanics in the study of the particle in a box. The rotational analogue is easily understood by considering molecules in a lattice where local barriers to rotation are too great for classical reorientation of the molecule to take place. The simplest example for us to consider here is where the rotation is one-dimensional—e.g., a methyl group rotor. Here it is sufficient to consider a single coordinate ϕ, used to describe the angular coordinate of the symmetrically distributed triangle of hydrogen

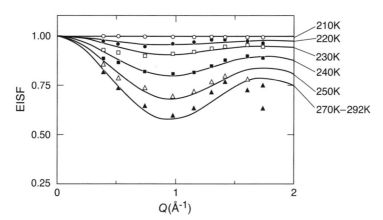

Figure 18. EISF for in-plane rotation of the nitric acid molecule intercalated in graphite. [Reprinted with permission from Simon et al. (1990), Phys Rev B, Vol. 41, Fig. 4, p.2393. © 1990 American Physical Society. URL: *http://link.aps.org/abstract/PRB/v41/p2390*]

atoms, around the central carbon atom (i.e., the C_{3v} axis of the CH_3 group). The Hamiltonian for this single particle motion is given by Mitchell et al. (2005):

$$\frac{h^2}{4\pi I}\frac{\partial^2}{\partial\phi^2} + \sum_{n\geq 1}\frac{V_{3n}}{2}\{1 + \cos[3n(\phi + \alpha_{3n})]\} = \frac{h^2}{4\pi I}\frac{\partial^2}{\partial\phi^2} + \sum_n b_n e^{i3n\phi} \quad (39)$$

where $I = 3mr^2$ is the moment of inertia for three hydrogen atoms of mass, m and displaced from the axis of rotation by the radius, r. The potential energy is expressed as a Fourier series, typically limited to $n \leq 2$, in which case experimental data are required to determine three parameters, V_3, V_6, and the relative phase, $\alpha = \alpha_6 - \alpha_3$. A basis set of free rotor wave functions can be used with the Hamiltonian and diagonalized to obtain a set of tunneling and vibrational (librational) energy levels and wave functions. In Figure 19 (from Johnson and Kearley 2000) the rotational potential is shown in the upper panel along with a schematic energy levels for a methyl rotor. Notice that each librational level is split into a singlet and a doublet, with symmetry species A and E, respectively. It is interesting to observe that there is an exponential decay of the tunnel splitting while there is a nearly linear increase in libration frequency as a function of barrier height are illustrated in the lower panel of Figure 19. The strength of the rotational potential at which the free rotor description (on the *left* of the lower panel) becomes tunneling is arbitrarily taken to be at about 2 meV (20 K). While tunneling data can be used to precisely define the PES, it is not true that this is a simple process once the rotor goes beyond a 1D type.

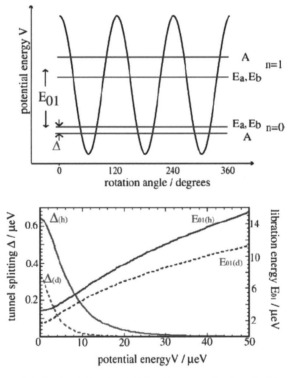

Figure 19. (upper panel) A simple rotational potential and corresponding tunnel split energy levels for the two lowest energy librational levels $n = 0, 1$. (lower panel) The tunnel splitting Δ and librational frequencies are shown as a function of the rotational hindering potential V for CH_3 and CD_3 with the solid and broken lines, respectively. [Used by permission of Annual Reviews, from Johnson and Kearley (2000), Ann Rev Phys Chem, 51, Fig. 1, p. 297-321]

Neutron diffraction and inelastic scattering in the study of the adsorbed films

Numerous neutron diffraction studies of adsorbed molecular films on or intercalated between the graphite basal planes, and on the lamellar halides, clays and MgO have been performed over the last three decades. The bulk of these studies were aimed at answering questions concerning the nature of 2D solids, solid-solid and solid-liquid phase transitions (especially KTHNY melting), layer-by-layer growth, preferential adsorption, and orientational ordering. Here we highlight only a few examples relevant to geochemistry and mineralogy, but encourage the interested reader to consult the conference proceedings (Sinha, 1980; Taub et al. 1991) and monograph (Bruch et al. 1997) for more information.

Graphite and carbonaceous materials

A surge of neutron (and subsequently X-ray) diffraction studies on the structure of adsorbed films on graphite followed the pioneering work of Passell and his coworkers (Kjems et al. 1976) in the 1970's. Using a simple difference technique, Passell was first to show that a non-surface specific probe like neutrons was excellent for the study of the structure and dynamics of adsorbed films. Sheets formed from compressed powders of crystalline graphite distributed under the trade names Grafoil (Union Carbide) and Papyex (Le Carbone Lorraine) were found to be ideal candidates for these "surface" studies. The largest concentration of neutron scattering studies of the structure and dynamics of these adsorbed phases has focused on adsorbed hydrocarbons. Much of the early work determined both the structure of these 2D solid phases and the nature of the melting transition for 2D films (Dash 1999; Hansen and Taub 1999). This is a special type of confined system since the behavior is nominally 2D. A concerted effort was made in the late 70's to the mid 80's to gather evidence to support the predictions by Kosterlitz and Thouless (1973), Halperin and Nelson (1979) and Young (1979)—the (KTHNY) theory—that melting in 2D proceeds via a two step, continuous process that involves a bond ordered state known as the "hexatic phase" that intervenes between the solid and usual isotropic liquid phase. Strandberg (1988) has written an excellent review of 2D melting and the reader is directed there for additional information. Perhaps the most extensively studied system from the stand point of continuous melting in 2D and of layering transitions involves the ethylene-on-graphite system (Kim et al. 1986; Larese et al. 1988b). Also relevant to interfacial or liquids in restricted geometries are the studies of surface melting and reentrant layering (Herwig et al. 2000). These studies are particularly important because they establish that interfacial solids can be stabilized at the adsorbate-film interface when additional liquid layers are adsorbed above them. Neutron scattering studies of layer-by-layer melting of rare gas solids by Larese (1993) and Phillips et al. (1993), and of alkane films (Arnold et al. 2002a,b), carboxylic acids (Bickerstaffe et al. 2004) and alkane-alcohol mixtures by Messe et al. (2005) have firmly established that this fascinating behavior appears in a wide range of fluid-solid interfacial systems.

MgO

Metal oxides are important compounds not only industrially but also as components of earth materials. This coupled with the fact that the next major component of the geosphere is water suggests that neutron studies of structure and dynamics of adsorbed films and interfacial water on metal oxides could act as a prototype for the study of more complex mineralogical systems. After adsorption on graphite perhaps the next most intensely studied system where 2D and interfacial phenomena discussed above has been investigated is magnesium oxide. While work by Coulomb et al. (1984) stimulated a number of papers in the 1980's, production of MgO powders in sufficient quantities to supply a large number of research groups was not available. Thanks to the recent interest in nanoscience and new synthetic methods for the production of large quantities of nanometer scale nanocubes of MgO a resurgence of the structure and dynamics of small molecular systems including methane and other n-alkanes

(Arnold et al. 2005, 2006b), hydrogen (Larese et al. 2006) and water have recently started to appear in the literature. One such study is a combined experimental and theoretical investigation of the rotational tunneling of methane on MgO (100) surfaces. Figure 20 shows the rotational tunneling spectrum for a monolayer film of methane measured using high resolution inelastic neutron scattering.

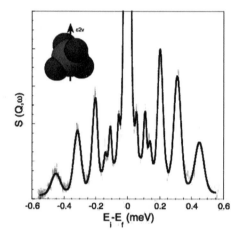

Figure 20. Inelastic neutron "rotational tunneling" spectrum collected at 1.5 K from a single layer of methane adsorbed to a MgO(100) surface. Analysis of this spectrum leads to the surprising result that methane sits in the "dipod" configuration, i.e., with its C_{2v} axis normal to the surface plane, as shown in the inset. [Reprinted with permission from Larese et al. (2001) Phys Rev Lett, Vol. 87, Fig. 1, p. 206102-2. © 2001 by the American Physical Society. URL: *http://link.aps.org/abstract/PRL/v87/e206102*]

SUMMARY AND FUTURE DIRECTIONS

There can be no disputing the fact that neutron diffraction and scattering have made a clear contribution to our current understanding of the structural and dynamical characteristics of liquid water and water containing dissolved ions at ambient conditions and to a somewhat lesser degree other state conditions involving a change in temperature and pressure. Indeed, a molecular-level understanding of how fluids (e.g., water, CO_2, CH_4, higher hydrocarbons, etc.) interact with and participate in reactions with other solid earth materials are central to the development of predictive models that aim to quantify a wide array of geochemical processes. The importance of the hydrogen-bond interaction, in water and other important hydrogenous fluids, has been highlighted as well as the sensitivity of the network to perturbations by a change in physical conditions or proximity to solute molecules and interfaces. Despite the large body of work that documents the nature of hydrogen bonding and associated interactions with its local surroundings, it is premature to assume that we have a complete understanding of the mechanisms that give rise to the particular properties exhibited by water and other simple molecular fluids. This is particularly true as one goes both above and below ambient conditions. For example, there is continuing discussion on the relation between the behavior of supercooled water, the structure of the amorphous ices, the behavior of molecular fluids at interfaces and the incorporation of simple molecules in hydrate clathrates. This is of particular interest since we have seen that nanoporous confinement of water at ambient conditions leads to structural and dynamical features that emulate the super-cooled state. In the context of natural systems, interrogation of fluids and fluid-solid interactions at elevated temperatures and pressures is an area requiring much more work, particularly for complex solutions containing geochemically relevant cations, anions, and other important dissolved species such CO_2 or CH_4. We have tried to describe a series of prototypical interfacial and surface problems using neutron scattering to stimulate the thinking of earth scientists interested applying some of these approaches to confined systems of mineralogical importance.

Our ability to predict the molecular-level properties of fluids and fluid-solid interactions relies heavily on the synergism between experiments such as neutron diffraction or inelastic neutron scattering and molecular-based simulations. Tremendous progress has been made in closing the gap between experimental observations and predicted behavior based on simulations due to improvements in the experimental methodologies and instrumentation on the one hand, and the development of new potential models of water and other simple and complex fluids on the other. For example there has been an emergence of studies taking advantage of advanced computing power that can accommodate the demands of *ab initio* molecular dynamics. On the neutron instrumentation side while much of the quasielastic work described above has been performed using instrumentation located at reactor based sources, the advent of 2nd generation spallation neutron sources like ISIS, new generation sources like the SNS at the Oak Ridge National Laboratory and the low repetition rate 2nd target station at ISIS offer significant opportunities for the study of interfacial and entrained liquids. At the very least, an improvement of the counting statistics by one to two orders of magnitude on many instruments such as vibrational and time-of-flight spectrometers at SNS will allow parametric studies of many systems which otherwise would be prohibitively time consuming. The extended-Q SANS diffractometer at SNS will offer very high intensity and unparalleled Q-range to extend the accessible length scale in the real space, from 0. 05 nm to150 nm. The backscattering spectrometer will provide very high intensity and excellent energy resolution through unprecedented range of energy transfers, thereby allowing simultaneous studies of translational and rotational diffusion components in various systems. The vibrational spectrometer with two orders of magnitude improvement in performance and the capability to perform simultaneous structural measurements should present exciting opportunities to and engender an entire new population of users in the neutron community.

ACKNOWLEDGMENTS

Research by DRC was sponsored by the Division of Chemical Sciences, Geosciences, and Biosciences, Office of Basic Energy Sciences (OBES), U.S. Department of Energy, under contract DE-AC05-00OR22725, Oak Ridge National Laboratory, managed by UT-Battelle, LLC. KWH and EM acknowledge support to SNS from the U.S. Department of Energy, under contract DE-AC05-00OR22725, Oak Ridge National Laboratory, managed by UT-Battelle, LLC. JZL gratefully acknowledges support from OBES, Division of Materials Sciences and Engineering under contract DE-AC05-00OR22725, and NSF under DMR-0412231. The authors would like to thank Drs. H.-R. Wenk, C. J. Benmore, A. A. Chialvo and J. M. Simonson for their constructive comments, R. Violet for help in assembling references for the chapter, and D. Cottrel for redrawing a number of the figures.

REFERENCES

Agamalian M, Drake JM, Sinha SK, Axe JD (1997) Neutron diffraction study of the pore surface layer of Vycor glass. Phys Rev E 55:3021-3027

Alba-Simionesco C, Dosseh G, Dumont E, Frick B, Geil B, Morineau D, Teboul V, Xia Y (2003) Confinement of molecular liquids: Consequences on thermodynamic, static and dynamical properties of benzene and toluene. Eur Phys J E 12:19-28

Ansell S, Neilson GW (2000) Anion-anion pairing in concentrated aqueous lithium chloride solution. J Chem Phys 112:3942-3944

Ansell S, Tromp RH, Neilson GW (1995) The solute and aquaion structure in a concentrated aqueous solution of copper(II) chloride. J Phys Condens Matter 7:1513-1524

Arnold T, Chanaa S, Cook R, Clarke SM, Larese, JZ (2006a) The structure of N-butane monolayer adsorbed on an MgO(110). Phys Rev B (in press)

Arnold T, Cook RE, Chanaa S, Clarke SM, Farinelli, PY, Larese JZ (2006b) Neutron scattering and thermodynamic investigations of thin films of n-alkanes adsorbed on MgO(100) surfaces. Phys Rev B (in press) doi: 10.1016/j.physb.2006.05.189

Arnold T, Cook RE, Larese JZ (2005) Thermodynamic investigation of thin films of ethane adsorbed on magnesium oxide. J Phys Chem B 109:8799-8805

Arnold T, Dong CC, Thomas RK, Castro MA, Perdigon A, Clarke SM, Inaba A (2002a) The crystalline structures of odd alkanes pentane, heptane, nonane, undecane, tridecane and pentadecane adsorbed on graphite at submonolayer coverages and from liquid. Phys Chem Chem Phys 4:3430-3435

Arnold T, Thomas RK, Castro MA, Clarke SM, Messe L, Inaba A (2002b) The crystalline structures of even alkanes hexane, octane, decane, dodecane and tetradecane monolayers adsorbed on graphite at submonolayer coverages and from liquid. Phys Chem Chem Phys 4:345-351

Badyal YS, Simonson JM (2003) The effects of temperature on the hydration structure around Ni^{2+} in concentrated aqueous solutions. J Chem Phys 119:4413-4418

Baker JM, Dore JC, Behrens P (1997) Nucleation of ice in confined geometry. J Phys Chem B 101:6226-6229

Bär NK, Ernst H, Jobic H, Kärger J (1999) Combined quasi-elastic neutron scattering and NMR study of hydrogen diffusion in zeolites. Magn Reson Chem 37:S79-S83

Bedzyk ML, Cheng L (2002) X-ray standing wave studies of minerals and mineral surfaces: principles and applications. Rev Mineral Geochem 49:221-266

Bée M (1988) Quasielastic Neutron Scattering. Adam-Hilger

Bellissent-Funel MC (1998) Is there a liquid-liquid phase transition in supercooled water? Europhys Lett 42:161-166

Bellissent-Funel MC (2001) Structure of supercritical water. J Molec Liq 90:313-322

Bellissent-Funel MC, Bosio L (1995) A neutron scattering study of liquid D_2O under pressure and at various temperatures. J Chem Phys 102:3727-3735

Bellissent-Funel MC, Bradley KF, Chen SH, Lal J, Teixeira J (1993) Slow dynamics of water molecules in confined space. Physica A 201:277-285

Bellissent-Funel MC, Chen SH, Zanotti J-M (1995) Single-particle dynamics of water molecules in confined space. Phys Rev E 51:4558-4569

Bellissent-Funel MC, Lal J, Bosio L (1992) Structural study of water confined in porous glass by neutron scattering. J Chem Phys 98:4246-4252

Bellissent-Funel MC, Teixeira J (1991) Dynamics of water studies by coherent and incoherent inelastic neutron scattering. J Molec Struc 250:213-230

Bellissent-Funel MC, Teixeira J, Bosio L, Dore JC (1989). A structural study of deeply supercooled water. J Phys Condens Matter 1(39):7123-7129

Benes NE, Jobic H, Verweij H (2001) Quasi-elastic neutron scattering study of the mobility of methane in microporous silica. Microporous Mesoporous Mater 43:147-152

Beta IA, Bohlig H, Hunger B (2004) Structure of adsorption complexes of water in zeolites of different types studied by infrared spectroscopy and inelastic neutron scattering. Phys Chem Chem Phys 6:1975-1981

Beta IA, Herve J, Geidel E, Bohlig H, Hunger B (2001) Inelastic neutron scattering and infrared spectroscopic study of furan adsorption on alkali-metal cation-exchanged faujasites. Spectrochem Acta A 57:1393-1403

Beta IA, Li JC, Bellissent-Funel MC (2003) A Quasi-elastic neutron scattering study of the dynamics of supercritical water. Chem Phys 292:229-234

Bickerstaffe A, Messe L, Clarke SM, Perdigon A, Cheah N, Inaba A (2004) Mixing behaviour of carboxylic acids adsorbed on graphite. Phys Chem Chem Phys 6:3545-3550

Bondarenko GV, Gorbaty YE (1973) Infrared spectra of v3 water-d1 at high pressures and temperatures. Dolk Phys Chem 210:369-371

Bosio L, Teixeira J, Bellissent-Funel MC (1989) Enhanced density fluctuations in water analyzed by neutron scattering. Phys Rev A 39:6612-6613

Botti A, Bruni F, Imberti S, Ricci MA, Soper AK (2004) Ions in water: The microscopic structure of a concentrated HCl solution. J Chem Phys 121:7840-7848

Botti A, Bruni F, Isopo A, Ricci MA, Soper AK (2002) Experimental determination of the site - site radial distribution functions of supercooled ultrapure bulk water. J Chem Phys 117:6196-6199

Botti A, Bruni F, Ricci MA, Soper AK (1998) Neutron diffraction study of high density supercritical water. J Chem Phys 109:3180-3184

Broseta D, Barré L, Vizika O, Shahidzadeh N, Guilbaud J-P, Lyonnard S (2001) Capillary condensation in a fractal porous medium. Phys Rev Lett 86:5313-5316

Brown GE, Henrich VE, Casey WH, Clark DL, Eggleston C, Felmy A, Goodman DW, Gratzel M, Maciel G, McCarthy MI, Nealson KH, Sverjensky DA, Toney MF, Zachara JM (1999) Metal-oxide surfaces and their interactions with aqueous solutions and microbial organisms. Chem Rev 99:77-174

Bruch LW, Cole MW, Zaremba E (1997) Physical adsorption: Forces and Phenomena. Intl Series Mono Chem No. 33. Oxford University Press

Bruni F, Ricci MA, Soper AK (1996) Unpredicted density dependence of hydrogen bonding in water found by neutron diffraction. Phys Rev B 54:11876-11879

Bruni F, Ricci MA, Soper AK (1998) Water confined in Vycor glass. I. A neutron diffraction study. J Chem Phys 109:1478-1485

Bruni F, Ricci MA, Soper AK (2001) Structural characterization of NaOH aqueous solution in the glass and liquid states. J Chem Phys 114:8056-8063

Chen SH, Gallo P, Bellissent-Funel MC (1995) Slow dynamics of interfacial water. Can J Phys 73:703-709

Chen SH, Liao C, Sciortino F, Gallo P, and Tartaglia P (1999) Model for single-particle dynamics in supercooled water. Phys Rev E 59:6708-6714

Chen SH, Teixeira J (1986) Structure and dynamics of low-temperature water as studied by scattering techniques. Adv Chem Phys 64:1-45

Chen SH, Teixeira J, Nicklow R (1982) Incoherent quasielastic neutron scattering from water in supercooled regime. Phys Rev E 26:3477-3482

Chialvo AA, Cummings PT (1998) Supercritical water and aqueous solutions: molecular simulation. *In*: Encyclopedia of Computational Chemistry. Scheyer PR (ed) Wiley & Sons, 4:2839-2859

Chialvo AA, Cummings PT, Simonson JM, Mesmer RE, Cochran HD (1998) Interplay between molecular simulation and neutron scattering in developing new insights into the structure of water. Ind Eng Chem Res 37:3021-3025

Chialvo AA, Simonson JM (2006) Ion association in aqueous LiCl solutions at high concentration: Predicted results via molecular simulation. J Chem Phys 124:154509-1-154509-9

Cole DR, Gruszkiewicz MS, Simonson JM, Chialvo AA, Melnichenko YB, et al. (2004) Influence of nanoscale porosity on fluid behavior. *In:* Water-Rock Interaction 1. Want R, Seal R (eds) p 735-739

Colmenero J, Arbe A, Alegria A, Monkenbusch M, Richter D (1999) On the origin of the non-exponential behavior of the α-relaxation in glass-forming polymer: incoherent neutron scattering and dielectric relaxation results. J Phys Condens Matter 11:A363-A370

Corsaro C, Crupi V, Longo F, Majolino D, Venuti V, Wanderlingh U (2005) Mobility of water in Linde type A synthetic zeolites: an inelastic neutron scattering study. J Phys Condens Matter 17:7925-7934

Coulomb J-P, Sullivan TS, Vilches OE (1984) Adsorption of Kr, Xe and Ar on highly uniform MgO smoke. Phys Rev B 30:4753-4760

Crupi V, Dianoux AJ, Majolino D, Migliardo P, Venuti V (2002a) Dynamical response of liquid water in confined geometry by laser and neutron spectroscopies. Phys Chem Chem Phys 4:2768-2773

Crupi V, Majolino D, Migliardo P, Veniti V (2002b) The puzzle of liquid water diffusive behavior: recent IQENS results. Physica A 304:59-64

Crupi V, Majolino D, Migliardo P, Venuti V (2002c) Neutron scattering study and dynamic properties of hydrogen-bonded liquids in mesoscopic confinement. 1. The water case. J Phys Chem B 106:10884-10894

Crupi V, Majolino D, Migliardo P, Venuti V, Dianoux AJ (2002d) Low-frequency dynamical response of confined water in normal and supercooled regions obtained by IINS. Appl Phys A 74:S555-S556

Crupi V, Majolino D, Migliardo P, Venuti V, Mizota T (2004a) Vibrational and diffusional dynamics of water in Mg50-A zeolites by spectroscopic investigation. Mol Phys 102:1943-1957

Crupi V, Majolino D, Migliardo P, Venuti V, Wanderlingh U, Mizota T, Telling M (2004b) Neutron scattering study and dynamic properties of hydrogen-bonded liquids in mesoscopic confinement. 2. The zeolitic water case. J Phys Chem B 108:4314-4323

Crupi V, Majolino D, Venuti V (2004c) Diffusional and vibrational dynamics of water in NaA zeolites by neutron and Fourier transform infrared spectroscopy. J Phys Condens Matter 16:S5297-S5316

Cunsolo A, Orecchini A, Petrillo C, Sacchetti F (2006) Quasieleastic neutron scattering investigation of the pressure dependence of molecular motions in liquid water. J Chem Phys 124:084503-1 – 084503-7

Danielewicz-Ferchmin I, Ferchmin AR (2004) Water at ions, biomolecules and charged surfaces. Phys Chem Liq 42:1-36

Dash JG (1999) History of the search for continuous melting. Rev Mod Phys 71:1737-1743

David F, Vokhmin V, Ionova G (2001) Water characteristics depend on the ionic environment. Thermodynamics and modelization of the aquo ions. J Molec Liq 90:45-62

Davidson AM, Mellot CF, Eckert J, Cheetham AK (2000) An inelastic neutron scattering and NIR-FT Raman spectroscopy study of chloroform and trichloroethylene in faujasites. J Phys Chem B 104:432-438

de Jong PHK, Neilson GW (1996) Structural studies of ionic solutions under critical conditions. J Phys Condens Matter 8:9275-9279

de Jong PHK, Neilson GW (1997) Hydrogen-bond structure in an aqueous solution of sodium chloride at sub- and supercritical conditions. J Chem Phys 107:8577-8585 de Siqueira AV, Lobban C, Skipper NT, Williams GD, Soper AK, Done R, Dreyer JW, Humphreys RJ, Bones JAR (1999) The structure of pore fluids in swelling clays at elevated pressures and temperatures. J Phys Condens Matter 11:9179-9188

Dierker SB, Wiltzius P (1991) Statics and dynamics of a critical binary fluid in a porous medium. Phys Rev Lett 66:1185-1188

Dimon P, Horn PM, Sutton M, Birgneau RJ, Moncton DE (1985) First-order and continuous melting in a two-dimensional system: Monolayer xenon on graphite. Phys Rev B 31:437-447

Dore JC (1984) Neutron diffraction studies of water in the normal and super-cooled liquid phase. J Physique C 7:C7-49-C7-64

Dore JC (1985) Structural studies of water by neutron diffraction. Water Sci Rev 1:3-92

Dore JC (1991) Structural studies of water and other hydrogen-bonded liquids by neutron diffraction. J Molec Struc 250:193-211

Dore JC (2000) Structural studies of water in confined geometry by neutron diffraction. Chem Phys 258:327-347

Dore JC, Garawi M, Bellissent-Funel MC (2004) Neutron diffraction studies of the structure of water at ambient temperatures, revisited [a review of past developments and current problems]. Molec Phys 102:2015-2035

Dore JC, Sufi MAM, Bellissent-Funel MC (2000) Structural change in D_2O as a function of temperature: The isochoric temperature derivative function for neutron diffraction. Phys Chem Chem Phys 2:1599-1602

Dore JC, Webber B, Hartl M, Behrens P, Hansen T (2002) Neutron diffraction studies of structural phase transformation for water-ice in confined geometry. Physica A 314:501-507

Dove MT (2002) An introduction to the use of neutron scattering methods in mineral sciences. Eur J Min 14:203-224

Dutta P, Sinha SK, Vora P, Nielson M, Passell L, Bretz M (1980) Neutron diffraction of melting on physisorbed monolayers of methane-d4 on graphite. *In*: Ordering in Two Dimensions. Sinha SK (ed) North Holland Pub, p 169-174

Egelstaff PA (1967) An Introduction to the Liquid State. Academic Press

Egelstaff PA (1992) Illustrations of radiation scattering data. NATO ASI Series C 379:29-44

Egelstaff PA, Polo JA, Root JH, Hahn LJ, Chen SH (1981) Structural rearrangements in low-temperature heavy water. Phys Rev Lett 47:1733-1736

Enderby JE (1983) Neutron scattering from ionic solutions. Ann Rev Phys Chem 34:155-185

Enderby JE (1995) Ion solvation via neutron scattering. Chem Soc Rev 24:159-168

Enderby JE, Cummings S, Herdman GJ, Neilson GQ, Salmon PS, Skipper N (1987) Diffraction and study of aqua ions. J Phys Chem 91:5851-5858

Faraone A, Chen SH, Fratini E, Baglioni P, Liu L, Brown C (2002) Rotational dynamics of hydration water in dicalcium silicate by quasielastic neuron scattering. Phys Rev E 65:040501-1–040501-4

Faraone A, Liu L, Chen SH (2003a) Model for the translation-rotation coupling of molecular motion in water. J Chem Phys 119:6302-6313

Faraone A, Liu L, Mou C-Y, Shih P-C, Brown C, Copley JRD, Dimeo RM, Chen SH (2003b) Dynamics of supercooled water in mesoporous silica matrix MCM-48-S. Eur Phys J E 12:S59-S62

Faraone A, Liu L, Mou C-Y, Shih P-C, Copley JRD, Chen SH (2003c) Translational and rotational dynamics of water in mesoporous silica materials: MCM-41-S and MCM-48-S. J Chem Phys 119:3963-3971

Fawcett WR (2004) Liquids, Solutions, And Interfaces – From Classical Macroscopic Descriptions To Modern Microscopic Details. Oxford

Fenter PA (2002) X-ray reflectivity as a probe of mineral-fluid interfaces: A user guide. Rev Mineral Geochem 49:150-216

Finney JL, Soper AK (1994) Solvent structure and perturbations in solutions of chemical and biological importance. Chem Soc Rev 23:1-10

Floquet N, Coulomb JP, Dufau N, Andre G (2004) Structure and dynamics of confined water in $AlPO_4$-5 zeolite. J Phys Chem B 108:13107-13115

Fratini E, Chen SH, Baglioni P, Bellissent-Funel MC (2001) Age-dependent dynamics of water in hydrated cement paste. Phys Rev E 64:020201-1–020201-4

Fratini E, Chen SH, Baglioni P, Cook JC, Copley JRD (2002) Dynamic scaling of quasielastic neutron scattering spectra from interfacial water. Phys Rev E 65:010201-1–010201-4

Frisken BJ, Cannell DS, Lin MY, Sinha SK (1995) Neutron-scattering studies of binary mixtures in silica gels. Phys Rev E 51:5866-5879

Fu H, Trouw F, Sokol PE (1999) A quasi-elastic and inelastic neutron scattering study of H_2 in zeolite. J Low Temp Phys 116:149-165

Gaballa GA, Neilson GW (1983) The effect of pressure on the structure of light and heavy water. Molec Phys 50:97-111

Gallo P, Rovere M (2003) Anomalous dynamics of confined water at low hydration. J Phys Condens Matt 15:7625-7633

Gay-Duchosal M, Powell DH, Lechner RE, Rufflé B (2000) QINS studies of water diffusion in Na-montmorillonite. Physica B 276-278:234-235

Geidel E, Lechert H, Döbler J, Jobic H, Calzaferri G, Bauer F (2003) Characterization of mesoporous materials by vibrational spectroscopic techniques. Micro Mesopor Mater 65:31-42

Glanville YJ, Pearce JV, Sokol PE, Newalker B, Komarneni S (2003) Study of H_2 confined in the highly ordered pores of MCM-48. Chem Phys 292:289-293

Gorbaty YE, Demianets YN (1983) The pair correlation functions of water at a pressure of 1000 bar in the temperature range 25-500 °C. Chem Phys Lett 100:450-454

Gorbaty YE, Kalinichev AG (1995) Hydrogen bonding in supercritical water. I. Experimental results. J Phys Chem 99:5336-5340

Gorbaty YE, Okhulkov AV (1994) High-pressure X-ray cell for studying the structure of fluids with the energy-dispersive technique. Rev Sci Inst 65:2195-2198

Goyal R, Fitch AN, Jobic H (2000) Powder neutron and X-ray diffraction studies of benzene adsorbed in zeolite ZSM-5. J Phys Chem B 104:2878-2884

Hahn RL, Narten AH, Annis BK (1986) Neutron scattering from solutions: The hydration of lanthanide and actinide ions. J Less-Common Metals 122:233-240

Hall PL, Ross DK (1981) Incoherent neutron-scattering functions for random jump diffusion in bounded and infinite media. Mol Phys 42:673-682

Halperin BI, Nelson DR (1978) Theory of two-dimensional melting. Phys Rev Lett 41:121-124

Hansen FY, Taub H (1999) The mechanism of melting in monolayer films of short and intermediate-length n-alkanes adsorbed on graphite. Inorg Mater 35:586-593

Hart RT, Benmore CJ, Neuefeind J, Kohara S, Tomberli B, Egelstaff PA (2005) Temperature dependence of isotopic quantum effects in water. Phys Rev Lett 94:047801-1 – 047801-4

Head-Gordon T, Hura G (2002) Water structure from scattering experiments and simulation. Chem Rev 102:2651-2670

Helm L, Foglia F, Kowall T, Merbach AE (1994) Structure and dynamics of lanthanide ions and lanthanide complexes in solution. J Phys Condens Matter 6:A137-A140

Hempelmann R (2000) Quasielastic Neutron Scattering and Solid State Diffusion (Oxford Series on Neutron Scattering in Condensed Matter). Oxford University Press

Henson NJ, Eckert J, Hay PJ, Redondo A (2000) Adsorption of ethane and ethene in Na-Y studied by inelastic neutron scattering and computation. Chem Phys 261:111-124

Herdman GJ, Neilson GW (1990) Neutron scattering studies of aqua-ions. J Molec Liq 46:165-179

Herwig KW, Fuhrmann D, Criswell L, Taub H, Hansen FY, Dimeo R, Neumann DA (2000) Dynamics of intermediate-length alkane films adsorbed on graphite. J Phys IV 10:157-160

Hewish NA, Neilson GW, Enderby JE (1982) Environment of Ca^{2+} ions in aqueous solvent. Nature 297:138-139

Horita H, Cole DR (2004) Chapter 9. Stable isotope partitioning in aqueous and hydrothermal systems at elevated temperatures. *In*: Aqueous Systems at Elevated Temperatures and Pressures: Physical Chemistry in Water, Steam and Hydrothermal Solutions. Palmer DA, Fernández-Prini R, Harvey AH (eds) Elsevier, p 277-319

Howell I, Neilson GW (1997) Ni^{2+} coordination in concentrated aqueous solutions. J Molec Liq 73,74:337-348

Howell I, Neilson GW, Chieux P (1991) Neutron diffraction studies of ions in aqueous solution. J Molec Struct 250:281-289

Hura G, Sorensen JM, Head-Gordon T (2000) A high-quality X-ray scattering experiment in liquid water at ambient conditions. J Chem Phys 113:9140-9148

Ichikawa K, Kameda Y, Yamaguchi T, Wakita H, Misawa M (1991) Neutron-diffraction investigation of the intramolecular structure of a water molecule in the liquid phase at high temperatures. Molec Phys 73:79-86

Idziak SHJ, Li Y (1998) Scattering studies of complex fluids in confinement. Curr Opin Colloid Interface Sci 3:293-298

Indris S, Heitjans P, Behrens H, Zorn R, Frick B (2005) Fast dynamics of H_2O in hydrous aluminosilicate glasses studied with neutron quasielastic scattering. Phys Rev B 71:064205-1-064205-9

Jobic H (1992) Molecular motions in zeolites. Spectrochem Acta A 48:293-312

Jobic H (2000a) Diffusion of linear and branched alkanes in ZSM-5. A quasi-elastic neutron scattering study. J Mol Calal A 158:135-142

Jobic H (2000b) Inelastic scattering of organic molecules in zeolites. Physica B 276:222-225

Jobic H, Bée M, Kärger J, Vartapetian RS, Balzer C, Julbe A (1995) Mobility of cyclohexane in a microporous silica sample: a quasi-elastic neutron scattering and NMR pulsed-field gradient technique study. J Membr Sci 108:71-78

Jobic H, Bée M, Pouget S (2000a) Diffusion of benzene in ZSM-5 measured by neutron spin-echo technique. J Phys Chem B 104:7130-7133

Jobic H, Czjzek M, van Santen RA (1992) Interaction of water with hydroxyl groups in H-mordenite: a neutron inelastic scattering study. J Phys Chem 96:1540-1542

Jobic H, Fitch AN, Combet J (2000b) Diffusion of benzene in NaX and NaY zeolites studied by quasi-elastic neutron scattering. J Phys Chem B 104:8491-8497

Jobic H, Kärger J, Bée M (1999) Simultaneous measurement of self- and transport diffusivities in zeolites. Phys Rev Lett 82:4260-4263

Jobic H, Paoli H, Méthivier A, Ehlers G, Kärger J, Krause C (2003) Diffusion of *n*-hexane in 5A zeolite studied by the neutron spin-echo and pulsed-field gradient NMR techniques. Micro Mesopor Mater 59:113-121

Jobic H, Tuel A, Krossner M, Sauer J (1996) Water interaction with acid sites in H-ZSM-5 zeolite does not form hydroxonium ions. A comparison between neutron scattering results and ab initio calculations. J Phys Chem 100:19545-19550

Johnson MR, Kearley GI (2000) Quantitative atom-atom potentials from rotational tunneling: their extraction and their use. Ann Rev Phys Chem 51:297-321

Kalinichev A (2001) Molecular simulations of liquid and supercritical water: Thermodynamics, Structure and hydrogen bonding. Rev Mineral Geochem 42:83-129

Kemner E, Overweg AR, Jayasooriya UA, Parker SF, de Schepper IM, Kearley GJ (2002) Ferrocene-zeolite interactions measured by inelastic neutron scattering. Appl Phys A 74:S1368-S1370

Kennedy D, Norman C (2005) What don't we know? Science 309:75-102

Kim HK, Zhang QM, Chan MH (1986) Experimental evidence of continuous melting of ethylene on graphite. Phys Rev Lett 56:1579-1582

Kittaka S, Takahara S, Yamaguchi T, Bellissent-Funel MC (2005) Interlayer water molecules in vanadium pentoxide hydrate. 8. Dynamic properties by quasi-elastic neutron scattering. Langmuir 21:1389-1397

Kjems J, Passell L, Taub H, Dash JG, Novaco AD (1976) Neutron scattering study of nitrogen adsorbed on basal plane oriented graphite. Phys Rev B 13:1446-1462

Knudsen KD, Fossum JO, Helgesen G, Bergaplass V (2003) Pore characteristics and water absorption in a synthetic smectite clay. J Appl Cryst 36:587-591

Kosterlitz JM, Thouless DJ (1973) Long range order and metastability in two dimensional solids and superfluids. J Phys C 6:1181

Larese JZ (1993) Multilayer argon films on graphite: structural and melting properties. Accounts Chem Res 26:353-360

Larese JZ (1997) Structure and dynamics of physisorbed phases. Solid State & Mat Sci 2:539-545

Larese JZ (1999) Neutron scattering investigations of the dynamics of thin films adsorbed on solid surfaces. Neutrons 57-68

Larese JZ, Frazier L, Arnold T, Hinde RJ, Ramirez-Cuesta AJ (2006) Direct observation of molecular hydrogen binding to magnesium oxide surfaces. Physica B (in press) doi: 10.1016/j.physb.2006.05.344

Larese JZ, Harada M, Passell L, Krim J, Satija S (1988a) Neutron scattering study of methane bilayer and trilayer films on graphite. Phys Rev B 37:4735-4742

Larese JZ, Passell L, Heidemann A, Richter D, Wicksted JP (1988b) Melting in two dimensions: The ethylene-on-graphite system. Phys Rev Lett 61:432-435

Larese JZ, y Marei DM, Sivia DS, Carlile CJ (2001) Tracking the evolution of interatomic potentials with high resolution inelastic neutron scattering. Phys Rev Lett 87:206102-1–206102-4

Leberman R, Soper AK (1995) Effect of high salt concentrations on water structure. Nature 378:364-366

Li JC, Ross DK, Howe LD, Stefanopoulos KL, Fairclough JPA, Heenan R, Ibel K (1994) Small-angle neutron scattering studies of the fractal-like network formed during desorption and adsorption of water in porous materials. Phys Rev B 49:5911-5917

Lin MY, Sinha SK, Drake JM, Wu X-I, Thiyagarajan P, Stanley HB (1994) Study of phase separation of binary fluid mixture in confined geometry. Phys Rev Lett 72:2207-2210

Line CMB, Winkler B, Dove MT (1994) Quasielastic incoherent neutron scattering study of the rotational dynamics of water molecules in analcime. Phys Chem Min 21:451-459

Liu L, Faraone A, Mou C-Y, Shih P-C, Chen SH (2004) Slow dynamics of supercooled water confined in nanoporous silica materials. J Phys Condens Matter 16:S5403-S5436

Loong C-K (2006) Inelastic scattering and applications. Rev Mineral Geochem 63:233-254

Loong CK, Price DL (1984) Hydrogen-bond spectroscopy of water by neutron scattering. Phys Rev Lett 53:1360-1363

Maddox SA, Gomez P, McCall KR, Eckert J (2002) Water mobility in Calico Hills tuff measured by quasielastic neutron scattering. Geophys Res Lett 29:Art. No. 1259, doi: 10.1029/2001Glv14167

Malikova N, Gadéne A, Marry V, Dubois E, Turq P (2006) Diffusion of water in clays on the microscopic scale: Modeling and experiment. J Phys Chem B 110:3206-3214

Malikova N, Gadéne A, Marry V, Dubois E, Turq P, Zanotti J-M, Longeville S (2005) Diffusion of water in clays – microscopic simulation and neutron scattering. Chem Phys 317:226-235

Mamontov E (2004) Dynamics of surface water in ZrO$_2$ studied by quasielastic neutron scattering. J Chem Phys 121:9087-9097

Mamontov E (2005) High-resolution neutron scattering study of slow dynamics of surface water molecules in zirconium oxide. J Chem Phys 123:024706-1 – 024706-9

Mamontov E, Kumzerov YA, Vakhrushev SB (2005a) Diffusion of benzene confined in the oriented nanochannels of chrysotile asbestos fibers. Phys Rev E 72:051502-1-051502-7

Mamontov E, Kumzerov YA, Vakhrushev SB (2005b) Translational dynamics of water in the nanochannels of oriented chrysotile asbestos fibers. Phys Rev E 71:061502-1 – 061502-5

Mansour F, Dimeo RM, and Peemoeller H (2002) High resolution inelastic neutron scattering from water in mesoporous silica. Phys Rev E 66:041307-1 – 041307-7

Mellot CF, Davidson AM, Eckert J, Cheetham AK (1998) Adsorption of chloroform in NaY zeolite: A computational and vibrational spectroscopy study. J Phys Chem B 102:2530-2535

Melnichenko YB, Schüller J, Richter R, Ewen B, Loong C-K (1995) Dynamics of hydrogen-bonded liquids confined to mesopores: a dielectric and neutron spectroscopy study. J Chem Phys 103:2016-2024

Melnichenko YB, Wignall GD, Cole DR, Frielinghaus H (2004) Density fluctuations near the liquid-gas critical point of a confined fluid. Phys Rev E 69:057102-1-057102-4

Melnichenko YB, Wignall GD, Cole DR, Frielinghaus H (2006) Adsorption of supercritical CO_2 in aerogels as studied by small-angle neutron scattering and neutron transmission techniques. J Chem Phys 124: 204711-1 – 204711-11

Melnichenko YB, Wignall GD, Cole DR, Frielinghaus H, Bulavin LA (2005) Liquid-gas critical phenomena under confinement: small-angle neutron scattering studies of CO_2 in aerogel. J Molec Liq 120:7-9

Messe L, Perdigon A, Clarke SM, Inaba A, Arnold T (2005) Alkane-alcohol mixed monlayers at the solid/liquid interface. Langmuir 21:5085-5093

Michielsen JCF, Bot A, van der Elsken J (1988) Small-angle X-ray scattering from supercooled water. Phys Rev A 38:6439-6441

Mitchell PCH, Parker SF, Ramirez-Cuesta AJ, Tomkinson J (2005) Vibrational Spectroscopy with Neutrons. World Scientific

Mitchell PCH, Tomkinson J, Grimblot JG, Payen E (1993) Bound water in aged molybdate alumina hydrodesulfurization catalysts – an inelastic neutron scattering study. Faraday Trans 89:1805-1807

Mitra S, Mukhopadhyay R (2003) Molecular dynamics using quasielastic neutron scattering technique. Curr Sci 84:653-662

Mitra S, Mukhopadhyay R, Pillai KT, Vaidya VN (1998) Diffusion of water in porous alumina: neutron scattering study. Solid State Commun 105:719-723

Mitra S, Mukhopadhyay R, Tsukushi I, Ikeda S (2001) Dynamics of water in confined space (porous alumina): QENS study. J Phys Condens Matter 13:8455-8465

Nair S, Chowdhuri Z, Peral I, Neumann DA, Dickinson LC, Tompsett G, Jeong H-K, Tsapatsis M (2005) Translational dynamics of water in a nanoporous layered silicate. Phys Rev B 71:104301-1-104301-8

Neilson GW, Adya AK (1997) Neutron diffraction studies on liquids. Ann Rep Prog Chem- Phys Chem, Sect C 93:101-145

Neilson GW, Adya AK, Ansell S (2002) Neutron and X-ray diffraction studies on complex liquids. Annu Rep Prog Chem- Phys Chem Sect C 98:273-322

Neilson GW, Broadbent RD (1990) The structure of Sr^{2+} in aqueous solution. Chem Phys Lett 167:429-431

Neilson GW, Broadbent RD, Howell I, and Tromp RH (1993) Structural and dynamical aspects of aqueous ionic solutions. J Chem Soc Faraday Trans 89:2927-2936

Neilson GW, Enderby JE (1989) The coordination of metal aquaions. Adv Inorg Chem 34:195-218

Neilson GW, Enderby JE (1996) Aqueous solutions and neutron scattering. J Phys Chem 100:1317-1322

Neilson GW, Mason PE, Ramos S, Sullivan D (2001) Neutron and X-ray scattering studies of hydration in aqueous solutions. Phil Trans R Soc London A 359:1575-1591

Neilson GW, Page DI, Howell WS (1979) A neutron diffraction study of the structure of heavy water at pressure using a new high-pressure cell. J Phys D: Appl Phys 12:901-907

Neilson GW, Skipper NT (1985) Potassium (1+) ion coordination in aqueous solution. Chem Phys Lett 114: 35-38

Neumann DA (2006) Neutron scattering and hydrogenous materials. Mater Today 9:34-41

Newsome JR, Neilson GW, Enderby JE (1980) Lithium ions in aqueous solution. J. Phys C-Solid State Phys 13:L923-L926

Ohtaki H, Radnai T (1993) Structure and dynamics of hydrated ions. Chem Rev 93:1157-1204

Paoli H, Méthivier A, Jobic H, Krause C, Pfeifer H, Stallmach F, Kärger J (2002) Comparative QENS and PFG NMR diffusion studied of water in zeolite NaCaA. Micro Mesopor Mater 55:147-158

Papadopoulos GK, Jobic H, Theodorou DN (2004) Transport diffusivity of N_2 and CO_2 in silicalite: coherent quasielastic neutron scattering measurements and molecular dynamics simulations. J Phys Chem B 108: 12748-12756

Petrillo C, Sacchetti F, Dorner B, Suck JB (2000) High-resolution neutron scattering measurement of the dynamic structure factor of heavy water. Phys Rev E 62:3611-3618

Phillips JM, Zhang QM, Larese JZ (1993) Why do vertical steps reappear in adsorption isotherms? Phys Rev Lett 71:2971-2974

Pitteloud C, Powell DH, Fischer HE (2001) The hydration structure of the Ni^{2+} ion intercalated in montmorillonite clay: a neutron diffraction with isotopic substitution study. Phys Chem Chem Phys 3: 5567-5574

Pitteloud C, Powell DH, Gonzalez MA, Cuello GJ (2003) Neutron diffraction studies of ion coordination and interlayer water structure in smectite clays: lanthanide(III)-exchanged Wyoming montmorillonite. Colloids Surf A 217:129-136

Pitteloud C, Powell DH, Soper AK, Benmore CJ (2000) The structure of interlayer water in Wyoming montmorillonite studied by neutron diffraction with isotropic substitution. Physica B 276-278:236-237

Postorino P, Tromp RH, Ricci MA, Soper AK, Neilson GW (1993) The interatomic structure of water at supercritical temperatures. Nature 366:668-670

Powell DH, Fischer HE, Skipper NT (1998) The structure of interlayer water in Li-montmorillonite studied by neutron diffraction with isotopic substitution. J Phys Chem B 102:10899-10905

Powell DH, Neilson GW, Enderby JE (1989) A neutron diffraction study of $NiCl_2$ in D_2O and H_2O. A direct determination of $g_{NiH}(r)$. J Phys-Condens Matt 1:8721-8733

Powell DH, Neilson GW, Enderby JE (1993) The structure of Cl^- in aqueous solutions: an experimental determination of $g_{ClH}(r)$ and $g_{ClO}(r)$. J Phys Condens Matter 5:5723-5730

Press W (1981) Single-Particle Rotations in Molecular Crystals. Springer Tracts in Modern Physics, Vol 92. Springer

Proffen T (2006) Analysis of disordered materials using total scattering and the atomic pair distribution function. Rev Mineral Geochem 63:255-274

Radlinski AP (2006) Small-angle neutron scattering and the microstructure of rocks. Rev Mineral Geochem 63:363-397

Ramsay JDF (1998) Surface and pore structure characterization by neutron scattering techniques. Adv Colloid Interf Sci 76-77:13-37

Ricci MA, Bruni F, Gallo P, Rovere M, Soper AK (2000) Water in confined geometries: experiments and simulations. J Phys Condens Matt 12:A345-A350

Ricci MA, Soper AK (2002) Jumping between water polymorphs. Phys A 304:43-52

Rosi NL, Eckert J, Eddaoudi M, Vodak DT, Kim J, O'Keeffe M, Yaghi OM (2003) Hydrogen storage in microporous metal-organic frameworks. Science 300:1127-1129

Schenkel R, Jentys A, Parker SF, Lercher JA (2004a) INS and IR and NMR spectroscopic study of C-1-C-4 alcohols adsorbed on alkali metal-exchanged zeolite X. J Phys Chem B 108:15013-15026

Schenkel R, Jentys A, Parker SF, Lercher JA (2004b) Investigation of the adsorption of methanol on alkali metal cation exchanged zeolite X by inelastic neutron scattering. J Phys Chem B 108:7902-7910

Simon Ch, Rosenman I, Batallan F, Rogerie J, Legrand JF, Magerl A, Lartigue C, Fuzellier H (1990) Measurement of defect mobility in a defect-mediated melting. Phys Rev B 41:2390-2397

Singwi KS, Sjölander A (1960) Diffusive motions in water and cold neutron scattering. Phys Rev 119:863-871

Skipper NT, Lock PA, Titiloye JO, Swenson J, Zakaria AM, Spencer Howells W, Fernadez-Alonso F (2006) The structure and dynamics of 2-dimensional fluids in swelling clays. Chem Geol 230:182-196

Skipper NT, Neilson GW, Cummings SC (1989) An X-ray diffraction study of aquated nickel (2+) and magnesium (2+) by difference methods. J Phys-Condens Matt 1:3489-3506.

Skipper NT, Smalley MV, Williams GD, Soper AK, Thompson CH (1995) Direct measurement of the electric double-layer structure in hydrated lithium vermiculite clays by neutron diffraction. J Phys Chem 99: 14201-14204

Skipper NT, Soper AK, McConnell JDC (1990) The structure of interlayer water in vermiculite. J Chem Phys 94:5751-5760

Skipper NT, Soper AK, Smalley MV (1994) Neutron diffraction study of calcium vermiculite: hydration of calcium ions in a confined environment. J Phys Chem 98:942-945

Soper AK (1996) Bridge over troubled water: the apparent discrepancy between simulated and experimental non-ambient water structure. J Phys Condens Matter 8:9263-9267

Soper AK (2000) The radial distribution functions of water and ice from 220 to 673 K and at pressures up to 400 MPa. Chem Phys 258:121-137

Soper AK (2005) Partial structure factors from disordered materials diffraction data: An approach using empirical potential structure refinement. Phys Rev B 72:104204(1)-104204(12)

Soper AK, Bruni F, Ricci MA (1997) Site-site pair potential functions of water from 25 to 400°C: Revised analysis of old and new diffraction data. J Chem Phys 106:247-254

Soper AK, Egelstaff PA (1981) The structure of liquid hydrogen chloride. Molec Phys 42:399-410

Soper AK, Phillips MG (1986) A new determination of the structure of water at 25°C. Chem Phys 107:47-60

Soper AK, Silver RN (1982) Hydrogen-hydrogen pair correlation function in liquid water. Phys Rev Lett 49: 471-474

Soper AK, Turner J (1993) Impact of neutron scattering on the study of water and aqueous solutions. Int J Modern Phys B 7:3049-3076

Sorensen JM, Hura G, Glaeser RM, Head-Gordon T (2000) What can X-ray scattering tell us about the radial distribution functions of water. J Chem Phys 113:9149-9161

Stanley HE, Teixeira J (1980) Interpretation of the unusual behavior of H_2O and D_2O at low temperatures: tests of a percolation theory. J Chem Phys 73:3404-3422

Stepanov AG, Shegai TO, Luzgin MV, Jobic H (2003) Comparison of the dynamics of *n*-hexane in ZSM-5 and 5A zeolite structures. Eur Phys J E 12:57-61

Stephens PW, Heiney PA, Birgeneau RJ, Horn PM, Moncton DE, Brown GE (1984) High resolution X-ray scattering study of the commensurate-incommensurate transition of monolayer Kr on graphite. Phys Rev B 29:3512-3532

Steytler DC, Dore JC, Wright CJ (1983) Neutron diffraction study of cubic ice nucleation in a porous silica network. J Phys Chem 87:2458-2459

Strandberg KJ (1988) Two dimensional melting. Rev Mod Phys 60:161-207

Swenson J, Bergman R, and Howells WS (2000) Quasielastic neutron scattering of two-dimensional water in a vermiculite clay. J Chem Phys 113:2873-2879

Swenson J, Bergman R, Longeville S (2001a) A neutron spin-echo study of confined water. J Chem Phys 115: 11299-11305

Swenson J, Bergman R, Longeville S, Howells WS (2001b) Dynamics of 2D-water as studied by quasi-elastic neutron scattering and neutron resonance spin-echo. Physica B 301:28-34

Swenson J, Jansson H, Howells WS, Longeville S (2005) Dynamics of water in a molecular sieve by quasielastic neutron scattering. J Chem Phys 122:084505-1-084505-7

Takahara S, Kittaka S, Kuroda Y, Yamaguchi T, Fujii H, Bellissent-Funel MC (2000) Interlayer water molecules in vanadium pentoxide hydrate, $V_2O_5 \cdot nH_2O$. Part 7. Quasi-elastic neutron scattering study. Langmuir 16:10559-10563

Takahara S, Nakano M, Kittaka S, Kuroda Y, Mori T, Hamano H, Yamaguchi T (1999) Neutron scattering study on dynamics of water molecules in MCM-41. J Phys Chem B 103:5814-5819

Takahara S, Sumiyama N, Kittaka S, Yamaguchi T, Bellissent-Funel MC (2005) Neutron scattering study on dynamics of water molecules in MCM-41. 2. Determination of translational diffusion coefficient. J Phys Chem B 109:11231-11239

Takamuku T, Yamagami M, Wakita H, Masuda Y, Yamaguchi T (1997) Thermal property, structure, and dynamics of supercooled water in porous silica by calorimetry, neutron scattering, and NMR relaxation. J Phys Chem B 101:5730-5739

Tassaing T, Bellissent-Funel MC (2000) The dynamics of supercritical water: A quasielastic incoherent neutron scattering study. J Chem Phys 113:3332-3337

Tassaing T, Bellissent-Funel MC, Guillot B, Guissani Y (1998) The partial pair correlation functions of dense supercritical water. Europhys Lett 42:265-270

Taub H, Torzo G, Lauter HJ, Fain SF (eds) (1991) Phase Transitions in Surface Films 2, Vol 267. NATO Adv Study Inst Series B: Physics. Plenum

Teixeira J, Bellissent-Funel MC, Chen SH, Dianoux AJ (1985) Experimental determination of the nature of diffusive motions of water molecules at low temperature. Phys Rev A 31:1913-1917

Thiessen WE, Narten AH (1982) Neutron diffraction study of light and heavy water mixtures at 25°C. J Chem Phys 77:2656-2662

Tromp RH, Neilson GW (1996) Neutron diffraction study of the hydration of ions in aqueous ion exchange resins. J Phys Chem 100:7380-7383

Tromp RH, Neilson GW, Soper AK (1992) Water-structure in concentrated lithium chloride solutions. J Chem Phys 96:8460-8469

Tromp RH, Postorino P, Neilson GW, Ricci MA, Soper AK (1994) Neutron diffraction studies of H_2O/D_2O at supercritical temperatures. A direct determination of $g_{HH}(r)$, $g_{OH}(r)$, and $g_{OO}(r)$. J Chem Phys 101: 6210-6215

Uffindell CH, Kolesnikov AI, Li JC, Mayers J (2000) Inelastic neutron scattering study of water in the subcritical and supercritical region. Phys Rev B 62:5492-5495

Venturini F, Gallo P, Ricci MA, Bizzarri AR, Cannistraro S (2001) Low frequency scattering excess in supercooled confined water. J Chem Phys 114:10010-10014

Venuti V, Crupi V, Majolino D, Migliardo P, Bellissent-Funel MC (2004) Neutron diffraction study of structure of water confined in a sol-gel silica glass. Physica B 350:E599-E601

Walrafen GE, Wang WH, Chu YC (1999) Raman spectra from saturated water vapor to the supercritical fluid. J Phys Chem B 103:1332-1338

Walton RI, Francis RJ, Halasyamani PS, O'Hare D, Smith RI, Done R, Humphreys RJ (1999) Novel apparatus for the *in situ* study of hydrothermal crystallizations using time-resolved neutron diffraction. Rev Sci Instr 70:3391-3396

Walton RI, Norquist A, Smith RI, O'Hare D (2002) Recent results from the *in situ* study of hydrothermal crystallizations using time-resolved X-ray and neutron diffraction methods. Faraday Disc 122:331-341

Wang J, Kalinichev AG, Kirkpatrick RJ (2006) Effects of substrate structure and composition on the structure, dynamics, and energetic of water at mineral-surfaces: A molecular dynamics modeling study. Geochim Cosmochim Acta 70:562-582

Warren BE (1941) X-ray diffraction in random layer lattices. Phys Rev 59:693-698

Webber JBW, Dore JC, Fischer H, Vuillard L (1996) Critical scattering by fluid cyclohexane in porous silica. Chem Phys Lett 253:367-371

Wernet Ph, Nordlund D, Bergmann U, Cavalleri M, Odelius M, Ogasawara H, Naslund LA, Hirsch TK, Ojamac L, Glatzel P, Petersson LGM, Nilsson A (2004) The structure of the first coordination shell in liquid water. Science 304:995-999

Wesolowski DJ, Ziemniak SE, Anovitz LM, Machesky ML, Bénézeth P, Palmer D.A. (2004) Solubility and surface adsorption characteristics of metal oxides. *In:* Aqueous Systems at Elevated Temperatures and Pressures: Physical Chemistry in Water, Steam and Hydrothermal Solutions. Palmer DA, Fernández-Prini, Harvey AH (eds) Elsevier, p 493-595

Wilding MC, Benmore CJ (2006) Structure of glasses and melts. Rev Mineral Geochem 63:275-311

Williams GD, Skipper NT, Smalley MV (1997) Isotope substitution of interfacial fluids in vermiculite clays. Physica B 234-236:375-376

Williams GD, Soper AK, Skipper NT, Smalley MV (1998) High-resolution structural study of an electrical double layer by neutron diffraction. J Phys Chem B 102:8945-8949

Williams Q, Hemley RJ (2001) Hydrogen in the deep Earth. Annual Rev Earth Planet Sci 29:365-418

Wilson JE, Ansell S, Enderby JE, Neilson GW (1997) Water structure around chloride ions in the presence of biologic macromolecule. Chem Phys Lett 278:21-25

Winter R (1993) Neutron and X-ray scattering of fluids at high pressure and high temperature. *In:* High Pressure Chemistry, Biochemistry and Materials Science. Winter R, Jonas J (eds) Springer, p 167-199

Yamaguchi T (1998) Structure of subcritical and supercritical hydrogen-bonded liquids and solutions. J Molec Liq 78:43-50

Yamanaka K, Yamaguchi T, Wakita H (1994) Structure of water in the liquid and supercritical states by rapid x-ray diffractometry using an imaging plate detector. J Chem Phys 101:9830-9836

Young AP (1979) Melting and the vector Coulomb gas in two dimensions. Phys Rev B 19:1855-1866

Zangi R (2004) Water confined to a slab geometry: a review of recent computer simulation studies. J Phys Condens Matt 16:S5371-S5388

Zanotti J-M, Bellissent-Funel MC, Chen SH (1999) Relaxational dynamics of supercooled water in porous glass. Phys Rev E 59:3084-3093

Reviews in Mineralogy & Geochemistry
Vol. 63, pp. 363-397, 2006
Copyright © Mineralogical Society of America

14

Small-Angle Neutron Scattering and the Microstructure of Rocks

Andrzej P. Radlinski

Geoscience Australia
GPO Box 378,
Canberra City, Australian Capital Territory 2601, Australia
e-mail: Andrzej.Radlinski@ga.gov.au

INTRODUCTION

The self-similarity of rocks on the macro-scale is well known—traditionally, the photographs of rock formations usually include a scale-defining object such as a coin, a hammer, a human silhouette, etc. We know now that rock self-similarity, expressed in the quantitative language of fractal geometry, is also ubiquitous in the micro-world as well. In fact, sedimentary rocks are some of the most extensive microstructural fractal systems found in nature. Much of the knowledge of self-similarity on the micro-scale has been accumulated over the last two decades using small-angle neutron scattering (SANS) and small-angle X-ray scattering (SAXS).

In the family of neutron scattering techniques, small-angle scattering has the lowest spatial resolution (Vogel and Priesmeyer 2006, this volume)—it cannot detect the position of individual atoms. It can, however (using two different experimental designs), explore the internal micro-architecture of the pore space over an impressive linear scale range (from nanometers to tens of micrometers). The technique is non-invasive and gives an average value for a given sample volume.

In the following, it is demonstrated how SANS can be used to explore the microstructure of sedimentary and igneous rocks and help gain insights into internal specific surface area, porosity, pore size distribution, mercury intrusion porosimetry, compaction, subsurface generation of oil and gas, adsorption of gases, imbibition of water, distribution of crystalline precipitates and the microstructural effects of heat treatment.

It is the author's intention to provide both a comprehensive introduction for newcomers to the subject and a reference text for those already familiar with small-angle scattering techniques. The article includes a review of theoretical results, selected examples, description of experimental procedures, examples of interpreted data for various types of rocks and references to original work.

BACKGROUND

Small-angle scattering (SAS) techniques

SANS (Small-Angle Neutron Scattering) and SAXS (Small-Angle X-ray Scattering), have been used for decades to study the geometry of supra-molecular objects. These objects can be suspended in solvents or aggregated into a solid phase. With modern small-angle scattering apparatus the size of objects accessible with SANS and SAXS extends from approximately 1 nm to 20 μm. There is an extensive literature on the theory and practice of SAS (for a review see Guinier et al. 1955; Glatter and Kratky 1982; Feigin and Svergun 1989; Espinat 1990; Lindner and Zemb 1991).

1529-6466/06/0063-0014$05.00   DOI: 10.2138/rmg.2006.63.14

SANS and SAXS are very similar techniques. There are subtle practical reasons for neutrons being, in general, more suitable for microstructural work on sedimentary rocks rather than X-rays. In the following, the emphasis is on SANS, but some examples of SAXS work are also given.

Rocks as natural fractal systems

Although geological samples (particularly coals) have been studied by SAXS since the 1930's, the advent of fractal geometry (Mandelbrot 1977, 1982) brought about unprecedented developments when the connection was made between fractals and the microstructure of heterogeneous surfaces (Pfeifer and Avnir 1983). Katz and Thompson (1985) used SEM and optical microscopy to demonstrate the fractal character of pore space in sandstones over length scales ranging from 0.1 to 100 μm. This was followed by a series of papers on the microstructure of shales, sandstones and carbonates, which established that the upper size limit for fractal behavior ranges from 5 to 100 μm (depending on rock lithology) (Krohn 1988, Thompson 1991). Other studies of fractal features of sandstone and limestone rocks include the works of Jacquin and Adler (1987) and Hansen and Skjeltrop (1988), who used optical microscopy to determine fractal volume and surface dimensions.

The discovery of the fractality of rock microstructure provided fresh impetus for SAS work. Bale and Schmidt (1984) derived the analytical form of the correlation function for surface fractals (micron-sized pores with fractal boundary surfaces) and applied it to the analysis of SAXS power law scattering data obtained for lignite coals. The expression for the correlation function was later refined by Mildner and Hall (1986) to account for the limited size of fractal objects and by Wong and Bray (1988) to explain finite scattering as the fractal dimension approaches the value of 3. Wong et al. (1986) used SANS to study the microstructure of sandstone, shale, limestone and dolomite samples over scales of 5 to 500 Å. They found non-universal power-law scattering—indicating the surfaces of sandstones and shales are fractal (which they attributed to varying clay content in the samples). Inspired by the unique microstructural properties of sedimentary rocks, Cohen (1987) presented a general theoretical model of various morphological regimes in sedimentary rocks. He attributed suppressed grain sintering and the formation of a rough grain-pore interface to a small pore-grain free energy in comparison to the grain boundary free energy, and pointed to the anti-sintering thermodynamic regime as the source of ubiquitous fractality of the pore-rock interface.

Geological applications of SAS

In the following decade, small-angle scattering techniques were integrated into mainstream petroleum geology and engineering (Radlinski et al. 1996a, 1999, 2000a, 2004a,b; Boreham et al. 2003) and coal geology (Reich et al. 1990; Radlinski and Radlinska 1999; McMahon et al. 2002; Prinz et al. 2004; Radlinski et al. 2004c). SAS research into various aspects of the microstructure of porous media, including rocks, has continued (e.g., Allen 1991; Broseta et al. 2001; Spalla et al. 2003), and there has been increased interest in the microstructure of igneous rocks (Lucido et al. 1988; Floriano et al. 1994; Kahle et al. 2004; Winkler et al. 2005).

Principle of SAS experiments

The principle of a pinhole SANS experiment is illustrated in Figure 1. A flux of monochromatic thermal neutrons propagating in the direction of their wavevector $\mathbf{k_0}$ is elastically scattered inside a sample of uniform thickness t and irradiated volume V. The magnitude of $\mathbf{k_0}$ is λ^{-1}, where λ is the neutron wavelength. One measures intensity dI scattered in direction $\mathbf{k}$, where by convention $\mathbf{k} - \mathbf{k_0} = \mathbf{s}$ and the quantity $\mathbf{Q} = 2\pi\mathbf{s}$ is called the scattering vector. It follows from Figure 1 that the magnitude of $\mathbf{s}$ is $2\sin\Theta/\lambda$. The magnitude of the scattering vector is thus related to neutron wavelength λ and the scattering angle 2θ by $Q = (4\pi/\lambda)\sin\theta$.

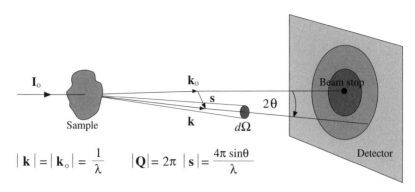

Figure 1. The principle of a pinhole SANS experiment.

The incident flux of the scattering radiation (particles) is denoted by Φ_0, i.e., $\Phi_0 = I_0/A$, where I_0 is the incident intensity (neutrons per second) and A is the beam cross sectional area at the sample position (Fig. 1). The scattered intensity monitored in the solid angle element $d\Omega$ targeted by the scattering vector $\mathbf{Q}$ can be expressed as

$$dI \propto \Phi_0 \frac{d\sigma}{d\Omega} d\Omega \tag{1}$$

where $d\sigma$ is the elemental scattering cross section. The quantity $d\sigma/d\Omega$ is called the differential cross section of scattering. It is the purpose of SAS experiment to measure $d\sigma/d\Omega$ in absolute units.

Scale range accessible using SANS and USANS instruments

The Bragg formula for diffraction, $\lambda = 2d\sin\theta$, shows that for a fixed λ, the angle of diffraction, θ, varies inversely with the distance between the diffracting lattice planes, d. For inorganic crystals, d and λ (for both X-rays and neutrons) are of the same order of magnitude (several Å) and the scattering angles are large. For materials with non-periodic structure, one is often interested in large-scale features (of the order of magnitude from tens of angstroms to a fraction of a millimeter). This requires acquisition of experimental data at small scattering angles.

The linear size of objects that contribute most to the scattering at a given Q-value is of the order of $1/Q$. For periodic structures the size is given by $2\pi/Q = \lambda/(2\sin\theta)$—it is another form of the Bragg law. For fractal (widely polydisperse non-periodic) systems, the size is approximately $2.5/Q \approx 0.4\lambda/(2\sin\theta)$ (Radlinski et al. 2000a). Thus, the precise linear size range accessible with SANS can be selected by choosing neutron wavelength λ and the range of the scattering angles. SANS and SAXS instruments are designed to maximize the Q-range and to minimize parasitic contributions to the scattering signal.

A SANS experiment is performed in a transmission geometry and measures the intensity of radiation scattered at angles very close to the direction of propagation of the incident beam. As the upper limit of 2θ in a SANS experiment is about $5°$, it follows that $\sin\theta \approx \theta$ and Q is simply a measure of the scattering angle in units of $Å^{-1}$. A typical range of neutron wavelengths available for a SANS experiment is $4 - 20$ Å.

The relationship between the scattering angle 2θ, scattering vector Q, and the linear size of objects accessible with various small-angle scattering instruments for two wavelengths (1.5 Å - close to wavelengths used in USANS and SAXS instruments, and 4 Å - close to the flux maximum for SANS instruments) is illustrated in Figure 2. At the smallest linear scales, Bragg

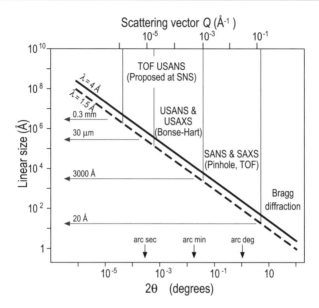

Figure 2. Linear scale range accessible with SANS and SAXS techniques. Q-values are calculated for $\lambda = 4$ Å.

diffraction covers the 2θ range from about $5°$ to nearly $180°$ (which corresponds to linear size range of about 1-20 Å). The angular range $4' < 2\theta < 5°$ (Q-range 10^{-1} to 10^{-3} Å^{-1}) is the traditional SANS (and SAXS) domain. This corresponds to the linear size range 20-3000 Å. Two types of SANS instruments (reactor based pinhole, e.g., Koehler 1986, Glinka et al. 1986, Lindner et al. 1992 and spallation source based time-of-flight, e.g., Thiyagarajan et al. 1997), operate in this range.

For the ultra-small angle (USANS and USAXS) region, where the scattering angles are in the range $6'' < 2\theta < 5'$ (Q-range 10^{-3} to 10^{-5} Å^{-1}) and the linear size ranges from 3000 Å to 30 μm, resolution of the incident and scattered beam is achieved by collimating and analyzing the beam using multiple reflections from perfect silicon crystals (Bonse and Hart 1965; Lambard and Zemb 1991; Agamalian et al. 1997; Hainbuchner et al. 2000). Practical USANS machines were developed only recently (Agamalian et al. 1997) and they have proved very useful in microstructural studies of geological materials (Radlinski et al. 1999, 2004a,b,c). There are several types of Bonse-Hart instruments—some have an option to sacrifice resolution for data acquisition speed (Mikula et al. 1988; McMahon and Treimer 1998). There is a proposal to construct a powerful TOF USANS instrument at the SNS facility in Oak Ridge, which theoretically could reach the minimum Q-value of nearly 10^{-6} Å^{-1} (linear size of 0.3 mm).

SUMMARY OF THEORETICAL RESULTS

Correlation function

The central concept in small-angle scattering is correlation function, $\gamma(r)$. For disordered (random) materials it plays the same role as the lattice structure in crystals. Following the notation of Debye and Bueche (1949) and Debye et al. (1957), the correlation function is defined as

$$\gamma(r) <\eta^2>_{Av} = <\eta_A\eta_B>_{Av} \tag{2}$$

where η_A, η_B are the local fluctuations in the physical property (dielectric constant for light, electronic density for X-rays, nuclear scattering length for neutrons) that provides a scattering contrast from some average value at points A and B a distance $\mathbf{r}$ apart; $<\eta^2>_{Av}$ is the average value of η^2. Defined as such, $\gamma(\mathbf{r})$ is confined within the limits $0 = \gamma(\infty) < \gamma(\mathbf{r}) < \gamma(0) = 1$.

According to the classical theory of scattering (Guinier and Fournet 1955), the intensity scattered by a unit volume is given by the integral:

$$I(Q) = \iint < \eta_A \eta_B >_{Av} \exp(ik\mathbf{sr}) d\tau_A d\tau_B \tag{3}$$

where $\mathbf{s}$ is the vector difference between unit vectors pointing in the direction $\mathbf{k}$ and $\mathbf{k_0}$, respectively, $|\mathbf{s}| = 2\sin\Theta$, $k = 2\pi/\lambda$ and $\mathbf{Q} = k\mathbf{s}$. Therefore

$$I(Q) = < \eta^2 >_{Av} \int \gamma(\mathbf{r}) \exp(i\mathbf{Qr}) d\tau \tag{4}$$

For isotropic media, where $\gamma(\mathbf{r})$ depends on the absolute value of $\mathbf{r}$ only and not on its direction, one gets:

$$I(Q) = 4\pi < \eta^2 >_{Av} \int_0^\infty r^2 \gamma(r) \frac{\sin(Qr)}{Qr} dr \tag{5}$$

This relationship shows that the scattering intensity within the entire Q-range is determined by Fourier transform of the correlation function. Conversely, the correlation function can be calculated by the inverse Fourier transform if the scattering intensity is known within a wide enough Q-range. At the heuristic level, Equations (4) and (5) indicate that the microstructural information accessible via the small-angle scattering technique is determined by the information content inherent in the correlation function.

Two phase approximation

For a wide range of substances, the SAS data for geological materials and porous media can generally be interpreted accurately using a two-phase approximation. In this approximation, the scattering volume is viewed as comprised of supra-molecular-size regions, each characterized by one of two possible values of the physical property that provides the scattering contrast. For instance, for porous media these two regions are the solid matrix and the pore space, respectively.

Two phase approximation is a simplification inherent in the SAS method and has been implicitly or explicitly employed for many years. Following Adler et al. (1990), one can describe a two phase (porous) medium in terms of a binary phase function $Z(\mathbf{x})$, where $Z(\mathbf{x}) = 1$ if $\mathbf{x}$ points to void and $Z(\mathbf{x}) = 0$ if $\mathbf{x}$ points to solid matrix. First two moments of $Z(\mathbf{x})$ yield the porosity, $\Phi = <Z(\mathbf{x})>$, and the correlation function:

$$\gamma(\mathbf{r}) = \frac{< Z(x)Z(\mathbf{x}+\mathbf{r}) - \Phi^2 >}{\Phi(1-\Phi)} \tag{6}$$

Equation (5) can now be written as

$$I(Q) = 4\pi(\Delta\rho)^2 \Phi(1-\Phi) \int_0^\infty r^2 \gamma(r) \frac{\sin(Qr)}{Qr} dr \tag{7}$$

where $(\Delta\rho)^2$ is the scattering contrast. Contrast calculations are discussed in detail in the following section. The correlation function can be expressed as the inverse Fourier transform of the scattering intensity data:

$$\gamma(\mathbf{r}) = \left[2\pi^2 (\Delta\rho)^2 \Phi(1-\Phi) \right]^{-1} \int_0^\infty Q^2 I(Q) \frac{\sin(Qr)}{Qr} dQ \tag{8}$$

Finally, the porosity Φ can be calculated from the invariant Q_{inv} (Porod 1952), defined as follows:

$$Q_{inv} = \int_0^\infty Q^2 I(Q) \ dQ = 2\pi^2 (\Delta\rho)^2 \Phi(1-\Phi) \qquad (9)$$

It is important to note the integration limits in Equations (7) and (8). The two way path from correlation function to the scattering intensity and back requires knowledge of γ in a wide range of r and of $d\sigma/d\Omega$ in a wide range of Q. At the very least, one has to make sure that the integrands in these equations have peaked within the limits of experimental data. The correlation function for porous media for larger features can be determined via the phase function $Z(\mathbf{x})$ and Equation (6) (using statistical analysis of microscopic images, including SEM; e.g., Ioannidis et al. 1996). For SAS data, due to experimental limitations at small Q-values, increased accuracy is achieved for smaller features (i.e., larger Q-values). Therefore for rock-like materials which may have very broad distribution of microstructural feature sizes, it is useful to combine SAS and microscopy data which overlap in the intermediate size range (Radlinski et al. 2004a).

The chemical composition of rocks can vary from near uniform (e.g., pure silica for sandstones) to complex mixtures of inorganic oxides and organic matter (e.g., for oil-bearing mudstones). The validity of two phase approximation for a particular type of rock and radiation needs to be verified on a case-by-case basis by direct calculation of the scattering length density.

Special cases

Form factor for simple shapes and scattering by polydisperse system of spheres. Traditional expression of the scattering cross section follows on from Equation (7):

$$I(Q) = 4\pi(\Delta\rho)^2 \Phi(1-\Phi) F(Q) \qquad (10)$$

where $F(Q) = \int_0^\infty r^2 \gamma(r) [\sin(Qr)/Qr] dr$ is the so-called form factor. There are analytical expressions for the form factor for simple geometrical objects like spheres, discs and parallelepipeds which can be specialized for limiting cases of 2D (flat discs) and 1D (needles) scattering objects. Feigin and Svergun (1987) and Espinat (1990) (in French) present excellent reviews of the subject.

Small-angle scattering curves for rocks often exhibit a negative power law, $I(Q) = AQ^{-m}$, where A is a constant and $3 < m < 4$. Schmidt (1982) has shown that power law scattering intensity can be explained theoretically if the scattering occurs on a particular polydisperse distribution $f(r)$ of randomly oriented independently scattering particles of any shape (provided that the particle shape distribution is independent of the distribution of particle dimensions, r):

$$f(r) \propto r^{(-2d+1-m)} \qquad (11)$$

where d is the particle dimensionality. For three dimensional particles, the theoretically possible range of m is $0 < m < 4$.

This work was published shortly before the concept of fractal geometry (Mandelbrot 1977, 1982) became widely known and applied to SAS in rocks (see following section). These results were later re-formulated to accord with fractal geometry (Schmidt 1989). Recently, Radlinski et al. (2004a,b,c) have demonstrated that the polydisperse spherical pore (PDSP) model can be used to accurately describe various aspects of the micro-architecture of sedimentary rocks, beyond the limitations of fractal models.

The PDSP model assumes that the pore space of a rock can be represented by a polydisperse distribution of independently scattering spheres. The form factor for a homogeneous sphere of radius r is:

$$F_{sph}(Qr) = \left[3\frac{\sin(Qr) - Qr\cos(Qr)}{(Qr)^3} \right]^2 \tag{12}$$

and the scattering intensity per unit volume of polydisperse spheres is given by (Guinier and Fournet 1955):

$$I(Q) = (\rho_1 - \rho_2)^2 \frac{\Phi}{\bar{V}_r} \int_{R_{min}}^{R_{max}} V_r^2 f(r) F_{sph}(Qr) dr \tag{13}$$

where $\bar{V}_r = \int_0^\infty V_r f(r) dr$ is the average pore volume and $f(r)$ is the probability function of the pore size distribution.

For numerical evaluation of $f(r)$, the scattering intensity can be expressed as

$$I(Q) = \sum_i IQ_{0i} \frac{\int_{R_{min}}^{R_{max}} V_r^2 F_{sph}(Qr) dr}{(R_{max,i} - R_{min,i})} \tag{14}$$

where R_{max} and R_{min} are the maximum and minimum pore radii, respectively, and

$$IQ_{0i} = \frac{(\rho_1 - \rho_2)^2 \Phi}{\bar{V}_r} f(r_i)(R_{max,i} - R_{min,i}) \tag{15}$$

Equation (14) can be solved numerically for $f(r)$ using the PRINSAS software (Hinde 2004). An analytical solution for $f(r)$ has been derived by Letcher and Schmidt (1966); also see page 232 of Feigin and Svergun (1987).

Total porosity (Φ) can be determined by summing the volume of all pores and dividing by the sample volume. The specific surface area for a probe size r can be calculated from the pore size distribution by summing the surface areas of all pores of radius larger than r and dividing by the sample volume:

$$\frac{S(r)}{V} = n_v \int_r^{R_{max}} A_r f(r') dr' \tag{16}$$

where the average number of pores per unit volume $n_v = \Phi / \bar{V}_r = I(0)/[(\rho_1 - \rho_2)^2 \bar{V}_r^2]$, $S(r)$ is the total surface area of pores with radius larger than r and $A_r = 4\pi r^2$.

Scattering by a sharp interface – Porod limit. The large-Q limit of the small-angle scattering domain is the region where atomic resolution has not been achieved, but where the observation scale is small and a well-defined interface appears to be smooth. This is the so-called Porod region (Porod 1951; Debye et al. 1957). At this limit, the differential scattering cross section for a two-phase system with a sharp interface is:

$$\frac{d\sigma}{d\Omega}(Q) = 2\pi(\rho_1 - \rho_2)^2 Q^{-4} \frac{S}{V} \tag{17}$$

where S is the total area of the interface inside the scattering volume V. If the contrast value is known, the intercept of the plot of $Q^4 I(Q)$ versus Q (the Porod plot) provides the value for the specific internal surface area, S/V.

This result is explicitly dependent on the existence of a sharp and smooth interface region, but (with subtle modifications discussed in detail by Auvray and Auroy 1991) applies to any shape of three dimensional scattering particle. Figure 3a shows the SAS curve calculated for a single sphere according to Equation (12). For large Q-values (the Porod limit) the oscillations merge into a Q^{-4}-like envelope. A small amount of polydispersity, expected in a physical system of real spheres, would average out the oscillations (as would the finite Q-resolution of

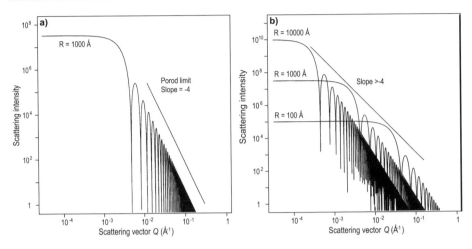

Figure 3. Results of numerical simulations of small-angle scattering for solid spheres: (a) scattering intensity for a single sphere of radius 1000 Å, (b) scattering curves for spheres of radii 10000, 1000 and 100 Å (after Radlinski et al. 2000a).

a SANS instrument). Figure 3b illustrates the effect of extensive polydispersity, similar to that theoretically discussed by Schmidt (1982). The slope of the scattering curve is now markedly larger than −4 (Radlinski et al. 2000a).

Scattering by rough surfaces: fractals and fuzzy interface. Correlation functions derived from microscopic data and power law SAXS and SANS curves obtained in the 1980's (see Background section) indicate that sedimentary rocks are fractal at microscopic scales. The early debate focused on whether rocks were mass fractals (akin to some colloidal systems) or surface fractals (as conceptually, there may be both surface and mass fractal structures present in a rock, Wong (2006)). A good discussion of small-angle scattering from fractals is presented in Schaefer et al. (1987), Martin and Hurd (1987) and Teixeira (1988).

Surface fractals are bulk objects with a rough surface—the roughness being scale invariant within a certain range of sizes. For surface fractals, the majority of the individual building blocks remain in the bulk. The surface area is proportional to r^{Ds}, where the surface fractal (Hausdorff) dimension D_s lies in the range $2 \leq D_s < 3$ and r is the linear scale (the length of the measuring stick). A good example of a surface fractal is planet Earth with its rough surface morphology for which D_s varies from 2 to 2.5, depending on geographical position and length scale.

For mass (volume) fractals the majority of building blocks are exposed on the surface. The volume and consequently, the mass of a mass fractal is proportional to r^{Dm} (where the mass fractal dimension D_m can be no larger than 3). An example is a river tributary system or a network of cracks in a solid. In the latter case the relative position of matter and void is reversed and the term "pore fractal" is sometimes used.

As rocks are physical objects, fractals related to rock microstructure can only exist from near-atomic scales (the building blocks can be reasonably expected to have linear size $r_o \leq 20$ Å) to an upper limit, termed a correlation length (denoted η for mass fractals and ξ for surface fractals, respectively). The correlation length is normally of the order of the grain size for a surface fractal and of the order of the largest pore diameter for a mass fractal. A specific form of the scattering law has been derived for both types of fractal objects. For mass fractals a good approximation for the correlation function for linear scales much longer than the size of the building block is (Freltoft et al. 1986):

$$\gamma(r) \propto r^{D_m - 3} \exp(-r/\eta) \tag{18}$$

where the power term is the correlation function for a perfect (mathematical) mass fractal (Schaefer and Keefer 1984) and the exponential term reflects the upper-size-limit-related decay of fractal properties. For Q-values much smaller than $1/r_o$, where r_o is the size of the building block, the scattering cross section for a single mass fractal particle occupying a volume v_0 and having average scattering length density ρ_s (Freltoft et al. 1986) is:

$$I(Q) = v_0(\rho_s - \rho_0)^2 \{1 + C_0\, Q^{-1}\Gamma(D_m - 1)\eta^{Dm-1} \times \tag{19}$$

$$\left[1 + (Q\eta)^2\right]^{\frac{1-D_m}{2}} \sin\left[(D_m - 1)\arctan(Q\eta)\right]\}$$

where ρ_0 is the scattering length density of embedding medium and C_0 is a constant. Furthermore, for $Q \ll 1/r_o$ the first term in curly brackets is insignificant in comparison to the second term—it only becomes significant in the large-Q limit of the small-angle scattering region (for full discussion see Freltoft el al. (1986)). A formulation of Equation (19) for a macroscopic sample of aerogels containing mass fractal particles of known density is given by Vacher et al. (1988). The limiting behavior of $I(Q)$ for sizes well below the correlation length, $\eta Q \gg 1$, is:

$$I(Q) \propto Q^{-Dm}\Gamma(D_m - 1) \times \sin\left[(D_m - 1)(\pi/2)\right], \quad D_m \le 3 \tag{20}$$

which is the $I(Q) = const\, Q^{-Dm}$ power law expected for mathematical mass fractals.

The correlation function for real surface fractals (Bale and Schmidt 1984; Mildner and Hall 1986) has a form similar to its counterpart for mass fractals:

$$\gamma(r) = \left(1 - C(r/\xi)^{3-D_S}\right)\exp(-r/\xi) \tag{21}$$

where the first term is the correlation function for perfect surface fractals (Bale and Schmidt 1984). The exponential term describes the decay of fractal properties as the length scale approaches the upper limit ξ, $C = N_0/[4\Phi(1-\Phi)V]$, V is the scattering volume, Φ is porosity and N_0 is a constant equal to the total surface area S separating the two phases in the sample when $D_s = 2$. The Fourier transform of $\gamma(r)$ gives:

$$I(Q) = AQ^{-1}\Gamma(5 - D_S)\xi^{5-D_S}\left[1 + (Q\xi)^2\right]^{\frac{D_S-5}{2}} \sin\left[(D_S - 1)\arctan(Q\xi)\right] \tag{22}$$

where $A = \pi N_0(\Delta\rho)^2$.

Allen (1991) proposed a modified expression for the correlation function for surface fractals (Eqn. 21) based on the assumption that the fractal surface is self-affine and that the surface area per unit volume, $(S/V)_r$, scales with the roughness scale, r, as follows:

$$(S/V)_r = (S/V)_0(r/\xi)^{2-D_S} \tag{23}$$

Equation (23) relates the specific surface area at any length scale, $(S/V)_r$, to its "smooth" value, $(S/V)_0$, measured with the roughness length scale ξ (where the correlation length ξ represents the upper limit for surface fractal properties). In this notation the constant A in Equation (22) is $A = \pi\xi^{Ds-2}(S/V)_0(\Delta\rho)^2$. For $D_s = 2$, $(S/V)_0 = N_0$, as expected.

The limiting behavior for sizes well below the correlation length, $\xi Q \gg 1$, is given by the expression derived by Bale and Schmidt (1984) for unconstrained surface fractals:

$$I(Q) = A\, Q^{D_s - 6}\Gamma(5 - D_s) \times \sin\left[(D_s - 1)(\pi/2)\right], \quad 2 < D_s \le 3 \tag{24}$$

where $A = \pi N_0(\Delta\rho)^2$. This is an $I(Q) = const\, Q^{Ds-6}$ power law which reduces to the Porod limit for $D_s = 2$.

The specific form of the small-angle scattering law, $I(Q) = constant \times Q^{-m}$, has been derived by many authors for both ideal and real surface fractals as well as mass fractals. It is important to note that transitional effects near the upper and lower limit of the fractal region may distort the scattering curve over a significant Q-range (Freltoft et al. 1986; Allen 1991). It is usually assumed that in order to justify the notion of a fractal, the appropriate geometrical properties should extend over at least one order of magnitude of the length scale (see discussion in Radlinski et al. 1996a).

Exponent m is D_m for mass fractals and $6-D_s$ for surface fractals. Since the allowed values for fractal dimensions are restricted ($2 < D_s \leq 3$ for surface fractals and $D_m \leq 3$ for mass fractals), the power law exponent is $m < 3$ for mass fractals and $3 < m \leq 4$ for surface fractals, where $m = 4$ for smooth surfaces (Porod law). This allows for a clear distinction between the two types of fractals using SAS techniques.

There is another class of scattering curve, for which the power law exponent is less than the Porod limit value of -4. Such a situation arises when a non-fractal porous medium has a fuzzy interface, i.e., characterized by a value of the scattering length density monotonously varying over a distance corresponding to the interface thickness. This can be achieved by coating the surface with a thin molecular layer. For fuzzy interfaces the scattering law has been theoretically predicted to have the form $I(Q) = const\, Q^{-m}$, where $4 < m < 5$ (Schmidt et al. 1991; McMahon et al. 2001).

The extended regions of power-law scattering curves (where the exponent $m = 1$ or $m = 2$), are observed for scattering on (non-fractal) one-dimensional particles (needles) or two-dimensional particles (platelets), respectively (Feigin and Svergun 1987; Espinat 1990). Radlinski et al. (1996b) show an example of SANS and SAXS scattering on a molecular system undergoing a thermally-driven 1D ($m = 1$) $\rightarrow$ 2D ($m = 2$) $\rightarrow$ 3D ($m = 4$) transition.

CALCULATIONS OF THE SCATTERING LENGTH DENSITY AND CONTRAST

General expressions

A knowledge of contrast is crucial for quantitative interpretation of absolutely calibrated SANS (and SAXS) results. In geological applications, a value for contrast is needed to determine rock porosity, internal specific surface area and other related quantities. Analysis of contrast values constitutes an integral part of preparations for a SAS experiment and often determines which instruments need to be used in order to best address the problem at hand. For this reason we provide a comprehensive discussion of the subject, including three worked examples of contrast calculation.

Rocks are somewhat transparent to neutrons and to a much lesser extent, X-rays. Therefore, these two particular types of radiation can be used for scattering experiments on geological samples. The physical property responsible for neutron scattering is the coherent scattering amplitude (nuclear potential)—for X-rays it is the electron density (electric charge). The strength of radiation-matter interaction that influences the contrast term, $(\Delta\rho)^2$, in Equation (7), is dependent on the average of all the nuclei in a pseudo-molecule whose chemical composition reflects the average composition of each of the two rock phases: the solid phase (rock matrix) and the fluid phase (the content of the pore space). The neutron scattering length density for each single phase of pseudo-molar mass M is:

$$\rho_n = \frac{N_A d}{M} \sum_j p_j \left(\sum_i s_i b_i \right)_j \tag{25}$$

where $N_A = 6.022 \times 10^{23}$ is the Avogadro's number, d is density (in g/cm^3), s_i is the proportion by number of nucleus i in the compound j, p_j is the proportion by molecular number of the compound j in the mixture and b_i is the coherent scattering amplitude for nucleus i.

For X-ray coherent (also called Rayleigh or elastic) scattering, the single-phase factor in the contrast term is simply:

$$\rho_{el} = I_e \rho_e = \frac{N_A d}{M} N_e I_e \tag{26}$$

where ρ_e is the electron density (i.e., number of electrons per unit volume), $I_e = e^2/(mc^2) = 2.82 \times 10^{-13}$ cm is the coherent scattering amplitude of a single electron, N_e is the number of electrons per one supra-molecule (of composition as per the double summation in Eqn. 25), M is the molecular weight of one supra-molecule and d is bulk density (in g/cm^3).

Scattering length density for common rock constituents

General data. Table 1 lists the coherent scattering amplitude, b (Sears 1990; Dianoux and Lander 2002), atomic mass (in atomic units) and the number of electrons, Z, for 15 elements, including those commonly found in inorganic and organic rocks. Density and molar weights for most common naturally abundant inorganic minerals are compiled in Table 2. Table 3 lists the neutron and X-ray scattering length densities for these minerals.

Compositional data obtained by X-ray fluorescence for two types of inorganic matter—Palaeozoic shale (Radlinski et al. 1996a) and ash from a Bowen Basin Permian coal (Radlinski et al. 2004c)—are presented in Table 4. Calculated crystalline density for these two rocks is 2.98 g/cm^3 for shale and 3.23 g/cm^3 for ash, whereas the measured density of the shale sample is 2.4 g/cm^3, i.e., 81% of the crystalline value. This 19% difference cannot be accounted for by sample porosity, as the porosity is only 5%. It is likely that at least some of the oxides comprising the rock are in an amorphous state (and are less densely packed than the crystalline state). The scattering length density (SLD) calculated from Equation (25) is 4.47×10^{10} cm^{-2} for shale (crystalline) and 4.79×10^{10} cm^{-2} for ash (crystalline). Taking into account the ~20% decrease in density for naturally abundant inorganic matter, these values are likely to be about 3.6×10^{10} cm^{-2} for shale and 3.75×10^{10} cm^{-2} for ash.

Table 1. Average coherent scattering amplitude of neutron scattering for naturally abundant isotopes, atomic mass and number of electrons for 15 elements commonly found in inorganic and organic rocks.

Element	b (10^{-12} cm)	Atomic mass (au)	Atomic No.
H	−0.3739	1.00794	1
D	0.6671	2.0	1
C	0.6646	12.011	6
N	0.936	14.007	7
O	0.5803	15.9994	8
Mg	0.5375	24.3050	12
Al	0.3449	26.982	13
Si	0.4149	28.0855	14
P	0.513	30.974	15
S	0.2847	32.066	16
K	0.371	39.0983	19
Ca	0.490	40.078	20
Ti	−0.330	47.867	22
Fe	0.954	55.845	26
Zr	0.716	91.224	40

Table 2. Density and molar weight for common naturally abundant inorganic minerals, water and heavy water. Abbreviations used: cr – crystalline, am – amorphous.

Mineral	Formula	Density (g/cm³)	Molecular Wt. (g/mole)
Quartz	SiO_2	2.7; 2.2 (fused)	60.085
Corundum	Al_2O_3	4.0	101.96
Hematite	Fe_2O_3	5.2	159.69
Anatase	TiO_2	3.9	79.865
Periclase	MgO	3.6	40.304
Lime	CaO	3.25-3.38	56.077
	K_2O	2.32	94.196
Zircon	$ZrSiO_4$	4.72 cr, 4.0 am	183.3071
Water	H_2O	1.0	17.999
Heavy water	D_2O	1.105	19.999

Table 3. Neutron, ρ_n, and X-ray, ρ_{el}, scattering length densities (SLD) for common minerals, water and heavy water. The value of $\Sigma s_i b_i$ (Eqn. 9) is shown in the last column.

Mineral	Formula	SAXS SLD ρ_{el} (10^{11} cm⁻²)	SANS SLD ρ_n (10^{10} cm⁻²)	$\Sigma s_i b_i$ (10^{-12} cm)
Quartz	SiO_2	2.29	4.264	1.5755
Corundum	Al_2O_3	3.332	5.743	2.4307
Hematite	Fe_2O_3	4.205	7.156	3.6489
Anatase	TiO_2	3.152	2.443	0.8306
Periclase	MgO	3.034	6.014	1.1178
Lime	CaO	2.816	3.817	1.0703
	K_2O	1.924	1.962	1.3223
Zircon	$ZrSiO_4$	2.702	—	—
Water	H_2O	0.94	−0.56	−0.1675
Heavy water	D_2O	0.94	6.4	1.9145

Table 4. Comparison of chemical composition for coal ash and typical shale (in wt% units).

Mineral	SiO_2	Al_2O_3	Fe_2O_3	TiO_2	CaO	K_2O	MgO	balance	total
Ash	49	25	13	1.3	1.0	1.0	—	9.7	100
Shale	58	13.4	6.7	-	3.1	3.1	2.4	13.3	100

SLD for neutrons. The values of ρ_n (for neutrons) are presented graphically in Figure 4. On the left hand side, open arrows indicate SLD values for individual inorganic minerals calculated using crystalline densities (Tables 2 and 3). Heavy arrows indicate SLDs of particular interest: for (partly) amorphous shale and ash, void, and light and heavy water. For sedimentary rocks void represents pore space (which is typically filled with water). Therefore, the typical contrast value is given by the square of the difference between the SLD for the shale and the SLD for the water.

Due to the large coherent scattering amplitude associated with deuterium (Table 1), the scattering length density for heavy water is very large. For instance, by saturating a rock sample with a mixture of water and heavy water of appropriate proportions one can match the scattering length density of the rock matrix and remove the matrix-void scattering contrast. Contrast matching (as well as contrast enhancement) by spatially selective deuteration is

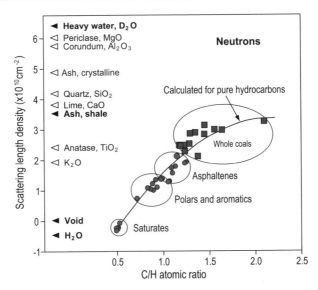

Figure 4. Neutron scattering length density for coals, hydrocarbons and common minerals.

arguably the most important feature unique to neutron scattering and has been widely employed in numerous studies. The most celebrated applications of contrast matching are in polymer science and biology. In the context of rock microstructure, contrast matching has been recently used by Broseta et al. (2001) to study capillary condensation of water in the pores of sandstone and by Snook et al. (2002), using d-toluene to determine closed porosity in coal chars used as fuel in industrial blast furnaces.

On the right hand side of Figure 4 we show values of SLD for organic compounds: whole coals (comprised of various solid organic macerals), fractions of crude oil (asphaltenes, polars and aromatics, and saturates) and pure hydrocarbons. For organic compounds, the value of SLD is largely determined by the atomic ratio of two dominant elements in the organic matter—carbon and hydrogen. As there is little neutron contrast between the solid organic and solid inorganic components of the rock, for rocks containing both organic and inorganic solid matter (e.g., hydrocarbon source rocks, ashy coals) the dominant scattering contrast is between the solid matrix and the pore space. Importantly, therefore, neutrons perceive sedimentary rocks as two-phase scattering systems, independent of the organic matter content.

At depths and temperatures corresponding to the oil generation window, for hydrocarbon source rocks the solid organic matter will gradually decompose into viscous bitumen, crude oil and eventually crack to light hydrocarbons and gas. This is accompanied by an increase in volume, leading to displacement of formation water with liquid hydrocarbons - a process known as primary migration. Due to the changes in pore fluid composition that take place during primary migration, hydrocarbon generation and expulsion in source rocks can be observed using SANS (Fig. 4).

SLD for X-rays. Values of scattering length density for X-rays, ρ_{el}, are plotted in Figure 5. It is apparent that ρ_{el} is strongly correlated with the density of the scattering material. Importantly, the matrix-void scattering contrast for organic matrix is about 25% of that of inorganic matrix. If the rock is purely organic (e.g., ash-free coal) or purely inorganic (e.g., sandstone), the system remains two-phase and the structural information obtained from both SAXS and SANS is identical. For rocks with both a significant organic and inorganic component, however, the

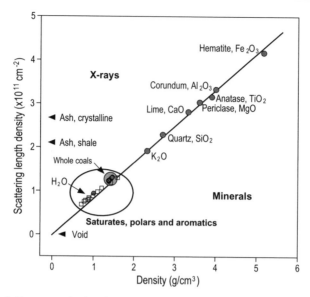

Figure 5. X-ray scattering length density for coals, hydrocarbons and common minerals.

system is two-phase for SANS, three-phase for SAXS and the scattering data obtained with the two techniques are no longer equivalent (Radlinski et al. 1996a).

Scattering lengths shown in Figure 5 have been calculated for X-ray photon energy of the incident primary beam which is greater than the excitation energy of inner electronic shells. When the absorption edge of an atom is close to the X-ray photon energy, the dispersion corrections to the coherent scattering amplitude can be up to 50% of the value given by Equation (26). With synchrotron X-ray sources it is possible to tune X-ray energy to the absorption edge of selected atoms, which results in enhanced contrast variation.

Selected examples

SANS - scattering length density for corundum. Corundum is one of the most abundant minerals in the Earth's crust. For Al_2O_3, Equation (25) reduces to one term only (there is no summation over *j*) and index *i* assumes values from 1 (for Al) to 4 (2, 3 and 4 for the three naturally abundant isotopes of oxygen). Coherent scattering amplitudes for naturally abundant isotopes are well known (Sears 1990, Dianoux and Lander 2002). Aluminium has one stable isotope, ^{27}Al ($s_1 = 2$, $b_1 = 0.3449 \times 10^{-12}$ cm) and oxygen has three: ^{16}O (natural abundance 99.76%, atomic mass 15.99491), ^{17}O (natural abundance 0.04%) and ^{18}O (natural abundance 0.20%). The three oxygen isotopes exhibit three coherent scattering amplitudes (*b*): $b_2 = b(^{16}O) = 0.5803 \times 10^{-12}$ cm, $b_3 = b(^{17}O) = 0.578 \times 10^{-12}$ cm and $b_4 = b(^{18}O) = 0.584 \times 10^{-12}$ cm. For a mineral with a naturally occurring isotopic distribution, the three values of b_i ($i = 1, 2, 3$) can be substituted with the abundance-weighted average value of $b_{ave} = 0.5803 \times 10^{-12}$ cm. An average atomic mass of 15.9994 can now be used (Table 1), and the range of index *i* in Equation (25) is reduced to (1, 2). Substituting $b_2 = b_{ave}$, $s_2 = 3$, and using a density of 4.0 g/cm³ and a molar mass of 101.96 g (Table 2), one can calculate from Equation (9) the scattering length density for crystalline corundum: $\rho_n = 5.743 \times 10^{10}$ cm⁻².

SANS - Scattering length density for coal. Typical results of chemical (ultimate) analysis of a coal are shown in Table 5: the amount of ash and elements H, C, N, S and O are all expressed in weight %. Two sets of numbers are given for each element - one is a percentage of dry mass with ash (dry), and the other is dry ash free (daf).

Table 5. Typical results of chemical analysis of a coal (Ammonate whole coal, vitrinite reflectance Ro = 1.28%). All numbers are in wt%.

Ash dry	H dry	H daf	C dry	C daf	N dry	N daf	S dry	S daf	O dry	O daf
11.63	4.63	5.24	79.4	89.849	1.35	1.53	0.64	0.72	2.35	2.659

Table 6. Atomic proportion (s_i of Eqn. 25) for five elements in a pseudo-molecule of dry ash free Ammonate whole coal. Note that daf weight percents add up to 100%.

Element	H	C	N	S	O
wt%	5.24	89.849	1.53	0.72	2.659
s	5.198	7.481	0.109	0.0225	0.166

To calculate ρ_n from Equation (25), assume the organic matrix of coal is composed of identical pseudo-molecules of composition $H_tC_uN_vS_xO_y$, where the generally non-integer numbers t, u, v, x, and y are the atomic proportions (i.e., numbers of atoms in a pseudo-molecule, s_i of Eqn. 25) for hydrogen, carbon, nitrogen, sulfur and oxygen, respectively. The atomic proportion for each chemical element is defined as the ratio of its dry ash free weight percent to its atomic mass. Table 6 lists atomic proportions calculated using data from Tables 5 and 1.

Because of the way atomic proportions are defined, the molar weight (M) of the molecule $H_tC_uN_vS_xO_y$ is 100 g ($M = \Sigma_i$ (atomic proportion)$_i$ × (atomic mass)$_i$). The scattering length density, ρ_n, can now be calculated from Equation (25) by substituting $N_A = 6.022 \times 10^{23}$, $d = 1.47$ g/cm^3 (obtained from direct measurement on a coal sample), $M = 100$ g and the s_ib_i value obtained for every element using data from Tables 1 and 6. The result is $\rho_n = 2.86 \times 10^{10}$ cm^{-2}.

SAXS - scattering length density for zircon. Zircons are sub-millimeter size crystals used for U-Pb dating of geological systems. Zircon crystals are too small for SANS measurements with currently available instruments (generally less than 1 mm across), but the geometry and volumetric fraction of radiation damaged (metamict) regions in zircons can be determined using SAXS utilizing the amorphous-crystalline contrast (Radlinski et al. 2003).

Consider a zircon crystal of composition $ZrSiO_4$. Both Zr (atomic mass 91.224, 40 electrons) and Si (atomic mass 28.0855, 14 electrons) have only one naturally abundant stable isotope. For the three oxygen isotopes, the average value of atomic mass is 15.9994 and the number of electrons per atom is 8 (Table 1). Therefore, the molar mass of $ZrSiO_4$ is 91.224 + 28.0855 + 4 × 15.9994 = 183.3071 grams, and the total number of electrons in one molecule is 40 + 14 + 4 × 8 = 86. The scattering length density, ρ_{el}, for crystalline zircon can now be calculated from Equation (26) by substituting $I_e = e^2/(mc^2) = 2.82 \times 10^{-13}$ cm, $N_A = 6.022 \times 10^{23}$, $M = 183.3071$ g, $N_e = 86$ and the crystalline density $d = 4.27$ g/cm^3 (Table 2). The result is $\rho_{el} = 3.757 \times 10^{11}$ cm^{-2}. For amorphous (metamict) zircon the only difference is the density, which now is $d = 4.0$ g/cm^3 (Table 2). The resulting scattering length density is $\rho_{el} = 3.184 \times 10^{11}$ cm^{-2}.

HOW A SANS EXPERIMENT IS UNDERTAKEN

Measurement of the absolute scattering cross section

For a pinhole SAS instrument (schematic diagram shown in Fig. 1), the signal registered by each element (pixel) of a two dimensional detector (whose position is defined by the scattering vector $\mathbf{Q}$ and the solid angle element $d\Omega$), is the number of counts $I(\mathbf{Q})$. $I(\mathbf{Q})$ is proportional to the number of neutrons (or X-ray photons) scattered by the illuminated volume of a sample of surface area A and thickness t:

$$I(\mathbf{Q},d\Omega) = I_0 A t T \frac{d\sigma}{d\Omega} d\Omega E(\mathbf{Q},\lambda) + \text{background} \qquad (27)$$

where λ is the wavelength, T is sample transmission, I_0 is the incident beam intensity and $E(\mathbf{Q}, \lambda)$ is the quantum efficiency of the detector element. The background term reflects electronic noise associated with the electronic detection system and external radiation (including high-energy cosmic radiation).

The purpose of an SAS experiment is to determine (for a number of samples) the value of the differential scattering cross section, $d\sigma/d\Omega$, in absolute units of cm^{-1}. In laboratory practice this is often done in the following sequence:

1. Measure the thickness and transmission for all samples

2. Run a standard sample of known t, T and $d\sigma/d\Omega$ in absolute units

3. Measure the dark current, i.e., background signal, with a platelet of cadmium for SANS (or lead for SAXS) placed in the sample position

4. If a sample is placed in a cell, measure the scattering from the empty cell

5. Measure the scattering from a series of samples using the same slits used for the standard sample

6. At the end of experiment re-run the dark current measurement

Absolutely calibrated data enable quantitative comparison with theoretical models. Leading SAS laboratories develop and calibrate scattering standards, including inter-laboratory calibration and SANS-SAXS cross-calibration (e.g., Ibel and Wright 1980, Wignall and Bates 1987, Russell et al. 1988). For an SAS experiment conducted in the manner described above, the scattering data (i.e., $I(\mathbf{Q},d\Omega)$) for a standard sample are used to determine the value of $Ad\Omega E(\mathbf{Q},\lambda)$ from Equation 27 (since I_0 is continuously monitored and the background scattering is measured independently). The result is then used to calculate $d\sigma/d\Omega$ in absolute units for every measured sample.

By following the above procedure it is also possible to process data from the first sample before running subsequent samples. Furthermore, useful results can be obtained even if SAS data for some of the samples were not acquired or were corrupted. A series of SANS experiments may take several days to complete, and there is always a possibility of unexpected reactor shut down (which is the most common cause of incomplete data acquisition).

Measurement of transmission

Transmission (T) is measured in a separate experiment by measuring the total intensity of radiation with (I^{tot}), and without (I_0^{tot}) a sample in the beam ($T = I^{tot} / I_0^{tot}$). One method used to measure transmission is to insert a total neutron counter a short distance downstream from the sample position to measure the signal with and without the sample in the beam. Leading SANS and SAXS laboratories have well tested procedures for transmission measurements.

Sample thickness

In the case of a liquid contained in a cell or a self-supporting solid, the measurement of sample thickness is trivial. Geological samples, however (especially those originating from exploration wells), are seldom available as an oriented core. The most common type of rock sample available from exploration drilling is "drill cuttings," which need to be crushed to coarse powders for neutron scattering experiments. The grain size of the crushed sample must represent a balance between average orientation over the sample volume (the grains cannot be too large) and the need to minimize the internal surface area associated with grain surfaces (the grains cannot be too small). Furthermore, for absolute calculations of rock properties one needs to take account of inter-granular porosity. Radlinski et al. (2004b) used resin-fixed powders of a narrow grain size distribution (0.355 – 0.475 mm) to fill the inter-granular porosity. The volume associated with the inter-granular porosity can then be determined both by calculation and by comparison with scattering data for solid samples. Spalla et al. (2003) used saturation with solvents to eliminate the need for thickness measurements in SAXS work on powders deposited onto a Kapton sheet. For powdered samples, to avoid significant contribution from scattering from the surface of individual grains, the grain size should be at least 10 times larger than the maximum linear distance accessible with the scattering apparatus ($\sim 1/Q_{min}$). Ideally, the minimum acceptable grain size should be determined experimentally.

MICROSTRUCTURE OF ROCKS REVEALED BY SANS - EXAMPLES

Sandstones

Scale range of fractal and Euclidean microstructure. Sandstones are composed predominantly of quartz with small amounts of interstitial clay and/or mica minerals. Subsurface sandstones constitute reservoir rocks for both water and hydrocarbons and have been extensively studied—especially in engineering applications related to porosity, permeability and fluid flow. As discussed previously, in the 1980's sandstone microstructure has been recognized to have a non-universal surface fractal character. The range over which the fractal behavior in sandstones has been observed by SAS was extended to about 20 μm using a then newly constructed low-background USANS instrument (Agamalian et al. 1997; Radlinski et al. 2000b). The observation limit was further extended to about 0.5 mm by combining SANS, USANS and backscattering scanning electron microscopy (BSEM) data (Radlinski et al. 2004a).

Figure 6 shows the absolute scattering intensity for a sample of sandstone presented on a log-log scale. Over 4.5 orders of magnitude on the length scale (from $2.5/Q = 1$ nm to 50 μm), the pore-matrix interface is a surface fractal ($D_s = 2.47$). Over the corresponding Q-range, the scattering intensity varies over 17 orders of magnitude (from 10^{-3} cm^{-1} to 10^{14} cm^{-1}). Note

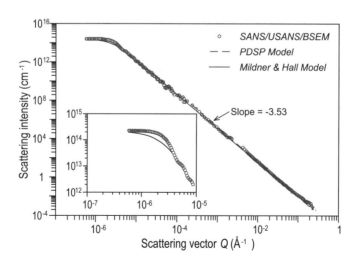

Figure 6. Absolutely calibrated SANS scattering intensity for a coarse sandstone, after background subtraction of 6×10^{-3} cm^{-1} (after Radlinski et al. 2004a).

that intensity values over 10^{11} cm^{-1} are inferred from the Fourier transform of the Backscatter Electron Microscopy (BSEM)-determined correlation function (Eqn. 5). For pore sizes larger than $2.5/Q = 50$ μm, the scattering curve flattens out, indicating a transition to a Euclidean scattering regime.

This particular rock is composed of 98% quartz, 1.5% mica and small amounts of dickite and montmorillonite (Radlinski et al. 2004a). The average grain diameter is approximately 250 μm and the porosity is 0.181. The value of porosity calculated from the invariant (Eqn. 9) agrees with an independently measured value to within 5%. The correlation length (ξ) (obtained by a reasonably good fit to the Mildner and Hall model; Eqn. 22) is 35 μm. The fit to the PDSP (polydisperse spheres) model, Equation (14), (which uses a form factor for homogeneous spheres, (Eqn. 12) and, therefore, is not constrained by an analytical form for the long-range cut-off of the correlation function) is very good. The model provides values for the specific surface area and pore size distribution (see Fig. 7), and gives an upper limit for surface fractal behavior of about 55 μm. Using this model, the volumetric fraction of total porosity associated with surface fractal geometry (Φ_{frac}/Φ_{total}) is approximately 62%.

These results demonstrate that fractal-like surface roughness can extend for a distance of approximately one fifth of the grain size diameter into the pore space, entirely controlling the specific surface area associated with molecular probe sizes and constitutes the majority of the pore volume. Only the largest pores (over 35 to 55 μm in diameter) are described by Euclidean geometry. NMR measurements of the decay of nuclear transverse magnetization in water-saturated samples indicate the existence of two scattering regimes - fractal and Euclidean (Radlinski et al. 2004a). These yield values of ξ of 35 μm and $\Phi_{frac}/\Phi_{total} = 47\%$ (which are in good agreement with neutron scattering data).

Synthetic mercury intrusion porosimetry curve. From the pore size distribution (see Fig. 7) and the known relationship between the capillary pressure and the pore size, it is possible to construct a synthetic mercury intrusion porosimetry (MIP) curve and compare it with the measured MIP curve. Figure 8 shows such a comparison and indicates that for this particular sample of sandstone, the ratio of the pore body size (measured by neutron scattering) to the pore throat size (measured by MIP) is about 3.5 and is independent of the pore size (Radlinski et al. 2004a).

Wetting mechanism for fractal pores. Broseta et al. (2001) published the first SAS study of the mechanism of wetting for the fractal pore-matrix interface in a rock (a Vosges sandstone with a porosity of 17%). The authors took advantage of the universal applicability of two-phase approximation for (organic matter free) sandstones: SANS and USAXS data were combined to calculate $d\sigma/d\Omega$ for dry rock samples and their fractal dimension ($D_s = 2.68$) was determined over three decades of the length scale. In order to observe size-specific invasion of pores by water, samples were exposed to a contrast-matched mixture of water and heavy water at several controlled vapor pressures. As the smallest pores imbibe water first, a shift of the large-Q cut-off of the fractal scattering regime towards smaller Q-values was observed with increased water vapor pressure. This was followed by Q^{-4} Porod scattering (from the smooth interface of water-filled small pores) at large Q-values. Wetting behavior was dominated by the capillary wetting regime (smallest pores filled with water first), with limited influence from the substrate-controlled wetting regime (covering a surface with a thin film of a uniform thickness) at higher water vapor pressures. The work of Broseta et al. (2001) constitutes the first direct visualization (in the Fourier reciprocal space) of the dynamics of fluid imbibition within a system of fractal pores.

In summary, SANS can be a useful technique in the analysis of fluid behavior in confined geometries (e.g., Lin et al. 1994a,b; Melnichenko et al. 2005). This application is further discussed by Cole et al. (2006, this volume).

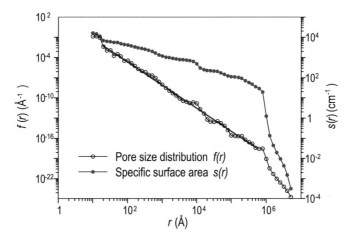

Figure 7. Distribution of pore size ($f(r)$) and specific surface area ($s(r)$) for a coarse sandstone. Straight line corresponds to $f(r) = const \times r^{-(D+1)}$ with $D = 2.49 \pm 0.03$ (after Radlinski et al. 2004a).

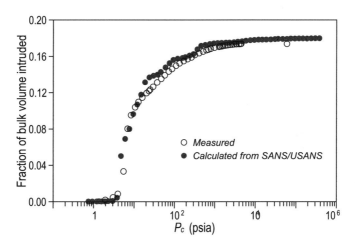

Figure 8. Measured and calculated mercury intrusion porosimetry curve for a coarse sandstone (after Radlinski et al. 2004a).

Mudstones: hydrocarbon source rocks

Principle of observation of hydrocarbon generation. Hydrocarbons are generated in organic-rich clastic rocks (e.g., mudstones) at elevated temperatures over geologic time. At the onset of bitumen generation, the organic matter expands and migrates through the pore space of the rock. This process can be detected by SANS/USANS due to the difference in contrast between the generated hydrocarbons and the pore water (Fig. 4) and has been studied for both natural (Radlinski et al. 1996a) and artificial (Radlinski et al. 2000a) hydrocarbon source rocks.

Figure 9 shows a schematic variation of SANS/USANS intensity with depth for three selected pore sizes r (corresponding to $Q \approx 2.5/r$). Two scenarios are considered – one for an organic-rich rock (in which the pore space eventually becomes saturated with generated

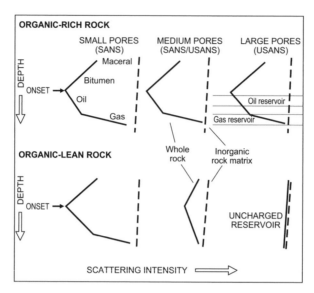

Figure 9. A schematic representation of the variation of SANS/USANS intensity with depth (at three fixed *Q*-values) for an organic-rich effective source rock (top) and an organic-lean rock (bottom) (after Radlinski et al. 2004b).

hydrocarbons and hydrocarbon expulsion ensues) and another for an organic-lean rock (for which only a fraction of the pore space is invaded by generated hydrocarbons). It is assumed that the organic matter is finely dispersed throughout the rock volume. In geochemical terms, an organic-rich rock is characterized by a TOC values greater than 2% and an organic-lean rock by a TOC values of less than 2% (where TOC is the weight percent of total organic carbon present in the rock).

Both subsurface temperature and pressure increase with depth. Under increasing pressure, the pores are expected to gradually compact. For an inorganic rock matrix (or organic-matter-free rock) of fixed lithology, the scattering intensity at a fixed *Q*-value decreases slightly with depth (as indicated by the broken lines). For rocks containing organic matter, increasing subsidence and thermal maturation results in the invasion of pore space with hydrocarbons generated from the organic source material—initially with bitumen, which then progressively cracks into lighter hydrocarbons (and eventually into gas). Using the corresponding scattering length density values for these various hydrocarbons (Fig. 4) one can schematically represent the scattering intensity versus depth as shown in Figure 9 with solid lines. Note that the scattering intensity for an organic-matter-free rock is always greater than for a rock of the same lithology containing dispersed organic matter.

The onset of oil generation occurs at a depth where maximum bitumen saturation (minimum scattering intensity) is observed for small pores. The onset of oil expulsion occurs at a depth where maximum bitumen saturation is observed for the largest pores (about 20 μm for typical shales; Radlinski et al. 1999). Only organic-rich rocks are capable of generating sufficient volumes of hydrocarbons to both fully saturate the pore space and to allow hydrocarbon expulsion from the source rock—thus becoming effective sources for hydrocarbon accumulations.

A basin-wide study of hydrocarbon generation and expulsion. These ideas were used to perform a comprehensive analysis of SANS and USANS data obtained for 165 potential source rocks of Late Jurassic - Early Cretaceous age, recovered from nine exploration wells drilled in the Browse Basin, offshore Western Australia (Radlinski et al. 2004b). This study demonstrated that conclusions drawn from SANS and USANS data are consistent with results obtained using traditional geochemical methods and can be used to independently calibrate and refine source rock generation/expulsion scenarios derived from geochemical modeling.

Scattering patterns form the Browse Basin rock samples are fractal-like, and the PDSP model (Eqn. 14) was used for numerical analysis. Figure 10 shows an example of the results: the variation of pore number density with depth (compaction) for four selected pore sizes, calculated for the giant Brewster gas accumulation. Figure 11 illustrates a comparison of USANS results with the predictions for oil and gas generation based on traditional geochemical thermal maturity indicators and thermal history analysis of the Brewster-1A well.

Coal

Coals as source rocks for hydrocarbons. Coals (in a form of fine powders and carbon blacks) were among the first substances for which classical SAS patterns were observed (Krishnamurti 1930; Warren 1934). Coal also provided one of the first examples of a surface fractal microstructure (Bale and Schmidt 1984) and has been studied extensively by SAXS and SANS—often in relation to industrial processes (Lin et al. 1978; Winans and Thiyaga-rajan 1988; Johnston et al. 1993; Snook et al. 2002; McMahon et al. 2002; Prinz et al. 2004; Radlinski et al. 2004c).

In the following we provide examples of SANS and USANS results pertaining to natural (untreated) coal. Despite its origin from higher plant debris (thermally altered in anoxic condi-tions) and the atomic composition being dominated by carbon and hydrogen, the microstructure of the coal—pore space interface is surprisingly similar to that of inorganic sedimentary rocks (Radlinski and Hinde 2002). The coal matrix is sensitive to thermal treatment, however, and in natural conditions exhibits two types of thermally induced coal-specific phenomena:

1. At small scales, the internal specific surface area gradually decreases with increasing rank and a microstructure comprising stacked lamellae of polyaromatic hydrocarbon sheets spaced at an average repeat distance of about 25 Å is formed (while retaining a surface fractal structure at larger scales).

2. In the narrow temperature range corresponding to the onset of hydrocarbon generation, the pore space microstructure is subject to a marked rearrangement.

These two points are illustrated in Figure 12, which shows the scattering intensity versus depth (for pore size $2.5/Q = 25$ nm) and the pore number density versus depth (calculated using the PDSP model) for a series of coal samples (which have undergone increasing thermal maturation with depth) derived from a petroleum exploration well. A discussion of the formation of the lamellar structure and the evidence that it can become intercalated with clay minerals (based on comparison of SAXS and SANS data) is given in Radlinski et al. (2004c).

Adsorption of gases. The magnitude of the internal specific surface area (SSA) in a coal determines the amount of gas (like CH_4 and CO_2) that can be adsorbed as a surface monolayer. Figure 13 shows a compilation of SSA data obtained for coals and vitrinite macerals (for both solid samples oriented in-bedding-plane and standard pellets) of different rank using the N_2 adsorption method and SANS (Radlinski et al. 2004c). As neutrons can detect both effective (open) and non-effective (closed) porosity, and the presence of connate (remnant) water slightly increases the scattering contrast and decreases the number of sites available for gas adsorption, the SANS-determined SSA values are systematically higher than those derived from the N_2 adsorption method. Both methods, however, give remarkably similar trends with respect of rank (expressed as vitrinite reflectance).

Total porosity. The total porosity of coal versus rank (expressed as % of carbon in dry ash free coal sample) is presented in Figure 14 against the backdrop of the world trend determined

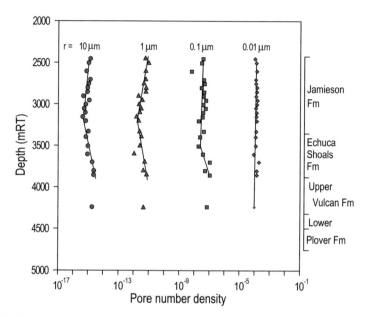

Figure 10. Variation of the pore number density for four selected pore sizes versus depth. Potential hydrocarbon source rock samples were recovered from the Brewster-1A well drilled on the giant Brewster gas accumulation, Browse Basin, Australia (after Radlinski et al. 2004b).

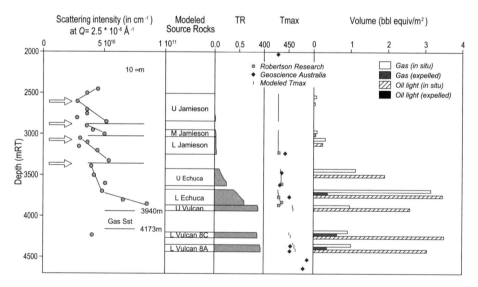

Figure 11. Comparison of USANS scattering intensity trends with modeled kerogen transformation (TR – Transformation Ratio) and *in situ* gas/oil generation and expulsion derived from geochemistry and thermal history analysis for Brewster-1A (after Radlinski et al. 2004b).

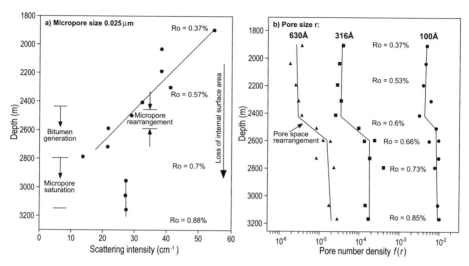

Figure 12. SANS results for coals from the Eastern View Sequence (Pelican-5 well, Bass Basin, Australia) showing (a) variation of the scattering intensity versus depth (for mean micropore size 25 nm ± 12 nm) and (b) pore number density calculated from SANS data using the PDSP model for micropore sizes 100, 316 and 630 Å (after Boreham et al. 2003).

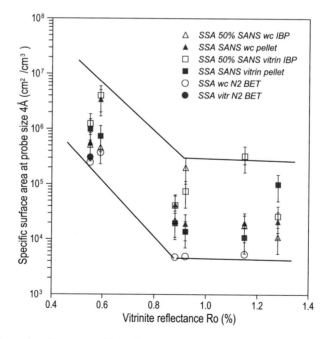

Figure 13. Comparison between specific surface area for coals of different ranks extrapolated to probe diameter 4 Å from SANS and nitrogen adsorption techniques.

by an independent method (Berkowitz 1979). The agreement is remarkably good and illustrates the utility of the PDSP model in interpreting SAS data for sedimentary rocks.

Igneous rocks and opals

Microstructure of igneous rocks. SANS work on igneous rocks was pioneered by Lucido et al. (1985). They studied alkaline basaltic rocks from Sicily and New Zealand and concluded that neutron scattering originated from light-colored small particles (leucocratic ocelli with an average diameter of around 190 Å) embedded in the rock—indicating the system had undergone rapid cooling near the critical condition. SANS data from a number of volcanic and plutonic rocks from South Africa, Morocco and Sardinia were interpreted as surface fractal scattering on droplets precipitated during spinodal decomposition of magma cooled at or near the critical condition (Lucido et al. 1988, 1991).

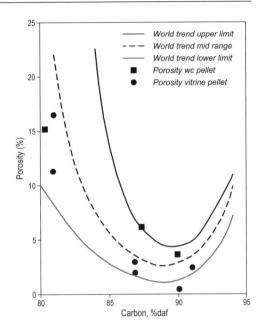

Figure 14. Porosity of coals and vitrinites calculated from SANS/USANS data for pore size range 2.5 nm to 10 μm (after Radlinski et al. 2004c).

Kahle et al. (2004) performed a combined SANS and microscopic study of various types of volcanic rocks: basalt, rhyolite, phonolitic pumice, phonolite and rhyolitic obsidian. Some of the rocks were thermally treated up to a temperature of 1900 K. The composition of samples was determined using microprobe and X-ray fluorescence analysis. The SANS patterns were fractal-like for basalt (due to voids with fractal surfaces) and Porod-like for pumice and phonolite (due to pores with smooth surfaces—except for thermally treated samples at length scales below 300 Å). The authors calculated specific surface area for these rocks and discussed the effect of heat treatment on rock microstructure. The results for pumice were consistent with the previous work of Floriano et al. (1994). For rhyolitic obsidian (which has no porosity) scattering occurs on crystalline precipitates of FeO embedded in the rock matrix. This matrix is composed predominantly of SiO_2 and Al_2O_3.

The work of Kahle et al. (2004) is significant as it combines a quantitative interpretation of absolutely calibrated SANS data with a verification of structural models using SEM, optical microscopy and microprobe analysis. Figure 15 shows SANS data for basalt samples obtained by melting basalt powder at 1400 K for 20 minutes (samples 8 and 9) and for 3 hours (samples 10 and 11). For samples thermally treated for 20 minutes, the scattering is very similar to SANS of as-received solid basalt samples and indicates a fractal structure ($D_s \approx 2.5$-2.7) over the length range from 50 Å to at least 3000 Å. For samples thermally treated for 3 hours, however, the scattering intensity decreases by roughly one order of magnitude. Slopes in the range −4.25 to −4.15 indicate the presence of a fuzzy rather than fractal interface. SEM images for the two types of samples (three different magnifications for each type) illustrate the loss of porosity upon heat treatment (Fig. 16). This type of work provides insights into the microstructural consequences of thermal processing of rocks in earth-interior-like conditions and complements the neutron imaging work of Kahle et al. (2004) (Winkler et al. 2005; also see Winkler 2006, this volume).

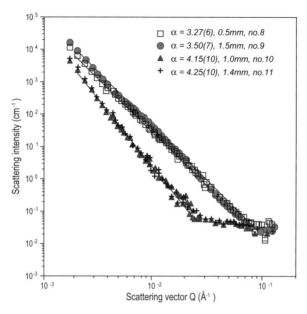

Figure 15. SANS data for 4 basalt samples thermally treated at 1400 K. Samples 8 and 9 were annealed for 20 minutes and samples 10 and 11 were annealed for 3 hours. [Reproduced with permission of E. Schweizerbart'sche Verlagsbuchhandlung (*http://www.schweizerbart.de*, from Kahle et al. (2004).]

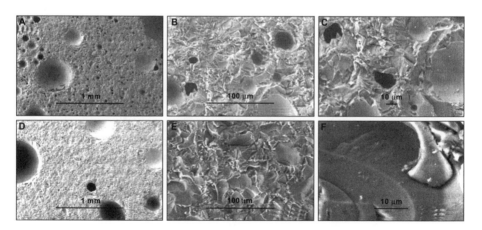

Figure 16. SEM images for basalt sample 8 (top) and sample 10 (bottom) showing the surface structure at various length scales. Magnification: A ×101, B ×1000, C ×1500, D ×100, E ×1010, F ×5000. [Reproduced with permission of E. Schweizerbart'sche Verlagsbuchhandlung (*http://www.schweizerbart.de*, from Kahle et al. (2004).]

Microstructure of opals. A precious opal of gem quality has been used for many years as a calibration standard for the D11 SANS facility at ILL (Ibel and Wright 1980). This particular application has led to a more extensive SANS study of a range of precious opals and potch opals from Australia, Brazil and Spain, as well as synthetic opals, flint nodules and a chalcedony band in an agate geode (Graetsch and Ibel 1997). Opals consist of closely packed non-crystalline silica spheres, with interstices filled with water and non-crystalline silica cement. The authors relate the SANS scattering patterns for these minerals in the Porod region to the type of stacking, the size distribution of the silica spheres, the amount of intersitial water and the chemical composition of the cement. Chalcedony and flint are composed of submicroscopic quartz fibers, and their SANS patterns are distinctly different (they also exhibit significantly less incoherent background scattering than opal). Figure 17 illustrates the SANS spectra of microcrystalline opals and quartz minerals (flint and chalcedony).

Clays

Due to their importance in civil engineering and in drilling-mud used in exploration drilling, clay/water mixtures have been studied extensively (including microstructural investigations using SAXS and SANS). Critical issues in the study of clays are the permeation of water and associated clay-swelling. For example, a mixture of water and sodium montmorillonite (a smectite) can form a thixotropic gel composed of up to 10 g of water per gram of clay.

The building blocks of clay minerals are flat, hydrophilic platelets about 10 Å thick with diameters in the range of hundreds to thousands of Å. In an aqueous environment, the clay platelets form sheet-like stacked layers intercalated with aqueous regions (pores). Typical dimensions in water-clay mixtures range from 10 Å for the thickness of a single layer, tens to thousands of Å for repeatable packages of stacked layers to hundreds of Å to tens of micrometers for clay platelet diameters and pore sizes (Knudsen et al. 2003). In order to fully describe these structures it is usually necessary to acquire data in a Q-range combining SANS (SAXS) and USANS (USAXS).

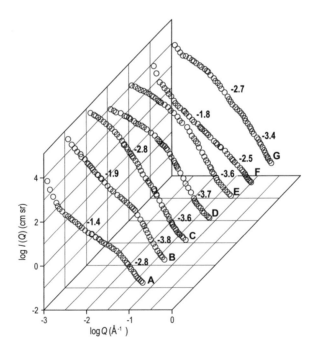

Figure 17. Radially averaged SANS data for microcrystalline opals and quartz minerals. A-C: opal-CT; D, E: opal-C; F: flint; G-chalcedony. Numbers refer to the slopes of the linear parts of the scattering curves. [Reproduced with kind permission of Springer Science and Business Media, from Graetsch and Ibel (1977)]

Morvan et al. (1994) reviewed early SANS and SAXS work on clays and presented results of SAXS and USAXS studies of suspensions of synthetic clay, laptonite (in water, up to 10% (w/w)) and of sodium montmorillonite clay (both in water (up to 20% (w/w) and in 1 M and 2 M NaCl (1.7% (w/w)). Figure 18 illustrates textures observed for montmorillonite samples dispersed in pure water and in brine and in swollen montmorillonite. Figure 19 shows calculated scattering curves for three different microstructures found in laptonite and montmorillonite.

Allen (1991) used SANS to test a microstructural model for compacted London clay and Oxford clay. His model is similar to that presented in Figure 18c, where the interlayer pores are called type I, pores between the closely inter-twined strands of lamellar stacks of different orientation are called type II and large, water containing pores are called type III. According to his analysis, type I pores are too small to be observed by SANS, a type II pore structure is a volume fractal and a type III pore surface is a surface fractal. Allen (1991) used the contrast-matching SANS method to observe swelling of London clay and Oxford clay and conjectured on the exchange of H_2O to D_2O in various pore types.

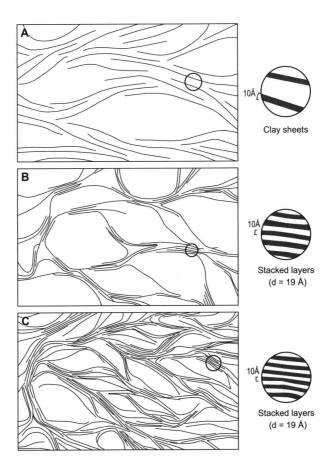

Figure 18. Three types of microstructural textures observed in montmorillonite clays. A: biphasic (nematic + holes), observed in sodium montmorillonite dispersed in pure water (Q^{-3} behavior at small Q-values); B: collapsed nematic + holes, observed in sample A after brine was added (Q^{-2} behavior at small Q-values); C: binding of dense sediments, observed in montmorillonite swollen in contact with brine (Q^{-3} behavior at small Q-values). [Reproduced from Morvan et al. (1994), with permission from Elsevier.]

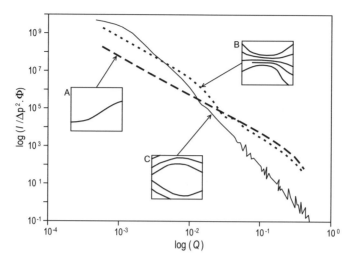

Figure 19. Small-angle scattering intensity (normalized to the contrast value, $\Delta\rho^2$, and clay volume fraction Φ) for three structural features in clays. A: water suspension of randomly oriented discs of radius 5000 Å and thickness 10 Å; B: water suspension of a mixture of randomly oriented discs of dimensions as in A and stacks of discs of radius 5000 Å and thickness 200 Å; C: polydisperse spherical pores filled with water in swollen clays. [Reproduced from Morvan et al. (1994), with permission from Elsevier.]

Knudsen et al. (2003) used SANS to obtain structural information about the synthetic clay Na-fluorohectorite. Oriented dehydrated samples (of porosity 41%) were prepared from suspension by applying uniaxial pressure (up to 26 MPa) at 120 °C. The microstructure of these samples at two orientations (parallel and perpendicular to applied stress) was assessed by SANS at various temperatures and water saturation. The obtained anisotropy ratio was 2:1, and the scattering curves were interpreted along the lines of a two-dimensional Debye model (Debye et al. 1957; Hall et al. 1983). Clay platelets were oriented preferentially in the direction perpendicular to applied stress. The relative pore water content in dry, normal and wet conditions was determined. The dynamics of water transport in a wet sample was studied using D_2O substitution and it was concluded that a slow restructuring of the internal surface may be taking place during H_2O/D_2O exchange on the time scale of about 10 hours.

Itakura et al. (2005) analyzed anisotropic multiple scattering USANS and SANS results obtained from two reconstituted samples of a natural kaolinitic soil used for containing industrial sludge waste. Samples were prepared from a slurry-like soil-water mixture uniaxially compressed at 400 kPa and 800 kPa. The authors demonstrate that the specific internal surface area obtained from multiple scattering analyses of water-saturated soils is consistent with the results of the Brunauer-Emmett-Teller (BET) method for air-dried fractions, thus providing a method for measuring the surface area of thick samples of fluid-saturated porous media.

MULTIPLE SCATTERING

The effect of multiple scattering on SANS curves

In the above we have assumed that each incident neutron has been scattered no more than once inside the specimen. In this approximation, the scattering cross section is independent of the neutron wavelength. Scattering, however, is a statistically random process and the probability of multiple scattering (MS) is finite but small (even for samples much thinner than the neutron mean free path). As shown in Figure 20, MS manifests itself as broadening

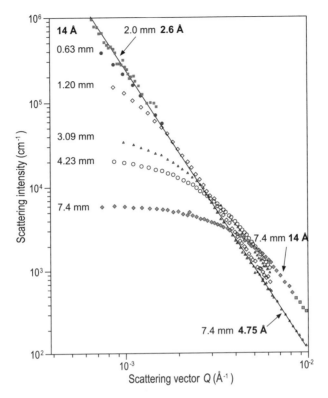

Figure 20. Multiple scattering SANS curves for a coarse sandstone measured using several neutron wavelengths for samples of various thicknesses. Note that distortion from the single scattering cross section (shown with solid line) is most accentuated for thick samples and long wavelengths (A.P. Radlinski and P. Lindner, unpublished).

(washing out) of the scattering curve. Its influence is particularly pronounced in strongly scattering materials such as rocks, clays, cements, ceramics, etc, for which the mean free path can be of the order of millimeters. The effect of MS is strongest in the small-Q region (for large scattering particle sizes) and MS probability depends on neutron wavelength as λ^2. Consequently, USANS and long wavelength SANS data for strongly scattering materials (including rocks) are susceptible to MS effects (Fig. 20).

In most cases, multiple scattering is an unwanted complication. A typical preventive approach is to prepare several samples of the same material of varying thicknesses to verify experimentally that there is no variation in the shape of the scattering curve (e.g., Radlinski et al. 1999). However, there is usually a practical limit to how thin a sample can be made. In some cases, SANS and USANS work is purposely undertaken on thick samples and the resulting MS patterns used to extract microstructural information.

Neutron scattering regimes

The formulation of neutron scattering regimes dates back to Weiss (1951) and has been more recently reviewed by Berk and Hardman-Rhyne (1988) and Krueger et al. (1991). The phase shift υ that a plane wave undergoes in traversing a scattering particle of radius R is $\upsilon = 2\Delta\rho R\lambda$, where $\Delta\rho = \rho_{particle} - \rho_{matrix}$ and ρ for each phase is given by Equation (25). Three scattering regimes are defined. The case of $\upsilon \ll 1$ is the usual SANS diffraction regime, where the scattered intensity is described by the Born approximation. The opposite case of $\upsilon \gg 1$ represents the multiple refraction regime, where the scattered intensity is derived from geometrical optics (von Nardroff 1926) and depends only upon the difference in index of refraction between the two phases (which itself is proportional to λ^2). Finally, the case of $\upsilon \approx 1$ is the MS regime most commonly encountered in SANS experiments for which the results can neither be described by a conventional SANS mechanism nor by multiple refraction.

SANS work utilizing multiple scattering (MS)

There have been a number of approaches to the MS problem. The literature is extensive and a non-exhaustive list of references is given. Weiss (1951) provided a general discussion and analysis of SANS data for several metallic powders and carbon black with a particle size of about 0.1 mm; Schelten and Schmatz (1980) derived analytical expressions (in kinematic approximation) for MS effects and calculated the single scattering cross section from MS patterns; Sabine and Bertram (1999) simplified the numerical scheme of Schelten and Schmatz and applied the theory to USANS data for a hydrated cement paste; Berk and others published a series of papers on MS theory and analysed SANS data for ceramics (Berk and Hardman-Rhyne 1986, 1988; Allen and Berk 1994); Mazumder and Sequeira (1992) worked on the theory of MS in statistical media; Krueger et al. (1991) applied MS to study the evolution of pore size distribution in alumina upon sintering; Maleyev (1995) considered theoretically MS from fractals; Bertram (1996) used computer simulations to investigate correlation effects in a system of closely packed polydispersed spheres; Šaroun (2000) developed a numerical scheme for MS USANS results and tested it on simulated data for polydispersed spheres; Bertram (2004) developed a method for extracting a single scattering differential cross section from MS USANS patterns based on matching Fourier transforms, which was subsequently applied to investigate the structure of oil-bearing and synthetic rock (Connolly et al. 2006).

NEUTRONS OR X-RAYS?

SANS and SAXS provide complementary information about the microstructure of rocks. Selection of a particular experimental technique should be based on careful analysis of the system to be studied, as the choice of neutrons or X-rays (or both) depends both on the type of rock and the microstructural problem at hand. Important issues to consider are the contrast, scale range of the inhomogeneities of interest, the sample size and the size of sample region to be investigated.

Neutrons have the ability to penetrate thick samples, offer a very wide range of scattering length density for contrast matching (or enhancement) and SANS data can be interpreted using a two-phase approximation for rocks with mixed organic-inorganic solid matrices. SAXS instruments offer high spectral resolution (small $\Delta\lambda/\lambda$) and can be used with small samples. SAXS data are unaffected by the incoherent scattering background. Synchrotron based SAXS instruments have a high beam intensity, which is very useful for low contrast samples. Some synchrotron-based SAXS instruments offer tight focus down to a 20 μm beam diameter.

CONCLUDING REMARKS

Collectively, modern SANS and USANS instruments provide access to microstructural features in rocks between 1 nm and 20 μm in size. Theoretical methods for reduction and interpretation of experimental data are well developed and numerous examples of applications are available in the literature. In recent years geological applications of SANS and USANS have matured to the point that they have became tools used to address geological questions rather than just a means to study individual rock samples.

Progress over the last two decades owes much to the theoretical advances made in relation to rock fractality, to the availability of absolutely calibrated SANS instruments and the development of a practical USANS instrument. With the proliferation of reactor based SANS and USANS instruments and the construction of new generation spallation neutron sources such as SNS, technological developments will play an important role in the future. For instance, the construction of a USANS instrument capable of detecting features up to 0.3 mm across is currently planned. Further developments in the use of grazing angle diffraction and reflectometry for studies of adsorbed molecules and the near-surface regions of rocks are possible.

The next frontier appears to be the ability to perform neutron scattering experiments at subsurface temperatures and pressures. High temperatures and pressures of the order of 100 GPa, characterize the Earth's lower mantle. These conditions can be replicated with laser-heated diamond anvil devices which hold very small volumes of rock. Such cells are currently being constructed for neutron scattering experiments at the new SNS facility (Parise 2006, this volume). Temperatures of up to 1300 °C and pressures up to 1 GPa (including a controlled axial component), simulating conditions down to a depth of approximately 50 km, are routinely used on large samples in rock physics experiments. A new field of study (*in situ* microstructural research in magmatic rocks) would present itself if these devices could be adapted to neutron diffraction and SANS experiments. A SANS cell capable of a maximum temperature of 400 °C and a pressure of 100 MPa would be sufficient to simulate conditions necessary to generate oil and gas from organic matter embedded in hydrocarbon source rocks and to study *in situ* the dynamics of hydrocarbon generation.

ACKNOWLEDGMENTS

Thanks are given to Steve Cadman and Alan Hinde for assistance in preparing this text and to J.S. Lin, D.R. Cole and H.R. Wenk for reviewing the manuscript. A.P. Radlinski publishes this chapter with the permission of the CEO, Geoscience Australia.

REFERENCES

Adler PM, Jacquin CG, Quiblier JA (1990) Flow in simulated porous media. Int J Multiphase Flow 16:691-712
Agamalian M, Wignall GD, Triolo R (1997) Optimization of a Bonse-Hart ultra-small-angle neutron scattering facility by elimination of the rocking-curve wings. J Appl Crystallogr 30:345-352
Allen AJ, Berk NF (1994) Analysis of small angle scattering data dominated by multiple scattering for systems containing eccentrically shaped particles or pores. J Appl Crystallogr 27:878-891
Allen AJ (1991) Time-resolved phenomena in cements, clays and porous rocks. J Appl Crystallogr 24:624-634
Auvray L, Auroy P (1991) Scattering by interfaces: variations on Porod's law. *In*: Neutron, X-ray and light scattering. Lindner P, Zemb Th (eds) Elsevier, p 199-221
Bale HD, Schmidt PW (1984) Small-angle X-ray scattering investigation of submicroscopic porosity with fractal properties. Phys Rev Lett 53:596-599
Berk NF, Hardman-Rhyne KA (1986) The phase shift and multiple scattering in small angle neutron scattering: application to beam broadening from ceramics. Physica B 136:218-222
Berk NF, Hardman-Rhyne KA (1988) Analysis of SAS data dominated by multiple scattering. J Appl Crystallogr 21:645-651
Berkowitz N (1979) An Introduction to Coal Technology. Academic Press
Bertram WK (1996) Correlation effects in small-angle neutron scattering from closely packed spheres. J Appl Crystallogr 29:682-685. See also Bertram WK (1988) Response to Pedersen's comment on 'Correlation effects in small-angle neutron scattering from closely packed spheres'. J Appl Crystallogr 31:489
Bertram WK (2004) Multiple scattering of USANS data using the method of matching Fourier transform. Presented at the Annual National Meeting of the American Crystallographic Association, Chicago 2004.

Bonse U, Hart M (1965) Small-angle X-ray scattering by spherical particles of polystyrene and polyvinyltoluene. Appl Phys Lett 7:238-240

Boreham CJ, Blevin JE, Radlinski AP, Trigg KR (2003) Coals as a source of oil and gas: a case study from the Bass Basin, Australia. Aust Pet Production Explor Assoc J 117-147

Broseta D, Barré L, Vizika O (2001) Capillary condensation in fractal porous medium. Phys Rev Lett 86: 5313-5316

Cohen MH (1987) The morphology of porous sedimentary rocks. *In*: Physics and chemistry of porous media II", AIP Conference Proceedings 154. Banavar JR, Koplik J, Winkler KW (eds) Am Inst Physics, p 1-16

Cole DR, Herwig KW, Mamontov E, Larese J (2006) Neutron scattering and diffraction studies of fluids and fluid-solid interactions. Rev Mineral Geochem 63:313-362

Connolly J, Bertram W, Barker J, Buckley C, Edwards T, Knott R (2006) Comparison of the structure on the nanoscale of natural oil-bearing and synthetic rock. J Petr Sci Eng 53(3-4):171-178

Debye P, Bueche AM (1949) Scattering by an inhomogeneous solid. J Appl Phys 20:518-525

Debye P, Anderson HR Jr, Brumberger H (1957) Scattering by an inhomogeneous solid. II. The correlation function and its application. J Appl Phys 28:679-683

Dianoux AJ, Lander G (eds) (2002) Neutron Data Booklet. Institut Laue-Langevin, Neutrons for Science

Espinat D (1990) Application des techniques de diffusion de la lumière, des rayons X et des neutrons à l'étude des systèmes colloïdaux. Rev l'Institut Français du Pétrole 45(6):1-131

Feigin LA, Svergun DI (1987) Structure Analysis by Small-angle X-ray and Neutron Scattering. Plenum Press

Floriano MA, Venezia AM, Deganello G, Svensson EC, Root JH (1994) The structure of pumice by neutron diffraction. J Appl Crystallogr 27:271-277

Freltoft T, Kjems JK, Sinha SK (1996) Power-law correlations and finite-size effects in silica particle aggregates studied by small-angle neutron scattering. Phys Rev B 33:269-275

Glatter O, Kratky O (eds) (1982) Small Angle X-ray Scattering. Academic Press

Glinka CJ, Rowe JM, LaRock JG (1986) The small-angle neutron scattering spectrometer at the National Bureau of Standards. J Appl Crystallogr 19:427-439

Graetsch H, Ibel K (1997) Small angle neutron scattering of opals. Phys Chem Minerals 24:102-108

Guinier A, Fournet G, Walker CB, Yudowitch KL (1955) Small-angle Scattering of X-rays. John Wiley and Sons

Hainbuchner M, Villa M, Kroupa G, Bruckner G, Baron M, Amenitsch H, Seidl E, Rauch H (2000) The new high resolution ultra-small-angle neutron scattering instrument at the High Flux Reactor in Grenoble. J Appl Crystallogr 33:851-854

Hall PL, Mildner DFR, Borst RL (1983) Pore size distributions of shaly rock by small angle neutron scattering. Appl Phys Lett 43:252-254

Hansen JP, Skjeltrop AP (1988) Fractal pore space and rock permeability implications. Phys Rev B 38:2635-2638

Hinde AL (2004) PRINSAS – a Windows-based computer program for the processing and interpretation of small-angle scattering data tailored to the analysis of sedimentary rocks. J Appl Crystallogr 37:1020-1024

Ibel K, Wright A (1980) An opal standard for very low momentum transfers in neutron small angle scattering. ILL Internal Scientific Report 80IB45S

Ioannidis MA, Kwiecien MJ, Chatzis I (1996) Statistical analysis of the porous microstructure as a method of estimating reservoir permeability. J Pet Sci Eng 16:251-261

Itakura T, Bertram WK, Knott RB (2005) The nanoscale structural response of a natural kaolinitic clayey soil subjected to uniaxial compression. Appl Clay Sci 29:1-14

Jacquin CG, Adler PM (1987) Fractal porous media. II: Geometry of porous geological structures. Transp Porous Media 2:571-596

Johnston PR, McMahon P, Reich MH, Snook IK, Wagenfeld HK (1993) The effect of processing on the fractal pore structure of Victorian brown coal. J Coll Interf Sci 155:146-151

Kahle A, Winkler B, Radulescu A, Schreuer J (2004) Small-angle neutron scattering study of volcanic rocks. Eur J Mineral 16:407-417

Katz AJ, Thompson AH (1985) Fractal sandstone pores: implications for conductivity and pore formation. Phys Rev Lett 54:1325-1328

Knudsen KD, Fossum JO, Helgesen G, Bergaplass V (2003) Pore characteristics and water absorption in a synthetic smectite clay. J Appl Crystallogr 36:587-591

Koehler WC (1986) The national facility for small-angle neutron scattering. Physica B 137:320-329

Krishnamurti P (1930) Studies in X-ray diffraction. Part I: The structure of amorphous scattering. Part II: Colloidal solutions and liquid mixtures. Indian J Phys 5:473-500

Krohn CE (1988) Sandstone fractal and Euclidean pore volume distributions. J Geophys Res 93(B4):3286-3296

Krueger S, Long GG, Page RA (1991) Characterization of the densification of alumina by multiple small-angle neutron scattering. Acta Crystallogr A47:282-290

Lambard J, Zemb Th (1991) A triple-axis Bonse-Hart camera used for high-resolution small-angle scattering. J Appl Crystallogr 24:555-561

Letcher JR, Schmidt PW (1966) Small-angle x-ray scattering determination of particle-diameter distributions in poly-disperse suspensions of spherical particles. J Appl Phys 37:649-655

Li JC, Ross DK, Howe LD, Stefanopoulos KL, Fairclough JPA, Heenan R, Ibel K (1994a) Small-angle neutron scattering studies of the fractal-like network formed during desorption and adsorption of water in porous materials. Phys Rev B 49:5911-5917

Lin JS, Hendricks RW, Harris LA, Yust CS (1978) Microporosity and micromineralogy of vitrinite in a bituminous coal. J Appl Crystallogr 11:621-625

Lin MY, Sinha SK, Drake JM, Wu X-I, Thiyagarajan P, Stanley HB (1994b) Study of phase separation of binary fluid mixture in confined geometry. Phys Rev Lett 72:2207-2210

Lindner P, Zemb Th (eds) (1991) Neutron, X-ray and Light Scattering. Elsevier

Lindner P, May RP, Timmins PA (1992) Upgrading of the SANS instrument D11 at the ILL. Physica B 180&181:967-972

Lucido G, Caponetti E, Triolo R, (1985) Preliminary small-angle neutron scattering experiments on magmatic rocks to detect critical phenomena. Mineral Petrogr Acta XXIX:133-138

Lucido G, Triolo R, Caponetti E (1988) Fractal approach in petrology: Small-angle neutron scattering experiments with volcanic rocks. Phys Rev B 39:9742-9745

Lucido G, Caponetti E, Triolo R, (1991) Fractality as a working tool for petrology: small-angle neutron scattering experiments to detect critical behaviour of magma. Geologica Carpathica 42:85-91

Maleyev SV (1995) Small-angle neutron scattering in fractal media. Phys Rev B 52:13163-13168

Mandelbrot BB (1977) Fractals: Form, Chance and Dimension. Freeman

Mandelbrot BB (1982) The Fractal Geometry of Nature. Freeman

Martin JE, Hurd AJ (1987) Scattering from fractals. J Appl Crystallogr 20:61-78

Mazumder S, Sequeira A (1992) Multiple small-angle scattering from a statistical medium. J Appl Crystallogr 25:221-230

Mc Mahon PJ, Treimer W (1998) The geometric origin of asymmetric small angle neutron scattering from an elastically bent double crystal diffractometer. J Crystallogr Res Technol 33:625-636

McMahon PJ, Snook IK, Treimer W (2002) The pore structure in processed Victorian brown coal. J Coll Interf Sci 251:177-183

McMahon PJ, Snook I, Smith E (2001) An alternative derivation of the equation for small angle scattering from pores with fuzzy interfaces. J Chem Phys 114:8223-8225

Melnichenko YB, Wignall GD, Cole DR, Frielinghaus H, Bulavin LA (2005) Liquid-gas critical phenomena under confinement: small angle neutron scattering studies of CO_2 in aerogel. J Mol Liquids 120:7-9

Mikula P, Lukas P, Eichhorn F (1988) A new version of a medium-resolution double-crystal diffractometer for the study of small-angle neutron scattering (SANS). J Appl Crystallogr 21:33-37

Mildner DFR, Hall PL, (1986) Small-angle scattering from porous solids with fractal geometry. J Phys D Appl Phys 19:1535-1545

Morvan M, Espinat D, Lambard J, Zemb T (1994) Ultrasmall- and small-angle X-ray scattering of smectite clay suspensions. Colloids Surf A 82:193-203

von Nardroff R (1926) Refraction of X-rays by small particles. Phys Rev 28:240-246

Parise JB (2006) High pressure studies. Rev Mineral Geochem 63:205-231

Pfeifer P, Avnir D (1983) Chemistry of noninteger dimensions between two and three. I. Fractal theory of heterogeneous surface. J Chem Phys 79:3558-3565

Porod G (1951) Die Röntgenkleinwinkelstreuung von dichtgepackten kolloiden Systemen. I Teil. Kolloid Z 124:83-114

Porod G (1952) Die Röntgenkleinwinkelstreuung von dichtgepackten kolloiden Systemen. II Teil. Kolloid Z 125:51-57, 108-122

Porod G (1982) General Theory. *In*: Small angle X-ray scattering. Glatter O, Kratky O (eds) Academic Press, p 17-51

Prinz D, Pyckhout-Hintzen W, Littke R (2004) Development of meso- and macroporous structure of coals with rank as analysed with small angle neutron scattering and adsorption experiments. Fuel 83:547-556

Radlinski AP, Boreham CJ, Wignall GD, Lin JS (1996a) Microstructural evolution of source rocks during hydrocarbon generation: A small-angle scattering study. Phys Rev B 53:14152-14160

Radlinski AP, Barré L, Espinat D (1996b) Aggregation of n-alkanes in organic solvents. J Mol Structure 383: 51-56

Radlinski AP, Radlinska EZ, Agamalian M, Wignall GD, Lindner P, and Randl OG (1999) Fractal geometry of rocks. Phys Rev Lett 82:3078-3081

Radlinski AP, Radlinska EZ (1999) The microstructure of pore space in coals of different rank: a small angle scattering and SEM study. *In*: Coalbed Methane: Scientific, Environmental and Economic Evaluation. Mastalerz M, Glikson M, Golding SD (eds) Kluvier Scientific Publishers, p 329-365

Radlinski AP, Boreham CJ, Lindner P, Randl OG, Wignall GD, Hope JM (2000a) Small angle neutron scattering signature of oil generation in artificially and naturally matured hydrocarbon source rocks. Org Geochem 31:1-14

Radlinski AP, Radlinska EZ, Agamalian M, Wignall GD, Lindner P, and Randl OG (2000b) The fractal microstructure of ancient sedimentary rocks. J Appl Crystallogr 33:860-862

Radlinski AP, Hinde AL (2002) Small angle neutron scattering and petroleum geology. Neutron News 13: 10-14

Radlinski AP, Claoue-Long J, Hinde AL, Radlinska EZ, Lin JS (2003) Small angle X-ray scattering measurement of the internal micro-structure of natural zircon crystals. Phys Chem Minerals 30:631-640

Radlinski AP, Ioannidis MA, Hinde AL, Hainbuchner M, Baron M, Rauch H, Kline SR (2004a) Angstrom-to-millimeter characterization of sedimentary rock microstructure. J Colloid Interf Sci 274:607-612

Radlinski AP, Kennard JM, Edwards DS, Hinde AL, Davenport R (2004b) Hydrocarbon generation and expulsion from Early Cretaceous source rocks in the Browse Basin, North West Shelf, Australia: a SANS study. Aust Pet Production Explor Assoc J 2004:151-180

Radlinski AP, Mastalerz M, Hinde AL, Hainbuchner M, Rauch H, Baron M, Lin JS, Fan L, Thiyagarajan P (2004c) Application of SAXS and SANS in evaluation of porosity, pore size distribution and surface area of coal. Int J Coal Geology 59:245-271

Reich MH, Russo SP, Snook IK, Wagenfield HK (1990) The application of SAXS to determine the fractal properties of porous carbon-based materials. J Colloid Interface Sci 135:353-362

Russell TP, Lin JS, Spooner S, Wignall GD (1988) Intercalibration of small-angle X-ray and neutron scattering data. J Appl Crystallogr 21:629-638

Sabine TM, Bertram WK (1999) The use of multiple-scattering data to enhance small-angle neutron scattering experiments. Acta Crystallogr A55:500-507

Šaroun J (2000) Evaluation of double-crystal SANS data influenced by multiple scattering. J Appl Crystallogr 33:824-828

Schaefer DW, Keefer KD (1984) Fractal geometry of silica condensation polymers. Phys Rev Lett 53:1383-1386

Schaefer DW, Wilcoxon JP, Keefer KD, Bunker BC, Pearson RK, Thomas IM, Miller DE (1987) Origin of porosity in synthetic materials. *In*: Physics and chemistry of porous media II", AIP Conference Proceedings 154, Banavar JR, Koplik J, Winkler KW (eds) Am Inst Physics, p 63-80

Schelten J, Schmatz W (1980) Multiple-scattering treatment for small-angle scattering problems. J Appl Crystallogr 13:385-390

Schmidt PW (1982) Interpretation of small-angle scattering curves proportional to a negative power of the scattering vector. J Appl Crystallogr 15:567-569

Schmidt PW (1989) Use of scattering to determine the fractal dimension. *In*: The fractal approach to heterogeneous chemistry. Avnir D (ed) John Wiley and Sons, p 67-79

Schmidt PW, Avnir D, Levy D, Hohr A, Steiner M, Roll A (1991) Small angle x-ray scattering from the surfaces of reversed phase silicas: power-law scattering exponents of magnitudes greater than four. J Chem Phys 94:1474-1479

Sears VF (1990) Coherent neutron scattering amplitudes. *In*: Neutron Scattering at the High Flux Isotope Reactor. Mook HA, Nicklow RM (eds) Oak Ridge National Laboratory, p 29-30

Snook I, Yarovsky I, Hanley HJM, Lin MY, Mainwaring D, Rogers H, Zulli P (2002) Characterization of metallurgical chars by small angle neutron scattering. Energy Fuels 16:1009-1015

Spalla O, Lyonnard S, Testard F (2003) Analysis of the small-angle intensity scattered by a porous and granular medium. J Appl Crystallogr 36:338-347

Teixeira J (1988) Small-angle scattering by fractal systems. J Appl Crystallogr 21:781-785

Thiyagarajan P, Epperson JE, Crawford RK, Carpenter JM, Klippert TE, Wozniak DG (1997) The time-of-flight small angle neutron diffractometer (SAD) at IPNS, Argonne National Laboratory. J Appl Crystallogr 30:280-293

Thompson AH (1991) Fractals in rock physics. Annu Rev Earth Planet Sci 19:237-262

Vacher R, Woignier T, Peloue J, Courtens E (1988) Structure and self-similarity of silica aerogels. Phys Rev B 37:6500-6503

Vogel SC, Priesmeyer H-G (2006) Neutron production, neutron facilities and neutron instrumentation. Rev Mineral Geochem 63:27-57

Warren BE (1934) X-ray diffraction study of carbon black. J Chem Phys 2:551-555

Weiss RJ (1951) Small angle scattering of neutrons. Phys Rev B 83:379-389

Wignall GD, Bates FS (1987) Absolute calibration of small-angle neutron scattering data. J Appl Crystallogr 20:28-40

Winans RE, Thiyagarajan P (1988) Characterization of solvent-swollen coal by SANS. Energy Fuels 2:356-358

Winkler B (2006) Applications of neutron radiography and neutron tomography. Rev Mineral Geochem 63: 459-471

Winkler B, Kahle A, Hennion B (2006) Neutron radiography of rocks and melts. International Conference on Neutron Scattering, Sydney, November 2005. Physica B (in press)

Wong Pz, Howard J, Lin JS (1986) Surface roughening and the fractal nature of rocks. Phys Rev Lett 57: 637-640

Wong Pz, Bray AJ (1988) Porod scattering from fractal surfaces. Phys Rev Lett 60:1344

Wong Pz (2006) Studies of the fractal nature of sedimentary rocks by small angle scattering and other techniques. International Conference on Neutron Scattering, Sydney, Australia, November 2005. Physica B (in press)

Reviews in Mineralogy & Geochemistry
Vol. 63, pp. 399-426, 2006
Copyright © Mineralogical Society of America

15

Neutron Diffraction Texture Analysis

Hans-Rudolf Wenk

Department of Earth and Planetary Science
University of California
Berkeley, California, 94720, U.S.A.
e-mail: wenk@berkeley.edu

INTRODUCTION

An intrinsic property of polycrystalline materials is the orientation distribution of crystallites. In some cases this distribution is random, yet often there is preferred orientation of crystallites relative to macroscopic axes that may have been attained during a deformation process. Many rocks—metamorphic, igneous as well as sedimentary—display non-random orientation distributions that are the cause for anisotropy of macroscopic physical properties. Interpretation of textures in materials has to rely on a quantitative description of orientation characteristics. Two types of preferred orientations need to be distinguished: The *shape preferred orientation* (or often abbreviated SPO) describes the orientation of grains with anisotropic shape. The *lattice preferred orientation* (LPO) or "texture" refers to the orientation of the crystal lattice. (LPO is an unfortunate term since the lattice does not always uniquely describe the crystal orientation, as in the trigonal mineral quartz with a hexagonal unit cell. "Crystallographic preferred orientation," CPO, would be more appropriate). Shape and crystal orientation can be correlated, such as in sheet silicates with a flaky morphology in schists, or fibers in fiber reinforced ceramics. In other cases they are not. In a rolled cubic metal or a plastically deformed quartzite, the grain shape depends on the deformation and is not directly related to the crystallography.

Many methods have been used to determine preferred orientation. Geologists have applied extensively the petrographic microscope equipped with a Universal stage to measure the orientation of morphological and optical directions in individual grains (e.g., Wahlstrom 1979; Wenk 1985). More recently electron diffraction, both with transmission (TEM) and scanning electron microscopes (SEM) have been used to measure orientation of crystals (e.g., Schwarzer and Weiland 1988; Randle and Engel 2000). In this case, the location of a grain can be determined, which permits to correlate microstructures, neighbor relations and texture. Also from individual orientation measurements the orientation distribution can be determined unambiguously. Grain statistics are generally limited to at best a few thousand grains and generally much less, even if many spots are recorded on a polycrystalline sample.

Other diffraction techniques rely on averages of diffraction signals over a polycrystalline sample. In this case spatial information is lost and in addition some ambiguity is introduced about the orientation distribution because normal diffraction effects are always centrosymmetric (Friedel's law) but an orientation relation is not (Matthies 1979). With averaging diffraction techniques statistics are highly improved and crystal orientations are weighted according to grain size.

In this context neutron diffraction plays an important role: For most materials, absorption is negligible compared to X-rays (Fig. 1). Large samples, 1-10 cm in diameter, of roughly spherical shape can be measured. Because the diffraction signal averages over large volumes rather than surfaces grain statistics are even better than with conventional X-rays. Figure 2 compares (0001) pole figures for quartz, one measured with the Universal stage and the other

1529-6466/06/0063-0015$05.00
DOI: 10.2138/rmg.2006.63.15

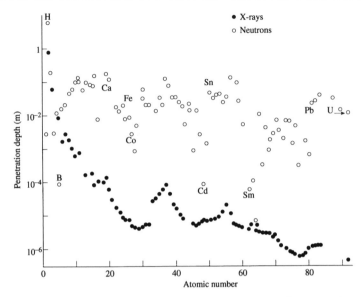

Figure 1. Absorption of neutrons and X-rays. The penetration depth corresponds to the thickness when the intensity has been reduced to 40%. Wavelength is 0.14 nm.

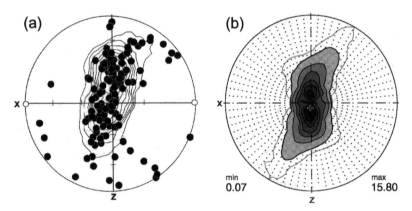

Figure 2. (0001) Pole figures of quartzite measured (a) on one hundred grains with a Universal stage petrographic microscope and (b) a neutron diffractometer with monochromatic radiation, averaging over approximately one million grains (Ghildiyal et al. 1999).

averaged over a million grains by neutron diffraction (Ghildiyal et al. 1999). The limited statistics of Universal stage measurements is obvious. Furthermore for quartz only *c*-axes can be measured with the Universal stage, whereas other crystallographic directions may be just as important to assess deformation characteristics. Figure 3 illustrates similar statistical limitations for calcite marble where pole figures were measured with an X-ray pole figure goniometer in reflection geometry on the surface of a slab (Fig. 3a) and by neutron diffraction on a sample cube (Fig. 3b) (Wenk et al. 1984). The X-ray pole figure shows an irregular pattern, whereas the neutron pole figure displays a symmetrical distribution, representative of the bulk orientation features of the sample. The low absorption has other advantages: Intensity corrections are generally unnecessary and environmental stages (heating, cooling,

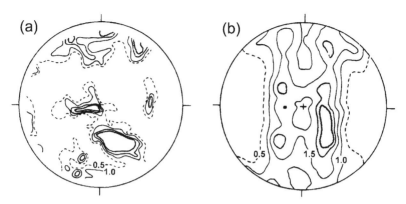

Figure 3. Comparison of (0006) pole figures of calcite for experimentally deformed marble. (a) Measured by X-ray diffraction in reflection geometry. (b) Measured by monochromatic neutron radiation in Jülich (Wenk et al. 1984). Equal area projection.

straining) can be used for *in situ* observation of texture changes. Neutron diffraction was first applied to textures in 1953 by Brockhouse to investigate magnetic structures, though rather unsuccessfully. Yet, during the last twenty years, neutron texture analysis has become firmly established in earth and materials science and has emerged as a favorite technique for many applications. This Chapter will describe the various experimental techniques, survey methods of quantitative data analysis and illustrate neutron texture analysis with examples.

Concepts of texture analysis, including texture representation, experimental techniques and interpretation, have been reviewed in books (e.g., Wenk 1985; Kocks et al. 2000). For earlier reviews of neutron diffraction texture analysis see also Brokmeier (1997, 1999), Feldmann (1989), Schaefer (2002) and Wenk (1994). For texture representations this Chapter will only use pole figures that represent the directional distribution of lattice plane normals {*hkl*} relative to sample coordinates. All pole figures shown use equal area projection of the orientation sphere and display pole densities in multiples of a random distribution (m.r.d.). The integral over a pole figure is 1 m.r.d.

EXPERIMENTAL TECHNIQUES

Neutron diffraction texture analysis relies on Bragg's law that stipulates that neutron waves *reflect* on lattice planes if the condition $2d_{hkl} \sin\theta = \lambda$ is satisfied. In a polycrystalline sample a detector at a particular orientation relative to the incident neutron beam only records signals from lattice planes that satisfy the reflection condition. In a textured sample the overall signal intensity changes if the sample is rotated relative to the detector and, if several detectors are available, each detector records different intensities and differently oriented crystals. From these intensity variations for different lattice planes *hkl* the orientation distribution can be obtained.

For X-ray techniques, whether in reflection or transmission geometry, the incident beam must not leave the specimen during rotations for a straightforward interpretation of intensity variations and proper volume/absorption/defocusing corrections need to be applied during data analysis. By contrast, for neutrons it is advantageous if the specimen does not leave the beam during rotations, so that the same volume is investigated at all times.

The elastic scattering of thermal neutrons by a crystal consists of two components, nuclear and magnetic scattering (see Parise 2006, this volume). *Nuclear scattering* is due to interactions between the neutron and the atomic nuclei and yields diffraction effects with equivalent information as X-ray scattering on electrons, but magnitudes of the scattering lengths are different and therefore diffraction peaks have different relative intensities. With neutrons signals from light elements are of similar magnitude as those from heavy ones. Also different isotopes can be distinguished. Scattering amplitudes of X-rays decrease with *d*-spacing, whereas those of neutrons do not. This improves the capability to measure low *d*-spacing reflections but their intensity is still lower because of thermal vibration and Lorentz polarization effects. *Magnetic scattering*, due to a dipole interaction between the magnetic moments of nucleus and shell electrons, is weaker (Harrison 2006, this volume). In materials with magnetic elements (e.g., Mn, Fe) peaks may occur in the diffraction pattern that are solely due to magnetic scattering and with those one can measure magnetic pole figures. They do display preferred orientation of magnetic dipoles in component crystals. If no magnetic superstructures are present, the magnetic contribution is, with presently available instrumentation, very difficult to separate from the nuclear scattering. Some examples will be shown later in this chapter.

Neutron diffraction texture studies are done either at reactors with a constant flux of thermal neutrons, or with pulsed neutrons at spallation sources. The wavelength distribution of moderated thermal neutrons is a broad spectrum with a peak at 1–2 Å (Vogel and Priesmeyer 2006, this volume). The low absorption and high penetration of neutrons relative to X-rays was mentioned. This is an expression of the weak interaction of neutrons with matter, which has the disadvantage that scattering is weak and long counting times are required.

A conventional neutron texture experiment at a reactor source uses monochromatic radiation produced with single crystal monochromators. A goniometer rotates the sample to explore the entire orientation range, analogous to an X-ray pole figure goniometer. Such texture measurements are routinely conducted at Chalk River (Canada), Geesthacht (Germany), LLB (France) and NIST (USA). To improve counting efficiency *position-sensitive detectors* have been applied that record a 2θ spectrum with many peaks simultaneously. Examples of such facilities are at ILL (instruments D1B and D20) and Jülich (Julios). With the advent of pulsed neutron sources it has become popular to use *polychromatic* neutrons and detectors that can identify the energy of neutrons by measuring the time of flight (TOF). Dedicated TOF diffractometers for texture research are at JINR, Dubna (SKAT), IPNS (GPPD) and LANSCE (HIPPO).

Monochromatic neutrons

A conventional texture experiment uses monochromatic radiation. With a Cu (111) or graphite (0002) monochromator wavelengths λ = 1.289 Å and λ = 2.522 Å are often selected. The detector is aligned relative to the incident beam at the angle 2θ for a selected lattice plane *hkl* to satisfy Bragg's law. The intensities are measured in different sample directions by rotating the sample around two axes with a goniometer (e.g., φ and χ), to cover the entire orientation range, often in 5° × 5° increments (Fig. 4). Intensities are directly proportional to pole densities. This method is analogous to that for an X-ray pole figure goniometer but with the advantage that defocusing corrections are not necessary. Figure 5 shows a calcite (0006) pole figure of experimentally deformed limestone that was measured at Geesthacht with this method.

It is also possible to use *position-sensitive detectors* which record intensities along a ring (1 dimension, 1D) rather than at a point. The ring can be mounted on a dif-

Figure 4. Eulerian cradle with rotation axes φ and χ used in a neutron pole figure goniometer (Hoefler et al. 1988).

fractometer so that it either records for a single reflection a whole range of lattice orientations at once (usually covering the diffractometer coordinates χ, Juul Jensen and Leffers 1989) or

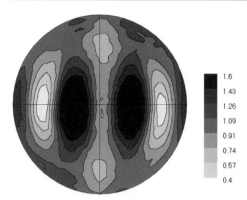

Figure 5. Calcite 0006 pole figure of experimentally deformed limestone, measured with monochromatic neutrons at GKSS, Geesthacht. Equal area projection, linear contours, pole densities in m.r.d. (Compare with Fig. 12).

so that it records a continuous 2θ range (Bunge et al. 1982). The latter is particularly interesting because it permits one to record many pole figures simultaneously and opens the possibility to deconvolute spectra that is valuable in the case of overlapping peaks. The geometry of such a system used at ILL is illustrated in Figure 6. Note that different positions on the detector record differently oriented lattice planes (Bunge et al. 1982). Figure 7 is a graphic representation of 72 spectra measured with a position sensitive detector at different sample orientations for the same experimentally deformed calcite limestone illustrated in Figure 5 (Wenk 1991). The relative changes in peak intensities are due to texture.

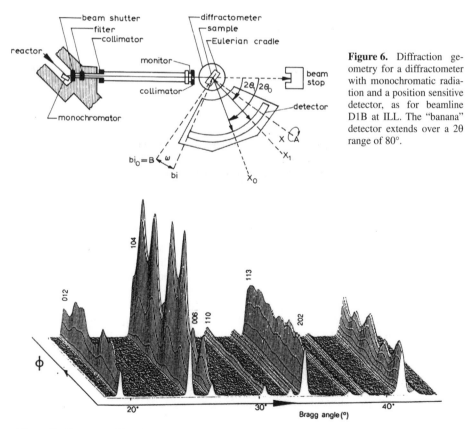

Figure 6. Diffraction geometry for a diffractometer with monochromatic radiation and a position sensitive detector, as for beamline D1B at ILL. The "banana" detector extends over a 2θ range of 80°.

Figure 7. 72 neutron diffraction spectra measured on an experimentally deformed calcite polycrystal (limestone) with a position sensitive detector at ILL with monochromatic neutrons; changes in intensity as a function of φ are due to texture (Wenk 1991).

Polychromatic time-of-flight neutrons

Another method to measure a spectrum simultaneously is at a fixed detector position but with *polychromatic* neutrons and a detector system that can identify the energy of neutrons, e.g., by measuring the time of flight (TOF). Figure 8 is a representation of Bragg's law for monochromatic and polychromatic neutrons for aluminum. For monochromatic neutrons with a wavelength $\lambda = 1.5$ Å a θ-scan (A) can be performed over the θ-range of interest by moving the detector, or the θ-span can be measured simultaneously with a 1D-position sensitive detector. There are peaks at each intersection of line A with the Bragg curves for lattice planes *hkl*. For polychromatic neutrons a whole range of wavelengths between λ_{min} and λ_{max} is available (line B in Fig. 8). A detector at a fixed scattering angle θ (e.g., 45°) records a whole *d*-spectrum for each sample orientation.

Since neutron scattering is weak, it is efficient to make better use of resources by building instruments with multiple

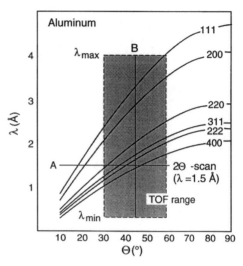

Figure 8. Bragg's law for low order diffractions of aluminum. A illustrates a 2θ scan with a conventional goniometer and monochromatic radiation. B gives the coverage of a continuous spectrum at a fixed detector position. With a position sensitive detector a large 2θ range is measured simultaneously (A) and with an energy sensitive or TOF detector a large wavelength range is covered (B). Both can be combined as with a 2D position sensitive single crystal diffractometer (Fig. 14).

detectors as with SKAT at JINR (24 detectors, Ullemeyer et al. 1998), GPPD at IPNS (14 detectors, Xie et al. 2003) and HIPPO at LANSCE (30 detectors, Wenk et al. 2003). On GPPD detectors are arranged at different angles in the plane of incident and diffracted beam. At SKAT they are positioned in a ring (bank) at right angles to the incident beam (Fig. 9). For HIPPO detectors are arranged on five banks at 2θ angles 10°, 20°, 40°, 90° and 145° (Fig. 19 in Vogel and Priesmeyer 2006, this volume) and each one records reflections from differently oriented lattice planes *(hkl)*. The 10° and 20° banks are not used for texture experiments because of poor resolution. The pole figure coverage for 40°, 90° and 145° banks with 30 detectors is illustrated in Figure 10a.

Thus the advantage of pulsed polychromatic neutrons and a detector system that can measure the time of flight (TOF) of neutrons and discriminate their energies is, that whole spectra with many Bragg peaks can be recorded simultaneously. With TOF neutrons and a multi-detector system, fewer sample rotations are necessary to perform quantitative texture analysis. For typical texture investigations with HIPPO, rotation around a single axis is sufficient, which eliminates the need for a 2-circle goniometer and simplifies the construction of environmental cells to measure textures at non-ambient conditions. Rotating the sample around a single axis perpendicular to the incident beam to several positions (0°, 45°, 67.5°, and 90° have been established), provides $4 \times 30 = 120$ spectra for the subsequent analysis. If detectors are at different θ angles, their resolution is different, which is illustrated in Figure 11 for calcite limestone. High angle detectors (e.g., 145°) have excellent resolution but intensities are weak, particularly at large *d*-spacings. Low angle detectors (e.g., 40°) have poor resolution but good counting statistics. The relative intensity differences between *hkl*'s for different detectors of a bank are indicative of texture (compare with Fig. 7). Combining information from all detectors provides measuring times down to a few minutes per sample. Table 1 compares some features of TOF spectrometers that are used for texture analysis.

Figure 9. SKAT texture diffractometer at JINR-Dubna with 3 circle goniometer and large ring with detector panels. Person mounting a sample is for scale (courtesy K. Ullemeyer).

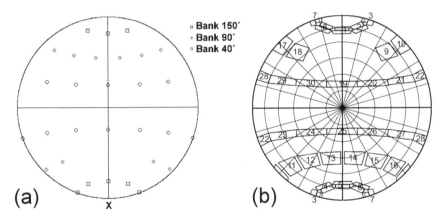

Figure 10. HIPPO pole figure coverage with 30 detector panels, distributed over 3 banks. (a) Point locations of detector centers. X is incident neutron beam. (b) Actual size and shape of detectors. Circles are at 15° intervals. Equal area projection.

The reliability of various neutron texture measurement techniques has been evaluated by circulating a textured polycrystalline calcite standard sample among 15 different neutron diffraction facilities (Wenk 1991; Lutterotti et al. 1997; Von Dreele 1997; Walther et al. 1995; Wenk et al. 2003). They include reactors with monochromatic radiation and point detectors, reactors with position sensitive detectors, pulsed reactors and spallation sources with TOF neutrons. In general textures measured on the same sample at different facilities agree very closely (some examples are shown in Fig. 12). For pole figures with strong diffraction intensities, standard deviations from the mean are 0.04-0.06 m.r.d. with a spread of maxima values of 0.2 m.r.d. The spread is considerably larger for pole figures with weak diffraction intensities (0.4 m.r.d.). For weak diffraction peaks position sensitive detectors and TOF techniques have an advantage over single tube detectors with monochromatic neutrons since integrated rather than peak intensities can be determined which yields better counting statistics.

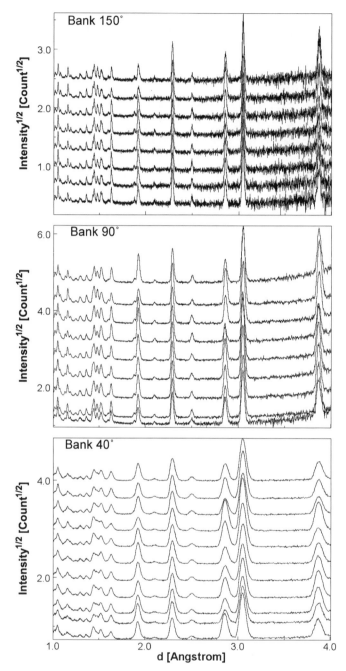

Figure 11. 30 diffraction spectra of calcite limestone, recorded simultaneously with the HIPPO diffractometer at LANSCE. Notice differences in counting statistics and resolution for the different detector banks. Relative peak intensity differences are due to texture (Wenk et al. 2002).

Table 1. Summary of features of neutron TOF spectrometers that are used for texture analysis.

Instrument	Flight path (m)	Flux at sample (neutrons s⁻¹ cm⁻²)	Number of detector panels
SKAT (JINR)	103	10^6	24
GPPD (IPNS)	20	3×10^6	14
HIPPO (LANSCE)	9	2.4×10^7	30

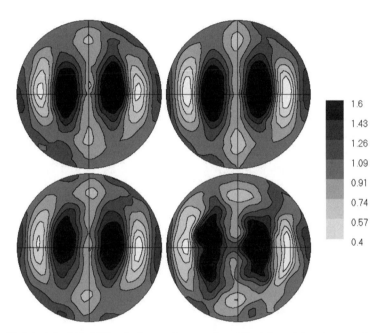

Figure 12. Selected calcite 0001 pole figures of an experimentally deformed limestone standard sample and used as a round robin to assess reliability of neutron diffraction texture measurements. Examples shown are pole figures recalculated from the OD based on measurements at four neutron diffraction facilities. (a) Conventional reactor with monochromatic neutrons (Julios at KFA, Jülich), (b) Reactor with monochromatic neutrons and position sensitive detector (D1B at ILL Grenoble) (Wenk 1991), (c) Pulsed reactor with TOF measurements, single peak extraction (SKAT at Dubna, Russia) (Walther et al. 1995), (d) Spallation neutrons with 30 detectors and OD determined with the Rietveld method (HIPPO at LANSCE) (Wenk et al. 2002). Pole figures are normalized so that densities are expressed in multiples of a uniform distribution (m.r.d.). Equal area projection, linear contours. Compare also with Figure 5.

Results from this round-robin experiment establish that neutron diffraction would clearly be the method of choice for texture measurements of bulk samples if it were more readily available. Quantitative texture information for the same sample obtained by different laboratories with neutron diffraction is much more reproducible than similar comparisons with conventional X-ray diffraction or electron microscopy. This is mainly because of the larger number of grains probed with neutrons, negligible surface preparation effects, and the absence of instrument dependent defocusing and absorption corrections.

Special techniques

This section will briefly discuss some neutron scattering techniques that were explored for special texture applications but have not yet become routine procedures.

2D position sensitive detectors for monochromatic radiation are available at some facilities (e.g., instrument D19 at ILL) and are mainly used for single crystal studies because they cover a significant portion of reciprocal space. This is quite attractive for texture analysis since the 2D diffraction pattern reveals portions of Debye rings for several Bragg peaks and intensity variations along rings can be analyzed for texture. An image is shown for illite clay in Figure 13a. Disadvantages are that the 2D recording is highly distorted in terms of diffraction angle and orientation space, rendering quantitative data extraction difficult. In this respect hard X-ray synchrotron diffraction images are far superior and obtained in a fraction of the time (Fig. 13b, Wenk et al. 2006a).

A similar technique has been investigated earlier by combining 2D position sensitive detectors and TOF (Wenk et al. 1991). With 2D position sensitive detectors as available at the single crystal diffractometers (SCD) of IPNS and LANSCE (Fig. 14), each detector location records a TOF spectrum and the 3D xyT data array can then be analyzed for texture. A time slice ($T = 6.93$ ms) displays Bragg lines of four diffraction peaks with intensity variations (Fig. 15a). Extracting xy intensities for $d = 2.845$ Å, corresponding to (0006) of calcite, provides an angular sector of $25° × 50°$ for a (0006) pole figure (Fig. 15b). While this technique is quite elegant in principle and has been tested on a few samples (aluminum and calcite), the data analysis is extremely complex due to distortions and non-linear corrections.

With strain neutron diffractometers, samples can be deformed and lattice strains can be recorded *in situ* at stress. Such facilities (e.g., ENGIN-X at ISIS, SMARTS at LANSCE, EP-SILON at Dubna) generally have two detectors, recording signals from lattice planes that are

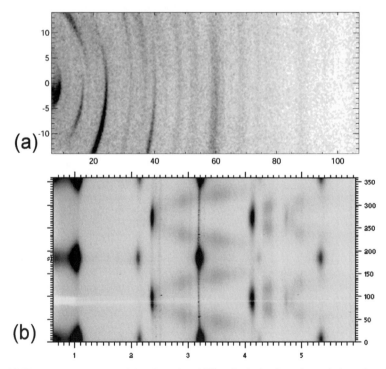

Figure 13. Texture measurements of the clay mineral illite, displaying intensity variations along Debye rings. (a) Position-sensitive neutron detector at D19, ILL; (b) "unrolled" synchrotron diffraction image, recorded with an image plate at HASY (Wenk et al. 2006a). Abscissa corresponds to 2θ angle.

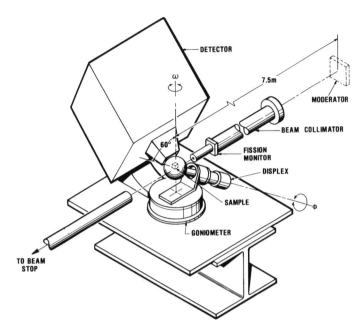

Figure 14. TOF 2D position sensitive detector for single crystal diffractometer (SCD) at LANSCE and IPNS.

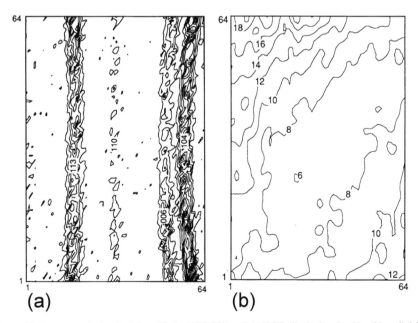

Figure 15. Texture analysis of calcite with the TOF-SCD at LANSCE, displaying the 64 × 64 *xy* division 2D detector. (a) Time slice of the *xyT* histogram with Bragg peaks appearing as high density lines. (b) Extraction of *xy* intensities for *d* = 2.845 A, corresponding to a segment of the calcite 006 pole figure (Wenk et al. 1991).

oriented perpendicular and parallel to the applied stress. This is generally not sufficient to obtain full texture information, but nevertheless some interesting results have been obtained for calcite (Schofield et al. 2003) and quartz that both undergo mechanical twinning. Polycrystalline quartz, compressed to 500 MPa at 500 °C shows systematic intensity changes of reflections that are sensitive to Dauphiné twinning (e.g., 10-12 and 20-11) and no changes with those that are not (e.g., 11-20) (Fig. 16). These intensity changes are due to texture and reveal that mechanical twinning in quartz initiates around 80 MPa at the conditions of the experiment and saturates around 500 MPa. If the texture of the starting material is known then texture changes during straining can be interpreted quantitatively based on structure factors and intensity changes of reflections.

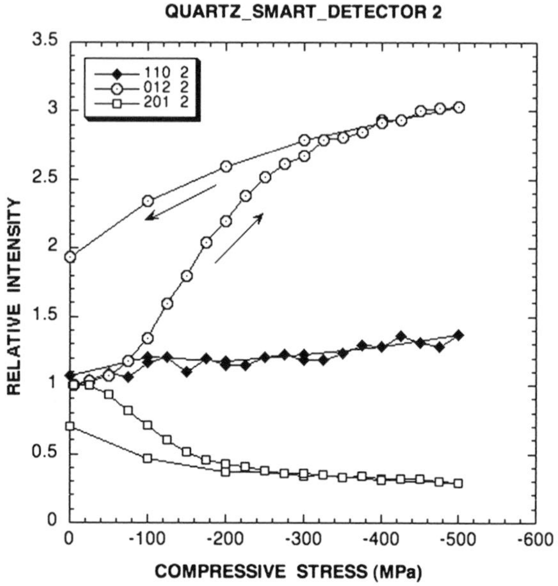

Figure 16. *In situ* observed intensity changes due to mechanical twinning, for several reflections as a sample of quartzite (novaculite) is compressed in the TOF SMARTS strain diffractometer at LANSCE. The detector records lattice planes that are perpendicular to the compression direction.

The neutron cross section of textured materials for certain energies is strongly direction dependent and intensities as function of energy can be used to estimate texture (Santisteban et al. 2006).

DATA ANALYSIS

The goal of texture measurements is to obtain a quantitative 3D orientation distribution (OD), often referred to as ODF (orientation distribution function) since it can be represented as a continuous mathematical function. The OD relates orientations of crystallites (with axes [100], [010] and [001]) to those of the sample (*x, y, z*) usually by means of three Euler angles ϕ_1, Φ, ϕ_2 (Fig. 17a). The 3D OD can not be directly measured with averaging diffraction techniques but methods are available to obtain the OD from measured pole figures (for a review see Kallend 2000). The OD, in terms of Euler angles ϕ_1, Φ, ϕ_2, can be viewed as a cylindrical distribution, with azimuth ϕ_1 and radial distance Φ corresponding to pole figure coordinates, and the axial distance ϕ_2 corresponding to crystal rotation (Wenk and Kocks 1987) (Fig. 17b, top). Pole figures are projections of the OD along complicated paths determined by crystal and sample geometry. A 001 pole figure is simply a projection along the cylinder axis. If several pole figures are measured, the 3D OD can be reconstructed, e.g., using tomographic methods and corresponding software packages are available (POPLA, Kallend et al. 1991; Beartex, Wenk et al. 1998 and others).

For converting relative diffraction intensities to experimental pole figures, it is necessary to apply scaling corrections for incident beam intensity and a subtraction of background intensity. If diffraction peaks are overlapped, they can be integrated and treated as overlapped pole figures (assigning appropriate structure factor weights), or they can be deconvoluted, fitting them with Gauss functions (e.g., Antoniadis et al. 1990; Merz et al. 1990; Larson and Von Dreele 2004). For both it is advantageous to have position sensitive detectors or use TOF to record continuous spectra. The experimental pole figures can be complete or incomplete and do not need to be normalized. Usually three or four incomplete pole figures are sufficient to determine the OD.

In the traditional approach ODs are determined from pole figure measurements of a *few* diffraction peaks *hkl* in *many* sample orientations. However, rather than to obtain the OD from discrete pole figures, one may want to use directly the diffraction spectra instead. In this case *many* diffraction peaks are used (a continuous spectrum) and thus only a *small* pole figure coverage is required. This corresponds in some ways to calculating the OD from "inverse pole figures" (Morris 1959), though exchange of crystal coverage for sample coverage is limited. The intensity of each diffraction peak in the spectrum also corresponds to a projection of orientation densities along a defined path (Fig. 17b, bottom).

Crystallographers have developed a technique to extract structural information from continuous diffraction spectra (Rietveld 1969; Young 1993; Larson and Von Dreele 2004) and this is the obvious method for TOF data where continuous spectra are measured anyway. An iterative combination of the crystallographic Rietveld profile analysis and OD calculation, proposed by Wenk et al (1994), was implemented for TOF data by Lutterotti et al. (1997) in the software MAUD (Materials Analysis Using Diffraction). It has since been refined (Matthies et al. 2005) and applied to many samples. The Rietveld texture analysis is particularly attractive for low symmetry compounds such as triclinic plagioclase (Xie et al. 2003) and polyphase materials (Wenk et al. 2001; Pehl and Wenk 2005) with many overlapping diffraction peaks. It enables efficient data collection and maximal use of measurements, and provides simultaneously structural and textural information about polycrystals, as well as a quantitative texture correction for crystal structure refinements of textured materials.

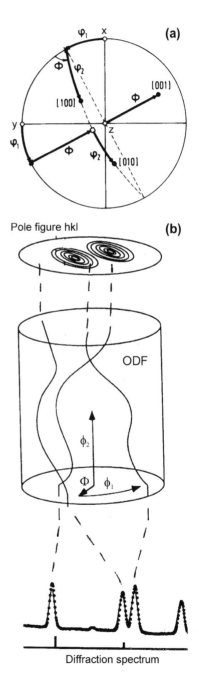

Figure 17. (*to the right*) (a) Euler angles ϕ_1, Φ, ϕ_2 define the orientation of a crystal with axes [100], [010] and [001] relative to a specimen coordinate system x, y, z. (b) A three-dimensional orientation distribution (OD), can be represented in a cylinder as function of Euler angles ϕ_1, Φ, ϕ_2. Pole figures are a projection of the OD (top). Also peak intensities of a diffraction spectrum are proportional to projections over OD paths (bottom).

It is pertinent to describe the Rietveld procedure for TOF neutron spectra in some detail and we use the LANSCE HIPPO diffractometer as an example. In the procedure all measured spectra from 30 detector banks and four rotations are taken as input, though only the information within a certain computation range (e.g., 0.8-3 Å) is generally used, yielding many data points (>200,000). Data from a detector bank are usually assumed to represent one point in orientation space (pole figure coverage in Fig. 10a). This is not strictly true because of the finite size of detectors. In HIPPO, for example, each detector has a size of 10-15° (depending on bank) and the actual pole figure coverage is illustrated in Figure 10b with polygons. Because of the detector geometry, the sharpest texture peaks that can be quantitatively represented are 10-15° in width, resulting in an optimal "texture resolution" of 25-30°. (Texture resolution is defined as the capability of quantitatively resolving two adjacent texture components). In principle the resolution could be increased by separating individual detector tubes but this would require much longer counting times and increase the computational effort enormously. For most deformation textures a 25-30° resolution is adequate.

With these diffraction data first instrumental, background, and phase parameters are refined. It is advantageous to calibrate instrumental parameters with a powder standard (such as Si) and obtain accurate values for instrumental peak aberrations that are convoluted with peak broadening due to the sample (e.g., crystallite size, microstrain and microstructure). From the spectra measured on the unknown sample for each detector a scale factor has to be refined to take account of detector efficiency and variations in absorption as neutrons pass through various components of the instrument. These scale factors may vary by a factor of 2 or more. Each detector is at a slightly different flight path, requiring a refinement of deviations from the average. Furthermore sample displacement parameters need to be refined to take account of the fact that the sample center as well as the goniometer rotation axis are usually not exactly aligned in the neutron beam. Next background parameters for each spectrum are determined. The background function is assumed to be a polynomial corrected with the incident spectrum (2nd to 4th order).

Once the background and instrument parameters are refined, crystallographic parameters such as lattice parameters, atomic coordinates and thermal parameters are determined. Now the texture refinement can start and for this several algorithms exist, most notably direct methods that solve the problem with tomography as alluded to above (Fig. 17b), and Fourier methods, generally referred to as "harmonic method" (Bunge 1965, 1982), where pole figures and OD are expressed as harmonic functions (hence ODF, see discussion above). The various methods, implemented in the Rietveld procedure, have recently been compared for TOF measurements of extruded aluminum with a strong and asymmetric texture (Matthies et al. 2005) and in Figure 18 001 pole figures recalculated from the OD are shown. The texture peak (above 1 m.r.d.) is similar in all pole figures but there are considerable differences for the region below 1 m.r.d. that constitutes nevertheless a large part of orientations. The tomographic method WIMV (Matthies and Vinel 1982), modified to permit irregular data coverage, provides smooth pole figures that are positive throughout (Fig. 18a). The harmonic method displays similar peaks, but considerable oscillations in the lower pole density regions, including slight negative deflections, for a harmonic expansion to $L_{max} = 12$ (Fig 18b). The pattern becomes considerably worse for $L_{max} = 16$ with many negative oscillations and exaggerated maxima (9.5 m.r.d.) (Fig. 18c).

We find that harmonic methods are adequate for a qualitative description of the main texture component. Due to the limited resolution of detectors and incomplete pole figure coverage, harmonic expansions beyond $L_{max} = 12$ introduce subsidiary oscillations, which are artifacts. The results from this analysis of aluminum illustrate the well-known limitations of the harmonic method: termination errors and lack of odd coefficients. Of the two currently available Rietveld programs that incorporate texture analysis, only MAUD (Lutterotti et al. 1997) includes discrete methods. GSAS (Von Dreele 1997) is limited to harmonic algorithms. Both do not incorporate any of the sophisticated approaches to extract odd coefficients in the harmonic expansion (e.g., Dahms and Bunge 1988; Van Houtte 1991). Even though

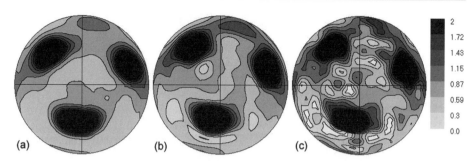

Figure 18. 100 pole figures recalculated from the OD of extruded aluminum, measured with the TOF diffractometer HIPPO and analyzing spectra with the Rietveld method. (a) Obtained with the tomographic method EWIMV, (b) Harmonic method with series expansion to $L_{max} = 12$, (c) series expansion to $L_{max} = 16$. Equal area projection, linear contour scale, pole densities in m.r.d. White regions are below zero (Matthies et al. 2005).

computationally straightforward, it turns out that limited detector resolution and irregular data coverage are much more serious for harmonic than discrete methods.

How much confidence can we have in a Rietveld refinement? A first indication is the value of a bulk R-factor that compares measured and refined values. However, particularly for 3D textures, such a single number is not adequate to reveal all possible shortcomings of the model. It is necessary to compare individual spectra and assess deviations. Figure 19 shows measured data (dots) and calculated spectrum (line) for calcite. A fit as illustrated here for a 90° detector is adequate. However, this is just one out of 40 90° detector spectra. An overall assessment is best provided by a "mapplot" that stacks all spectra and represents intensities with gray shades (Fig. 20). The lower part are measured spectra, the top part recalculated ones. Also here the similarity, both of peak intensities and their variation, as well as background intensities are very similar, giving us confidence that the refinement is good.

With multidetector systems such as HIPPO and data analysis with the Rietveld method neutron texture analysis has become a routine. An automatic sample changer can measure up

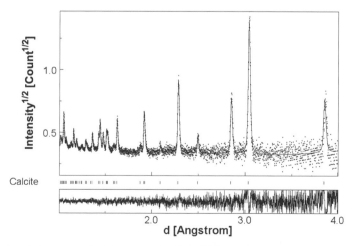

Figure 19. A typical spectrum of calcite measured with the TOF HIPPO diffractometer (150° bank detector panel). Dots are measured data and line is the fit obtained with the Rietveld code MAUD. Deviations (bottom) are minimal and largely due to counting statistics which is worst for large *d*-spacings.

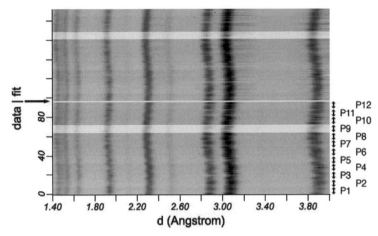

Figure 20. Limestone round robin standard measured with HIPPO and analyzed with the Rietveld method. Mapplot illustrating with gray shades intensity variations for 12 detectors of the 90° bank and 8 sample rotations. Bottom are measured data and top is the Rietveld fit. Note the excellent reproduction of background and peak intensities as well as peak positions. The wavy *d*-spacing variation is due uncertainties about the sample position relative to detectors (Wenk et al. 2002).

to 32 samples without intervention of the instrument scientist. Textures can be measured *in situ* at temperature, stress (CRATES) and pressure, opening the field to investigate texture changes in representative sample volumes, e.g., during phase transformations or recrystallization. But TOF multidetector systems have also their drawbacks. The average flightpaths from sample to detector has to be calibrated with an external standard of different geometry than the actual sample and thus needs to be refined, introducing uncertainty. Also the representative sample center may be displaced during rotations. Processing of the data recorded by the 30 detector panels on 720 detector tubes with 4,136,113 TOF channels is not only complex but intensities also are subject to electronic instabilities and may vary slightly over the duration of an experiment. And counting statistics for high angle detectors and large *d*-spacings that contain diagnostic reflections in many minerals is poor, with much uncertainty about spectral corrections. None of these limitations is critical and can generally be refined, but the investigator has to be aware of all the possible complications that may affect data quality.

APPLICATIONS

Compared with X-ray pole figure diffractometry and electron backscatter diffraction patters (EBSP), the number of texture analyses by neutron diffraction is small. The main reason for this is the limited access to neutron facilities and long delays in obtaining beamtime. For those who need texture results right now, neutron diffraction is not the best method. Another reason is the involved data processing procedure that requires expertise. To obtain an OD from pole figures or EBSP data is a matter of minutes, for neutron diffraction, particularly when using the Rietveld technique, it still takes hours or even weeks in the case of complex materials. Thus it is no surprise that for materials with strong and well separated diffraction peaks (cubic and hexagonal metals, quartz, calcite) conventional methods are adequate and there is no need to write proposals, wait for approval, travel to distant facilities, just to measure a few samples. Similarly the cost of neutrons does not justify experiments on such materials. Yet there are cases where neutrons have distinct advantages and, interestingly, many of these applications are in earth sciences. Neutrons are essential for low symmetry crystals (such as triclinic plagioclase: Wenk et al. 1985; Ullemeyer et al. 1994; Xie et al. 2003) and polyphase materials, including many rocks (Dornbusch et al. 1994; Siegesmund et al. 1994; Helming et al. 1996; Gastreich et al. 2000; Wenk et al. 2001; Leiss et al. 2002; Kurz et al. 2004; Ivankina et al. 2005; Pleuger et al. 2005) with closely spaced and partially overlapping diffraction peaks. For such materials conventional pole figure goniometry cannot be used.

Grain statistics

As was pointed out earlier, grain statistics can be important (Figs. 2 and 3) and neutron diffraction that analyzes large sample volumes has distinct advantages in characterizing bulk materials. This is particularly relevant for ore minerals that have been investigated extensively. Pyrite (Jansen et al. 1992; Siemes et al. 1993), chalcopyrite (Jansen et al. 1993, 1996a,b), pyrrhotite (Niederschlag and Siemes 1996), galena (Skrotzki et al. 2000) and hematite (Rosière et al. 2001; Guenther et al. 2002; Siemes et al. 2003; Hansen et al. 2004) are often coarse-grained and only analyzing large volumes provides adequate statistics. This is significant for estimating the anisotropy of physical properties such as magnetic susceptibility (Siemes et al. 2000; Fig. 21). Similar considerations apply to iron meteorites, where orientation relations between phases in Wiedmanstätten patterns, and volume fractions of orientation variants for bcc lamellae of kamacite are elegantly determined with neutron diffraction (Höfler et al. 1988; Fig. 22).

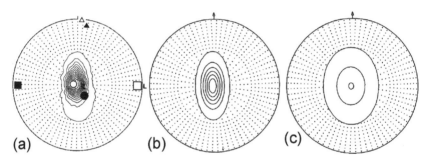

Figure 21. Measured and modeled 0003 pole figures of hematite from the Andrade mine, Brazil. (a) Experimental pole figure measured by neutron diffraction . Closed symbols are measured principal axes of magnetic susceptibility, open symbols are calculated axes. (b) Modeled pole figure based on a Bingham distribution to fit experimental pole densities and (c) modeled pole figure to fit measured susceptibilities (Siemes et al. 2000).

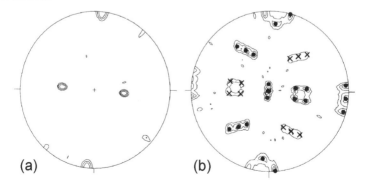

Figure 22. Pole figures of the Gibeon octahedrite meteorite measured by neutron diffraction at Jülich. This sample shows Widmanstätten patterns with intergrowth of fcc taenite and bcc kamacite. (a) (200) pole figure of taenite, (b) 200 pole figure of kamacite. Symbols indicate ideal orientations for a Nishiyama-Wassermann relationship between bcc and fcc phases. The variants with dots are emphasized over the variants with crosses (Höfler et al. 1988).

Statistical considerations also apply if textures are weak and large volumes are needed to establish significant patterns, as in the case of quartz that was deformed by a shock wave from a meteorite impact (Wenk et al. 2005), in fine-grained carbonates in deep sea sediments (Ratschbacher et al. 1994) and halite (Skrotzki et al. 1995). The good statistics was crucial in establishing anisotropy in calcite marble used as building materials (De Wall et al. 2000; Leiss and Weiss 2000; Siegesmund et al. 2000; Zeisig et al. 2002). This latter case is interesting since it represents a new application for texture analysis of rocks. Because of the anisotropy of thermal expansion of calcite, certain directions in textured marbles are most susceptible to microfracturing and spallation during seasonal temperature changes and the sectioning of slabs ought to take preferred orientation into account (Fig. 23).

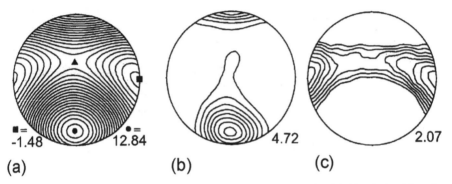

Figure 23. Fabric anisotropy in Carrara marble. (a) Dilatation coefficient a (in $10^{-6}\,C^{-1}$) calculated from the texture; (b) 0006 and (c) $11\bar{2}0$ pole figures measured by neutron diffraction on spherical samples of 30 mm diameter at Dubna, maxima (in m.r.d.) are indicated. Equal area projection (Leiss and Weiss 2000).

Polymineralic rocks

Increasingly textures determined by neutron diffraction are being used in the interpretation of geologic deformation histories (Leiss et al. 1999; Ullemeyer and Weber 1999; Leiss and Molli 2003; Pleuger et al. 2005), as well as the interpretation of seismic anisotropy in the crust (Ivankina et al. 2005; Ullemeyer et al. 2006). Preferred orientation of quartz has been of prime interest in petrofabric research, ever since the first pole figure was published by Schmidt (1925). Quartz textures have been measured in many rocks and the great variety of fabric types was classified by Sander (1950). Until recently most of this research relied on Universal stage measurements, which can only determine orientations of the c-axes. Also today most discussions of quartz textures emphasize (0001) pole figures. Neutron diffraction has been used to establish the full crystal orientation distribution that provides important additional geological information about the deformation history.

The case is granitic mylonites, rather common highly deformed quartz-bearing rocks. Neutron diffraction is preferred because these rocks contain several phases (mainly quartz, plagioclase feldspar and biotite mica), and because they are fairly coarse-grained. Only a volume average can provide adequate statistics. Pehl and Wenk (2005) investigated textures of mylonites from the Santa Rosa mylonite zone in southern California by TOF neutron diffraction, analyzing data with the Rietveld method. Figure 24 shows a spectrum with the Rietveld fit, documenting a large number of diffraction peaks, many of them overlapped. With the Rietveld method volume fractions could be refined (quartz: 38 vol%, plagioclase: 50 vol%, biotite 12 vol%), as well as orientation distributions for the three phases. The plagioclase texture is more or less random, biotite has a very strong texture. Only quartz will be discussed here. Figure 25a displays pole figures recalculated from the OD. They are projected on the foliation plane, with

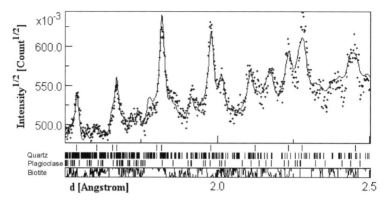

Figure 24. Rietveld fit (line) of a HIPPO TOF spectrum (dots) of granitic mylonite with three principal phases quartz, plagioclase and biotite (90° detector).

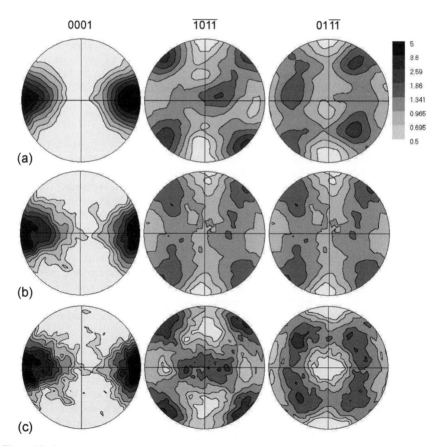

Figure 25. Quartz in mylonite of granitic composition from Palm Canyon, California. (a) Pole figures recalculated from the OD using neutron diffraction and Rietveld method. (b) Discretization of OD and applying Dauphiné twinning to all orientations. (c) Imposing compression in the center (pole of foliation) and inducing Dauphiné twinning to minimize elastic energy. Equal area projection, logarithmic contours, pole densities in m.r.d. (Pehl and Wenk 2005).

the lineation direction vertical. By analyzing many samples, the authors showed that this texture is fairly constant over 20 km and thus representative of tectonic deformation. The c-axis pattern with a maximum in the foliation plane at right angles to the lineation is typical of dynamic recrystallization; a-axes align preferentially in the lineation direction.

Of particular interest are the patterns for positive rhombs $\{10\overline{1}1\}$ and negative rhombs $\{01\overline{1}1\}$. They are different and furthermore have a lower symmetry than the orthorhombic $c = (0001)$ and $a = \{11\overline{2}0\}$ pole figures. The two rhombs are related by Dauphiné twinning, with host and twin related by a 2-fold rotation about the c-axis. Indeed mechanical twinning under stress and due to the large elastic anisotropy of quartz may have introduced this pattern. The effects of twinning can be modeled by first applying it to all orientations (Fig. 25b). This does not affect c- and a-axes, and positive and negative rhombs become equal. Now a compressive stress is applied perpendicular to the foliation (center of pole figure) and Dauphiné twinning is induced to minimize the elastic energy. It produces the texture pattern in Figure 25c. This pattern has a great similarity with the observed texture (Fig. 25a), suggesting that mechanical twinning did indeed occur and can thus be used as a paleopiezometer. Mechanical twinning in quartz rocks initiates around 80 MPa (Fig. 16). Deviations of rhombohedral pole figures from orthorhombic symmetry are attributed to a component of non-coaxial deformation.

Interestingly it has been found that with increasing deformation quartz-bearing mylonites become susceptible to the alkali silica reaction if used as aggregate in concrete and render such concrete very unstable (Monteiro et al. 2001). This is probably due to intracrystalline deformation of quartz and with neutron diffraction texture analysis the overall deformation state of the rock can be assessed (Wenk and Pannetier 1990) and its suitability as aggregate material established.

In situ experiments and phase transformations

So far all examples were investigated at ambient conditions. However, neutron diffraction offers the possibility to record texture changes *in situ* at temperature, pressure and stress. The HIPPO diffractometer at LANSCE is unique by the availability of furnaces, cryostats, pressure cells and straining devices that are compatible with texture measurements. The trigonal (α)-hexagonal (β)-trigonal phase transformation of quartz in quartzite is used to illustrate this capability. A cylinder of mylonitic quartzite, 1 cm in diameter, is heated in a vacuum furnace. At 300 °C the texture pattern is roughly trigonal, with a c-axes maximum and three concentrations of rhombs (Fig. 26a). Upon heating to 625 °C the texture becomes hexagonal with no changes to c-axes or a-axes (Fig. 26b). The phase transformation is displacive and reversible, with only minor distortions of bonds. Upon cooling each crystal could choose between two orientation variants, related by a 180° rotation about the c-axis. It turns out that there is a perfect memory and the texture reverts exactly to the initial pattern (Fig. 26c). The cause of this texture memory is not yet understood but qualitatively attributed to the effect of stresses imposed by neighboring grains.

Similar heating experiments have been performed on metals documenting texture changes during recrystallization and a less perfect texture memory in the hcp→bcc→hcp transformation of zirconium (Wenk et al. 2004) and the bcc→fcc→bcc transformation of iron (Wenk et al. 2006b). Such *in situ* texture measurements may help us to better understand anisotropy changes during phase transformations and particularly variant selection in martensitic transitions. In titanium Bhattacharyya et al. (2006) could document that during the phase transformation preexisting bcc grains grow, replacing hcp orientations.

To end our discussion of examples we go to low temperature. With neutron diffraction it is possible to investigate textures of ice that is not only relevant on earth but also for the outer planets where high pressure forms of ice control their rheology. For neutron diffraction it is necessary to use deuterated ice to avoid strong inelastic scattering (see also Kuhs and

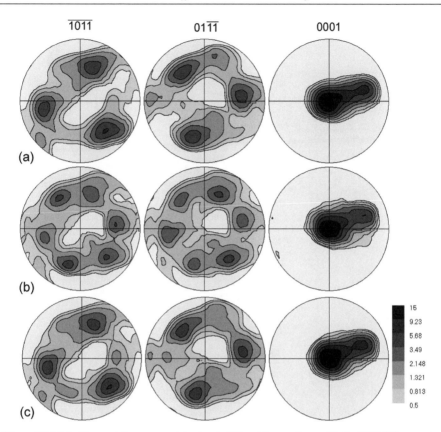

Figure 26. Mylonitic quartzite, measured *in situ* at different temperatures with the TOF diffractometer HIPPO. (a) 300 °C, (b) 625 °C (above the hexagonal phase transformation), (c) 300 °C (after cooling). Note the perfect texture memory. Equal area projection, logarithmic contour scale, pole densities in m.r.d.

Hansen 2006, this volume). The experimental study of Bennett et al. (1994) documented for hexagonal ice I pole figures with *c*-axes parallel to the compression direction (Fig. 27a) and for rhombohedral high pressure ice II *c*-axes perpendicular to the compression direction, indicating different deformation mechanisms (Fig. 27b).

Magnetic textures

Harrison (2006, this volume) discussed the unique magnetic scattering of neutrons. This was the original incentive for Brockhouse (1950) to use neutrons for texture analysis. There have been several attempts to determine magnetic textures, i.e., the orientation of magnetic domains (e.g., Henning et al. 1981; Zink et al. 1994, 1997; Birsan et al. 1996). As was outlined by Bunge (1989) different situations may arise in magnetic materials: In demagnetized ferromagnetic materials such as cubic iron magnetic moments may be oriented randomly in any of the six <100> directions. In this case the crystallographic and magnetic texture are identical. A weak magnetic field produces a preferential alignment of moments along certain <110> directions destroying the cubic crystal symmetry. Birsan et al. (1996) conducted an interesting study on silicon steel. The 110 pole figure of the demagnetized material, measured by neutron diffraction (Fig. 28a), corresponds to a typical rolling texture. A difference pole figure subtracting the demagnetized pole figure from the pole figure with an applied magnetic field in the rolling direction (center) is shown in Figure 28b. The magnetic part of neutron

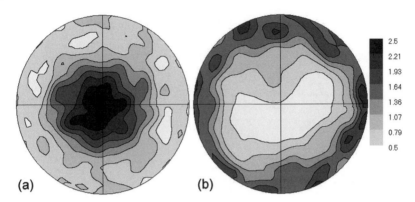

Figure 27. (0001) pole figures of ice deformed experimentally in axial compression (compression direction in center). (a) Hexagonal ice I. (b) Rhombohedral high pressure polymorph ice II. Equal area projection, linear scale, pole densities in m.r.d. (Bennett et al. 1994).

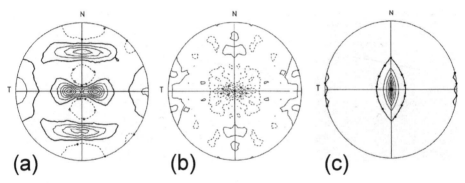

Figure 28. Magnetic silicon steel investigated by neutron diffraction. (a) (110) pole figure of demagnetized steel, (b) difference pole figure (magnetized, 150 mA-demagnetized) illustrating magnetic scattering, (c) magnetic texture of the magnetized sample obtained by modeling (Birsan et al. 1996).

scattering has a minimum in the center and concentrations near the transverse direction. It can be used to determine, through modeling, the magnetic pole figure, i.e., the alignment of magnetic moments (Fig. 28c) which turns out to be more or mess parallel to the {100} maxima of easy magnetization and the magnetic field. In high magnetic fields the magnetization is no longer related to a crystallographic direction and thus becomes independent of the crystallographic texture.

To our knowledge there are no applications to minerals but studies of magnetic textures of ferrimagnetic mineral ores such as manganite and magnetite could reveal informative distributions to assess paleomagnetic properties of rocks.

New possibilities

Classical applications of neutron diffraction texture analysis to coarse materials, improving grain statistics, and to low-symmetry polyphase materials were discussed. Also mentioned were *in situ* experiments at high and low temperature. There is a wealth of other possibilities to use neutron diffraction to characterize direction-dependent properties. Texture patterns may be compared with strain pole figures that display elastic deformation (Wang et al. 2002, 2003), either caused by applied stress or expressing residual stress. Such directional stress measurements are barely explored for geological materials (Darling et al.

2004; Schofield et al. 2003; Daymond 2006, this volume). This opens exciting possibilities for quantitative modeling, combining textures, lattice strain, stored energy (Hayakawa et al. 1997) and microstructure to better understand material-forming processes (Agnew et al. 2003; Brown et al. 2003, 2005). Much of this is promoted by metallurgists but similar theories are also applied to understand deformation of rocks (for a review e.g., Wenk 1998).

COMPARISON OF METHODS AND RECOMMENDATIONS

The optimal choice of pole-figure measurements depends on many variables, such as availability of equipment, material to be analyzed, and data requirements.

Neutron diffraction is advantageous for determination of complete pole figures in coarse-grained aggregates and, in principle, allows determination of magnetic pole figures. The advantage of neutrons is that bulk samples rather than surfaces are measured, that coarse grained materials can be characterized, that environmental cells (heating, cooling, straining) are available and that resolution in d is much better than with conventional X-ray pole figure goniometers in reflection geometry where peak broadening occurs because of defocusing effects with sample tilt. This makes it possible to measure complex composites with many closely spaced diffraction peaks. With position sensitive detectors, and particularly with TOF, continuous spectra can be recorded; shifts in peak positions can be used for residual stress determination and intensities to extract simultaneously texture information, for example with the Rietveld method.

For routine metallurgical practice and many other applications in materials science and geology, *X-ray diffraction* with a pole figure goniometer in back-reflection geometry is generally adequate. It is fast, easily automated, inexpensive both in acquisition and maintenance. X-ray diffraction texture analysis done on a flat surface is restricted to fine-grained materials (<1 mm) that are homogeneous within the plane. Back reflection provides only incomplete pole figures, but this drawback can be overcome by data analysis. Pole figures can only be measured adequately if diffraction peaks are sufficiently separated. In geological samples and ceramics, X-ray diffraction is therefore generally limited to single phase aggregates of orthorhombic or higher crystal symmetry.

Electron diffraction with a *TEM* is most tedious but provides valuable information about dislocation microstructures and, at least two-dimensionally, about interaction between neighbors and about heterogeneities within grains (e.g., Schwarzer and Weiland 1988). This is important data to interpret deformation processes.

Electron Backscatter Diffraction Patterns (EBSP or EBSD), measured with the *SEM* and produced on polished surfaces, can be used to determine local orientation and this technique has become an important addition to the methods of texture measurements (Lloyd et al. 1991; Randle and Engel 2000). While such measurements do not provide direct information on dislocation microstructures, they allow for determination of local orientation correlations that are important in the study of recrystallization or of stress concentrations on which failure may occur. It can be used to produce orientation maps (OIM, Orientation Imaging Microscopy, Adams et al. 1993). Determination of individual grain orientations has advantages for OD calculations because of the absence of ambiguity, which is inherent in pole figures. With the possibility of automation this technique has become comparable in expense and effort to X-ray diffraction analysis. It remains confined to samples of fairly high crystallinity and moderate deformation. Figure 29 compares pole figures of quartz obtained by EBSP (Fig. 29a) and neutron diffraction (Fig. 29b). They are virtually identical (Kunze et al. 1994) but for the neutron data, measured on a 1cm cube, they represent the actual data, for the EBSD results they had to be arbitrarily smoothed with a 15° filter to obtain the same pole densities.

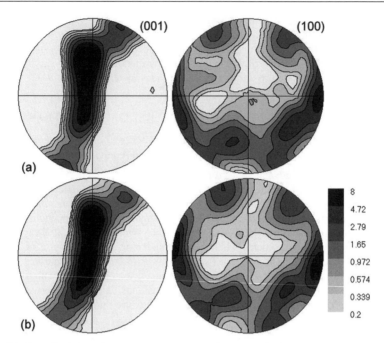

Figure 29. Naturally deformed mylonitic quartzite. Pole figures obtained from (a) EBSP measurements in the SEM and arbitrary smoothing with a 15° Gauss filter and (b) by neutron diffraction with a monochromatic source at Jülich and recalculated from the OD. Equal area projection, linear contours, pole densities in m.r.d. (Kunze et al. 1994).

A new technique is to use *hard X-rays* ($\lambda \sim 0.1$ Å) produced in a *synchrotron* and record 2D diffraction images with image plates or CCD cameras. Such high energy X-rays can penetrate several millimeters of thickness and thus have become comparable to neutrons as far as volume averages are concerned (Wenk and Grigull 2002). Synchrotron X-rays have the advantage of very high intensity. It is likely that they will replace neutrons for some applications, particularly if only grain size and resolution are issues. Similar Rietveld techniques are applied to obtain quantitative texture information from 2D images as from 1D neutron spectra (Lonardelli et al. 2005).

This Chapter has reviewed methods of texture analysis by neutron diffraction, an increasingly important application of neutron scattering. Those interested in preferred orientation and how it relates to macroscopic anisotropy of physical properties as well as deformation history and rheology should consult the literature (e.g., Wenk 1985; Kocks et al. 2000; Karato and Wenk 2002).

ACKNOWLEDGMENTS

Neutron scattering was a wonderful opportunity to meet people from a wide variety of fields, starting with the first experiments in 1980 at Jülich with W. Schäfer and ILL with H.-J. Bunge and J. Pannetier, and more recently the close interaction at Los Alamos around HIPPO with K. Bennett, R. Von Dreele, A. Hurd and S. Vogel. I am grateful to reviewers for their comments, particularly Sven Vogel and Mark Daymond and contributions from H.-G. Brokmeier, W. Schäfer and K. Ullemeyer. Research was supported by several agencies, foremost NSF, DOE and IGPP.

REFERENCES

Adams BL, Wright SI, Kunze K (1993) Orientation imaging: The emergence of a new microscopy. Metall Trans 24A:819-831

Agnew SR, Tomé CN, Brown DW, Holden TM, Vogel SC (2003) Study of slip mechanisms in a magnesium alloy by neutron diffraction and modeling. Scripta Mater 48:1003-1008

Antoniadis A, Berruyer J, Filhol A (1990) Maximum-likelihood methods in powder diffraction refinements. Acta Cryst A46:692-711

Bennett K, Wenk H-R, Durham WB, Stern L, Kirby SH (1997) Preferred crystallographic orientation in the ice I→II transformation and the flow of ice II. Philos Mag A 76:413-435

Bhattacharyya D, Viswavathan GB, Vogel SC, Williams DJ, Venkatesh V, Fraser HL (2006) A study of the mechanism of α to β phase transformation by tracking texture evolution with temperature in Ti-6Al-4V using neutron diffraction. Scripta Mater 54:231-236

Birsan M, Szpunar JA, Tun Z, Root JH (1996) Magnetic texture determination using nonpolarized neutron diffraction. Phys Rev B 53:6412-6417

Brockhouse BN (1953) The initial magnetization of nickel under tension. Can J Phys 31:339-355

Brokmeier H-G (1997) Neutron diffraction texture analysis. Physica B 234-236:977-979

Brokmeier H-G (1999) Advantages and applications of neutron texture analysis. Textures Microstructures 33:13-34

Brown DW, Abeln SP, Blumenthal WR, Bourke MAM, Mataya MC, Tomé CN (2005) Development of crystallographic texture during high rate deformation of rolled and hot-pressed beryllium. Metall Mater Trans 36A:929-939

Brown DW, Bourke MAM, Clausen B, Holden TM, Tomé CN, Varma R (2003) A neutron diffraction and modeling study of uniaxial deformation in polycrystalline beryllium. Metall Mater Trans 34A: 1439-1449

Bunge H-J (1965) Zur Darstellung allgemeiner Texturen. Z Metallk 56:872-874

Bunge H-J (1982) Texture Analysis in Materials Science – Mathematical Methods. Butterworths

Bunge H-J (1989) Texture and magnetic properties. Textures Microstructures 11:75-91

Bunge H-J, Wenk H-R, Pannetier J (1982) Neutron diffraction texture analysis using a position sensitive detector. Textures Microstructures 5:153-170

Dahms M, Bunge H-J (1988) A positivity method for the determination of complete orientation distribution functions. Textures Microstructures 10:21-35

Darling TW, TenCate JA, Brown DW, Clausen B, Vogel SC (2004) Neutron diffraction study of the contribution of grain contacts to nonlinear stress-strain behavior. Geophys Res Lett 31, doi:10.1029/2004GL020463

Daymond MR (2006) Internal stresses in deformed crystalline aggregates. Rev Mineral Geochem 63:427-458

DeWall H, Bestmann M, Ullemeyer K (2000) Anisotropy of diamagnetic susceptibility in Thassos marble: a comparison between measured and modeled data. J Struct Geol 22:1761-1771

Dornbusch HJ, Skrotzki W, Weber K (1994) Development of microstructure and texture in high-temperature mylonites from the Ivrea Zone. *In*: Textures of Geological Materials. Bunge HJ et al. (Eds) Deutsch. Gesell. Metallkunde, p 187-202

Feldmann K (1989) Texture investigation by neutron time-of-flight diffraction. Textures Microstructures 10: 309-323

Gastreich M, Jansen E, Raith M, Kirfel A (2000) Polfigurmessungen mit Neutronen an einem Orthopyroxen-Sillimanit-Granulit. Z Kristall (Suppl.) 17: 73

Ghildiyal H, Jansen E, Kirfel A (1999) Volume texture of a deformed quartzite observed with U-stage and neutron diffractometery. Textures Microstructures 31:239-248

Günther A, Brokmeier H-G, Petrovsky E, Siemes H, Helming K, Quade H (2002) Mineral preferred orientation and magnetic properties as indicators of varying strain conditions in naturally deformed iron ore. Appl Phys A 74:1080-1082

Hansen A, Chadima M, Cifelli F, Brokmeier H-G, Siemes H (2004) Neutron pole figures compared with magnetic preferred orientations of different rock types. Physica B 350:120-122

Harrison RJ (2006) Neutron diffraction of magnetic materials. Rev Mineral Geochem 63:113-143

Helming K, Schmidt D, Ullemeyer K (1996) Preferred orientations of mica bearing rocks described by texture components. Textures Microstructures 25:211-222

Henning K, Wieser E, Betzl M, Feldmann K (1981) Magnetic texture (Magnetic pole figures). Proceedings of the 6th International Conference on Texture of Materials, Tokyo, Japan (The Iron and Steel Institute of Japan) Vol. 2:967

Höfler S, Will G, Hamm H-M (1988) Neutron diffraction pole figure measurements on iron meteorites. Earth Planet Sci Lett 90:1-10

Ivankina TI, Kern H, Nikitin AN (2005) Directional dependence of P- and S- wave propagation and polarization in foliated rocks from the Kola superdeep well: Evidence from laboratory measurements and calculations based on TOF neutron diffraction. Tectonophysics 407:25-42

Jansen EM, Brokmeier H-G Siemes H (1996) Neutron texture investigations on natural Mt. Isa chalcopyrite ore. Part II: Preferred orientation of chalcopyrite after different experimental deformation conditions. Textures Microstructures 28: 1-15

Jansen EM, Brokmeier H-G, Siemes H (1996) Neutron texture investigations on natural Mt.Isa chalcopyrite ore. Part I: Preferred orientation of one and the same chalcopyrite sample before and after experimental deformation. Textures Microstructures 26:167-179

Jansen EM, Merz P, Schaeben H, Schäfer W, Siemes H, Will G (1992) Determination of preferred orientation of pyrite in a chalcopyrite ore by means of neutron diffraction. Textures Microstructures 19:203-210

Jansen EM, Siemes H, Merz P, Schäfer W, Will G, Dahms M (1993) Preferred orientation of experimentally deformed Mt. Isa chalcopyrite ore. Mineral Mag 57:45-53

Juul Jensen DJ, Leffers T (1989) Fast texture measurements using position sensitive detector. Textures Microstructures 10:361-374

Kallend JS (2000) Determination of the orientation distribution from pole figure data. *In:* Texture and Anisotropy. Kocks UF, Tomé CN, Wenk H-R (eds) Cambridge University Press, p 102-125

Kallend JS, Kocks UF, Rollett AD Wenk H-R (1991) Operational texture analysis. Mater Sci Eng A132:1-11

Karato S-Y, Wenk H-R (eds) (2002) Plastic Deformation of Minerals and Rocks. Reviews in Mineralogy and Geochemistry, Volume 51. Mineralogical Society of America

Kocks U F, Tomé CN, Wenk H-R (2000) Texture and Anisotropy. Preferred Orientations in Polycrystals and Their Effect on Materials Properties, 2nd Paperback Edtn. Cambridge University Press

Kuhs WF, Hansen TC (2006) Time-resolved neutron diffraction studies with emphasis on water ices and gas hydrates. Rev Mineral Geochem 63: 171-204

Kunze K, Adams BL, Heidelbach F, Wenk H-R (1994) Local microstructural investigations in recrystallized quartzite using orientation imaging microscopy. Proc. ICOTOM 10, Mater Sci Forum 157-162:1243-1250

Kurz W, Jansen E, Hundenborn R, Pleuger J, Unzog W (2004) Microstructures and preferred orientations of omphacite in Alpine eclogites: implications for the exhumation of (ultra-) high pressure units. J Geodynamics 37:1-55

Larson AC, Von Dreele RB (2004) General Structure Analysis System (GSAS). Los Alamos National Laboratory Report LAUR 86-748

Leiss B, Gröger HR, Ullemeyer K, Lebit H (2002)Textures and microstructures of naturally deformed amphibolites from the northern Cascades, NW USA. *In:* Deformation Mechanisms, Rheology and Tectonics: Current Status and Future Perspectives. De Meer S, Drury MR, De Bresser JHP, Pennock GM (eds) Geological Society, London, Special Publications, 200:219-238

Leiss B, Molli G (2003) "High-Temperature" texture in naturally deformed calcite marble from the Alpi Apuane, Italy. J Struct Geol 25:649-658

Leiss B, Siegesmund S, Weber K (1999): Texture asymmetries as shear sense indicators in naturally deformed mono- and polyphase carbonates. Textures Microstructures 33:61-74

Leiss B, Weiss T (2000) Fabric anisotropy and its influence on physical weathering of different types of Carrara marbles. J Struct Geol 22:1737-1745

Lloyd GE, Schmidt N-H, Mainprice D, Prior DJ (1991) Crystallographic textures. Mineral Mag 55:331-345

Lonardelli I, Wenk H-R, Lutterotti L, Goodwin M (2005) Texture analysis from synchrotron diffraction images with the Rietveld method: dinosaur tendon and salmon scale. J Synchr Radiation 12: 354-360

Lutterotti L, Matthies S, Wenk H-R, Schultz AJ, Richardson J W (1997) Combined texture and structure analysis of deformed limestone from time-of-flight neutron diffraction spectra. J Appl Phys 81:594-600

Matthies S (1979) On the reproducibility of the orientation distribution function of texture samples from pole figures (ghost phenomena). Phys Stat Sol (b) 92:K135-138

Matthies S, Pehl J, Wenk H-R, Lutterotti L, Vogel S (2005) Quantitative texture analysis with the HIPPO TOF diffractometer. J Appl Cryst 38:462-475

Matthies S, Vinel GW (1982) On the reproduction of the orientation distribution function of textured samples from reduced pole figures using the concept of conditional ghost correction. Phys Stat Sol (b) 112:K111-114

Merz P, Jansen E, Schaefer E, Will G (1990) PROFAN-PC: a PC program for powder peak profile analysis. J Appl Cryst 23:444-445

Monteiro PJM, Shomglin K, Wenk H-R, Hasparyk NP (2001) Effect of aggregate deformation on alkali-silica reaction. ACI Materials J 98:179-183

Morris PR (1959) Reducing the effects of nonuniform pole distribution in inverse pole figure studies. J Appl Phys 30:595-596

Niederschlag E, Siemes H (1996) Influence of initial texture, temperature and total strain on the texture development of polycrystalline pyrrhotite ores in deformation experiments. Textures Microstructures 28:129-148

Parise JB (2006) Introduction to neutron properties and applications. Rev Mineral Geochem 63:1-25

Pehl J, Wenk H-R (2005) Evidence for regional Dauphiné twinning in quartz from the Santa Rosa mylonite zone in Southern California. A neutron diffraction study. J Struct Geol 27:1741-1749

Pleuger J, Froitzheim N, Jansen E (2005) Folded continental and oceanic nappes on the southern side of Monte Rosa (western Alps, Italy): Anatomy of a double collision suture. Tectonics 24, TC4013, doi.1029/2004TC001737

Popa NC, Balzar D (2001) Elastic strain and stress determination by Rietveld refinement: Generalization treatment for textured polycrystals for all Laue classes. J Appl Cryst 34:187-195

Rajmohan N, Hayakawa Y, Szpunar JA, Root JH (1997) Neutron diffraction for stored energy measurement in interstitial free steel. Acta Mater 45:2485-2494

Randle V, Engler O (2000) Introduction to Texture Analysis: Macrotexture, Microtexture and Orientation Mapping. Gordon and Breach Science Publishers

Ratschbacher L, Wetzl A, Brokmeier H-G (1994) A neutron goniometer study of the preferred orientation of calcite in fine-grained deep-sea carbonate. Sedimentary Geol 89:315-324

Rietveld HM (1969) A profile refinement method for nuclear and magnetic structures. J Appl Cryst 2:65-71

Rosière CA, Siemes H, Quade H., Brokmeier H-G, Jansen EM (2001) Microstructures, textures and deformation mechanisms in hematite. J Structural Geol 23:1429-1440

Sander B (1950) Einführung in die Gefügekunde der Geologischen Körper, Vol. 2. Springer

Santisteban JR, Edwards L, Stelmukh V (2006), Characterization of textured materials by TOF transmission. Physica B (in press)

Schäfer W (2002) Neutron diffraction applied to geological texture and residual stress analysis. Eur J Mineral 14:263-290

Schmidt W (1925) Gefügestatistik. Tschermaks Mineral Petrograph Mitt 38:392-423

Schofield PF, Covey-Crump SJ, Stretton IC, Daymond MR, Knight KS, Holloway RF (2003) Using neutron diffraction measurements to characterize the mechanical properties of polymineralic rocks. Mineral Mag 67:967-987

Schwarzer RA, Weiland H (1988) Texture analysis by the measurement of individual grain orientations.- Electron microscopical methods and application to dual phase steel. Textures Microstructures 8-9: 443-456

Siegesmund S, Helming K, Kruse R (1994), Complete texture analysis of a deformed amphibolite: comparison between neutron diffraction and U-stage data. J Struct Geol 16:131-142

Siegesmund S, Ullemeyer K, Weiss T, Tschegg EK (2000) Physical weathering of marbles caused by anisotropic thermal expansion. Int J Earth Sci 89:170-182

Siemes H, Klingenberg B, Rybacki E, Naumann M, Schaefer W, Jansen E, Rosiere CA (2003) Texture, microstructure and strength of hematite ores experimentally deformed in the temperature range 600 °C to 1100 °C and at strain rates between 10^{-4} and 10^{-6} s^{-1}. J Struct Geol 25:1372-1391

Siemes H, Zilles D, Cox SF, Merz P, Schaefer W, Will G, Schaeben H, Kunze K (1993) Preferred orientation of experimentally deformed pyrite measured by means of neutron diffraction. Mineral Mag 57: 29-43

Skrotzki W, Helming K, Brokmeier H-G, Dornbusch H-J, Welch P (1995) Textures in pure shear deformed rock salt. Textures Microstruct 24:133-141

Skrotzki W, Tamm R, Oertel C-G, Röseberg J, Brokmeier H-G (2000) Microstructure and texture formation in extruded lead sulfide. J Struct Geol 22:1621-1632

Ullemeyer K, Helming K, Siegesmund S (1994) Quantitative texture analysis of plagioclase. In: Textures of Geological Materials. Bunge HJ, Siegesmund S, Skrotzki W, Weber K (eds) DMG Informationsgesellschaft, p 93-108

Ullemeyer K, Siegesmund S, Rasolofosaon PNJ, Behrmann JH (2006) Experimental and texture-derived P-wave anisotropy from the TRANSALP traverse: An aid for the interpretation of seismic field data. Tectonophysics 414: 97-116

Ullemeyer K, Spalthoff P, Heinitz J, Isakov NN, Nikitin AN, Weber, K (1998) The SKAT texture diffractometer at the pulsed reactor IBR-2 at Dubna: experimental layout and first measurements. Nucl Instrum Methods A 412:80-88

Ullemeyer K, Weber K (1999) Lattice preferred orientation as an indicator of a complex deformation history of rocks. Textures Microstructures 33:45-60

Van Houtte P (1991) A method for the generation of various ghost correction algorithms – the example of the positivity method and the exponential method. Textures Microstructures 13:199-212

Vogel SC, Priesmeyer H-G (2006) Neutron production, neutron facilities and neutron instrumentation. Rev Mineral Geochem 63:27-57

Von Dreele RB (1997) Quantitative texture analysis by Rietveld refinement. J Appl Cryst 30:517-525

Wahlstrom EE (1979) Optical Crystallography. Wiley

Walther K, Ullemeyer K, Heinitz J, Betzl M, Wenk H-R (1995) TOF texture analysis of limestone standard: Dubna results. J Appl Cryst 28: 503-507

Wang YD, Peng RL, Wang X-L, McGreevy RL (2002) Grain-orientation-dependent residual stress and the effect of annealing in cold-rolled stainless steel. Acta Mater 50:1717-1734

Wang YD, Wang X-L, Stoica AD, Richardson JW, Lin Peng R (2003) Determination of the stress orientation distribution function using pulsed neutron sources. J Appl Cryst 36:14-22

Wenk H-R (1985) Preferred Orientation in Deformed Metals and Rocks: An Introduction to Modern Texture Analysis. edited by H. R. Wenk, pp. 11-47, Academic Press

Wenk H-R (1991) Standard project for pole figure determination by neutron diffraction. J Appl Crystallogr 24: 920-927

Wenk H-R (1994) Texture analysis with TOF neutrons. In: Time-of-Flight Diffraction at Pulsed Neutron Sources (JD Jorgensen and AJ Schultz. Edts.), Trans Am Cryst Assn 29: 95-108

Wenk HR (1999) A voyage through the deformed Earth with the self-consistent model. Model Simul Mater Sci Eng 7:699-722

Wenk H-R Grigull S (2003) Synchrotron texture analysis with area detectors. J Appl Crystallogr 36:1040-1049

Wenk H-R Lonardelli I, Vogel SC, Tullis J (2005) Dauphiné twinning as evidence for an impact origin of preferred orientation in quartzite: An example from Vredefort, South Africa. Geology 33:273-276

Wenk H-R Pehl J, Williams DJ (2002) Texture changes during the quartz $\alpha-\beta$ phase transition studied by neutron diffraction. Los Alamos National Laboratory report LA-14036-PR, 36-39

Wenk H-R, Bunge HJ, Jansen E, Pannetier J (1985) Preferred orientation of plagioclase - neutron diffraction and U-stage data. Tectonophysics 126:271-284

Wenk H-R, Cont L, Xie Y, Lutterotti L, Ratschbacher L, Richardson J (2001) Rietveld texture analysis of Dabie Shan eclogite from TOF neutron diffraction spectra. J Appl Crystallogr 34:442-453

Wenk H-R, Huensche I, Kestens L (2006b) *In situ* observation of texture changes in ultralow carbon steel. Mater Trans (in press)

Wenk H-R, Kern H, Schäfer W, Will G (1984) Comparison of x-ray and neutron diffraction in texture analysis of carbonate rocks. J Struct Geol 6:687-692

Wenk HR, Kocks UF (1987) The representation of orientation distributions. Metall Trans 18A:1083-1092

Wenk H-R, Larson AC, Vergamini PJ, Schultz AJ (1991) TOF of pulsed neutrons and 2d detectors for texture analysis of deformed polycrystals. J Appl Phys 70:2035-2040

Wenk H-R, Lonardelli I, Franz H, Nihei K, Nakagawa S (2006a) Texture analysis and elastic anisotropy of illite clay. Geophysics (in press)

Wenk H-R, Lonardelli I, Williams D (2004) Texture changes in the hcp-bcc-hcp transformation of zirconium studied *in situ* by neutron diffraction. Acta Mater 52:1899-1907

Wenk H-R, Lutterotti L, Vogel S (2003) Texture analysis with the new HIPPO TOF diffractometer. Nucl Instrum Methods A515:575-588

Wenk H-R, Matthies S, Donovan J, Chateigner D (1998) BEARTEX, a Windows-based program system for quantitative texture analysis. J Appl Cryst 31:262-269

Wenk H-R, Matthies S, Lutterotti L (1994) Texture analysis from diffraction spectra. In: Textures of Materials ICOTOM-10. Bunge HJ (ed) Switzerland: Trans. Tech. Pubs. 157-162:473-479

Wenk H-R, Pannetier J (1990) Texture development in deformed granodiorites from the Santa Rosa mylonite zone, southern California. J Struct Geol 12:177-184

Xie Y, Wenk H-R, Matthies S (2003) Plagioclase preferred orientation by TOF neutron diffraction and SEM-EBSD. Tectonophysics 370:269-286

Young RA (1993) The Rietveld Method. International Union of Crystallography, Oxford University Press

Zeisig A, Weiss T, Siegesmund S (2002) Thermal expansion and its control on the durability of marbles. In: Natural Stones, Weathering Phenomena, Conservation Strategies and Case Studies. Siegesmund S, Weiss T, Vollbrecht A (eds) Geological Society Special Publication 205:57-72

Zink U, Brokmeier H-G and Bunge HJ (1994) Determination of magnetic textures in ferromagnetics by means of neutron diffraction. Mater Sci Forum 157-162:251-256

Zink U, Bunge HJ, Brokmeier H-G (1997) Neutron texture measurement with applied magnetic field. Physica B 234-236:980-982

Reviews in Mineralogy & Geochemistry
Vol. 63, pp. 427-458, 2006
Copyright © Mineralogical Society of America

Internal Stresses in Deformed Crystalline Aggregates

Mark R. Daymond

Department of Materials Science and Engineering
Queens University
Kingston, Ontario, K7L 3N6, Canada
e-mail: daymond@me.queensu.ca

INTRODUCTION

This Chapter is an introduction to the various length scales of internal stresses, the causes of internal stresses and in particular their measurement by neutron diffraction. Some of the approaches used to model and interpret internal stresses are also introduced, and the use of these approaches to interpret deformation mechanisms illustrated through a number of examples.

In nearly every case when an engineering material is employed it will experience some stresses, if only those due to its own weight. Real materials are not the uniform continuum they are sometimes considered to be, and local differences in structure or properties mean that the local 'internal' stress is not necessarily equal to any external applied stress. These internal stresses are fundamental in controlling the deformation and failure of materials; they can have a considerable effect on material properties, including fatigue resistance, fracture toughness and strength. These stresses can vary greatly as a function of position within a body, due to the processes experienced during its production. Consequently, their measurement and interpretation is of considerable interest (e.g., Hutchings 1990), and great efforts have been employed over the years in developing accurate and precise measurement techniques. Such stresses can develop in a deformed material at many length scales and from many mechanisms. Fundamentally, internal stresses arise due to the elastic response of the material when an inhomogeneous distribution of non-elastic strains is imposed. These non-elastic strains could be due to plastic strain, precipitation, phase transformations, thermal expansion, etc. (Noyan and Cohen 1987). The origins in all cases thus come down simply to two aspects: heterogeneity and constraint. That is, when the various constituent parts of a component or material would, if unconstrained, exhibit different responses to the applied load (be that stress, temperature, electric field, etc.) the constraints imposed by the bulk of the surrounding material result in the build up of stresses between these constituent parts.

However, before the occurrence and impact of stresses in materials is discussed in some detail, the use of diffraction techniques for the measurement of stress is summarized, with an emphasis on neutron diffraction. Examples of the work carried out in the modeling of internal stresses will be reviewed in some detail, since it is the combination of experimental data with micromechanical modeling that has proven to be a highly fruitful area of research, providing great insight into the microscale deformation mechanisms which may be occurring. Considerable work has been carried out in this field in the study of the deformation of metals, and to a lesser extent in ceramics, and so it is here that the focus of the Chapter lies. However the Chapter also covers recent work applying these ideas to geological materials, and considers some of the particular issues which need to be addressed in applying the large body of work on engineering materials to geological systems.

1529-6466/06/0063-0016$05.00

DOI: 10.2138/rmg.2006.63.16

MEASUREMENT OF STRESS BY DIFFRACTION

One highly successful technique for making non-destructive measurements of stresses within engineering samples is diffraction, using either X-rays or other probes (Noyan and Cohen 1987). Conventional X-rays are limited to surface studies, while more penetrating radiation such as neutrons or synchrotron X-rays allow studies at greater depth in the material (Fitzpatrick and Lodini 2003). This has the attraction of allowing the study of bulk average properties, without the complications of the changing stress states and constraints which occur near the surface of a material. Many text books are available on the subject of the measurement of stresses by diffraction; while the aspects relevant to the study of deformation are summarized here, for more detailed and complete discussions the reader is referred either to the two books cited above or to, for example, Hauk (1997), Krawitz (2001) or Hutchings et al. (2005).

Length scales of stress

Those working in the measurement and interpretation of stresses typically classify the measured stresses according to the length scales over which they operate, relative to the microstructural repeat unit. The advent of new strain measurement techniques using EBSD (Wilkinson 1997) or synchrotron X-ray micro-diffraction (e.g., Martins et al. 2004) which probe materials at very short length scales are acting to blur these divisions somewhat, but the distinctions remain very useful. Macroscale or Type I stresses act over length scales at which the material can be considered to be a continuum, usually millimeters or larger. For example, the plastic deformation induced in the surface layers of a material by grinding or shot peening results in stresses in these layers because of the constraining effect of the bulk material below (Fig. 1); these stresses act over hundreds of micrometers. Other examples include the stresses in anything from bent bars and welded sheet metal, to the stresses between tectonic plates. Type I stresses are the stresses that would be of primary concern to a design engineer. A whole host of techniques, both destructive and non-destructive, are available for the measurement of such stresses, and a flavor of some of them can be found in the review by Ruud (1982). At the other

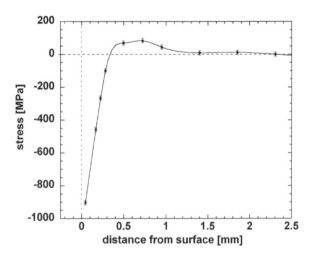

Figure 1. The stress profile generated by shot peening of a nickel plate, illustrating the Type I stress component in the direction in the plane of the surface of the plate. Plasticity is expected to have occurred in approximately the top 0.25 mm of material and generates compressive stresses near the surface with balancing tensile stresses deeper in the material. Data was obtained by the author on the ENGIN instrument at the ISIS pulsed neutron source, as part of the VAMAS TWA20 exercise (Webster 2001), and is previously unpublished. Lines are a guide to the eye.

end of the lengthscale, Type III stresses vary *within* the individual length scale of the microstructure, i.e., within the grain or crystallite. They may vary on the level of the atomic lattice arising from, for example, atomic vacancies, the distortion of the lattice caused by dislocations, or the strains at a semi-coherent precipitate interface. In contrast, Type II stresses vary on the intermediate length scale which is comparable to that of the microstructure repeat unit. For example, the different properties of the phases of a composite lead to so-called "interphase" Type II stresses (see below).

Figure 2 is a schematic illustrating the different length scales of stress discussed above, and how they relate to a microstructure. The stresses experienced at any given position in the material will be a summation of the various types described above. The discussion in this Chapter will consider mainly Type II stresses, with a more cursory treatment of Type I and III, because it is Type II stresses which are principally important in interpreting and understanding the response of the polycrystalline aggregate to applied load. However the various techniques for the measurement of internal stresses will be sensitive to the different types in different ways, and so some attention is given to the other types as necessary. In the literature a distinction is usually drawn between residual and internal stresses; the former are those which are self-equilibrating in a body which has no external forces or constraints acting on it. They will be additive to those stresses arising from any applied load; the internal stresses are thus a combination of the residual and those due to the applied stresses.

Experimental arrangements

Neutron diffraction can be used to measure components of the strain tensor directly from changes in crystal lattice spacing. When illuminated by radiation of wavelength similar to atomic lattice spacings, crystalline materials diffract this radiation as distinctive Bragg peaks. The stress experienced

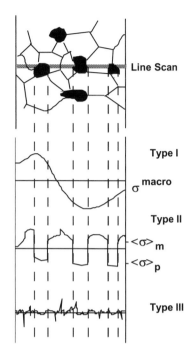

Figure 2. Schematic illustrating the length scales of stresses. The top part of the figure illustrates a microstructure, with different grains, and different phases (light/dark). The stresses that exist along the line scan shown are indicated in the lower part of the figure. Type I stress vary on a long length scale, while Type II stresses vary on the scale of the microstructure, as a function of phase (e.g. matrix $<\sigma>_m$ and particle $<\sigma>_p$, where "$<>$" indicates "average") as well as a function of grain orientation within the phase. Type III stresses vary within the microstructural element. The total stress at any given position in the material is the sum of the three stress contributions. [Reprinted from Fitzpatrick et al. (1997), Fig. 1, with permission of Elsevier].

by the material causes a change in the spacing of the atomic lattice. If the entire volume that is diffracting and contributing to a particular diffraction peak has a change in mean elastic strain, that diffraction peak moves. This is the case for Type I and Type II stresses. In contrast, if the average stress in the diffracting volume remains constant but the *distribution* of stresses changes, the shape and width of the peak changes. This effect on diffraction peaks is caused by Type III stresses, and by changes in the distribution of Type II stresses.

Determination of strain

The two classes of sources of neutron source, monochromatic and time-of-flight, can both be used for the measurement of strain. For a monochromatic wavelength source, differentiation of Bragg's law gives:

$$\varepsilon = \left(\frac{\Delta d}{d_0} \right) = -\Delta\theta \cot\theta_0 \tag{1}$$

where $\Delta\theta$ is in radians, and the values θ_0 and d_0 are those obtained in a stress-free sample of the material. Type I and Type II stresses both cause a change in position of the diffraction peak—it is this change which is measured and interpreted as strain, based on Equation (1). Type III stresses, on the other hand, cause a change in the width of the diffraction peak; their analysis is outside the scope of this review, but see e.g. (Warren 1990; Gubicza et al. 2004; Mittemeijer and Scardi 2004). To be confident of correctly determining the peak position, it is important that an appropriate peak shape is used in the peak fitting process, particularly in the case where changes in width may occur. As discussed elsewhere in this issue (Vogel and Priesmeyer 2006, this volume), typically Gaussian or pseudo-Voigt peak shapes are used (e.g., Hutchings et al. 2005).

The neutron diffraction instruments that are used for strain measurements are, in their principles of operation, very similar to general purpose powder diffraction instruments. A number of key differences exist however. Firstly, in contrast to conventional powder diffraction, but in the same way as for texture measurements, when measuring elastic strains it is important to note that the direction in which the strains are measured relative to the sample becomes a critical aspect of the data analysis. The direction in which strain is measured is along the scattering vector $\mathbf{Q}$, which is perpendicular to the diffracting planes shown as diagonal lines in Figure 3a. When measuring multiple diffraction peaks (each with different θ_0) in a particular direction in the sample, the sample/detector configuration is usually reoriented such that the scattering vector $\mathbf{Q}$ always lies in the same orientation relative to the sample. It should be noted that, in general, for each *hkl* diffraction peak a different strain will be determined, due to both differences in the elastic stiffnesses and to plastic anisotropy, as will be discussed below in the sections on intergranular strains. Secondly, such instruments are often used to make measurements to determine the internal stress occurring in large engineering components, or on samples contained within bulky sample environment facilities (required to impose temperature, stress etc. on the samples). As such there are certain physical constraints on the diffraction instrument, particularly the requirement of accurate physical positioning (typically ±0.05 mm) of large objects relative to the diffraction beam. The concepts involved behind the design of a neutron diffraction instrument are discussed in detail by Johnson and Daymond (2002).

At a time-of-flight (TOF) neutron scattering instrument, neutron pulses, each with a continuous range of velocities and therefore wavelengths, are directed at a specimen. The incident spectra are polychromatic, thus many lattice planes are recorded in each measurement. The scattering vectors $\mathbf{Q}$ for all reflections recorded in a given detector lie in the same direction, and thus measure the strain in that same direction. It should be emphasized again that each reflection is produced from a particular family of grains that are oriented such that a specific *hkl* plane diffracts to the detector. Strain can then be calculated from the shift in a given reflection:

$$\varepsilon_{hkl} = \frac{\Delta t}{t_0} = \frac{\Delta d}{d_0} = \frac{\Delta\lambda}{\lambda_0} \tag{2}$$

Figure 3 (*on facing page*). (a) Schematic illustrating the experimental arrangement for measurement of strains at a monochromatic neutron source. The strain is measured in the direction "Q" (e.g., Webster 2001). The diagonal hash lines indicate the measurement volume, and the orientation of the lattice planes which are diffracting. (b) Schematic illustrating the experimental arrangement for measurement of strains at a time-of-flight neutron source. The strain is measured simultaneously in the two directions "Q" (e.g., Webster 2001). The cross hash lines indicate the measurement volume, and the orientation of the two sets of lattice planes which are diffracting. (c) Schematic of the Engin-X beamline for strain measurement at ISIS. Components are: a) incident beam definition (guide and slits); b) neutron detectors; c) positioning table (x-y-z-θ); d) scattered beam definition (radial collimator); e) sample (sphere).

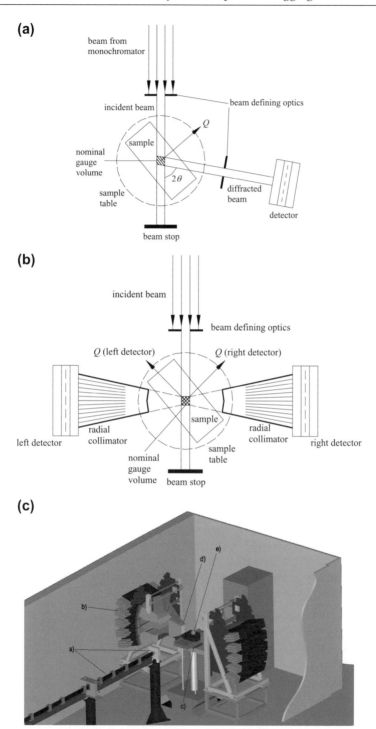

Figure 3. *caption on facing page*

where t is the time-of-flight, d is the lattice spacing and λ is the wavelength. However, since an entire diffraction spectrum with multiple diffraction peaks is obtained for each measurement direction, without the need for any reorientation of sample or detector, it is common practice in the measurement of Type I strains/stresses that instead of analyzing individual diffraction peaks, the strain is determined from the whole diffraction pattern using a Rietveld (1969) style analysis where a model of the crystallographic structure is used to create a simultaneous fit to the entire diffraction spectrum. In this case, strain is given by:

$$\varepsilon = \frac{\Delta a}{a_0} \tag{3}$$

where Δa is the change in lattice parameter relative to the lattice parameter a_0 of a stress-free or reference sample of the material. The strain determined by changes in the lattice parameter in this way from a conventional Rietveld refinement will be only slightly and subtly affected by any Type II stresses, hence its common use for the determination of Type I stresses. The impact of Type II stresses on Rietveld refinements and ways in which the effects can be incorporated and utilized are described by Daymond et al. (1997, 1999a). The diffraction peak observed at TOF sources has a distinct asymmetry, due to the way that the neutrons are produced. However this asymmetry can be easily characterized as an instrumental parameter, and thus an appropriate peak shape used in the fitting procedure (e.g., Ikeda and Carpenter 1985).

The uncertainty in strain u_ε is given by (e.g., Webster 2001):

$$u_\varepsilon^2 = \frac{u_d^2}{d^2} + \frac{u_{d_0}^2}{d_0^2} \tag{4}$$

where

$$u_d^2 = d^2 \left[\frac{u_\lambda^2}{\lambda^2} + u_\theta^2 \cot^2 \theta \right] \tag{5}$$

Equation (4) is the correct value of the overall uncertainty for the strain. However if the same value of d_0 is used to calculate a series of strain measurements, their *relative* uncertainties are more simply given by:

$$u_\varepsilon^2 = \frac{u_d^2}{d^2} \tag{6}$$

The impact of background and the instrument resolution on the strain accuracy obtained, and thus subsequently on the count times required for a given measurement is discussed by Withers et al. (2001).

Some readers may be familiar with the $\sin^2 \psi$ technique used in the measurement of stress by the diffraction of conventional X-rays (e.g., Noyan and Cohen 1987; Hauk 1997). Exactly the same methodology may be employed in neutron diffraction (Hauk 1997), however the imperative for it with conventional X-rays—namely that the small penetration depth means that strains can only be measured in reflection and not in transmission—is not present for neutrons. The technique is thus only sometimes used, since although it adds information and accuracy, it requires more experimental time.

By suitable use of slits and focusing mechanisms on both the incident and scattered beam, neutron diffraction can be used to determine the elastic strain *within a defined volume* in a poly-crystalline solid (shown as diagonal hash shading in Fig. 3a). By moving the sample relative to this volume (which is fixed by the geometry of the diffraction apparatus), it is possible to build up a map of the spatial variation of strain, and/or phases present. The minimum spatial resolution presently achievable by this technique with neutron diffraction is of the order of 1/3 or 1/2 mm, but experiments carried out with spatial resolutions of between 1 to 4 mm are more common. Careful spatial alignment of the beam-defining mechanisms and the sample are required when carrying out mapping of the variation of strains as a function of position in an object. This is now a common and well established technique for engineering components (e.g., Preuss et al. 2002; Edwards et al. 2005), and the technique has also been applied to the mapping of archaeological (Siano et al. 2006) and geological (Meredith et al. 1997; Scheffzük et al. 2005) samples. Some interesting work has looked at mapping the strains occurring around a crack tip under load (Sun et al. 2005), although the high spatial resolution required for this often makes such studies more suited to synchrotron X-rays (Withers et al. 2002), as discussed below. For a detailed description of the experimental methods and potential pitfalls involved in making spatial mapping measurements, which is outside the scope of this review, the reader should consult Webster (2001) or Hutchings et al. (2005).

Table 1. Neutron sources with dedicated strain measurement facilities.

Facility
AEC, South Africa
ANSTO, Australia
FRM-2, Germany
GKSS, Germany
HiFR, ORNL, USA
HMI, Germany
ILL, France
ISIS, UK
JAERI, Japan
JRC, Netherlands
KAERI, S. Korea
KUR, Japan
LANSCE, LANL, USA
LLB, France
MURR, USA
NIST, USA
NPI, Czech Republic
NRC, Canada
PSI, Switzerland
SNS, USA (under construction)

This section has introduced the key requirements of a neutron diffractometer used to measure stresses, which are similar for both monochromatic and time-of-flight neutron sources. These are a well defined incident beams which are critical to spatial mapping measurements; standard detectors appropriate for the type of source; a method to define where in the sample the scattered beam originates, which is dependent on the type of source; and a high capacity, high precision table capable of moving objects relative to the neutron beam. Typical capacities for such tables in modern instruments might be 500 kg, with 500 mm of travel, and 0.01 mm/100 mm precision. A schematic of one of the most recent stress measurement diffractometers, ENGIN-X (Santisteban et al. 2006a) is given in Figure 3c, and illustrates these components. A list of neutron sources which have neutron diffractometers dedicated to strain measurement is given in Table 1 (after Webster 2001). The GKSS, ISIS, LANSCE, SNS and PSI facilities have pulsed neutron strain measurement diffractometers, while the other facilities in Table 1 have monochromatic neutron strain measurement diffractometers. The benefit of the pulsed neutron sources for strain measurements is that, as already stated, a large number of diffraction peaks are easily obtained. The major benefit of the monochromatic sources is that if only one or a small number of diffraction peaks are required, data acquisition times are usually lower than at a pulsed source.

Calculation of stresses

So far we have considered only the measurement of strains. Stress and elastic strain are second rank tensors which are related through the elastic constants of a solid. Given the strain it is therefore possible to calculate the *stress* in the scattering volume provided the relevant material elastic constants are known. Determination of the full strain tensor requires measurements of the elastic strain along at least six independent directions. If the principal strain directions within the body are known this can be reduced, typically to three orthogonal directions. The conversion of strain to stress follows a slight modification of the well-known equations that are used for isotropic solid state mechanics (Hooke's law):

$$\sigma_{xx} = \frac{E_{hkl}}{\left(1+\nu_{hkl}\right)\left(1-2\nu_{hkl}\right)}\left[\left(1-\nu_{hkl}\right)\varepsilon_{xx} + \nu_{hkl}\left(\varepsilon_{yy} + \varepsilon_{zz}\right)\right] \tag{7}$$

where E_{hkl} and v_{hkl} represent the Young's modulus and Poisson's ratio respectively appropriate for the given *hkl* lattice reflection, and ε_{ii} and σ_{ii} represent the strain and stress respectively in the directions indicated. The choice of values for E_{hkl} and v_{hkl} is discussed in more detail in the section on diffraction elastic constants. It should be noted that Equation (7) can be used even when the strains are not in principal directions, provided that they are orthogonal. However in this case, the stresses determined will also not necessarily be principal stresses, i.e., the presence of any shear stresses is ignored. Typically principal directions are assumed based on sample/experiment geometry, with the actual experimental determination of principal direction (Krawitz and Winholtz 1994) being the exception, rather than the rule.

In the case of the Rietveld analysis of multiple diffraction peaks, typically the macroscopic elastic properties would be the appropriate ones to be used in Equation (7), as justified empirically by Daymond et al. (1997, 1999a), and somewhat more rigorously in (Daymond 2004). Authors have also suggested alternate multi-peak averaging approaches appropriate for the study of Type I stresses (Kamminga et al. 2000; Daymond 2004).

Comparison with some other techniques

A range of other scattering techniques are available for the measurement of strain, most obviously conventional X-rays (e.g., Noyan and Cohen 1987), but also including high energy synchrotron X-rays (e.g., Fitzpatrick and Lodini 2003) and EBSD (Wilkinson 1997). A major difference in comparison with conventional X-ray or electron diffraction is that the path length of thermal neutrons is much higher for most materials. Thus, while conventional X-rays and electrons probe the surface layers of materials, neutrons interact with the "bulk" of the material. This has a particular impact when considering the stress state found in phases or grains. The near surface will be in plane stress conditions, since the stress component normal to a free surface must be zero. This is likely to be very different from the stress state found within the depth of a material, where the constraint of surrounding grains plays a significant role in the development of internal stresses.

High energy synchrotron X-rays can, however, have a path length which, while typically smaller is certainly close to that of neutrons, particularly for lighter elements (Table 2). With the growing availability of high energy synchrotron sources in the last 10 years, the engineering neutron diffraction strain measurement community has rapidly taken on board the new technique. The first synchrotron X-ray diffractometers which will be dedicated at least part time to strain measurement are presently under commissioning or construction. The great benefit of synchrotron X-rays over neutrons is that a very high flux is obtained, typically allowing 1s data acquisition times, which should be compared to around 100 s at neutron facilities under optimum conditions. Synchrotron X-rays have a very low divergence compared to neutrons. While this has the advantage that it is easy to produce a small beam size if such is desired, it has the negative consequence that for large (20 μm+) grain size samples (see below) very few

Table 2. Approximate attenuation lengths (63% reduction in intensity) for laboratory X-rays, synchrotron X-rays and neutrons. Adapted from (Withers 2004).

Source	Energy (keV)	Wavelength (Å)	Approximate attenuation length (mm)				
			Al	**Ti**	**Fe**	**Ni**	**Cu**
Thermal neutrons	2.5×10^{-5}	1.80	96	18	8	5	10
Synchrotron X-rays	150	0.08	39	14	7	5	5
Synchrotron X-rays	60	0.21	13	3	1.1	0.8	0.7
Synchrotron X-rays	31	0.40	3.3	1	0.16	0.11	0.10
Lab. X-rays (Cu Kα)	8.05	1.54	0.076	0.011	0.004	0.023	0.021

grains within the sample are at the diffraction condition, which may mean that the diffraction pattern obtained does not represent a "powder average." In such cases it may be possible to "rock" the sample during data acquisition to bring more grains into the diffraction condition, or alternately to take advantage of this fact and simply measure these single grains within the polycrystal and thus to interpret their behavior directly (e.g., Martins et al. 2004) rather than via a polycrystal average. At the high X-ray energies typically used for such measurements, the wavelength of the X-rays is very small, resulting in a smaller scattering angle than found with neutrons. For example, an 80 keV X-ray has a wavelength of ~0.15 Å. For a steel (111) diffraction peak, this results in a scattering angle $2\theta = 4°$, compared with a 2θ close to 90° for thermal neutrons. This difference in scattering geometry has an impact on many experimental considerations, and is discussed in more detail in (Fitzpatrick and Lodini 2003). One result is that the depth at which a strain can be measured in an object is highly dependent on which component of strain is being measured relative to the surfaces, meaning that in much of the published literature only in-plane components of strain are reported. While this has an impact on "strain scanning" measurements, it is often not an issue for *in situ* loading experiments and such measurements have become increasingly popular where sample constraints allow (e.g., Daymond and Withers 1996b; Pyzalla et al. 2006).

Finally, we mention in passing the technique of pulsed neutron transmission strain measurement. The transmission spectrum of thermal neutrons through a polycrystalline sample displays sudden, well-defined increases in intensity as a function of neutron wavelength. These Bragg edges occur because for a given *hkl* reflection, the Bragg angle increases as the wavelength increases until 2θ is equal to 180°. At wavelengths greater than this critical value, no scattering by this particular *hkl* lattice spacing can occur, and there is a sharp increase in the transmitted intensity. The transmitted spectrum can be analyzed (Santisteban et al. 2002) and changes in the position of these sharp changes in intensity interpreted as changes in the elastic strain, with the scattering vector **Q** being along the direction of the neutron beam. The set-up in such experiments is simple; the sample is placed in a collimated pulsed neutron beam, and the detector is located in line behind the sample. By using a two-dimensional array of detectors it is possible to map the lattice spacing in simple geometry samples, producing images analogous to neutron radiography although with to-date only millimeter resolution. The technique has been applied to archaeological samples (Siano et al. 2006), and single crystals (Santisteban et al. 2006b) as well as to engineering structures (Santisteban et al. 2002). However the technique requires a white beam of neutrons from a pulsed source and at present is only available at a small number of facilities; the ENGIN-X instrument at ISIS is equipped with a transmission detector which can operate simultaneously with the conventional diffraction detectors.

INTERNAL STRESSES

Some of the example studies described below detail measurements and studies of residual stresses, that is those present in a material due to its previous thermo-mechanical processing. A common origin of such residual stresses is due to thermal misfit, that is differing thermal contractions/expansions between different parts of the sample after exposure to elevated temperatures. However, many studies published have considered not just residual stresses, but the generation of stresses during *in situ* loading. In this way, the deformation history of a single sample can be monitored without the ambiguity of making measurements on a suite of pre-deformed samples. In this case, typically a conventional hydraulic or screw driven loading frame is used, oriented appropriately relative to the diffraction scattering geometry. The majority of work to date in this area has considered simple uniaxial loading. The applied load is usually incremented, and held at constant macroscopic stress (or strain) while neutron data are collected, before the load or strain is again incremented. Typically the load frame is oriented in the neutron diffractometer such that the scattering vector lies either on the axis parallel to the loading direction or perpendicular to it, thus exploring the axial strains and the Poisson strains, i.e., parallel and perpendicular to the applied load respectively (see e.g., Fig. 3b). An example of such an apparatus and its use is described in detail by (Daymond and Priesmeyer 2002). Other additions to such *in situ* loading facilities might include simultaneous heating (Ma et al. 2005) or cooling (Oliver et al. 2004c) of the sample. More recent work has started to address the perceived requirement for loading under elevated confining pressures (Covey-Crump et al. 2006a).

The advantage of diffraction in the study of multiphase polycrystalline materials becomes clear. Since in such samples each crystalline phase will produce a different diffraction pattern, the lattice parameter (or individual peak position) for each phase can be determined and monitored during deformation. By monitoring the changes in the lattice parameter the relative strains (and hence stresses) in each phase can be determined, and compared with the bulk sample response. Thus during *in situ* loading, assuming that the individual phase elastic properties are known, the fraction of the applied load borne by each phase can be determined. The lattice strain is a purely elastic measure, whether the composite as a whole is deforming elastically or plastically. The plastic deformation is thus interpreted based on the nature of load transfer between phases (or grain orientations as will be discussed), and from changes in peak width. It should be noted that recent work (Poulsen et al. 2005) has suggested that even amorphous structures, with their very different scattering characteristics, are susceptible to some of the treatments described here, but this is very much "work in progress."

Interphase stresses

It is perhaps easiest to understand the origin of Type II internal stresses when we consider the deformation of a composite material. Man-made composites have become ubiquitous; materials that are produced by physical mixing and interpenetration of individual phases, designed to provide properties which are, overall, superior to those of the individual constituent phases. Man mimics nature in this case since most natural materials are in fact composites of some form. The understanding of the way that the properties of the individual phases combine to produce the properties of the composite, and the distribution and transfer of stress between the phases of a composite during deformation, has long been an area of interest due to the influence of these stresses on the properties of the bulk composite material. In a multiphase material, the stresses arising between the phases can be considered to be due to shape misfit between the phases and to have two general origins: load transfer arising from differences in the elastic properties of the two phases; and "shape misfits" caused by effects such as differential thermal expansion or differential plastic flow (e.g., Clyne and Withers 1993). Figure 4 shows a schematic illustrating the origin of interphase stresses due to thermal misfit strains.

During loading of a sample, differences in the elastic properties of the phases lead to internal stresses and the distribution of load between the phases. Subsequent plastic flow due to local differences in yield stress and hardening coefficient causes local reaction stresses and thus redistribution of the applied load. The various micromechanical effects often give multiphase materials improved properties relative to their constituent phase, but sometimes they may also be detrimental and therefore must be understood to allow the use of composites use in real components. Many neutron diffraction studies have demonstrated the load partitioning between phases occurring in multi-phase metal-based composite systems, typically in situations where one phase does not yield (e.g., Withers et al. 1989; Allen et al. 1992; Bourke et al. 1993; Withers and Clarke 1998), although a number of systems where both phases potentially exhibit plastic behavior have also been studied (e.g., Dunst and Mecking 1996; Carter and Bourke 2000).

Consider the data shown in Figure 5, which shows the elastic strains in a Cu-Mo composite under applied loading to a few percent plastic strains (Daymond et al. 1999b). The Mo constitutes 15 vol% of the composite as a whole and is made up of discrete approximately equiaxed particles in a continuous Cu matrix. As is common for such measurements the strains are shown relative to the starting unstressed value, that is the quoted strains are not

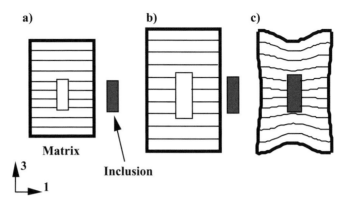

Figure 4. Schematic demonstrating the origin of thermal misfit strains, due to a rise in temperature on an idealized single reinforcement / matrix composite. Consider first an inclusion that fits exactly within a hole in the matrix. a) The inclusion is separated from the matrix. b) Both are heated, resulting in an expansion of the matrix and inclusion; the matrix expands more than the inclusion due to a higher coefficient of thermal expansion. c) The matrix and inclusion are joined back together; the reinforcement is in tension (pulled larger than it would otherwise be) and the matrix must have balancing compression.

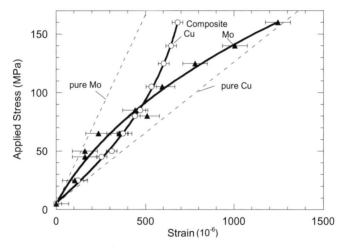

Figure 5. The internal phase strains measured by neutron diffraction in a Cu-Mo composite undergoing quasistatic tensile deformation. Lines are a guide to the eye. The nominal moduli for the pure phases is also shown (Used with permission. Fig. 3a from Daymond et al. 1999b).

absolute, since they do not take into account any initial thermal residual stresses produced during fabrication. The strains are the elastic strains parallel to the applied load as measured by diffraction. The experimental scatter shown (±30 µe in the Cu and ±70 µe in the Mo, where µe represents a strain of 10^{-6}) was determined from the average error in lattice parameter calculated from the statistics of the diffraction analysis. The larger errors in the Mo phase are due to the smaller scattering volume and concomitant poorer counting statistics.

For applied stresses below 70 MPa, the reinforcement and matrix exhibit a linear response, with apparent moduli (slope of applied stress vs. elastic strain) which are within the limits imposed by the single phase Young's moduli. However, the strains in the Mo reinforcement are larger than in a hypothetical pure Mo sample and the strains in the Cu matrix are smaller than a hypothetical pure Cu sample. From this we can infer that elastic load transfer occurs from the compliant Cu matrix to the stiffer Mo inclusions, demonstrating the effect of constraint simply in the elastic regime. At a stress of around 80 MPa, plasticity starts to occur in the Cu matrix, which has a lower yield stress than Mo. For a further given increment in applied stress the increase in elastic strain (i.e., stress) in the Cu is now smaller than in the elastic regime. Correspondingly, the increment in elastic strain in the plastically harder (non-yielding) Mo phase is larger than in the elastic regime. This is a classic demonstration of composite plastic load transfer (Clyne and Withers 1993).

Intergranular strains – elastic anisotropy

At a smaller length scale, we must take into account that for almost all materials the elastic stiffness of a single crystal (or crystallite) of material is dependent on its orientation relative to the applied load. This elastic anisotropy leads to a variation in the internal strains (and hence stresses) which are experienced by differently oriented grains in a polycrystal due to an applied load. The most well known result of this effect, is the requirement to use "diffraction elastic constants" or "plane specific stiffnesses" when making diffraction measurements of stress.

The elastic anisotropy of crystal structures in general is treated in many texts, for example, (Nye 1992). The treatment below summarizes that given by Nye, but considers only the high symmetry cubic and hexagonal structures. These make up the majority of engineering, though not geological materials. Using the conventional collapsed tensor notation (Nye 1992) of strain (ε), stress (σ), stiffness (C) and compliance (S) and the summation convention:

$$\varepsilon_i = S_{ik}\sigma_k \quad \text{and} \quad \sigma_i = C_{ik}\varepsilon_k \tag{8}$$

For a *single crystal* with cubic symmetry, it is found that

$$\frac{1}{E_{hkl}} = S_{11} - 2\left(S_{11} - S_{12} - \tfrac{1}{2}S_{44}\right)A_{hkl} \tag{9}$$

where E_{hkl} is the plane specific Young's modulus of the single crystal, and

$$A_{hkl} = \frac{\left(h^2k^2 + h^2l^2 + k^2l^2\right)}{\left(h^2 + k^2 + l^2\right)^2} \tag{10}$$

and thus has limiting values of $A_{h00} = 0$ and $A_{hhh} = \frac{1}{3}$. Hence the term $2(S_{11}-S_{12}-\frac{1}{2}S_{44})$ governs the anisotropy of cubic materials, that is the way that the stiffness varies as a function of *hkl*. It is often reported as the dimensionless "cubic anisotropy factor" $2(S_{11}-S_{12})/S_{44}$ $= 2C_{44}/(C_{11}-C_{12})$ with values in metals ranging from 0.71 in chromium, to 1.01 for the nearly isotropic tungsten, to 3.38 for copper (Hosford 1993). In the case of a *polycrystal*, the anisotropy inherent in the individual crystallite response given in Equation (9) will still be apparent. However, in the same way that the apparent moduli of the phases in the two phase composite described in the previous section were altered by the elastic load transfer and brought closer together than the individual phase moduli, the values of E_{hkl} in the polycrystal will not be as extreme as that seen in the single crystal. That is, while in copper we expect the plane specific modulus E_{200} of the *polycrystal* (determined by examining the response of the {200} type diffraction peak—the diffraction elastic constant) to be lower than that of E_{111}, in practice the difference between the two values will be smaller than that given by Equation (9) for a *single crystal*. The determination of the plane specific diffraction elastic constant of a polycrystal requires the use of the modeling approaches described in the next section. We note in passing however, that a linear variation with modulus as a function of A_{hkl} is still expected to hold for an untextured cubic polycrystal (Bollenrath et al. 1967), i.e., for a random polycrystal:

$$\frac{1}{E_{hkl}} = S - S'A_{hkl} \tag{11}$$

where S and S' are related to S_{11}, S_{12}, S_{44} via the polycrystal averaging method used (see section on models below). For the majority of cubic metals $E_{111} > E_{200}$ however there are some exceptions where the anisotropy is in the opposite sense such as chromium and niobium, and this is also the case for some inorganic phases, such as NaCl.

Due to the elastic anisotropy of the individual crystallites, the presence of texture has an influence on the observed elastic response of a polycrystalline aggregate. This is discussed in more detail in the section covering diffraction elastic constants.

For hexagonal structures, as would be expected given the lower symmetry, a more complicated expression holds and three parameters are required to describe the variation in elastic stiffness as a function of *hkl*. The variation is solely as a function of angle relative to the "*c*" axis, with the elastic stiffness being isotropic within the "*a*" plane. For hexagonal structures the appropriate expression is

$$\frac{1}{E_{hkl}} = \left(1 - L^2\right)^2 S_{11} + L^4 S_{33} + \left(1 - L^2\right) L^2 \left(2S_{13} + S_{44}\right) \tag{12}$$

where L is the cosine of the angle between the unit vector of the *hkl* plane normal and the hexagonal *c*-axis. As symmetry is reduced further, more and more parameters are required to describe the variation of elastic stiffness with crystallographic direction, for example see Nye (1992).

Intergranular strains – plastic anisotropy

As slip occurs preferentially on certain slip systems (plastic anisotropy), plastic relaxations lead to a further level of misfit stresses which is dependent on grain orientation. In effect one can consider the polycrystal as a composite in which every orientation of grain has a slightly different yield stress, thereby producing a highly complex stress-strain response. While long known of (Greenough 1947, 1952) these strains due to plastic anisotropy became of interest to the neutron diffraction community (MacEwen et al. 1983) more recently, in particular because of the potential to validate micromechanical plasticity models (see below). In Figure 6 the response of the lattice reflections with the five largest *d*-spacing are plotted against the applied stress for a single phase austenitic (*fcc*) stainless steel undergoing tensile loading (Daymond et al. 1997). Also shown is the 0.2% strain yield limit for plasticity, commonly used to determine the onset of macroscopic plasticity (e.g., Dieter 1986). It is worth re-iterating that each line represents the response of a family of grains within the polycrystal which are oriented such that the given *hkl* lattice plane is parallel to the loading direction. As expected, the initial response is elastic (linear), as described in the previous section. Deviations from linearity of the individual plane responses occur close to the onset of macroscopic plasticity. Once plastic deformation occurs, the yield of preferentially oriented grains relative to their neighbors causes strain redistribution, and there is a strong divergence from the hitherto linear response. In direct analogy to the description of the generation of interphase stresses in the section above, as subsets of grains become plastic, they do not accumulate elastic load at the same rate as they did when they where elastic, causing changes in the partitioning between the different grain orientations for a given incremental load increase. The first diffraction peaks that show evidence of the onset of plasticity in this case were the 531 (not shown) and 331 reflections, which showed an upward inflection at about 200 MPa (Clausen et al. 1999). These grain families thus play the part of plastically "soft" directions, with corresponding load transfer to the grains that are still deforming elastically, such as the 200 grains. However in contrast to the Cu-Mo example given above, at around 260 MPa this latter grain family also starts to

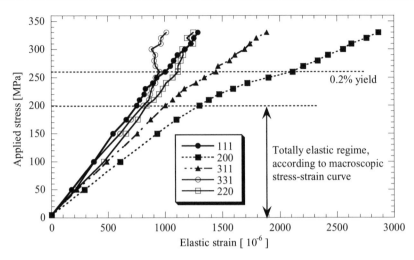

Figure 6. The internal strains measured on different diffraction peaks in a stainless steel undergoing quasistatic tensile deformation. Lines are a guide to the eye. (Used with permission of American Institute of Physics, Fig, 2 in Daymond et al. 1997)

show yield behavior. At this point, model predictions indicate that the majority of the grains have become plastic and accordingly the lattice strain response is again linear with increment in applied stress, although with slightly different gradients to that seen in the elastic regime. The origin of the plastic anisotropy comes from the anisotropy of slip in an *fcc* crystal structure. In *fcc* Fe (and hence in steel) slip occurs on the {111} planes and in the <110> directions. The analysis of the impact of this fact on the deformation of single crystals is well documented in text books (e.g., Dieter 1986), and long established; the concept is typically first met through the introduction of the Schmid factor, which in a single crystal relates the applied stress to a resolved shear stress along the particular slip system. During loading of a single crystal those slip systems that first reach the critical resolved shear stress required to initiate slip will yield first. The situation is more complicated in the case of a polycrystal and we cannot simply determine the order that differently oriented grains will reach a stress sufficient to cause yield. This is because the stress experienced by any particular grain orientation is not the applied stress, for two reasons. Firstly, due to elastic anisotropy the grains oriented, for example, with the highly elastically compliant {200} direction parallel to the applied load actually experience a lower stress than the elastically stiff {111} direction. Secondly, the surrounding grains provide a constraint resulting in a non-uniaxial stress state being experienced by any individual grain within the polycrystal aggregate. Hence to quantitatively describe the evolution of strains such as those shown in Figure 6, we require a model of polycrystal plasticity (see below).

Due to the elastic and plastic anisotropy of the individual crystallites, the presence of crystallographic texture can have an influence on the observed intergranular strains generated in the polycrystalline aggregate. Firstly, texture can alter the relative magnitude of the stresses experienced by differently oriented grain families, thereby leading to changes in how plasticity initiates in the different grain families. In *fcc* materials, where only one family of slip system operates, this effect is relatively small and is typically outweighed by the residual stresses and hardening (caused by dislocation structures) generated by the deformation which produced the texture in the first place (Daymond et al. 2000). In lower symmetry materials however, texture can have a major effect on which type of slip system can operate and hence on the internal strains generated; the effect is even more strongly evidenced when a unidirectional deformation mode such as twinning can operate (Agnew et al. 2003; Oliver et al. 2004b, 2005).

Some recent *in situ* studies have investigated the influence of cyclic loading on interphase or intergranular stresses. In this case the strain development can simply be monitored through multiple consecutive cycles (Lorentzen et al. 2002) however this ties the maximum cycling rate to the data acquisition rate and thus, given finite experimental time, the maximum number of cycles which can be studied. A greater number of cycles can be studied by interrupting rapid cyclic loading at various stages of the fatigue life to then measure the strain development around a single cycle (Korsunsky et al. 2004). Short cycle times can also be dealt with through the use of stroboscopic data acquisition techniques (Daymond and Withers 1996a) which sacrifice time resolution as a function of cycle number to increase time resolution *within* the cycle.

Finally, the intergranular strains which arise due to plasticity have been used to provide a semi-quantitative "fingerprint" of the macroscopic plastic deformation undergone by a material. The approach is to fit the differences in strains observed for the various diffraction peaks to a simple single parameter model. By comparison with calibration data, obtained for example from a uniaxial loading test, a correlation can be drawn between the single parameter fit to the diffraction data and the macroscopic plastic strain. The approach has been used for cubic and hexagonal materials with some success (Daymond et al. 1997, 1999a; Korsunsky et al. 2002).

Intergranular strains – perpendicular to the applied load

In the direction perpendicular to a uniaxially applied load, initially elastic Poisson strains are observed, with the various grain families being gradually brought into compression for an applied tensile stress. Just as each lattice plane has a different diffraction elastic constant, it will also will have a different Poisson's ratio (e.g., Daymond and Bouchard 2006). Deviations from linearity occur in the Poisson grain family strain response at applied stresses corresponding to macroscopic plasticity, however it is considerably harder to explain these nonlinearities in terms of simple "load sharing" arguments. To explain part of the reason for this, let us consider a cubic material. If we consider a (110) plane oriented with its plane normal *parallel* to the applied load, rotating the plane around the (110) normal simply alters the plane which is perpendicular to the loading direction. This has little effect on the strains which are measured in the parallel direction, since the influence is only via interaction with neighboring grains. If a sufficiently large population of grains is monitored we can expect that any effects will average out and thus the dispersion of strains in the population of grains oriented with [110] parallel to the loading direction is small. However, if the (110) plane normal is oriented *perpendicular* to the applied load, rotating about the [110] axis brings a variety of different lattice planes into alignment with the direction of loading—in fact it is possible to have either the (111)-extreme stiffness type or (200)-extreme compliance type planes oriented axially, while maintaining the [110] axis transverse to the load. The response of a given (110) plane perpendicular to the loading direction will thus be highly dependent on which plane is oriented parallel to the loading direction; in fact this can affect both the elastic Poisson and the plastic response of the observed diffraction peak (Oliver et al. 2004a). This effect was also observed in model calculations (Clausen et al. 1998) which predicted, for *fcc* materials, firstly a larger standard deviation of the perpendicular lattice strain response than of the parallel lattice strain in the elastic regime for elastically anisotropic materials, and secondly a drastic increase in the standard deviation of the perpendicular lattice strain in the plastic regime for both elastically isotropic and anisotropic materials. Thus the dispersion of strains in the direction perpendicular to the applied load is very large, and will be highly dependent on the exact population of grains present. Some subsets of the grain population may actually show a decrease in the transverse compressive strain, possibly even becoming tensile in the transverse direction (Oliver et al. 2004a) for an applied tensile stress. Even a small weighting of one level of grain orientation over another, e.g. preferred (111 parallel, 110 perpendicular) over (200 parallel, 110 perpendicular) will alter the mean strain observed in a diffraction peak perpendicular to the applied load. Both experimental and model results will be sensitive to this, because a relatively small number of grains contributed to a given diffraction peak.

In order to address these complexities in interpreting strains occurring in directions away from the primary loading axis, a number of authors have started to measure—in analogy to texture measurements—"strain pole figures" where the strain of a given reflection is measured for all orientations in a sample (e.g., Larsson et al. 2004). Some authors have gone further and calculated a "stress orientation distribution function" (SODF) based on such strain pole figures (Wang et al. 2003). However, because of the extra complexity compared to texture measurements (since a strain tensor must be determined, not just a crystal orientation), it is necessary to use some model of how strain is distributed between differently oriented grains in order to calculate the SODF, and it is thus not a direct experimental measure; Wang et al. (2003) use an elastic Eshelby model (see below).

MODELING OF PHASE INTERACTIONS AND INTERNAL STRESSES

Elastic models

Bulk aggregate properties. The evaluation of mean elastic properties for polycrystals or multi-phase materials is one of the more important problems in micromechanics and several approaches have been used over the years. Some of the more significant ones are discussed below. The key issue is one of determining the response of the aggregate, based on the properties of its constituent phases or grains. For engineering materials the drive is typically the desire to design optimum material (composite) properties for a given application. For geological materials it may be because we have made measurements on a multiphase system which has a particular geometric microstructure and wish to apply the results to a differently structured system. A number of models of varying complexity have been developed to model either the "bulk" properties based on the properties of the constituent phases, or of interpreting the behavior of constituent phases obtained by measurement.

In the Voigt (1910) approximation, in a composite body subjected to a stress at its boundaries all the composite elements (whether they are phases or grains with different orientations) are subjected to the same uniform strain, which is thus the macroscopic composite strain. We then expect obtain stress discontinuities at the boundaries of phases. In contrast, in the Reuss (1929) approximation it is assumed that the stress in the phases of the composite material is equal to the average stress applied to the material, resulting in strain discontinuities at phase boundaries. Using these approaches it is possible to develop expression for the bulk elastic properties, based on those of the constituent phases, see Simmons and Wang (1971) or Fitzpatrick and Lodini (2003). However these models are typically considered only as bounds for the real behavior since, based on energetic grounds (Hill 1952), they represent the extremes for a purely elastically deforming material. Nonetheless, for the determination of composite properties, where the constituent phases are elastically isotropic, they are easy to use and the aggregate modulus of a composite can be very simply determined. The so called "Hill-Neerfield average", which is obtained by taking the arithmetic mean of the Reuss and Voigt approaches is often used as a first estimate of material behavior for approximately isotropic systems (e.g., Noyan and Cohen 1987), and is typically found to be in good agreement with more complex models, at least in the case of untextured polycrystals.

Hashin and Shtrikman (1962, 1963) have also derived elastic properties for multiphase materials, making no prescriptions about the geometry of the phases, assuming only that the individual phases are isotropic, homogenous and well distributed. They defined the elastic behavior of the material in terms of the strain energy stored in the material subjected to uniform strains or stresses. By minimization of either potential or complementary potential energy under suitable constraints, two bounds on behavior can be produced which are more restrictive than the Voigt / Reuss limits. The Hashin and Shtrikman model is often used to check more complicated models' predictions of the elastic moduli of multi-phase systems because, since the model is generally to be considered to come close to the limit of how closely the two bounds for composite behavior can be brought when the only given information is elastic properties and volume fraction. The Hashin and Shtrikman bounds for isotropic materials are:

$$K_{HS} = \left\{ \sum_i \frac{c_i}{3K_i + 4\mu} \right\}^{-1} \sum_j \frac{c_j K_j}{3K_j + 4\mu} \tag{13}$$

$$\mu_{HS} = \left\{ \sum_i \frac{c_i}{6\mu_i(K + 2\mu) + \mu(9K + 8\mu)} \right\}^{-1} \sum_j \frac{c_j \mu_j}{6\mu_j(K + 2\mu) + \mu(9K + 8\mu)}$$

where c_i is the volume fraction of the ith phase (or grain family), K_i and μ_i are the bulk and shear modulus of the ith phase respectively and:

if $K = \max\{K_i\}$, $\mu = \max\{\mu_i\}$ we obtain the upper bound

but if $K = \min\{K_i\}$, $\mu = \min\{\mu_i\}$ we obtain the lower bound.

The Young's modulus can be calculated from K and μ for either bound using the conventional isotropic relation:

$$E_{HS} = \frac{9K_{HS}\mu_{HS}}{3K_{HS} + \mu_{HS}} \tag{14}$$

For example, the elastic response of an approximately equi-volume olivine-magnesiowüstite composite under uniaxial compression has been shown to lie within relatively tight Hashin and Shtrikman bounds (Fig. 7), based on neutron diffraction measurements of the internal strains borne by each phase (Schofield et al. 2003). Covey-Crump et al. (2006b) have shown how the strain borne by the phases in a halite-calcite composite, starts off roughly equidistant between the Voigt and Reuss limits during initial loading, and then deviates from linearity tending towards the Voigt limit as plasticity occurs in the halite phase (Fig. 8). The initial formulation was later extended (Watt 1980; Watt and Peselnick 1980) to materials of lower symmetry (except triclinic) for which, naturally, rather more complex versions of Equation (13) arise.

Both the advantage and disadvantage of the modeling approaches given above is that no description of the microstructure is included. Many more complicated models which do take into account geometrical arrangements have been developed in the prediction of composite elastic properties; some such as e.g., the Shear Lag Model are reviewed in Clyne and Withers (1993), while others, such as the Eshelby model, are discussed in the context of composites below.

Diffraction elastic constants. While the determination of average aggregate properties based on individual components is an important area of study, the calculation of diffraction elastic constants (i.e., the plane specific Young's modulus) from the single crystal elastic stiffnesses is of great interest to the diffraction community. The same models described in the previous section can be applied to this task. For example, for the Voigt model we see that all grains in the polycrystal aggregate will experience the same strain (which must be the average strain), hence all *hkl* orientations will have the same lattice strain in a given direction, and thus the moduli can be found by the approach given in the previous section. For the Reuss model, since all grains experience the same (applied) stress, the appropriate elastic modulus is simply that described in Equation (9). The same modeling approaches can also be employed to determine the appropriate modulus in the Poisson's direction (e.g., Noyan and Cohen 1987).

The most common model used in the calculation of diffraction elastic constants of untextured polycrystalline materials however is the Kröner (1958, 1961) model. This approach

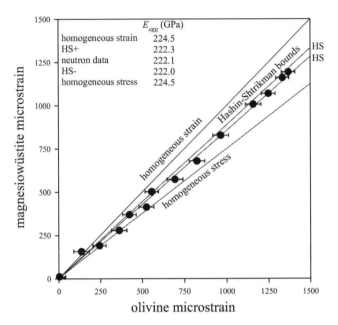

Figure 7. The relationship between the phase average elastic strains parallel to the direction loading, of the phases of an elastically isotropic 54% olivine + 46% magnesiowüstite composite during elastic deformation, from Schofield et al. (2003). Also shown are the theoretical bounds on the behavior predicted by homogeneous stress (Reuss) and homogeneous strain (Voigt) conditions, and the Hashin-Shtrikman bounds, which the neutron diffraction data lies between.

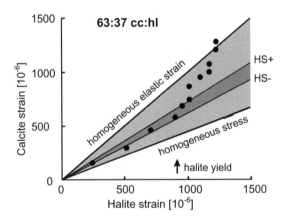

Figure 8. Comparison of the calcite and halite phase average strains parallel to the direction of loading at various increasing applied stresses, for a 63% calcite/37% halite (by volume) composite. Also shown is the strain partitioning predicted during the elastic phase of the deformation assuming homogeneous stress (Reuss) and homogeneous strain (Voigt), and the region that lies between the Hashin-Shtrikman (HS±) bounds. While deformation is elastic, the system is well described by the HS bounds, but once plasticity occurs in the halite, deviation from linearity occurs indicating load redistribution from the halite to the calcite phase. [Used wtih permission from Elsevier, from Covey-Crump et al. (2006b), Fig. 1b.]

utilizes the Eshelby approach (see below) so is often also referred to as a self-consistent analysis. By assuming random texture and spherical grains it is possible to derive analytic expressions for the diffraction elastic constants. These expressions are relatively complex, for example see Fitzpatrick and Lodini (2003). Here just the final result for cubic symmetry materials is given:

$$G^3 + \alpha G^2 + \beta G + \gamma = 0 \qquad (15)$$

where

$$\alpha = (9A_1 + 4A_2)/8$$
$$\beta = A_2(3A_1 + 12A_3)/8$$
$$\gamma = 3A_1A_2A_3/8$$
$$1/A_1 = 3(S_{11} + S_{12}); \qquad 1/A_2 = S_{44}; \qquad 1/A_3 = 2(S_{11} - S_{12})$$

The solution to Equation (15) is substituted into Equation (16):

$$E_{hkl} = \frac{\omega}{1 - 2\omega(1 - 5A_{hkl})t_{44}}; \qquad \nu_{hkl} = \frac{2G\left[1 + \omega(1 - 5A_{hkl})t_{44}\right] - \omega}{2G\left[1 - 2\omega(1 - 5A_{hkl})t_{44}\right]} \qquad (16)$$

where A_{hkl} is given by Equation (10) and

$$t_{44} = \frac{(G - A_2)(3A_1 + 6G)}{2G\left[8G^2 + G(9A_1 + 12A_2) + 6A_1A_2\right]}; \qquad \omega = \frac{9A_1G}{3A_1 + G} \qquad (17)$$

The anisotropy of the individual crystallites means that texture *does* have an effect on the observed diffraction elastic constants. Thus, more recently, in order to determine diffraction elastic constants workers have employed Finite Element crystal plasticity approaches (Wern et al. 2002), or self-consistent approaches (see below) which are more complex than the Kröner model. Such models have the advantage of being able to include the influence of texture on elastic properties, but the disadvantage of requiring numerical solution.

The reverse problem—determining the single crystal elastic constants given the measured polycrystal diffraction elastic constants—is tractable using appropriate modeling techniques. This reverse approach to calculating the single crystal stiffness is particularly important for many engineering systems e.g., TiAl where single crystals of sufficient size for individual testing simply cannot be fabricated. A similar issue would also arise for many geological materials. The results obtained by such methods match reasonably well with other reported data when alternative approaches are used. However, it should be noted that the results are sensitive to the quality of the data and the details of the modeling approach used. Using an appropriate approach is particularly important when the polycrystal exhibits texture. Dealing with this problem is outside of the scope of this paper; readers are referred, for example, to the review of techniques appropriate for untextured materials given in (Gnaeupel-Herold et al. 1998), and for a rigorous approach to dealing with the reverse problem in the case of textured materials to (Matthies et al. 2001). Some simple approaches are also described in (Howard and Kisi 1999).

Inelastic models

Composite models - Eshelby model. A number of avenues have been explored in attempting to include the effect of phase/grain geometry on the mechanical properties of polycrystalline materials, one of the most successful being the Eshelby or Mori-Tanaka method (Eshelby 1957, 1959; Mura 1987), where inclusions (phases or grains) are modeled as ellipsoids. The approach is to consider the elastic field about a single inclusion in an infinite isotropic matrix of the same elastic constants. The inclusion is "cut" from the unstressed matrix and then allowed to undergo a stress-free shape change. Its surface is then stressed such that it can be replaced and "welded" (i.e. no interfacial sliding allowed) into the original hole from which it was cut. The constraints are then removed, and an equilibrium with the matrix is attained when the inclusion reaches a particular strain. By building up combinations of such shape changes it is possible to simply model elastic and thermal loading of the matrix and inclusion. While such an exercise can be carried out for an arbitrarily shaped inclusions, Eshelby showed that in the special case of ellipsoidal inclusions, an analytical solution is possible. Since ellipsoids can, dependent on aspect ratio, vary from cylinders, to spheres to flat discs, this gives great flexibility in modeling composites. The approach has been used with considerable success in modeling composite mechanics. An accessible review of the method, including its application to the deformation of composites, and comparisons with neutron diffraction strain measurement is given by Withers et al. (1989). The Eshelby method has also been developed extensively e.g., to take account of arbitrary elastic anisotropy of the inclusion and of the host material and for porous elasticity (Levin and Alvarez-Tostado 2003).

As well as the obvious advantages of using the Eshelby technique for modeling elastic properties of composites which are gained from the fact that it incorporates the phase geometry in a simple manner, considerable success has also been achieved using various adaptations of the method to model plasticity in composites (Clyne and Withers 1993). For example Fitzpatrick et al. (1997; 2002) have shown how it is possible to use the internal strains measured by diffraction, in combination with the Eshelby method to separate out thermal, elastic and plastic contributions to the observed response of a two phase composite where one phase is elastic.

Composite models - finite element models. Finite element modeling has been shown to be an extremely powerful tool for investigating load sharing between phases in a composite. In this case, the microstructure of the composite is modeled using a finite element mesh, where the properties of the mesh are varied spatially to match the appropriate elasto-plastic properties of the composite. While a few attempts (e.g., Brockenbrough et al. 1991) have been made to model particular microstructures (by digitizing of actual micrographs) it is in particular the unit cell approach which has had greatest success. In this approach, the random distribution of phases is modeled as some uniform regular array, which can then be approximated using a unit cell approach, i.e., modeling a single reinforcement, but with boundary conditions that are appropriate to mimic an infinite regular array (e.g., Levy and Papazian 1990). This has obvious computational advantages, and seems to be a reasonable approximation in many cases. For example; Daymond et al. (1999b) have used the technique to interpret the shift from plasticity to creep on raising the temperature of Cu-Mo, and to separate the relative contributions of creep and plasticity in the deformation of a thermally ratcheting Al-SiC composite (Daymond and Withers 1996a), while Agrawal et al. (2003) used it to study the generation of tensile stress at the interface of a co-continuous Cu-Al_2O_3 composite due to thermal misfit strains.

Polycrystal plasticity models – Taylor and Sachs. Two of the simplest models for dealing with polycrystalline plasticity developed by Taylor (1938) and Sachs (1928) are—despite their age—still used. The Taylor (1938) model is analogous in many ways to the Voigt model for elasticity; the crystals are treated as rigid-plastic (i.e., no elastic deformation) and plastic strain is assumed to be homogenous, irrespective of grain orientation; this strain is equal to the average strain. The same assumption is made for strain rate. Thus stress is discontinuous across boundaries. In particular, it has been used with considerable success in the prediction of texture development in metals. The determination of yield in a grain is controlled by the usual Schmid relationship. In the Sachs (1928) model on the other hand, rigid plastic grains are assumed to all have the same stress, thus having strain discontinuities at grain boundaries. It is usually considered a lower bound for polycrystal models, while the Taylor model is considered an upper bound. Since these models treat grains as rigid plastic, neither are particularly appropriate to the study of internal strains in materials.

Based on the Eshelby method (described above) more complex schemes were introduced by Kröner (1961) and Budiansky and Wu (1962) and were significant advances in that, unlike the Sachs and Taylor models, they included the elastic anisotropy of the material. The approach taken was to model grains in the polycrystal as ellipsoids in an infinite medium. The medium has the average properties of the population of grains. No direct grain-to-grain interaction is accounted for—all interaction is accounted for via the medium properties. The coupling between the grains and the medium was modeled as purely elastic, and resulted in very low deformation heterogeneity. A significant advance was made by Hill/Hutchinson (next section) who then introduced an elasto-plastic coupling between grain and medium.

Polycrystal plasticity models – Hill/Hutchinson. As suggested in the description of the origin of Type II intergranular strains in steel and aluminum, if we are to explore the different responses of the variously oriented grain families, i.e., to understand the behavior of the individual diffraction peaks and the intergranular strains, we need a description which captures the interaction of the elastic and plastic properties of the polycrystalline aggregate. A popular approach in recent years has been the Hill (1965) self-consistent approach, first implemented by Hutchinson (1970). In this model a population of grains is chosen with a distribution of orientations and volume fractions that match the measured texture. Each grain in the model is treated as an ellipsoidal inclusion and is attributed anisotropic elastic constants and slip mechanisms characteristic of a single crystal of the material under study. Interactions between individual grains and the surrounding medium (which has properties of the average of all the grains) are performed using an elasto-plastic Eshelby (1959) self-consistent formulation. Since the properties of the medium derive from the average response of all the grains, it is initially undetermined and must be solved by iteration. Small total deformations are assumed, and usually no lattice rotation or texture development is incorporated. A number of groups have developed such models, some with slight modifications and developments compared to that described by Hutchinson. Details of the implementation of two examples of these more recent models can be found in Turner et al. (1995) and Clausen et al. (1998). Such models are typically termed elasto-plastic self-consistent (EPSC) models.

The single crystal elastic constants used in the model may be taken from the literature values, however there may be some variation with alloying (Dawson et al. 2001). Thus, the values are typically verified by comparison to the elastic experimental data, or if sufficiently high quality measurements are collected they can be calculated directly from the data in the elastic regime (e.g., Matthies et al. 2001). Considering the plastic flow law, the critical resolved shear stress and exponential hardening coefficients used are various, but in many calculations, an extended Voce law (Tomé et al. 2001) is used:

$$\tau = \tau_0 + \left(\tau_1 + \theta_1 \Gamma\right)\left(1 - \exp\left[\frac{-\theta_0 \Gamma}{\tau_1}\right]\right) \tag{18}$$

where Γ is the accumulated shear strain in the grain. The crystallographic shear flow stress τ in Equation (18) describes (in an average way) the resistance to activation that the deformation modes experience. The threshold value is τ_0, and it usually increases with deformation due to strain-hardening, which is here shown to follow a modified Voce law; τ_1 is the Voce stress where the hardening extrapolates to a zero value of accumulated shear and θ_0 is the athermal initial hardening rate (Kocks et al. 1998; Kocks and Mecking 2003). The final hardening rate θ_1 can be non-zero. Typically the hardening parameters in Equation (18) are fitted to give optimum agreement with the macroscopic stress-strain curves, since while single crystal values can be determined explicitly, the actual values found in the polycrystal are influenced by grain size, dislocations, precipitates etc.

Plastic slip is assumed to take place on the appropriate slip systems, for example for *fcc* crystal structures the twenty-four <110> {111} systems. The same set of yield and hardening values are initially applied to all systems and grains (i.e., for *fcc* structures there are only four fitting parameters in the model for a given test). However each individual slip system in a given grain is subsequently kept track of during modeling of the deformation, and the hardening curve followed appropriately according to the accumulated shear strain.

In order to provide reasonable comparisons with diffraction data a subset of the total population of grains used in the model is identified for each diffracting family, as defined by the condition of having an *hkl* plane-normal lying within, for example, a 5° cone around the nominal scattering direction. In many cases the angular span can be chosen to match the actual detector coverage used in the experiment. A population of grains is chosen to represent the texture—typically this might be of the order of 1000 to 5000 appropriately weighted grains (see below for further discussion as to required grain populations).

This model has been shown to be in good to excellent agreement in comparison with diffraction data for a range of metals, including *fcc* (Pang et al. 1998b; Clausen et al. 1999), *bcc* (Pang et al. 1998a; Daymond and Priesmeyer 2002), and *hcp* (Turner and Tomé 1994; Pang et al. 1997; Daymond and Bonner 2002) crystal structures. An example indicating the level of agreement which can be obtained is shown in Figure 9. A number of important issues remain in the implementation of these models, most significantly the way that interactions between the grain and matrix are fine tuned (e.g., Tomé 1999), and the way that interactions between different slip systems, i.e. latent hardening (Xu et al. 2006), are handled. Attention has also been applied more recently to the study of multiphase systems, where the interphase and intergranular stresses will superimpose in some manner (e.g., Gharghouri et al. 1999; Dye et al. 2001; Daymond and Priesmeyer 2002; Daymond et al. 2005; Daymond and Fitzpatrick 2006)

Polycrystal plasticity models – crystal plasticity finite element models. The other major class of model used to describe polycrystal plasticity is the finite element approach. In the

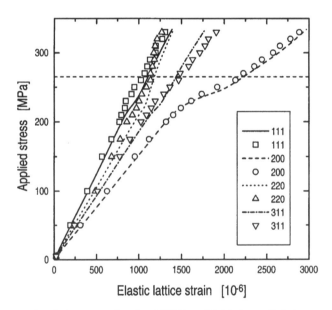

Figure 9. A comparison between experiment and EPSC model data for applied stress versus the elastic lattice strains parallel to the applied stress for several diffraction peaks in *fcc* steel. Experimental data corresponds to that shown in Figure 6. Symbols are measured and lines are calculated. The horizontal dotted line represents the macroscopic 0.2% yield limit. [Used with permission of Elsevier from Clausen et al. (1999), Fig. 5a.]

crystal plasticity finite element (CPFE) approach, a finite element mesh is used to represent a polycrystalline aggregate. This is different from the composite unit cell model when, typically, only two phases are modeled. In CPFE the constitutive behavior varies spatially over the FE mesh to a much greater extent, with the local mesh behavior chosen to simulate the elasto-plastic behavior of an aggregate of many single crystals. The method was originally developed for the prediction of texture evolution where it has proven extremely effective e.g. (Dawson and Marin 1998), and has more recently been applied to model the elastic strain response of grains during plastic deformation and compared with diffraction results e.g. (Dawson et al. 2001). The great advantage of the CPFE method is that it can explicitly account for the influence of neighboring grains, local neighborhood and grain-to-grain interaction. The major disadvantage of the CPFE method is the high computational overhead required to simulate a significant number of grains. For the calculation of mean grain family elastic strains, the CPFE method is of the order of 100 times slower than EPSC for the same grain population (Fonseca et al. 2006). Further, since nearest neighbor effects are explicitly accounted for in CPFE, this means that typically it is necessary to model a larger population of grains to get a reasonable polycrystal average than in the EPSC case. This therefore means that it is likely that while EPSC and CPFE approaches will agree well in predicting *average* elastic strain in a particular grain family, the *distribution* of strain around this mean is likely to be better physically represented by the CPFE approach than by the EPSC model, although to date there have been only one direct quantitative comparison between the approaches (Fonseca et al. 2006). While a large body of CPFE publications exist, with relevance to comparisons with neutron diffraction strain measurement, the work of Dawson et al. (2001, 2005) and Bate and Fonseca (2004) should be noted. Particular success has been achieved in the study of Al and in other cubic alloys. Some recent studies have considered mixed finite element—crystal plasticity models, whereby the FE code is used to model spatial variation while a (non-FE) crystal plasticity code model is used to provide the constitutive law for the FE model. Such multiscale models have been developed either using a decoupled (Tomé et al. 2001; Oliver et al. 2004a) approach, where the crystal plasticity code is run separately to the FE code, via a "look-up table", as well as through true coupled (Raabe et al. 2002; Daymond 2005) approaches, where the crystal plasticity code is embedded into the FE code.

Further examples of the use of polycrystal plasticity models to interpret deformation

Twinning and domain reorientation. While the intergranular stresses which develop during plastic deformation of *fcc* metals thus seem at least reasonably well understood, the trends of internal stress generation in lower symmetry materials are less well characterized. Part of the reason for this is that typically multiple deformation modes can potentially operate (e.g., basal slip, prism slip) in such materials. Further, observations vary considerably between different materials due to the sensitivity of the different deformation modes to factors such as unit cell aspect ratio, grain size and temperature. In addition, in some materials it is well known that mechanical twinning can be an important deformation mechanism. Since twinning is a unidirectional shear mechanism, in highly textured materials complex interactions between twinning and slip can occur depending on the sense of the imposed load relative to the texture (e.g., Kaschner et al. 2006). Since twinning is associated with an actual crystallographic rearrangement, as well as with changes in strain, twinning is associated with a change in intensity of diffraction peaks. A number of neutron diffraction studies of internal strain generation have been carried out on a range of materials that undergo twinning—predominantly *hcp* materials such as alloys based on Zr (Turner et al. 1995), Be (Brown et al. 2003), Mg (Gharghouri et al. 1999; Agnew and Duygulu 2005) and Ti (Cho et al. 2002), all of which have important technological applications. For example, Gharghouri et al. (1999) used their neutron diffraction results to suggest that both $\{10\bar{1}1\}$ and $\{10\bar{1}2\}$ twinning must be active modes in their particular Mg alloy. Such twinning reorientations in metals are highly analogous to the domain switching that occurs in a whole class of ferroelectric/ferroelastic ceramics (e.g., Cain et al. 1994; Rogan et al. 2003) and which been successfully studied using diffraction. Indeed, in some ways these are simpler systems to interpret than metals, since the domain switching is expected to be the only deformation mechanism present. For example, Rogan et al. (2003) were able to demonstrate and quantify the different amounts of domain switching occurring in each phase of two phase Pb(Zr,Ti)O$_3$ (i.e., where both rhombohedral and tetragonal forms are present) under mechanical loading. Such mechanical twinning is also highly relevant to the study of a range of minerals, most obviously calcite, quartz and spinels.

Effect of temperature. By probing the internal stresses during deformation as a function of temperature using neutron diffraction, and comparing the results obtained with micromechanical models, it is possible to elucidate the effect of temperature on mechanisms. For example, Daymond et al. (1999b) studied the influence of raised temperature on the internal stresses generated in a Cu-Mo particulate composite, interpreting the observed strains using a unit cell finite element model and demonstrating the influence of local creep diffusion as the temperature was raised. An example of the study of the influence of temperature on intergranular strains in an *fcc* stainless steel is given in (Daymond and Bouchard 2006). A slip based polycrystal plasticity model was in good agreement with the experimental data up to 0.4 of melting point (T_m). By 0.49 T_m the model was still in reasonably good quantitative agreement, but by 0.55 T_m agreement was at best qualitative. Oliver et al. (2004b) have examined the influence of temperature on the relative contributions of slip and twinning modes as a function of temperature in a magnesium alloy; this is significant because the critical resolved shear stress required for twinning is much less temperature sensitive than that for slip.

Phase transformations. Neutron diffraction has long been used for the study of phase transformations, i.e. where the crystallographic phase changes either by a diffusive or displacive mechanism. Particular interest in the crystallography physics community has focused on trans-formations caused either by a change of temperature (Redfern 2006, this volume) or magnetic field (Von Dreele 2006, this volume). Geologists have used the technique to study the influence of pressure (Parise 2006, this volume). Given the plethora of work we give a few example here of those studies where internal stresses have specifically been of interest. Particularly with relevance to engineering applications, there has been much work on Shape Memory Alloys (SMAs). These are an important class of engineering materials in that the phase transforma-tions are affected by both temperature and stress, making them potentially useful in a range of actuator and sensor applications. Due to the highly non-linear nature of their response, there is a strong drive to obtain models to assist in their implementation in engineering design. SMAs owe their unusual mechanical behavior to a martensitic transformation. The advantage of study-ing the transformation using neutron diffraction is that, by suitable analysis of the data, it is pos-sible to determine the volume fraction, the preferential selection of crystallographic variants or orientations (Wenk 2006, this volume) and the development of internal strains, as the transfor-mation progresses as a function of applied load (see Fig. 10). Examples of such studies, include work on NiTi (Vaidyanathan et al. 1999), CuAlZnMn (Šittner et al. 2002) and FePd (Oliver et al. 2003). For example, Oliver et al. (2003) showed that internal stresses played a significant role in driving the reversal of martensite variant changes when the applied stress was removed. Other examples of studies of the interaction of internal stresses with phase transformations include studies on steels (Oliver et al. 2002) and ceramics (Üstündag et al. 1995).

Particular application of the techniques to geological materials and systems

Engineers have long sought to develop constitutive models for material behavior which can be applied in predicting the performance and lifetime of components. There is a direct analogy here with those working in rock mechanics, who require such models to describe the elastic and plastic properties of geological materials, though admittedly the length scales and timescales are somewhat different. Thus, an understanding of the elastic properties of rocks plays an important role in the interpretation of seismological data, while models of the plastic and visco-plastic

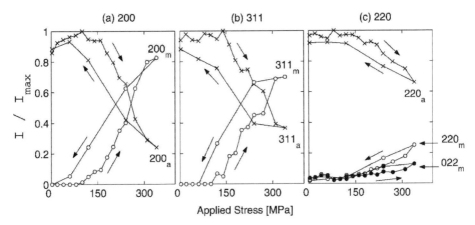

Figure 10. Variation of integrated peak intensities with applied stress in the direction parallel to the applied stress, during stress-induced transformation at 24 °C in FePd. Crosses are austenite diffraction peaks; circles are martensite diffraction peaks with the same Miller indices, as labeled for each plot. Intensities are normalized with respect to the maximum intensity of the appropriate austenite reflection. Arrows indicate the loading and unloading curves. [Used wtih permission of Elsevier from Oliver et al. (2003), Fig. 6.]

properties are applied in the thermo-mechanical modeling of the lithosphere (Schofield et al. 2003). It should be emphasized that the majority of systems that are of interest to geologists are in fact composite structures (polymineralic rocks) which thus have mechanical properties dependent on the individual phases, and the geometric arrangement of these phases and their interfaces. The ability to predict the behavior of such composites based on the properties of their constituent phases would thus be of great significance. The application of the techniques described above, which have been well established in the study of engineering materials to geological systems is relatively new, but a small number of studies are in the literature.

Pintschovius et al. (2000) considered the residual stresses resulting from the plastic deformation of monomineralic rock—demonstrating that even quite small stresses could be successfully measured. *In situ* studies of the loading of monomineralic rocks have looked at the elastic response of different diffraction planes (Scheffzük et al. 1998) as well as the effect of microcracking on the response of diffraction planes under nominally elastic loading (Meredith et al. 2001). The techniques described earlier in this paper have been applied to spatial mapping studies of Type I stresses (e.g., Meredith et al. 1997), and their variation near a previously shock loaded interface (Scheffzük et al. 2005). An interesting group of studies considering the elastic strain partitioning in polyphase rocks (Frischbutter et al. 2000; Covey-Crump et al. 2001, 2003) have demonstrated the limits of validity of the simple elastic models detailed in this paper in describing the phase response. Some of this work is reviewed by Schofield et al. (2003). Darling et al. (2004) used neutron diffraction to examine the internal strain state in sandstones which exhibit a non-linear, hysteretic relation between stress and macroscopic strain (i.e., measured by an extensometer). Whilst without a constraining pressure or elevated temperature (or both) the possibility of studying plasticity in geological systems is limited to certain materials systems such as halite, there has been some initial work in this area (Covey-Crump et al. 2006b). Recent work has looked at the way that introducing a confining pressure can further widen the scope of *in situ* loading studies (Dobson et al. 2005; Covey-Crump et al. 2006a). Finally, "man-made" rocks are also susceptible to theses techniques with, for example, Schulson et al. (2001) using neutron diffraction to monitor the development of internal strain within the calcium hydroxide phase of hardened Portland cement paste as the material cooled, with the effect attributed to the thermal mismatch between calcium hydroxide and calcium silicate hydrate phases.

There are a number of complexities which need to be considered when we are considering making measurements on geological systems, compared to the wide range of studies on engineering materials. We summarize these aspects here; some are discussed in more detail by Schofield et al. (2003).

Grain size. Typical studies of polycrystalline materials have assumed that a powder approximation can be made, that is sufficient grains/crystallites are present in the diffracting volume to be able to use statistical arguments in interpreting the deformation. Recent developments at neutron and in particular synchrotron X-ray facilities have pushed the techniques towards the study of a small number of individual grains (Martins et al. 2004), but this area of research is still in its infancy in terms of providing data for direct comparison and validation of models. Putting an exact figure on the number of grains that should be present in the measurement volume is not simple, since it will depend on the characteristics of the diffraction instrument, and in particular on the divergence of the incident beam and the angular coverage of the detector, both of which affect how many of the grains in the volume will contribute to the diffraction pattern. Based on model observations, it is typically considered that several thousand grains are required in the gauge volume for results to be reasonably statistically significant—typically 1% or less of the grains within the measurement volume will contribute to the diffraction signal. Hence obtaining results that are statistically meaningful requires many grains, with Clausen et al. (2003) suggesting around 10,000 based on their modeling studies. Reliable diffraction measurements are usually carried out on several hundred thousand grains. Geological materials often have large grain sizes which may complicate this requirement, particularly for spatial mapping studies. In such cases techniques to increase the number of grains which can contribute are often employed, for example carrying out a small angular "rocking" of the sample about the principal measurement vector, or spatial scanning along an axis where no strain variation is expected. If a smaller number of grains are present, it may still be possible to carry out measurements, but a significantly increased scatter is likely to be observed in the results due to the variation in number of sampled grains, and hence to the significance of local neighborhood on the measured strains.

Crystal structure and symmetry. Geological materials often have a low symmetry, compared to many engineering materials which are either cubic or hexagonal. This reduced crystal symmetry results in an increase in the number of peaks—in extreme cases such as monoclinic very few fully separate (i.e., non-overlapping) peaks are available, making a multipeak analysis preferable. Further it should be noted that a decrease in symmetry increases the required count times, although this is not particularly significant in comparison to the long count times experienced for some Ti and Co based engineering alloys due to the particular neutron scattering/absorption properties of these materials.

When crystal structures are very close, such as ortho- and clino-pyroxenes, the similar structures result in very similar diffraction patterns. As the diffraction patterns become more and more similar, distinguishing the phases requires higher resolution instruments and increased data collection times. However, even when phases are close together in structure, advanced data analysis strategies can be employed in some cases (e.g., Stone et al. 1999). Similar problems with peak overlap and difficulty in distinguishing phases may occur if many phases are present, or phases are present in very small quantities. Typically measurement will not give accurate strains if less than a few percent by volume of a phase is present (unless its scattering is significantly stronger than the other phases). If more than four or five phases are present, one would also expect significant overlap in the diffraction patterns which would make separating out the elastic strains in the individual phases difficult. However, it should be noted that quite typically a measurement accuracy in strain of 10^{-4} and sometimes 5×10^{-5} is obtained. For an atomic lattice of 0.3 nm, this corresponds to monitoring the average atomic lattice spacing to an accuracy of better than 0.03 pm.

Elemental issues. Unlike X-rays which have scattering that decreases more or less monotonically as atomic number increases, the scattering of neutrons varies—for our purposes—more or less randomly across the periodic table (Bacon 1975). Of particular note H has a very large incoherent scattering (i.e., creates a significant background signal), while the presence of isotopes of certain elements most noticeably Gd, B and Cd cause significant neutron absorption. If significant quantities of these elements are present in the material it is not possible to transmit neutrons through more than a small thickness of material.

CONCLUSIONS

Neutron diffraction is an extremely effective probe for the measurement of internal stresses in polycrystalline materials. Measurements of diffraction peak position can be correlated to changes in the lattice parameter and hence interpreted as an elastic strain. With a knowledge of the appropriate elastic stiffness, stress can be calculated from these strains. The changes in stress state, in particular its variation between phases or as a function of crystallographic orientation allow an interpretation of the deformation mechanisms that are operating. Models of polycrystal deformation are particularly helpful in interpreting such experimental data.

The techniques of strain measurement by neutron diffraction and synchrotron x-ray diffraction have thus been extensively used in the study of engineering materials. They have led directly to insights into the influence of processing methods on residual stress development and into the interpretation of deformation mechanisms operating in materials. There seems considerable potential for the techniques to be applied to geological materials, an area that is at present just starting to be explored.

ACKNOWLEDGMENTS

This work has benefited greatly from various collaborations and interactions with my colleagues. In particular I would like to thank Dr. Mike Fitzpatrick, of the Open University, UK, and the manuscript reviewers for useful comments on the Chapter. In addition, many members of the neutron diffraction stress measurement community worked together closely in the period 1996-2001 under the auspices of VAMAS TWA-20, resulting in the publication of an ISO Technology Trends Assessment (Webster 2001). This Chapter incorporates the recommendations made by that group.

This work benefited from the funding provided by the Canadian National Science and Engineering Research Council, and by the Canadian Foundation for Innovation through a Canada Research Chair in Mechanics of Materials.

REFERENCES

Agnew SR, Duygulu O (2005) Plastic anisotropy and the role of non-basal slip in magnesium alloy AZ31B. Int J Plasticity 21:1161-1193
Agnew SR, Tome CN, Brown DW, Holden TM, Vogel SC (2003) Study of slip mechanisms in a magnesium alloy by neutron diffraction and modelling. Scripta Mater 48:1003-1008
Agrawal P, Conlon K, Bowman KJ, Sun CT, Cichocki FR, Trumble KP (2003) Thermal residual stresses in co-continuous composites. Acta Mater 51:1143-1156
Allen AJ, Bourke M, Dawes S, Hutchings MT, Withers PJ (1992) The analysis of internal strains measured by neutron diffraction in Al/Sic metal matrix composites. Acta Metall Mater 40:2361-2373
Bacon GE (1975) Neutron Diffraction, 3rd ed. Oxford
Bate PS, Fonseca JQd (2004) Texture development in the cold rolling of IF steel. Mater Sci Eng A 380:365-377
Bollenrath F, Hauk V, Müller EH (1967) On calculation of polycrystalline elasticity constants from single crystal data. Z Metallkd 58:76-82

Bourke MAM, Goldstone JA, Shi N, Allison JE, Stout MG, Lawson AC (1993) Measurement and prediction of strain in individual phases of a 2219Al/TiC/15p-T6 composite during loading. Scripta Mat 29:771-776

Brockenbrough JR, Suresh S, Wienecke HA (1991) Deformation of MMCs with continuous fibers: geometrical effects of fiber distribution and shape. Acta Metall Mater 39:735-752

Brown DW, Bourke MAM, Clausen B, Holden TM, Tome CN, Varma R (2003) A neutron diffraction and modeling study of uniaxial deformation in polycrystalline beryllium. Metall Mater Trans A 34A:1439-1449

Budiansky B, Wu TT (1962) Theoretical prediction of plastic strains in polycrystals. *In: Proceedings of the 4th U. S. National Congress of Applied Mechanics,* University of California, Berkeley, California. Rosenberg RM (ed) Vol. **1**, p 1175-1185

Cain MG, Bennington SM, Lewis MH, Hull S (1994) Study of the ferroelastic transformation in zirconia by neutron-diffraction. Philos Mag B 69:499-507

Carter DH, Bourke MA (2000) Neutron diffraction study of the co-deformation behaviour of beryllium-aluminium composites. Acta Mater 48:2885-2900

Cho JR, Dye D, Conlon KT, Daymond MR, Reed RC (2002) Intergranular strain accumulation in a near-alpha titanium alloy during plastic deformation. Acta Mater 50:4847-4864

Clausen B, Leffers T, Lorentzen T (2003) On the proper selection of reflections for the measurement of bulk residual stresses by diffraction methods. Acta Mater 51:6181-6188

Clausen B, Lorentzen T, Bourke MAM, Daymond MR (1999) Lattice strain evolution during uniaxial tensile loading of stainless steel. Mater Sci Eng 259:17-24

Clausen B, Lorentzen T, Leffers T (1998) Self-consistent modelling of the plastic deformation of fcc polycrystals. Acta Mater 46:3087-3098

Clyne TW, Withers PJ (1993) An Introduction to Metal Matrix Composites. Cambridge University Press

Covey-Crump SJ, Holloway RF, Schofield PF, Daymond MR (2006a) A new apparatus for measuring mechanical properties at moderate confining pressures in a neutron beamline. J Appl Cryst 39:222-229

Covey-Crump SJ, Schofield PF, Daymond MR (2006b) Using neutrons to investigate strain partitioning between the phases during plastic yielding of calcite + halite composites. Physica B *in press*

Covey-Crump SJ, Schofield PF, Stretton IC (2001) Strain partitioning during the elastic deformation of an olivine-mangesiowustite aggregate. Geophys Res Lett 28:4647-4650

Covey-Crump SJ, Schofield PF, Stretton IC, Knight KS, Ismail WB (2003) Using neutron diffraction to investigate the elastic properties of anisotropic rocks: results from an olivine + orthopyroxene mylonite. J Geophys Res - Solid Earth 108(B2): Art. No. 2092

Darling TW, TenCate JA, Brown DW, Clausen B, Vogel SC (2004) Neutron diffraction study of the contribution of grain contacts to nonlinear stress-strain behavior. Geophys Res Lett 31:Art. No. L16604

Dawson P, Boyce D, MacEwen S, Rogge R (2001) On the influence of crystal elastic moduli on computed lattice strains in AA-5182 following plastic straining. Mater Sci Eng A 313:123-144

Dawson PR, Boyce DE, Rogge RB (2005) Correlation of diffraction peak broadening to crystal strengthening in finite element simulations. Mater Sci Eng A 399:13-25

Dawson PR, Marin EB (1998) Computational mechanics for metal deformation processes using polycrystal plasticity. Adv Appl Mech 34:77-169

Daymond MR (2004) The determination of a continuum mechanics equivalent elastic strain from the analysis of multiple diffraction peaks. J Appl Phys 96:4263-4272

Daymond MR (2005) A combined finite element and self-consistent model; validation by neutron diffraction strain scanning. Mater Sci Forum 495:1019-1024

Daymond MR, Bonner NW (2002) Lattice strain evolution in IMI 834 under applied stress. Mater Sci Eng A 340:263-271

Daymond MR, Bouchard PJ (2006) Elastoplastic deformation of 316 stainless steel under tensile loading at elevated temperatures. Metall Mater Trans A 37A:1863-1873

Daymond MR, Bourke MAM, Von Dreele RB (1999a) Use of Rietveld refinement to fit an hexagonal crystal structure in the presence of elastic and plastic Anisotropy. J Appl Phys 85:739-747

Daymond MR, Bourke MAM, Von Dreele RB, Clausen B, Lorentzen T (1997) Use of Rietveld refinement for residual stress measurements and the evaluation of macroscale plastic strain from diffraction spectra. J Appl Phys 82:1554-1562

Daymond MR, Fitzpatrick ME (2006) Effect of cyclic plasticity on internal stresses in a Metal Matrix Composite. Metall Mater Trans A 37A:1977-1986

Daymond MR, Hartig C, Mecking H (2005) Interphase and intergranular strains in a composite with both phases undergoing plastic deformation. Acta Mater 53:2805-2813

Daymond MR, Lund C, Bourke MAM, Dunand DC (1999b) Elastic phase-strain distribution in a particulates reinforced Metal-Matrix Composite deforming by slip or creep. Metall Mater Trans 30A:2989-2997

Daymond MR, Priesmeyer HG (2002) Elastoplastic deformation of ferritic steel and cementite studied by neutron diffraction and self-consistent modelling. Acta Mater 50:1613-1626

Daymond MR, Tomé CN, Bourke MAM (2000) Measured and predicted intergranular strains in textured austenitic steel. Acta Mater 48:553-564

Daymond MR, Withers PJ (1996a) A new stroboscopic neutron diffraction method for monitoring materials subjected to cyclic loads: thermal cycling of metal matrix xomposites. Scripta Mater 35(6):717-720

Daymond MR, Withers PJ (1996b) A synchrotron radiation study of transient internal strain changes during the early stages of thermal cycling of MMCs. Scripta Mater 35(10):1229-1234

Dieter GE (1986) Mechanical Metallurgy. McGraw-Hill

Dobson D, Mecklenburgh J, Alfe D, Wood I, Daymond MR (2005) A new belt-type apparatus for neutron-based rheological measurements at gigapascal pressures. High Press Res 25:107-118

Dunst D, Mecking H (1996) Analysis of experimental and theoretical rolling textures of two phase titanium alloys. Z Metallkd 87:498-507

Dye D, Stone HJ, Reed RC (2001) A two phase elastic-plastic self-consistent model for the accumulation of microstrains in Waspalloy. Acta Mater 49:1271-1283

Edwards L, Bouchard PJ, Dutta M, Wang DQ, Santisteban JR, Hiller S, Fitzpatrick ME (2005) Direct measurement of the residual stresses near a "boat-shaped" repair in a 20 mm thick stainless steel tube butt weld. Int J Press Vessels Piping 82:288-298

Eshelby JD (1957) The determination of the elastic field of an ellipsoidal inclusion, and related problems. Proc Roy Soc A241:376-396

Eshelby JD (1959) The elastic field outside an ellipsoidal inclusion. Proc Roy Soc London Ser A 252:561-569

Fitzpatrick ME, Hutchings MT, Withers PJ (1997) Separation of macroscopic, elastic mismatch and thermal-expansion misfit stresses in Metal-Matrix Composite quenched plates from neutron-diffraction measurements. Acta Mater 45:4867-4876

Fitzpatrick ME, Lodini A (2003) Analysis of Residual Stress by Diffraction using Neutron and Synchrotron Diffraction. Taylor and Francis

Fitzpatrick ME, Withers PJ, Baczmanski A, Hutchings MT, Levy R, Ceretti M, Lodini A (2002) Changes in the misfit stress in an Al/SiC$_p$ metal matrix composite under plastic strain. Acta Mater 50:1031-1040

Fonseca JQd, Bate PS, Oliver EC, Daymond MR, Withers PJ (2006) Modelling and measuring intergranular stresses during the early stages of plasticity. Acta Mater, submitted

Frischbutter A, Neov D, Scheffzuk C, Vrana M, Walther K (2000) Lattice strain measurements on sandstones under load using neutron diffraction. J Struct Geol 22:1587-1600

Gharghouri MA, Weatherly GC, Embury JD, Root J (1999) Study of the mechanical properties of Mg-7.7at% Al by *in situ* neutron diffraction. Philos Mag A 79:1671-1695

Gnaeupel-Herold T, Brand PC, Prask HJ (1998) Calculation of single-crystal elastic constants for cubic crystal symmetry from powder diffraction data. J Appl Cryst 31:929-935

Greenough GB (1947) Residual lattice strains in plastically deformed metals. Nature 160:258-260

Greenough GB (1952) Quantitative X-ray diffraction observations on strained metal aggregates. Prog Metal Phys 3:176-219

Gubicza J, Nam NH, Balogh L, Hellmig RJ, Stolyarov VV, Estrin Y, Ungar T (2004) Microstructure of severely deformed metals determined by X-ray peak profile analysis. J Alloys Compd 378:248-252

Hashin Z, Shtrikman S (1962) A variational approach to the theory of the elastic behaviour of polycrystals. J Mech Phys Sol 10:343-352

Hashin Z, Shtrikman S (1963) A variational approach to the theory of the elastic behaviour of multiphase materials. J Mech Phys Sol 11:127-140

Hauk V (1997) Structural and Residual Stress Analysis by Nondestructive Methods. Elsevier

Hill R (1952) The elastic behaviour of a crystalline aggregate. Prof Phys Soc London A 65:349-354

Hill R (1965) Continuum micro-mechanics of elastoplastic polycrystals. J Mech Phys Sol 13:89-101

Hosford WF (1993) The Mechanics of Crystals and Textured Polycrystals. Oxford

Howard CJ, Kisi EH (1999) Measurement of single crystal elastic constants by neutron diffraction from polycrystals. J Appl Cryst 32:624-633

Hutchings MT (1990) Neutron diffraction measurement of residual stress fields - the answer to the engineers' prayer? Nondestr Test Eval 5:395-413

Hutchings MT, Withers PJ, Holden TM, Lorentzen T (2005) Introduction to the Characterization of Residual Stress by Neutron Diffraction. Taylor and Francis

Hutchinson JW (1970) Elastic-plastic behaviour of polycrystalline metals and composites. Proc R Soc London A 319:247-272

Ikeda S, Carpenter JM (1985) Wide-energy-range, high-resolution measurements of neutron pulse shapes of polyethylene moderators. Nucl Instrum Methods Phys Res A239:536-544

Johnson MW, Daymond MR (2002) An optimum design for a time-of-flight neutron diffractometer for measuring engineering stresses. J Appl Cryst 35(1):49-57

Kamminga JD, Keijser THd, Mittemeijer EJ, Delhez R (2000) New methods for diffraction stress measurement: a critical evaluation of new and existing methods. J Appl Cryst 33:1059-1066

Kaschner GC, Tomé CN, Beyerlein IJ, Vogel SC, Brown DW, McCabe RJ (2006) Role of twinning in the hardening response of zirconium during temperature reloads. Acta Mater 54(11):2887-2896

Kocks UF, Mecking H (2003) Physics and phenomenology of strain hardening: the FCC case. Progr Mater Sci 48:171-173

Kocks UF, Tomé CN, Wenk H-R (1998) Texture and Anisotropy. Cambridge

Korsunsky AM, Daymond MR, James KE (2002) The correlation between plastic strain and anisotropy strain in aluminium alloy polcrystals. Mater Sci Eng A 334:41-48

Korsunsky AM, James KE, Daymond MR (2004) Intergranular stresses in polycrystalline fatigue: diffraction measurement and self-consistent modelling. Eng Fracture Mechanics 71 (4-6):805-812

Krawitz AD (2001) Introduction to Diffraction in Materials Science and Engineering. John Wiley and Sons

Krawitz AD, Winholtz RA (1994) Use of position-dependent stress-free standards for diffraction stress measurements. Mater Sci Eng A 185(1-2):123-130

Kröner E (1958) Zur Behandlung des Quantenmechanischen Vielteilchenproblems mit Hilfe von Mehrteilchenfunktionen. Z Physik 151:504-518

Kröner E (1961) Zur Plastischen Verformung des Vielkristalls. Acta Metall 9:155-161

Larsson C, Clausen B, Holden TM, Bourke MAM (2004) Measurements and predictions of strain pole figures for uniaxially compressed stainless steel. Scripta Mater 51:571-575

Levin VM, Alvarea-Tostado JM (2003) Eshelby's formula for an ellipsoidal elastic inclusion in anisotropic poroelasticity and thermoelasticity. Int J Fracture 119:4-2:L79-L82

Levy A, Papazian JM (1990) Tensile properties of short fibre-reinforced SiC/Al composites, part II. Finite element analysis. Metall Trans 21A:411-420

Lorentzten T, Daymond MR, Clausen B, Tomé CN (2002) Lattice strain evolution during cyclic loading of stainless steel. Acta Mater 50(6):1627-1638

Ma S, Brown D, Bourke MAM, Daymond MR, Majumdar BS (2005) Microstrain evolution during creep of a high volume fraction superalloy. Mater Sci Eng A 399:141-153

MacEwen SR, Faber J, Turner APL (1983) The use of time-of-flight neutron diffraction to study grain interaction stresses. Acta Metall 31:657-676

Martins RV, Margulies L, Schmidt S, Poulsen HF, Leffers T (2004) Simultaneous measurement of the strain tensor of 10 individual grains embedded in an Al tensile sample. Mater Sci Eng A 387:84-88

Matthies S, Priesmeyer HG, Daymond MR (2001) On the diffractive determination of elastic single crystal constants using polycrystalline samples. J Appl Cryst 34:585-601

Meredith PG, Knight KS, Boon SA, Wood IG (2001) The microscopic origin of thermal cracking in rocks: an investigation by simultaneous time-of-flight neutron diffraction and acoustic emission monitoring. Geophys Res Lett 28:2105-2108

Meredith PG, Wood IG, Knight KS, Boon SA (1997) *In situ* measurement of strain partitioning during rock deformation by neutron diffraction imaging. J Conf Abstracts 2:50

Mittemeijer EJ, Scardi P (2004) Diffraction Analysis of the Microstructure of Materials. Springer

Mura T (1987) Micromechanics of Defects in Solids. Nijhoff

Noyan IC, Cohen JB (1987) Residual Stress - Measurement by Diffraction and Interpretation. Springer-Verlag

Nye JF (1992) Physical Properties of Crystals. OUP

Oliver EC, Daymond MR, Withers PJ (2004a) Interphase and intergranular stress generation in carbon steels. Acta Mater 52:1937-1951

Oliver EC, Daymond MR, Withers PJ (2004b) Neutron diffraction study of extruded magnesium during cycling and elevated temperature loading. Materials Sci Forum 490-491:257-262

Oliver EC, Daymond MR, Withers PJ (2005) Effects of texture and anisotropy on intergranular stress development in zirconium. Mater Sci Forum 495-497:1553-1558

Oliver EC, Mori T, Daymond MR, Withers PJ (2003) Neutron diffraction study of stress induced martensite transformation and variant change in FePd. Acta Mater 51:6453-6464

Oliver EC, Mori T, Daymond MR, Withers PJ (2004c) Stress-induced martensitic transformation and variant change in an Fe-Pd shape memory alloy. Mater Sci Eng A 378:328-332

Oliver EC, Withers PJ, Daymond MR, Ueta S, Mori T (2002) Neutron diffraction study of stress induced martensitic transformation in TRIP steel. Appl Physics A 74:1143-1145

Pang JWL, Holden TM, Mason TE (1997) *In situ* generation of intergranular strains in Zircaloy under uniaxial loading. Acta Mater 47(2):373-383

Pang JWL, Holden TM, Mason TE (1998a) The development of intergranular strains in a high-strength steel. J Strain Analysis 33:373-383

Pang JWL, Holden TM, Mason TE (1998b) *In situ* generation of intergranular strains in an Al7050 alloy. Acta Mater 46:1503-1518

Parise JB (2006) High pressure studies. Rev Mineral Geochem 63:205-231

Pintschovius L, Prem M, Frischbutter A (2000) High precision neutron diffracion measurements for the determination of low level residual stresses in a sandstone. J Struct Geol 22(11-12):1581-1585

Poulsen HF, Wert JA, Neufeind J, Honkimaki V, Daymond MR (2005) Measuring strain distributions in amorphous materials. Nature Mater 1:33-36

Preuss M, Pang JWL, Withers PJ, Baxter GJ (2002) Inertia welding nickel-based superalloy: Part II. Residual stress characterization. Metall Mater Trans A 33:3227-3234

Pyzalla A, Camin B, Buslaps T, Di Michiel M, Kaminski H, Kottar A, Pernack A, Reimers W (2006) Simultaneous tomography and diffraction analysis of creep damage. Science 308 (5718):92-95

Raabe D, Zhao Z, Park SJ, Roters F (2002) Theory of orientation gradients in plastically strained crystals. Acta Mater 50:421-440

Redfern SAT (2006) Neutron powder diffraction studies of order-disorder phase transitions and kinetics. Rev Mineral Geochem 63:145-170

Reuss A (1929) Berechung der Fließgrenze von Mischkristallen auf Grund der Plastizitätsbedingung für Einkristalle. Z angew Math Mech 9:49-58

Rietveld HM (1969) A profile refinement method for nuclear and magnetic structures. J Appl Cryst 2:65-71

Rogan RC, Üstündag E, Clausen B, Daymond MR (2003) Texture and strain analysis of the ferroelastic behavior of Pb(Zr,Ti)O₃ by *in situ* neutron diffraction. J Appl Phys 93:4104

Ruud CO (1982) A review of selected non-destructive methods for residual stress measurement. NDT Int 15: 15-23

Sachs Z (1928) Zur Ableitung einer Fließbedingung. Z Ver Dtsch Ing 72:734

Sanitsteban JR, Oliver EC, Daymond MR, Alianelli L, Edwards L. (2006b) Tensile deformation of a Cu mosaic crystal along the <110> direction studied by time of flight neutron transmission. Mater Sci Eng A, in press

Santisteban JR, Daymond MR, Edwards L, James JA (2006a) ENGIN-X: a third generation neutron strain scanner. J Appl Cryst, in press

Santisteban JR, Edwards L, Fitzpatrick ME, Steuwer A, Withers PJ, Daymond MR, Johnson MW, Rhodes N, Schooneveld EM (2002) Strain imaging by Bragg edge neutron transmission. Nucl Instrum Methods A 481:765-768

Scheffzük C, Frischbutter A, Walther K (1998) Intracrystalline strain measurements with time-of-flight neutron diffractoin: application to a Cretaceous sandstone from the Elbezone (Germany). Schriftenreihe fur Geowissenschaften 6:39-48

Scheffzük C, Walther K, Frischbutter A, Eichhorn F, Daymond MR (2005) Residual strain and texture measurements using neutron TOF diffraction on a dolomite-anhydrite rock and a quartz-dunnite compound. Solid State Phen 105:61-66

Schofield PF, Covey-Crump SJ, Stretton IC, Daymond MR, Knight KS, Holloway RF (2003) Using neutron diffraction measurements to characterize the mechanical properties of polymineralic rocks. Mineral Mag 67:967-987

Schulson EM, Swainson IP, Holden TM (2001) Internal stress within hardened cement paste induced through thermal mismatch: Calcium hydroxide versus calcium silicate hydrate. Cement Concrete Res 31:1785-1791

Siano S, Bartoli L, Santisteban JR, Kockelman W, Daymond MR, Miccio M, Marinis Gd (2006) Non-destructive investigation of Picenum bronze artefacts using neutron diffraction. Archaeometry 48:77-96

Simmons G, Wang H (1971) Single Crystal Elastic Constants and Calculated Aggregate Properties: a Handbook. MIT Press

Šittner P, Lukáš P, Neov D, Daymond MR, Novák V, Swallowe GM (2002) Stress induced martensitic transformation in CuAlZnMn polycrystals investigated by two *in situ* neutron diffraction techniques. Mater Sci Eng A A324:225-234

Stone HJ, Holden TM, Reed RC (1999) On the generation of microstrains during the plastic deformation of Waspalloy. Acta Mater 47:4435-4448

Sun YN, Choo H, Liaw PK, Lu YL, Brown DW, Bourke MAM (2005) Neutron diffraction studies on lattice strain evolution around a crack-tip during tensile loading and unloading cycles. Scripta Mater 53(8): 971-975

Taylor GI (1938) Plastic strain in metals. J Instrum Methods 62:307-324

Tomé CN (1999) Self-consistent polycrystal models: A directional compliance criterion to describe grain interactions. Modell Simul Mater Sci Eng 7:723-738

Tomé CN, Maudlin PJ, Lebensohn RA, Kaschner GC (2001) Mechanical response of zirconium - I. Derivation of a polycrystal constitutive law and finite element analysis. Acta Mater 49:3085-3096

Turner PA, Christodoulou N, Tomé CN (1995) Modelling of the mechanical response of rolled Zircaloy-2. Int J Plasticity 11:251-265

Turner PA, Tomé CN (1994) A study of residual stresses in Zircaloy-2 with rod texture. Acta Mater 42:4143-4153

Üstündag E, Subramanian R, Dieckmann E, Sass SL (1995) In situ formation of metal-ceramic microstructures in the Ni-Al-O system by partial reduction reactions. Acta Metall Mater 43:383-389

Vaidyanathan R, Bourke MAM, Dunand DC (1999) Phase fraction, texture and strain evolution in superelastic NiTi and NiTi-TiC composites investigated by neutron diffraction. Acta Mater 47:3353-3366

Vogel SC, Priesmeyer H-G (2006) Neutron production, neutron facilities and neutron instrumentation. Rev Mineral Geochem 63:27-57

Voigt W (1910) Lehrbruch der Krystallphysik. Teubner

Von Dreele RB (2006) Neutron Rietveld refinement. Rev Mineral Geochem 63:81-98

Wang YD, Wang X-L, Stoica AD, Richardson JW, Peng RL (2003) Determination of the stress orientation distribution function using pulsed neutron sources. J Appl Cryst 36:14-22

Warren BE (1990) X-Ray Diffraction. Dover Publications

Watt JP (1980) Hashin-Shtrikman bounds on the effective elastic moduli of polycrystals with monoclinic symmetry. J Appl Phys 51(3):1520-1524

Watt JP, Peselnick L (1980) Clarification of the Hashin-Shtrikman bounds on the effective elastic moduli of polycrystals with hexagonal, trigonal, and tetragonal symmetries. J Appl Phys 51(3):1525-1531

Webster GA (2001) Polycrystalline Materials -- Determination of Residual Stresses by Neutron Diffraction. International Standards Organization, Report # TTA-3

Wenk H-R (2006) Neutron diffraction texture analysis. Rev Mineral Geochem 63:399-426

Wern H, Kock N, Maas T (2002) Selfconsistent calculation of the x-ray elastic constants of polycrystalline materials for arbitrary crystal symmetry. Mater Sci Forum 404:127-132

Wilkinson AJ (1997) Methods for determining elastic strains from electron backscatter diffraction and electron channelling patterns. Mater Sci Technol 13:79-84

Withers PJ (2004) Depth capabilities of neutron and synchrotron diffraction strain measurement instruments. I The maximum feasible path length. J Appl Crystall 37:596-606

Withers PJ, Clarke AP (1998) A neutron diffraction study of load partitioning in continuous Ti/SiC composites. Acta Mater 46:6585-6598

Withers PJ, Daymond MR, Johnson MW (2001) The accuracy of diffraction peak location. J Appl Cryst 34:737-743

Withers PJ, Preuss M, Webster PJ, Hughes DJ, Korsunsky AM (2002) Residual strain measurement by synchrotron diffraction. Mater Sci Forum 404:1-10

Withers PJ, Stobbs WM, Pedersen OB (1989) The application of the Eshelby method of internal stress determination for short fibre metal matrix composites. Acta Metall Mater 37:3061-3084

Xu F, Holt RA, Daymond MR (2006) Development of internal strains during uni-axial deformation in textured Zircaloy-2: Part 2 modelling. Acta Mater, submitted

Reviews in Mineralogy & Geochemistry
Vol. 63, pp. 459-471, 2006
Copyright © Mineralogical Society of America

17

Applications of Neutron Radiography and Neutron Tomography

Bjoern Winkler

Institut fuer Geowissenschaften / Abt. Kristallographie
Universitaet Frankfurt (Main)
Senckenberganlage 30, D-60054 Frankfurt, Germany
e-mail: b.winkler@kristall.uni-frankfurt.de

INTRODUCTION

In radiographic methods, the attenuation of an incident beam on passing through an object is used to study the internal structure of this object. The availability of techniques to "look inside" large, optically intransparent objects without having to destroy them is obviously appealing and has had a profound impact in many fields, most clearly in medicine, but also in condensed matter studies. The development of radiography started with Röntgen's discovery of X-rays in 1895, where the absorption of X-rays by bones were demonstrated publicly in 1896 by obtaining a radiograph of a hand. The medical applications were obvious, and already in World War I, mobile X-ray units were routinely deployed. The main shortcoming of X-ray radiography in early medical applications is that often it cannot be used to distinguish different kinds of soft tissue and that the two dimensional projections are sometimes difficult to interpret. This changed dramatically with the development of "computerized axial tomography" for which the Nobel prize was awarded in 1979 to Hounsfield and Cormack. Computed tomography, as it is now frequently called, is the three-dimensional reconstruction of the interior of an object from many radiographs taken at different angles, and this combination of data sets enables to distinguish between different soft tissues, such as kidneys and liver.

Medical computed X-ray tomography scanners were used soon afterwards to investigate minerals and rocks, for example in an early study of chondritic meteorites (Arnold et al. 1983). These had a comparatively poor spatial resolution of ~1 mm. As there is no need for low radiation doses in material science studies, and as the penetrating power of high energy X-rays is larger and higher spatial resolutions can be obtained, dedicated scanners for material science have been developed. Commercially available "table top" scanners now offer spatial resolution of less than 1 μm. More recently, the advantages of synchrotron radiation (monochromatic, collimated and extremely brilliant beams) have been exploited in tomography. For example, the CT system at SPring-8 has a spatial resolution of about 1 μm (Uesugi et al. 2001).

With the availability of neutron beams for research in the early 1950's, a similar development of neutron radiography and tomography occurred. This was, in part, driven by the need to investigate objects for which X-ray radiography could not be employed, namely for the investigation of reactor fuel assemblies, which are intransparent to X-rays. A brief history of the development of neutron radiography, together with a list of general references has recently been presented by Berger (2004).

Neutron radiography and neutron tomography are now mature techniques, and this chapter gives a brief overview over some aspects relevant to their application in earth sciences. An example of a neutron radiograph and a neutron tomograph of a rock is shown in Figure 1.

The main advantage of using neutrons as a probe in comparison to other radiographic and

1529-6466/06/0063-0017$05.00   DOI: 10.2138/rmg.2006.63.17

Figure 1. Photograph (left), neutron radiograph (center) and computed tomograph (right) of a pegmatite. The diameter of the sample is 35 mm. The linear attenuation coefficient of hydrogen-rich mica grains is very different from those of the surrounding anhydrous minerals, and hence the mica grains can easily be distinguished. Image processing software can then be used to e.g., determine the volume fraction of mica in the sample. These measurements and image processing were performed at the PSI, Switzerland.

tomographic techniques is that due to the high penetrating power of neutrons, large samples can be investigated with little or no radiation damage, and that light elements can be detected in an environment dominated by heavy elements. Clearly, like any other techniques, neutron radiography has its strengths and limitations, and a large number of complementary approaches, such as X-ray, synchrotron or γ-ray tomography, nuclear magnetic resonance tomography or proton radiography are now available. In the earth sciences, neutron radiography and neutron tomography have only rarely been used up to now, and it is hoped that the present outline will stimulate researchers to exploit the unique capabilities of these techniques.

FUNDAMENTAL ASPECTS

The principle of radiographic and tomographic methods is to measure the attenuation of an incident beam along a line through an object as shown in Figure 2.

For a parallel and monochromatic beam the attenuation measured by a detector element is given by

$$-\ln\left(\frac{I}{I_0}\right) = \int \mu(x,y)ds \tag{1}$$

where I_0 and I are the incident and the detected beam intensity, $\mu(x,y)$ is the linear attenuation coefficient at (x,y) and the integration is along the straight line from the final aperture to the detector. The attenuation is due to scattering and absorption. The very different interaction mechanisms of neutrons and X-rays with matter result in very different μ and are the origin of the distinct capabilities and limitations of neutron and X-ray radiography.

The scattering of neutrons is conventionally separated into "coherent" and "incoherent" scattering. The coherent scattering cross section, σ_{icoh}, the incoherent scattering cross section, σ_{inc}, and the absorption cross section, σ_{abs}, of elements are tabulated in standard reference works (see e.g., Dianoux and Lander 2002 and references cited therein). All scattering and

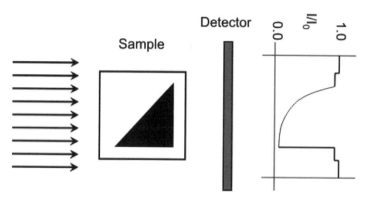

Figure 2. 2D schematic view of a neutron radiographic experiment. In this simplified drawing, a parallel neutron beam is used to illuminate a sample. The position sensitive detector records the attenuation of the beam after transmission through an inhomogeneous sample. The measured intensity is a projection of the distribution of the linear attenuation coefficients within the sample.

absorption processes contribute to the attenuation of the transmitted beam and the linear attenuation coefficient μ is given by

$$\mu = \sum \frac{\sigma_i \rho_i N_A}{m_{\text{molar},i}} \qquad (2)$$

where $\sigma_i = \sigma_{\text{coh},i} + \sigma_{\text{inc},i} + \sigma_{\text{abs},i}$ is the sum of the coherent and incoherent scattering and absorption cross section of the ith element (usually given in barn, where 1 barn = 10^{-24} cm^2), ρ_i is the density of the ith element (in g/cm^3), and $m_{\text{molar},i}$ the molar mass of element i. For example, the normalized transmitted intensity I/I_0 for thermal neutrons (with wave length λ = 1.78 Å) after transmission of 1 cm of pure aluminum (ρ = 2.7 g/cm^3, m_{molar} = 26.9815 g/mol, σ = 0.239 barn) is 0.985. The cross section, and hence the penetrating power of neutrons of a given wave length, depends on the neutron energy, this will be discussed below.

From Figure 2 it is evident that the experiment will depend on the characteristics of the neutron beam (brilliance, divergence, spectral distribution) and the properties of the detector. Both aspects will be discussed below.

Neutron beams

Depending on the type of experiment, neutron radiography can be performed essentially with all available neutron sources, including portable ones. In one special case, namely in "neutron-induced autoradiography" the neutron beam is only used to activate certain elements in the sample under investigation, so that the emission can then be measured by placing the object on a detector (generally an image plate, see below). This approach has, for example, been used to investigate paintings and other cultural heritage objects in order to reveal hidden details with a non-destructive method (Fischer et al 1999; Schillinger 1999). In geoscientific applications, neutron-induced autoradiography have been used to map the distribution of a number of elements in rock samples (Flitsiyan 1998; Romer et al. 2005).

In practice, the overwhelming number of radiographic and tomographic studies are performed at dedicated stations at neutron scattering facilities. It is therefore of interest to establish, for which purposes which of the different radiography/tomography stations are best suited. The neutron flux and the divergence of the incident beam are parameters which need to be taken into account. A well-established criterion to characterize the beam divergence is the L/D-ratio, where L is the length of the flight path from a pinhole with diameter D (typically a few cm) to the sample. High L/D-ratio therefore indicate a small beam divergence. The experimental determination of the L/D-ratio is standardized (see references in Berger 2004) as it is by no means straightforward to compute it for neutron guides. L/D-ratios vary from ~50 up to values >10,000 (Schillinger 1999; Schillinger et al. 2000).

Cold neutrons are generally guided to the neutronography station by a slightly curved neutron guide. This has the advantage that the background is much reduced, as γ-radiation and fast neutrons travel in straight lines and hence do not reach the sample. At other installations the radiography station is located at the end of a flight tube facing a cold neutron source. At such stations, the spectral distribution of the neutron beam is significantly broader compared to that at the end of a guide.

Depending on the specific installation, the flux of thermal neutrons from beam ports can be significantly higher than for cold neutrons. Currently, the most intense beam for neutronographic investigations is available at the Institute Laue Langevin in Grenoble, where a thermal neutron flux of ~ 3×10^9 (neutrons cm^{-2} s^{-1}) illuminates the sample at the end of a 15 m flight path. At other sources the available flux is generally one to three orders of magnitude smaller.

It is also possible to use fast neutrons with energies of a few MeV (Nordlund et al. 2001; Sanami et al. 2001; Bücherl et al. 2004). These have a significantly greater penetrating power than thermal neutrons, which is necessary if, for example, very hydrogen-rich compounds are investigated.

Detectors

There is a wide variety of detectors available for neutron imaging work, which differ from the detectors often used in diffractometers and spectrometers, since point or linear detectors (such as ^{3}He-counters) are inappropriate. An informative summary on currently used detectors has recently been presented by Lehmann et al. (2004).

The choice of a detector for a specific experiment should be based on considering the following aspects: (1) the spatial resolution of the detector assembly is one of the limiting factors determining the size of the smallest features which can be observed. (2) For dynamic imaging, the time resolution is important and is given by the number of images which can be collected and processed in a given period of time with a sufficient signal-to-noise ratio for the observation of non-periodic processes. For periodic processes, "stroboscopic" techniques can be used, these are discussed below. (3) The dynamic range and linearity of the detector is important if accurate intensity values are essential. This is always very important for neutron tomographic studies. (4) The sensitive area of the detector limits the size of objects which can be studied by neutron tomography but is not as crucial for radiography, where individual images can be combined or where large objects can be "scanned." (5) Quantitative studies require digital data—this is one of the main draw-backs of film-based methods.

For those detectors which are based on the detection of photons (i.e., X-ray film, image plates or video cameras), converters (scintillators) are employed, in which an incident neutron triggers a reaction so that photons can be measured. Neutron converters are based on elements which absorb neutrons and emit secondary radiation. This secondary radiation is then absorbed by a photo stimulable phosphor, and can then be detected optically or leads to the formation of a latent image. Typical converter materials are ^{6}Li-doped ZnS, Gd_2O_3 or Gd_2O_2S:Tb. The choice of the thickness of the converter is a compromise, as the neutron detection efficiency increases with thickness, but the spatial resolution decreases (Baechler et al. 2002).

For neutron radiographic work, photographic X-ray film used to be the most commonly used detector, but it has now, with few exceptions, been replaced by imaging plates (see below). Films still play a role in routine quality control, such as in the testing of components for aviation and space technology, where rigorous and standardized procedures have been established, and where the advantages, such as the good spatial resolution (20-50 µm), outweigh the disadvantages, such as the non-linear response of the film, the necessary wet-chemical processing and that the data is initially only analog.

Image plate detectors for X-rays use Eu-doped BaFBr layers as a storage phosphor dispersed in an organic binder supported on a polymer sheet. An incident photon will excite an electron into a metastable state, where it is trapped, and hence a latent image is formed. If the image plate is later scanned by a red laser, a recombination process is induced, during which blue light is emitted. Such image plates can be adapted to neutron detection by adding gadolinium (usually as Gd_2O_3) to the phosphor layer. On neutron capture, the gadolinium releases a conversion electron, which will then create a storage center (Thoms et al. 1997; Schlapp et all 2002). The resolution of an image plate is slightly worse than that of photographic film (~50–100 µm). Image plates replace film, as they have a higher sensitivity, high linearity and a higher dynamic range. Also, the intensity data is immediately available in a digital form. They are generally used as "off-line IP," i.e., the IP-holder is detached from the scanner. This allows for custom made dimensions of the detector, but off-line IPs cannot be used efficiently for neutron tomography, as the read-out and erasure procedure cannot be fully automated.

Video cameras based on conventional pick up tube technology (e.g., orthicon or vidicon tubes) are rapidly replaced by CCD-based cameras. Most modern CCD based cameras are cooled by Peltier elements in order to increase the signal-to-noise ratio. With the exception of facilities with a very high flux, cameras can be used for the observation of processes with a time resolution of a few tenths of a second. At very high flux facilities, a time resolution of a few ten µs can be obtained in very favorable conditions. For example, the interaction of a water jet with a metal melt has been recorded with "high-frame" neutron radiography with 1125 frames/s (Sibamoto et al. 2002). However, short exposure times with a poor signal to noise ratio, the limited dynamic range available in high speed cameras, and the need to use image intensifiers generally restrict the available resolution to ~200-500 µm for studies of fast processes. The improvement of the spatial and temporal resolution of CCD-cameras is an active field of research (Schillinger 1999; Koerner et al. 2000; Schillinger et al. 2005) and significant improvements have been achieved in the last years. As the CCDs are sensitive to radiation, they cannot be exposed to the direct beam. Hence, the camera is housed in a light-tight and radiation-shielded box and aligned at 90° to the neutron flight path. The camera sees an image of the neutron converter on a mirror placed at 45° to the neutron beam and the optical axis of the camera. Video cameras are often used in conjunction with image intensifiers, which, while increasing the sensitivity, significantly constrain the achievable spatial resolution.

The development of new detectors is an active area of research. The characteristics of amorphous silicon flat panels have recently been discussed by Lehmann and Vontobel (2004). They seem to offer improvements at least in two areas, size and resistance against radiation damage. While CCD-based detectors have an active area of a few cm^2 only, amorphous silicon flat panels with an active area of about 20×30 cm^2 are available. This is of interest for those experimental stations in which the beam is collimated so that a large area is illuminated. In the neutronography station GENRA-3 at the GKSS facility in Geesthacht, Germany, for example, the beam illuminates an area of ~ 45×45 cm^2. Other new developments include CMOS-based detectors (Lehmann and Vontobel 2004).

NEUTRON RADIOGRAPHY

Neutron radiography is an established technique exploited commercially in materials science, e.g., for testing of parts in aviation and space technology. Neutron radiography has been used to determine water contents and diffusion in rocks and related compounds, such as bricks and concrete (Nordlund et al. 2001; El-Ghany El Abd and Milczarek 2004; Masschaele et al. 2004). Standard static neutronography is, however, of little interest in the earth sciences, as neutron tomographic experiments provide much more information and these will be discussed below. Neutron radiography can also be used for the observation of processes. In the context of *in situ* observations, this is of interest, as sample environments, such as high temperature furnaces, can be constructed so as to be nearly transparent to the neutron beam (Kahle et al 2003b).

Dynamic neutron radiography

For the observation of processes, two distinct methods are available, namely "dynamic neutron radiography" for non-periodic and "stroboscopic neutron radiography" for periodic processes. The latter will only briefly be introduced, as there are so far no geoscience-related applications. The observation of non-periodic processes at non-ambient conditions, especially at high temperatures, is very promising and two examples will be mentioned here.

The first example is "falling sphere" experiments. Radiographic studies of falling sphere experiments (Fig. 3) have widely been used in the earth sciences, as they present a unique opportunity to study densities and viscosities of melts at non-ambient conditions. At synchrotrons, falling sphere experiments in high pressure cells have been used to study melts at high pressure (e.g., Mori et al. 2000, Suzuki et al. 2002). These experiments are very challenging, and care has to be taken to avoid systematic errors, e.g., due to convection, non-uniform heating, ill-defined starting time of the sphere motion, or short falling times. At least for ambient pressure, high temperature investigations, neutrons offer a viable alternative, where most of these experimental problems can be avoided (Kahle et al. 2003a). Also, dynamic neutron radiography allows to observe sphere falls over large distances (~10 cm) and for a large range of velocities (~0.05-10 cm/min).

Large sample volumes also allow to use falling sphere experiments to investigate inhomogeneous melts with a significant fraction of crystallites. This would be problematic with alternative approaches, such as rotational viscosimetry (Dingewell 1995). However, first attempts indicated that possibly the main problem will be to ensure that the crystalline fraction neither floats to the top of the melt nor sinks to the bottom over the time required for the measurements, i.e., for a few hours.

The large illuminated sample volume also allows to study other processes. In a preliminary study, it has been shown that melt dynamics can be monitored (Fig. 4). Here again, large samples can be used and observation times can range from a few seconds to a few hours.

Further obvious applications for dynamic neutronography include the observation of bubble formation, of hydrogen diffusion, and of changes in interfaces between melts with different densities under shear.

COMPUTED NEUTRON TOMOGRAPHY

The images obtained from neutron radiographic investigations are two dimensional projections of the distribution of linear attenuation coefficients in the sample. In analogy to the developments in X-ray tomography, the advent of sufficiently powerful computers and suitable detectors allowed neutron tomographic investigations. Computed neutron tomography is now an established technique in the material sciences.

Due to its immense importance in diagnostic medicine, all aspects of the data processing required to obtain tomographic images from radiographs have been studied extensively and the most often used algorithms are well documented (e.g., Kak 1988). Also, due to the large market for data analysis software, a variety of commercial and public domain software packages have been developed. As the data analysis in tomographic experiments is quite independent of the ap-

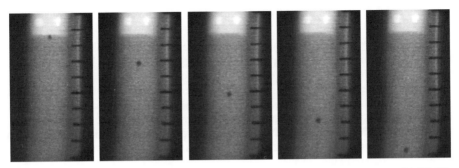

Figure 3. Time dependent neutron radiography. Here a sphere made of strongly absorbing material (Hf, Gd, Ta) falls in a silicate melt at high temperatures ($T > 1000$ °C). From the speed the viscosity of the melt can be determined. The distance between the markers on the left of each frame is 1 cm. Exposure times were 1/10 s. Total falling times range from a few minutes to an hour, so that several thousand positions of the sphere s function of time can be determined. These measurements were performed at the LLB, France.

$t = 0$ s   10 s   28 s   42 s   58 s   95 s

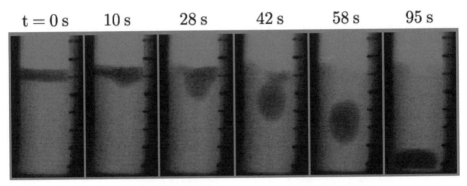

Figure 4. Dynamics of two immiscible melts. In this feasibility study, a brass plate was put on top of a silicate melt. After melting, the brass then descended through the silicate melt. These observations can be used to "calibrate" numerical models. These measurements were performed at the LLB, France.

plication, it is comparatively easy for a novice to obtain useful images if the main technical aspects are understood. Hence, in the following, only a few technical details will be mentioned.

If scattering and the energy dependence of the linear attenuation coefficient is neglected, tomography is the reconstruction of the distribution of the linear attenuation coefficients, $\mu(x,y)$ in the two-dimensional case described by Equation (2) or, in the more general case $\mu(\mathbf{x})$, where $\mathbf{x}$ describes a location in three dimensions. From the projections $\int \mu(\mathbf{x})ds$, which can be measured for different straight paths s, the distribution of $\mu(\mathbf{x})$ must then be found by reconstruction. There are several techniques available, with the most important ones being (1) "filtered back projection," which relies on the Radon transform, (2) Fourier reconstruction and (3) algebraic reconstruction. The accuracy of these three approaches should be very similar. Alternative approaches have been explored by several authors, e.g., use of maximum entropy methods (Zawisky et al. 2004).

The reconstruction of the distribution $\mu(\mathbf{x})$ is, of course, limited by the resolution of the individual images. It can be shown that if the detector would be the limiting factor, the maximal resolution could be achieved if the transmission is measured for $\pi/2 \times$ (number of pixels /line) sample orientations. In practice, due to the beam divergence and other limiting factors, it is not possible to achieve this optimal resolution. Instead the required rotation by 180° is covered in steps of slightly less than 1°, so that ~200-400 images are collected. Depending on the available flux and detector, a full data set can be collected in a few hours.

Tomography relies on accurate intensity measurements, and this implies that the influence of inhomogeneities or fluctuations in the beam intensity and differences in the efficiency of detector elements have to be minimized and corrected for. Hence, corrections with "flat field" (open beam) frames and "dark images" with a closed beam shutter are performed. An internal scaling can be obtained by comparing parts of the image which are not obscured by the sample during a rotation. The software required for the transformation of the individual frames to a three-dimensional rendering of the structure is commercially available from several companies. Fast computers with large (a few GB) random access memory are required for the processing of high resolution images.

Applications of neutron computed tomography in the material sciences are very wide ranging. In geosciences, there are comparatively few applications. Neutron tomography is very well suited to investigate the distribution and intergrowth of minerals (Figs. 5 and 6).

This is of interest, as the description of the textures of coarse grained igneous rocks is essential for an understanding of their formation process. Tomography can efficiently provide information of grain sizes and pore systems (de Beer et al. 2004). It can also be used to visualize the flow of liquids in bulk rocks (Wilding et al. 2005; Dierick et al. 2005).

A paleontological study visualized fossilized leaves of a cornifer in an eocene fossil assemblage , with neutron tomography (Abele H, Ballhausen H, Cywinski R, Dawson M, Francis J, Gähler R, Stephens R, Van Overberghe A, pers. com.) (Fig. 7). This indicated a close relationsship of the fossil cornifer to present day "monkey puzzle trees" (araucaria araucana).

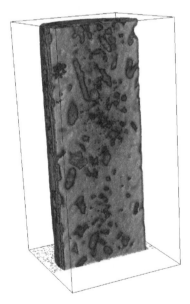

Figure 5. Photograph (left) and slice through a computed tomograph (right) of a basanite, a volcanic rock of felsic composition with an aphanitic texture. The surfaces of the pores are covered with water-containing zeolites. Image processing software can be used to determine the volume fraction of the pore system and to investigate if this is an open or closed pore system. These measurements and image processing were performed at the PSI, Switzerland.

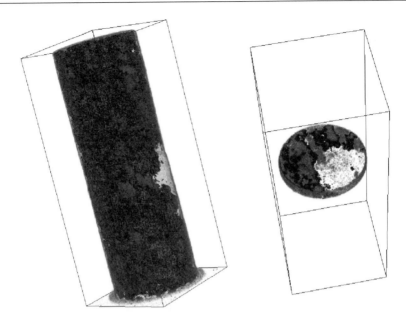

Figure 6. In a garnet-mica schist, individual mineral species can be identified in the tomogram (left). The slice on the right shows the intergrowth of garnet with the other minerals in the sample. Garnet is anhydrous and hence much more transparent to neutrons than the hydrous minerals such as mica. These measurements and image processing were performed at the PSI, Switzerland.

From a practical point of view, most tomographs are obtained from objects whose largest dimension is smaller than the detector edge and which can be fully illuminated by the neutron beam. Hence, currently available set-ups limit the edge lengths of the object under investigation to about 20 cm. However, it is possible to perform a tomography of so-called "regions-of-interest" of objects larger than the detector or beam with additional computational effort (Schillinger et al. 2004).

In the simplified description given above, a monochromatic incident beam has been assumed. For a polychromatic neutron beam, a formally correct description would have to incorporate the energy dependence of the linear attenuation coefficient, and hence $\mu(\mathbf{x})$ would have to be replaced by $\mu(\mathbf{x},E)$. For cold and thermal neutrons, the absorption cross section is generally proportional to $1/v$, i.e., the higher the neutron energy (the shorter the wavelength and the higher the velocity v) the higher the penetration power. This may lead to so-called "beam hardening," i.e., the mean energy of the transmitted spectrum is shifted

Figure 7. Tomography of fossilized leaves of a conifer enclosed in a sediment. This image was produced at the ILL (Grenoble) by a collaboration of researchers from the University of Leeds and the University of Heidelberg (Abele H, Ballhausen H, Cywinski R, Dawson M, Francis J, Gähler R, Stephens R, Van Overberghe A. pers. comm.). A comparison with leaves from present day plants indicates a close relationship to the monkey puzzle tree. Such studies may help to determine the evolution of leaf morphology and of the distribution of plants.

to higher energies with respect to the incident spectrum. This may lead to problems in the tomographic analysis of materials containing strong neutron absorbers, and may lead to artifacts in edge regions or for weakly attenuating regions, surrounded by strongly attenuating regions (Schillinger 1999). In addition, the exponential attenuation can also be violated due to incoherent scattering, especially in the presence of elements with a large incoherent and a small absorption cross section. Finally, coherent elastic scattering of neutrons may lead to a discontinuity in the energy dependence of μ if the lattice spacing is large enough as to lead to Bragg scattering. Then, the attenuation coefficient for neutrons with a wave length shorter than that corresponding to the Bragg edge will be larger than the attenuation coefficient for long wavelength neutrons. These effects may become significant in detailed investigations and the development of algorithms to correct for such effects is a current field of research (Hassanein et al. 2005; Kasperl et al. 2005).

FURTHER EXPERIMENTAL APPROACHES

Stroboscopic techniques

Periodic processes can be observed with neutron stroboscopic imaging. The principle is straightforward: a shutter is installed between detector and sample and synchronized with the repetition rate of the process. The shutter must be able to open and close on a short time scale relative to the time required to complete one period. The shutter is opened for a short time for a given part of the process, closed for the remainder, and opened again when the cycle repeats. The detector is therefore illuminated only for a fraction of the cycle. The detector must be able to accumulate information so that after a few hundred passes a reasonable signal-to-noise ratio is obtained. Then, the synchronization is changed, so that the shutter is open at a different fraction of the periodic process. Current technology allows to observe periodic processes with a time resolution of a few hundred microseconds. Technical aspects of this approach, which has been very successfully applied to study combustion engines and fuel injection systems, have recently been discussed by Brunner et al. (2005) and Schillinger et al. (2005).

Phase contrast imaging

Phase contrast imaging is well known in optical microscopy, where it allows to resolve features below the diffraction limit. For this development, a Nobel prize was awarded to F. Zernike in 1953. Phase radiography with neutrons has been first been demonstrated recently by Allman et al. (2000).

While a detector placed very close to the object will register the absorption contrast only, a detector placed in the near-field region will register the influence of the refractive index on the propagation of the neutron wave. This leads to an enhancement in contrast. However, the demands on the spatial coherence are stringent for phase contrast neutronography. To achieve this spatial coherence, pin holes with a diameter of a few tenth mm are used. This increases the L/D-ratio, but also leads to a drastic decrease of the flux at the sample. A number of papers have now been published (e.g., McMahon et al. 2003; Kardjilov et al. 2004) which have demonstrated, that polychromatic neutron beams can be used for phase-contrast radiography. Due to the increased incident intensity in comparison to monochromatic beams, this leads to a significant decrease in measurement times.

While phase contrast measurements can be used to only enhance the contrast, quantitative studies in which the variation of the beam intensity along the beam direction is measured and analyzed with a transport-of-intensity equation are possible. This then allows to distinguish between those materials, which have the same linear attenuation coefficient but a different refractive index (Lehmann et al. 2005).

Neutron interferometry

Neutron interferometry, based on the use of large perfect crystals as beam splitters, is a well established technique, mainly employed to study properties of neutrons (Bonse and Rauch, 1979). Neutron interferometry can also be used to image objects and recently, an interferometry-based neutron phase contrast tomography with a spatial resolution of about 50 μm has been developed by Zawisky et al. (2004). However, the limited space for a sample environment and the complexity of such experiments will very likely limit the applicability of this approach for earth science related studies.

COMPARISON TO X-RAY TOMOGRAPHY

The main differences between X-ray- and neutron-based investigations are due to the difference in the interaction between photons and neutrons with matter, and due to the very different beam characteristics of X-ray and synchrotron radiation and neutron beams. Neutrons can detect light elements, such as hydrogen, in a matrix of heavy elements, and that makes them ideally suited to study water in an anhydrous matrix. The brilliance of synchrotron beams, however, is several orders of magnitude larger than that of neutron beams, and hence for the investigation of very small samples with very high resolution, X-ray and synchrotron radiation will always be preferable. A list of linear attenuation coefficients for thermal neutrons and 100 keV electrons and a comparison of X-ray and neutron tomographic images of an ammonite and an eclogite have recently been published by Vontobel et al. (2005).

Further comparative studies emphasize that in many cases it will be beneficial to have a variety of radiations available and to be able to compare radiographs or tomographs (e.g., Masschaele et al. 2004). This is possible at some facilities. For example, the new ANTARES facility at the FRM-II in Munich allows to use cold and thermal neutrons, γ-radiation, and X-rays from a 320 kV tube (in a quasi-parallel beam geometry, as the X-ray source is 12 m away from the sample). As only the converter or detector is changed, while the sample is unmoved, images obtained with the different radiations can be superimposed or subtracted without further manipulation (Schillinger et al. 2006).

SUMMARY AND OUTLOOK

Neutron radiography and neutron tomography are techniques which are mature and readily accessible. They can be used to study the interior of large, complex objects with a resolution of ~100 μm, and observe non-periodic processes with a temporal resolution of a fraction of a second. Neutron radiographic investigations have already been carried out at high temperatures. There have been no radiographic investigations at high pressures up to now and there are several problems which need to be overcome in the construction of large volume autoclaves suitable for such studies. There have also been no tomographic investigations at high temperatures, although it should be straightforward to adopt available furnaces for such experiments. Time resolved studies of systems relevant to the earth sciences are also in their infancy, yet it is conceivable that dynamic neutron radiography will play a prominent role here in the future.

The relatively large number of neutron radiography stations, which are available at nearly all neutron scattering facilities (e.g., FRM-II in Munich, PSI in Villigen, NIST in Gaithersburg, ILL in Grenoble) allows comparatively easy access to these techniques and hence, especially in conjunction with complementary techniques such as γ-ray, synchrotron or X-ray radiography and tomography, open new and exciting possibilities to study the interior of and processes in complex materials of relevance to the earth sciences.

REFERENCES

Allman B, Mcmahon P, Nugent K, Paganin D, Jacobson D, Arif M, Werner S (2000) Phase radiography with neutrons. Nature 408:158-159

Arnold JR, Testa JP, Friedman PJ, Kambic GX (1983) Computed tomographic analysis of meteorite inclusions. Science 219:383-384

Baechler S, Kardjilov N, Dierick M, Jolie J, Kühne G, Lehmann E, Materna T (2002) New features in cold neutron radiography and tomography. Part I: thinner scintillators and a neutron velocity selector to improve the spatial resolution. Nucl Instr Methods Phys Res A491:481-491

Berger H (2004) Advances in neutron radiographic techniques and applications: a method for nondestructive testing. Appl Radiat Isot 61:437-442

Bonse U, Rauch H (eds) (1979) Neutron Interferometry. Claredon Press

Brunner J, Engelhardt M, Frei G, Gildemeister A, Lehmann E, Hillenbach A, Schillinger B (2005) Characterization of the image quality in neutron radioscopy. Nucl Instr Methods Phys Res A542:123-128

Bücherl T, Kutlar E, Lierse von Gostomski C, Calzada E, Pfister G, Koch D (2004) Radiography and tomography with fast neutrons at the FRM-II - a status report. Appl Radiat Isot 61:537-540

de Beer FC, Middleton MF, Hilson J (2004) Neutron radiography of porous rocks and iron ore. Appl Radiat Isot 61:487-495

Dianoux AJ, Lander G (eds) (2002) Neutron Data Booklet. Institut Laue Langevin

Dingwell DB (1995) Viscosity and Anelasticity of Melts. *In:* Mineral Physics and Crystallography. A Handbook of Physical Constants. Ahrens TJ (ed) American Geophysical Union Reference Shelf 2, p 209- 217

Dierick M, Vlassenbroeck J, Masschaele B, Cnudde V, van Hoorebeke L, Hillenbach A (2005) High-speed neutron tomography of dynamic processes. Nucl Instrum Methods Phys Res A542:296-301

El-Ghany El Abd A, Milczarek JJ (2004) Neutron radiography study of water absorption in porous building materials: anomalous diffusion analysis. J Phys D Appl Phys 37:2305-2313

Fischer CO, Gallagher M, Laurenze C, Schmidt C, Slusallek K (1999) Digital imaging of autoradiographs from paintings by Georges de La Tour (1593-1652). Nucl Instrum Methods Phys Res A424:258-262

Flitsiyan E (1998) Use of neutron-activation techniques for studying elemental distributions: Applications to geochemistry. J Alloys Compd 275-277:918-923

Hassanein R, Lehmann E, Vontobel P (2005) Methods of scattering corrections for quantitative neutron radiography. Nucl Instrum Methods Phys Res A542:353-360

Kahle A, Winkler B, Hennion B (2003a) Is Faxen's correction function applicable to viscosity measurements of silicate melts with the falling sphere method? J Non-Newtonian Fluid Mech 112:203-215

Kahle A, Winkler B, Hennion B, Boutrouille P (2003b) High-temperature furnace for dynamic neutron radiography. Rev Sci Instrum 74:3717-3721

Kak AC (1988) Principles of Computerized Tomographic Imaging. IEEE Press

Kardjilov N, Lehmann E, Steichele E, Vontobel P (2004) Phase-contrast radiography with a polychromatic neutron beam. Nucl Instrum Methods Phys Res A527:519-530

Kasperl S, Vontobel P (2005) Application of an iterative artifact reduction method to neutron tomography. Nucl Instrum Methods Phys Res A542:392-398

Koerner S, Lehmann E, Vontobel P (2000) Design and optimization of a CCD-neutron radiography detector. Nucl Instrum Methods Phys Res A454:158-164

Lehmann EH, Lorenz K, Streichele E, Vontobel P (2005) Non-destructive testing with neutron phase contrast imaging. Nucl Instrum Methods Phys Res A542:95-99

Lehmann E, Vontobel P (2004) The use of amorphous silicon flat panels as detector in neutron imaging. Appl Radiat Isot 61:567-571

Lehmann EH, Vontobel P, Frei G, Brönnimann C (2004) Neutron imaging-detector options and practical results. Nucl Instrum Methods Phys Res A531:228-237

Masschaele B, Dierick M, van Hoorebeke L, Cnudde V, Jocobs P (2004) The use of neutrons and monochromatic X-rays for non-destructive testing in geological materials. Environ Geol 46:486-492

McMahon PJ, Allman BE, Jacobson DL, Arif M, Werner SA, Nugent KA (2003) Quantitative phase radiography with polychromatic neutrons. Phys Rev Lett 91:145502

Mori S, Ohtani E, Suzuki A (2000) Viscosity of the albite melt to 7 GPa at 2000 K. Earth Planet Sci Lett 175:87-92

Nordlund A, Linden P, Por G, Solymar M, Dahl B (2001), Measurements of water content in geological samples using transmission of fast neutrons. Nucl Instrum Methods Phys Res A462:457-462

Romer RL, Heinrich W, Schröder-Smeibidl B, Meixner A, Fischer CO, Schulz C (2005) Elemental dispersion and stable isotope fractionation during reactive fluid-flow and fluid immiscibility in the Bufa del Diente aureole, NE-Mexico: evidence from radiographies and Li, B, Sr, Nd, and Pb isotope systematics. Contrib Mineral Petrol 149:400-429

Sanami T, Baba M, Saito K, Yamazaki T, Miura T, Ibara Y, Taniguchi S, Yamadera A, Nakamura T(2001) Fast-neutron profiling with an image plate. Nucl Instrum Methods Phys Res A458:720-728

Schillinger B (1999) Neue Entwicklungen zu Radiographie und Tomographie mit thermischen Neutronen und zu deren routinemäßigem Einsatz. Ph.D. thesis. Technische Universität München, München

Schillinger B, Abele H, Brunner J, Frei G, Gähler R, Gildemeister A, Hillenbach A, Lehmann E, Vontobel P (2005) Detection systems for short-time stroboscopic neutron imaging and measurements on a rotating engine. Nucl Instrum Methods Phys Res A542:142-147

Schillinger B, Calzada E, Mühlbauer M, Schulz M (2006) Multiple radiation measurements at the radiography facility ANTARES. FRM-II Progress Report 2006, TU München, München

Schillinger B, Kardjilov N, Kuba A (2004) Region of interest tomography of bigger than detector samples. Appl Radiat Isot 61:561-565

Schillinger B, Lehmann E, Vontobel P (2000) 3D neutron computed tomography: requirements and applications. Physica B 276-278:59-62

Schlapp M, von Seggern H, Massalovitch S, Ioffe A, Conrad H, Brueckel T (2002) Materials for neutron-image plates with low γ-sensitivity. Appl Phys A74[Suppl] S109-S111

Sibamoto Y, Kukita Y, Nakamura H (2002) Visualization and measurement of subcooled water jet injection into high-temperature melt by high-frame-rate neutron radiography. Nucl Technol 139:205-220

Suzuki A, Ohtani E, Funakoshi K, Terasaki H, Kubo T (2002) Viscosity of albite melt at high pressure and high temperature. Phys Chem Minerals 29:159-165

Thoms M, Lehmann MS, Wilkinson C (1197) The optimization of the neutron sensitivity of image plates. Nucl Instrum Methods Phys Res A384:457-462

Uesugi K, Suzuki Y, Yagi N, Tsuchiyma A, Nakano T (2001) Development of high spatial resolution X-ray CT system at BL47XU in SPring-8. Nucl Instrum Methods Phys Res A467-A468:853-856

Vontobel P, Lehmann E, Carlson WD (2005) Comparison of X-ray and neutron tomography investigations of geological materials. IEEE Trans Nuclear Sci 52:338-341

Wilding M, Lesher CE, Shields K (2005) Applications of neutron computed tomography in the geosciences. Nucl Instrum Methods Phys Res A542:290-295

Zawisky M, Bonse U, Dubus F, Hradil Z, Rehacek J (2004) Neutron phase contrast tomography on isotope mixtures. Europhys Lett 68:337-343